$$\binom{5}{2} : \frac{5!}{2!\,3!}$$

$$\binom{5}{3} : \frac{5!}{3!\,2!}$$

$$\binom{5}{4} : \frac{5!}{4!}$$

Mathematical Structures
for Computer Science

Fourth Edition

Mathematical Structures for Computer Science

Fourth Edition

Judith L. Gersting
University of Hawaii at Hilo

W. H. Freeman and Company
New York

To Adam and Jason, two beloved discrete
structures, now grown and nonisomorphic

Acquisitions Editor:	Richard J. Bonacci; Michelle Russel Julet
Marketing Manager:	Kimberly Manzi
Project Editor:	Penelope Hull
Text Designer:	Diana Blume
Cover Designer:	Salem Krieger
Illustrator:	Vantage Art, Burmar Graphics
Production Coordinator:	Paul W. Rohloff
Illustration Coordinator:	Bill Page
Compositor:	Black Dot
Manufacturer:	R R Donnelley & Sons, Harrisonburg

Library of Congress Cataloging-in-Publication Data

Gersting, Judith L.

Mathematical structures for computer science / Judith L. Gersting.—4th ed.

p.   cm.

Includes index.

ISBN 0-7167-8306-1 (alk. Paper)

1. Mathematics.   2. Mathematical models.   3. Computer science—Mathematics.

I. Title

QA39.2.G47   1998

004'.01'51—dc21

98-29149
CIP

Printed in the United States of America

Second printing, 1999

W. H. Freeman and Company
41 Madison Avenue, New York, NY 10010
Houndmills, Basingstoke RG21 6XS, England

Contents in Brief

Contents

Preface

The importance of logical thinking, the power of mathematical notation, and the usefulness of abstractions are contributions of the discrete mathematics course that remain fundamental to the computer science curriculum. In this Fourth Edition of *Mathematical Structures for Computer Science,* I have tried to make these ideas even more accessible to the student.

Content Changes

Although the chapter titles are unchanged from the Third Edition, there are nonetheless a number of changes.

- The sections on propositional and predicate logic in Chapter 1 have been rewritten to use a "natural" deduction system rather than an axiomatic (Hilbert) system. The old system had the advantage of few axioms and rules to keep track of, but also had the disadvantage that these axioms and rules were not particularly intuitive and required some imagination on the part of the student to apply. The new system consists of no axioms and many rewriting and inference rules; these are presented in a series of tables. Although it may be a disadvantage to have so many rules, I would in no way imagine asking students to memorize these rules; I would rather expect them to rely on the tables until these rules become second nature. The increase in the number of rules is more than offset by the fact that the rules are quite mechanical to apply, and students find it much easier to construct proof sequences.
- Chapters 5 and 6 on graphs and trees have been reorganized to place more emphasis on trees.
- Sections have been reorganized to allow more flexibility in order of presentation. In Chapter 1, all of propositional logic is presented before any predicate logic material, so propositional logic can be covered in its entirety and predicate logic postponed or eliminated, as the instructor wishes. Similarly, optional "applications" topics have been put into separate sections:

Although these topics demonstrate to students the usefulness of the material forming the core of the course, there is not time in a one-semester course, much less a one-quarter course, to cover them all. With the new arrangement, instructors should find it easier to pick and choose among them.

Pedagogy

Practice problems, Chapter Objectives, Section Review, Chapter Terminology Review, Chapter Self-Test, and On-the-Computer exercises have been retained. The following features have been added.

- Each chapter has a "motivating" opening problem, which appears later in the chapter as an exercise.
- Many tables have been used throughout to summarize information or to present step-by-step approaches to a problem.
- There are Reminder notes throughout; these contain little points that I find I have harped on in class over the many years I have taught this course. (For example, "To prove that something isn't true, try proof by contradiction.")
- Algorithms have been rewritten in a language-neutral pseudocode.
- New examples and exercises have been added. In response to users' requests, I've tried to aim most of the new exercises at an intermediate level, not too easy and not too difficult.

Answers to all Practice problems are given at the back of the book, as are answers to all starred Exercises, which are approximately one-quarter of the total Exercises.

Supplements

A complete **Solutions Manual** is available to instructors from the publisher.

Acknowledgments

The reviewers of this edition provided a number of helpful comments. Thank you to

Manjit Bhatia, Bowie State University
Herbert B. Enderton, University of California, Los Angeles
John Geske, Kettering University
Tom Head, Binghamton University
Margaret Johnson, Stanford University
Mark Ramras, Northeastern University
Olga Shumsky, Northwestern University
Ellen M. Stenson, Stenson Educational Consulting
George Trapp, West Virginia University
William J. Weber, Southeast Missouri State University

Without the support of my family, my book-writing efforts wouldn't be possible. This time, special thanks to my husband, John, for his skill with the scanner!

Note to the Student

As you go through this book, you'll encounter many new terms and ideas. Try reading with pencil and paper at hand and work the Practice problems as you encounter them. They are intended to reinforce or clarify some new terminology or method just introduced; answers are given at the back of the book. Pay attention also to the Reminders that point out common pitfalls or provide helpful hints.

You may find at first that the thought processes required to solve the exercises in the book are new and difficult. Your most important attribute for success will be perseverance. When you read a problem and don't see how to solve it right away, don't give up! Mull it over in your head for a day or two, be sure you understand all the terminology used in the problem, play with some ideas. Mathematical results don't spring fully formed from the foreheads of mathematical geniuses. Well, maybe from mathematical geniuses, but for the rest of us, it takes work, patience, false starts, and **perseverance.**

Enjoy the experience!

Mathematical Structures for Computer Science

Fourth Edition

Formal Logic

Chapter Objectives

After studying this chapter, you will be able to:

- Use the formal symbols of propositional logic.
- Find the truth value of an expression in propositional logic.
- Construct formal proofs in propositional logic, and use such proofs to determine the validity of English language arguments.
- Use the formal symbols of predicate logic.
- Find the truth value in some interpretation of an expression in predicate logic.
- Use predicate logic to represent English language sentences.
- Construct formal proofs in predicate logic, and use such proofs to determine the validity of English language arguments.
- Understand how the programming language Prolog is built on predicate logic.
- Mathematically prove the correctness of programs that use assignment statements and conditional statements.

You have been selected to serve on jury duty for a criminal case. The attorney for the defense argues as follows:

If my client is guilty, then the knife was in the drawer. Either the knife was not in the drawer or Jason Pritchard saw the knife. If the knife was not there on October 10, it follows that Jason Pritchard did not see the knife. Furthermore, if the knife was there on October 10, then the knife was in the drawer and also the hammer was in the barn. But we all know that the hammer was not in the barn. Therefore, ladies and gentlemen of the jury, my client is innocent.

Question: Is the attorney's argument sound? How should you vote?

It's much easier to answer this question if the argument is recast in the notation of formal logic. Formal logic strips away confusing verbiage and allows us to concentrate on the underlying reasoning being applied. In fact, formal logic—the subject of this chapter—provides the foundation for the organized, careful method of thinking that characterizes any reasoned activity—a criminal investigation, a scientific experiment, a sociological study. In addition, formal logic has direct applications in computer science. The last two sections of this chapter explore a programming language based on logic and the use of formal logic to verify the correctness of computer programs. Also, circuit logic (the logic governing computer circuitry) is a direct analog of the statement logic of this chapter. We shall study circuit logic in Chapter 7.

Section 1.1   Statements, Symbolic Representation, and Tautologies

Formal logic can represent the statements we use in English to communicate facts or information. A **statement** (or **proposition**) is a sentence that is either true or false.

EXAMPLE 1   Consider the following:

    a. Ten is less than seven.
    b. How are you?
    c. She is very talented.
    d. There are life forms on other planets in the universe.

Sentence (a) is a statement because it is false. Because item (b) is a question, it cannot be considered either true or false. It has no truth value and thus is not a statement. Sentence (c) is neither true nor false because "she" is not specified; therefore (c) is not a statement. Sentence (d) is a statement because it is either true or false; we do not have to be able to decide which.

Connectives and Truth Values

In English, simple statements are combined with connecting words like *and* to make more interesting compound statements. The truth value of a compound statement depends on the truth values of its components and which connecting words are used. If we combine the two true statements "Elephants are big" and "Baseballs are round," we would consider the resulting statement, "Elephants are big and baseballs are round," to be true. In this book, as in many logic books, capital letters near the beginning of the alphabet, such as A, B, and C, are used to represent statements and are called **statement letters**; the symbol $\wedge$ is a **logical connective** representing *and*. We agree, then that if A is true and B is true, $A \wedge B$ (read "A and B") should be considered true.

PRACTICE 1[1]   **a.** If A is true and B is false, what truth value would you assign to $A \wedge B$?
    b. If A is false and B is true, what truth value would you assign to $A \wedge B$?
    c. If A and B are both false, what truth value would you assign to $A \wedge B$?

    The expression $A \wedge B$ is called the **conjunction** of A and B, and A and B are called the **conjuncts** of this expression. Table 1.1 summarizes the truth value of $A \wedge B$ for all possible truth values of the conjuncts A and B. Each row of the table represents a particular truth value assignment to the statement letters, and the resulting truth value for the compound expression is shown.

[1]Answers to practice problems are in the back of the text.

Another connective is the word *or*, denoted by the symbol ∨. The expression $A \vee B$ (read "A or B") is called the **disjunction** of A and B, and A and B are called the **disjuncts** of this expression. If A and B are both true, then $A \vee B$ would be considered true, giving the first line of the truth table for disjunction (see Table 1.2).

A	B	$A \wedge B$
T	T	T
T	F	F
F	T	F
F	F	F

Table 1.1

A	B	$A \vee B$
T	T	T
T	F	
F	T	
F	F	

Table 1.2

PRACTICE 2   Use your understanding of the word *or* to complete the truth table for disjunction, Table 1.2.   ●

Statements may be combined in the form "if statement 1, then statement 2." If A denotes statement 1 and B denotes statement 2, the compound statement would be denoted by $A \rightarrow B$ (read "A implies B"). The logical connective here is **implication**, and it conveys the meaning that the truth of A implies or leads to the truth of B. In the implication $A \rightarrow B$, A stands for the **antecedent** statement and B stands for the **consequent** statement.

The truth table for implication is less obvious than that for conjunction or disjunction. To understand its definition, let's suppose your friend remarks, "If I pass my economics test, then I'll go to the movie Friday." If your friend passes the test and goes to the movie, the remark was true. If your friend passes the test but doesn't go to the movie, the remark was false. If your friend doesn't pass the test, then—whether he or she goes to the movie or not—you could not claim that the remark was false. You would probably want to give the benefit of the doubt and say that the statement was true. By convention, $A \rightarrow B$ is considered true if A is false, regardless of the truth value of B.

PRACTICE 3   Summarize this discussion by writing the truth table for $A \rightarrow B$.   ●

The **equivalence** connective is symbolized by ↔. The expression $A \leftrightarrow B$ is shorthand for $(A \rightarrow B) \wedge (B \rightarrow A)$. We can write the truth table for equivalence by constructing, one piece at a time, a table for $(A \rightarrow B) \wedge (B \rightarrow A)$, as in Table 1.3. From this truth table, $A \leftrightarrow B$ is true exactly when A and B have the same truth value.

A	B	$A \rightarrow B$	$B \rightarrow A$	$(A \rightarrow B) \wedge (B \rightarrow A)$
T	T	T	T	T
T	F	F	T	F
F	T	T	F	F
F	F	T	T	T

Table 1.3

The connectives we've seen so far are called **binary connectives** because they join two expressions together to produce a third expression. Now let's consider a **unary connective**, a connective acting on one expression to produce a second expression. **Negation** is a unary connective. The negation of A—symbolized by A'—is read "not A."

PRACTICE 4   Write the truth table for A'. (It will require only two rows.)   ●

Table 1.4 summarizes the truth values for all of the logical connectives. This information is critical to an understanding of logical reasoning.

A	B	$A \wedge B$	$A \vee B$	$A \to B$	$A \leftrightarrow B$	A'
T	T	T	T	T	T	F
T	F	F	T	F	F	
F	T	F	T	T	F	T
F	F	F	F	T	T	

Table 1.4

Because of the richness of the English language, words that have different shades of meaning are nonetheless represented by the same logical connective. Table 1.5 shows the common English words associated with various logical connectives.

REMINDER:
A only if *B*
means *A* → *B*.

English Word	Logical Connective	Logical Expression
and; but; also; in addition; moreover	Conjunction	$A \wedge B$
or	Disjunction	$A \vee B$
If A, then B. A implies B. A, therefore B. A only if B. B follows from A. A is a sufficient condition for B. B is a necessary condition for A.	Implication	$A \to B$
A if and only if B. A is necessary and sufficient for B.	Equivalence	$A \leftrightarrow B$
not A It is false that A ... It is not true that A ...	Negation	A'

Table 1.5

Suppose that $A \to B$ is true. Then, according to the truth table for implication, the consequent, B, can be true even though the antecedent, A, is false. So while the truth of

A leads to (implies) the truth of B, the truth of B does not imply the truth of A. The phrase "B is a necessary condition for A" to describe $A \to B$ simply means that if A is true, then B is necessarily true, as well. "A only if B" describes the same thing, that A implies B.

EXAMPLE 2

The statement "Fire is a necessary condition for smoke" can be restated as "If there is smoke, then there is fire." The antecedent is "there is smoke," and the consequent is "there is fire." ●

PRACTICE 5

Name the antecedent and consequent in each of the following statements. (*Hint:* Rewrite each statement in if–then form.)

a. If the rain continues, then the river will flood.
b. A sufficient condition for network failure is that the central switch goes down.
c. The avocados are ripe only if they are dark and soft.
d. A good diet is a necessary condition for a healthy cat. ●

EXAMPLE 3

Expressing the negation of a statement must be done with care, especially for a compound statement. Table 1.6 gives some examples. ●

Statement	Correct Negation	Incorrect Negation
It will rain tomorrow.	It is false that it will rain tomorrow. It will not rain tomorrow.	
Peter is tall and thin.	It is false that Peter is tall and thin. Peter is not tall or he is not thin. Peter is short or fat.	Peter is short and fat. Too strong a statement. Peter fails to have both properties (tallness and thinness) but may still have one property.
The river is shallow or polluted.	It is false that the river is shallow or polluted. The river is neither shallow nor polluted. The river is deep and unpolluted.	The river is not shallow or not polluted. Too weak a statement. The river fails to have either property, not just fails to have one property.

Table 1.6

PRACTICE 6

Which of the following represents A' if A is the statement "Julie likes butter but hates cream"?

a. Julie hates butter and cream.
b. Julie does not like butter or cream.
c. Julie dislikes butter but loves cream.
d. Julie hates butter or likes cream. ●

We can string statement letters, connectives, and parentheses (or brackets) together to form new expressions, as in

$$(A \rightarrow B) \wedge (B \rightarrow A)$$

Of course, just as in a computer programming language, certain *syntax rules* (rules on which strings are legitimate) prevail; for example,

$$A \,))\wedge \wedge \rightarrow BC$$

would not be considered a legitimate string. An expression that is a legitimate string is called a **well-formed formula**, or **wff**. In order to reduce the number of parentheses required in a wff, we stipulate an order in which connectives are applied. This *order of precedence* is:

1. connectives within parentheses, innermost parentheses first
2. $'$
3. $\wedge, \vee$
4. $\rightarrow$
5. $\leftrightarrow$

This means that the expression $A \vee B'$ stands for $A \vee (B)'$, not $(A \vee B)'$. Similarly, $A \vee B \rightarrow C$ means $(A \vee B) \rightarrow C$, not $A \vee (B \rightarrow C)$. However, we'll often use parentheses anyway, just to be sure there is no confusion.

In a wff with a number of connectives, the connective to be applied last is the **main connective**. In

$$A \wedge (B \rightarrow C)'$$

the main connective is $\wedge$. In

$$((A \vee B) \wedge C) \rightarrow (B \vee C')$$

the main connective is $\rightarrow$. Capital letters near the end of the alphabet, such as P, Q, R, and S, are used to represent wffs. Thus P could represent a single statement letter, which is the simplest kind of wff, or a more complex wff. We might represent

$$((A \vee B) \wedge C) \rightarrow (B \vee C')$$

as

$$P \rightarrow Q$$

if we want to hide some of the details for the moment and only concentrate on the main connective.

Wffs composed of statement letters and connectives have truth values that depend on the truth values assigned to their statement letters. We write the truth table for any wff by building up the component parts, just as we did for $(A \rightarrow B) \wedge (B \rightarrow A)$. The main connective is addressed in the last column of the table.

EXAMPLE 4   The truth table for the wff $A \vee B' \rightarrow (A \vee B)'$ is given in Table 1.7. The main connective, according to the rules of precedence, is implication.   ●

A	B	B'	$A \vee B'$	$A \vee B$	$(A \vee B)'$	$A \vee B' \to (A \vee B)'$
T	T	F	T	T	F	F
T	F	T	T	T	F	F
F	T	F	F	T	F	T
F	F	T	T	F	T	T

Table 1.7

If we are making a truth table for a wff that contains n different statement letters, how many rows will the truth table have? From truth tables done so far, we know that a wff with only one statement letter has two rows in its truth table, and a wff with two statement letters has four rows. The number of rows equals the number of true–false combinations possible among the statement letters. The first statement letter has two possibilities, T and F. For each of these possibilities, the second statement letter has two possible values. Figure 1.1a pictures this as a two-level "tree" with four branches showing the four possible combinations of T and F for two statement letters. For n statement letters, we extend the tree to n levels, as in Figure 1.1b. The total number of branches then equals 2^n. The total number of rows in a truth table for n statement letters is also 2^n.

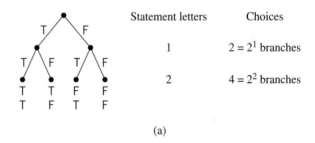

(a)

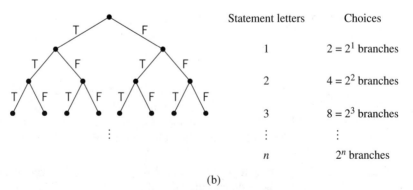

(b)

Figure 1.1

A	B	C
T	T	T
T	T	F
T	F	T
T	F	F
F	T	T
F	T	F
F	F	T
F	F	F

Table 1.8

This tree structure also tells us how to enumerate all the T–F combinations among the n statement letters when setting up a truth table. If we read each level of the tree from bottom to top, it says that the T–F values for statement letter n (which will compose the last column of the truth table) alternate, those for statement letter $n - 1$ alternate every two values, those for statement letter $n - 2$ alternate every four values, and so forth. Thus a truth table for three statement letters would begin as shown in Table 1.8. The values for statement letter C alternate, those for statement letter B alternate in groups of two, and those for statement letter A alternate in groups of four, resulting in something like a sideways version of the tree. (Reading the rows from the bottom up and using 1 for T and 0 for F shows that we are simply counting up from zero in binary numbers.)

PRACTICE 7   Construct truth tables for the following wffs.

    a. $(A \to B) \leftrightarrow (B \to A)$ (Remember that $C \leftrightarrow D$ is true precisely when C and D have the same truth value.)

    b. $(A \lor A') \to (B \land B')$

    c. $[(A \land B') \to C']'$

    d. $(A \to B) \leftrightarrow (B' \to A')$

Tautologies

A wff like item (d) of Practice 7, whose truth values are always true, is called a **tautology**. A tautology is "intrinsically true" by its very structure; it is true no matter what truth values are assigned to its statement letters. A simpler example of a tautology is $A \lor A'$; consider, for example, the statement "Today the sun will shine or today the sun will not shine," which must always be true because one or the other of these must happen. A wff like item (b) of Practice 7, whose truth values are always false, is called a **contradiction**. A contradiction is "intrinsically false" by its very structure. A simpler example of a contradiction is $A \land A'$; consider "Today is Tuesday and today is not Tuesday," which is false no matter what day of the week it is.

REMINDER:
A, B, C stand for single statement letters; P, Q, R, S stand for wffs.

Suppose that P and Q represent two wffs, and it happens that the wff $P \leftrightarrow Q$ is a tautology. If we did a truth table using the statement letters in P and Q, then the truth values of the wffs P and Q would agree for every row of the truth table. In this case, P and Q are said to be **equivalent wffs**, denoted by $P \Leftrightarrow Q$. Thus $P \Leftrightarrow Q$ states a fact, namely, that the particular wff $P \leftrightarrow Q$ is a tautology. Practice 7(d) has the form $P \leftrightarrow Q$, where P is the wff $(A \to B)$ and Q is the wff $(B' \to A')$, and $P \leftrightarrow Q$ was shown to be a tautology. Therefore, $(A \to B) \Leftrightarrow (B' \to A')$.

We shall list some basic equivalences, prove one or two of them by constructing truth tables, and leave the rest as exercises. We represent any contradiction by 0 and any tautology by 1.

Some Tautological Equivalences

1a. $A \lor B \Leftrightarrow B \lor A$	1b. $A \land B \Leftrightarrow B \land A$	(commutative properties)
2a. $(A \lor B) \lor C \Leftrightarrow$	2b. $(A \land B) \land C \Leftrightarrow$	(associative properties)
$A \lor (B \lor C)$	$A \land (B \land C)$	
3a. $A \lor (B \land C) \Leftrightarrow$	3b. $A \land (B \lor C) \Leftrightarrow$	(distributive properties)
$(A \lor B) \land (A \lor C)$	$(A \land B) \lor (A \land C)$	
4a. $A \lor 0 \Leftrightarrow A$	4b. $A \land 1 \Leftrightarrow A$	(identity properties)
5a. $A \lor A' \Leftrightarrow 1$	5b. $A \land A' \Leftrightarrow 0$	(complement properties)

Note that 2a allows us to write $A \lor B \lor C$ with no need for parentheses; similarly, 2b allows us to write $A \land B \land C$.

EXAMPLE 5   The truth table in Table 1.9a verifies equivalence 1a, the commutative property for disjunction, and that in Table 1.9b verifies 4b, the identity property for conjunction. Note that only two rows are needed for Table 1.9b because 1 (a tautology) cannot take on false truth values.

A	B	$A \lor B$	$B \lor A$	$A \lor B \leftrightarrow B \lor A$
T	T	T	T	T
T	F	T	T	T
F	T	T	T	T
F	F	F	F	T

(a)

A	1	$A \land 1$	$A \land 1 \leftrightarrow A$
T	T	T	T
F	T	F	T

(b)

Table 1.9

PRACTICE 8   Verify equivalence 5a.

The equivalences in the list are grouped into five pairs. In each pair, one equivalence can be obtained from the other by replacing $\land$ with $\lor$, $\lor$ with $\land$, 0 with 1, or 1 with 0. Each equivalence in a pair is called the **dual** of the other. Thus, 1a and 1b (commutativity of disjunction and commutativity of conjunction) are duals of each other. This list of equivalences appears in a more general setting in Chapter 7.

Two additional equivalences that are very useful are De Morgan's laws, named for the nineteenth-century British mathematician Augustus De Morgan, who first stated them.

De Morgan's Laws

$$(A \lor B)' \Leftrightarrow A' \land B' \qquad \text{and} \qquad (A \land B)' \Leftrightarrow A' \lor B'$$

Each is the dual of the other. De Morgan's laws help in expressing the negation of a compound statement, as in Practice 6.

Suppose that P and Q are equivalent wffs. Then P can be replaced by Q in any wff R containing P, resulting in a wff R_Q that is equivalent to R.

EXAMPLE 6   Let R be $(A \to B) \to B$, and let P be $A \to B$. From Practice 7(d), P is equivalent to $Q = B' \to A'$. Replacing P with Q in R, we get $R_Q = (B' \to A') \to B$. The truth tables for R and R_Q are shown in Tables 1.10 and 1.11, which show that R and R_Q are equivalent.

A	B	$A \to B$	$(A \to B) \to B$
T	T	T	T
T	F	F	T
F	T	T	T
F	F	T	F

Table 1.10

A	B	A'	B'	$B' \to A'$	$(B' \to A') \to B$
T	T	F	F	T	T
T	F	F	T	F	T
F	T	T	F	T	T
F	F	T	T	T	F

Table 1.11

Logical Connectives and Programming

The logical connectives AND, OR, and NOT are available in many programming languages, in which AND takes precedence over OR. These connectives, in accordance with the truth tables we have defined, act on combinations of true or false expressions to produce an overall truth value. Such truth values provide the decision-making capabilities fundamental to the flow of control in computer programs. Thus, at a conditional branch in a program, if the truth value of the conditional expression is true, the program will next execute one section of code; if the value is false, the program will next execute a different section of code. If the conditional expression is replaced by a simpler, equivalent expression, the truth value of the expression and hence the flow of control of the program is not affected, but the new code is easier to understand and may execute faster.

EXAMPLE 7   Consider a statement in a computer program that has the form

```
if ((outflow > inflow) and not ((outflow > inflow) and (pressure < 1000)))
    do something;
else
    do something else;
```

Here the conditional expression has the form

$$A \land (A \land B)'$$

where A is "outflow > inflow" and B is "pressure < 1000." This expression can be simplified by replacing some wffs with equivalent wffs.

$$
\begin{aligned}
A \land (A \land B)' &= A \land (A' \lor B') && \text{(De Morgan's Laws)} \\
&= (A \land A') \lor (A \land B') && \text{(tautology 3b)} \\
&= 0 \lor (A \land B') && \text{(tautology 5b)} \\
&= (A \land B') \lor 0 && \text{(tautology 1a)} \\
&= A \land B' && \text{(tautology 4a)}
\end{aligned}
$$

The statement form can therefore be written as

```
if ((outflow > inflow) and not (pressure < 1000))
    do something;
else
    do something else;
```

An Algorithm

To test whether a wff is a tautology, we can always write its truth table. For n statement letters, 2^n rows will be needed for the truth table. Suppose, however, that the wff has implication as its main connective, so that it has the form $P \rightarrow Q$ where P and Q are themselves wffs. Then we can use a quicker procedure than constructing a truth table to determine whether $P \rightarrow Q$ is a tautology. We assume that $P \rightarrow Q$ is *not* a tautology, and see whether this leads to some impossible situation. If it does, then the assumption that $P \rightarrow Q$ is not a tautology is also impossible, and $P \rightarrow Q$ must be a tautology after all.

To assume that $P \rightarrow Q$ is not a tautology is to say that it can take on false values, and, by the truth table for implication, $P \rightarrow Q$ is false only when P is true and Q false. By assigning P true and Q false, we determine possible truth values for the wffs making up P and Q. We continue assigning the truth values so determined until all occurrences of statement letters have a truth value. If some statement letter is assigned both true and false values by this process, we have an impossible situation, so the wff $P \rightarrow Q$ must be a tautology. Otherwise, we have found a way to make $P \rightarrow Q$ false, and it is not a tautology.

What we have described is a set of instructions—a procedure—for carrying out the task of determining whether $P \rightarrow Q$ is a tautology. This procedure can be executed by mechanically following the instructions; in a finite amount of time, we will have the answer. In computer science terms, the procedure is an *algorithm*.

Definition: Algorithm
An **algorithm** is a set of instructions that can be mechanically executed in a finite amount of time in order to solve some problem.

Algorithms constitute the very heart of computer science, and we shall have more to say about them throughout this book. You are probably already aware that the major task in writing a computer program for solving a problem consists of devising an algorithm (a procedure) to produce the problem solution.

Algorithms are often described in a form that is a middle ground between a purely verbal description in paragraph form (as we gave above for deciding if $P \rightarrow Q$ is a tautology) and a computer program (that, if executed, would actually carry out the steps of the algorithm) written in a programming language. This compromise form to describe algorithms is called **pseudocode**. An algorithm written in pseudocode should not be hard to understand even if you know nothing about computer programming. The only thing to note about the pseudocode used in this book is that lines preceded by double slashes (//) are explanatory comments, not part of the algorithm itself.

Following is a pseudocode form of the algorithm to determine whether $P \rightarrow Q$ is a tautology.

ALGORITHM   TautologyTest

TautologyTest(wff P; wff Q)
//Given wffs P and Q, decides whether the wff $P \rightarrow Q$ is a tautology.

  //Assume $P \rightarrow Q$ is not a tautology
  P = true   //assign T to P
  Q = false   //assign F to Q
  repeat
      for each compound wff already assigned a truth value,
          assign the truth values determined for its components
  until all occurrences of statements letters have truth values

  if some letter has two truth values **then** //contradiction, assumption false
          write ("$P \rightarrow Q$ is a tautology.")

  else   //found a way to make $P \rightarrow Q$ false
          write ("$P \rightarrow Q$ is not a tautology.")

  end if

end TautologyTest

The algorithm first assigns the truth values "true" to P and "false" to Q, consistent with the assumption that $P \to Q$ is not a tautology. The algorithm then enters a *loop*, where a sequence of steps is repeated until some condition is met. Within the loop, truth assignments continue to be made to smaller and smaller components of the original P and Q until all occurrences of individual statement letters have truth values. Then the algorithm tests whether a contradiction has occurred, and writes out the information about whether $P \to Q$ is a tautology.

EXAMPLE 8

Consider the wff $(A \to B) \to (B' \to A')$. This matches the pattern needed in order to use algorithm *TautologyTest*, namely $P \to Q$, where P is $A \to B$ and Q is $B' \to A'$. Following the algorithm, we first assign truth values

$A \to B$ true   and $B' \to A'$ false

Moving on to the loop, the assignment of false to the compound statement $B' \to A'$ determines the further assignments

B' true   and A' false

or

B false   and A true

Now working with P, A true and $A \to B$ true determines the assignment

B true

At this point all occurrences of statement letters have truth values, as follows:

$$
\begin{array}{cc}
A{-}T \; B{-}T & B{-}F \; A{-}T \\
\underline{(A \to B)} & \to \underline{(B' \to A')} \\
T & F
\end{array}
$$

This terminates the loop. In the final step of the algorithm, B now has an assignment of both T and F, so the algorithm decides that $(A \to B) \to (B' \to A')$ is a tautology. Actually, we learned this earlier (in Practice 7(d)) by building a truth table. ●

REMINDER:
Algorithm *TautologyTest* applies only when the main connective is →.

Algorithm *TautologyTest* decides whether wffs of a certain form are tautologies, namely those where the main logical connective is →. However, the process of building a truth table and then examining all the truth values in the final column constitutes an algorithm to decide whether an arbitrary wff is a tautology. This second algorithm is therefore more powerful because it solves a more general problem, but algorithm *TautologyTest* is usually faster for those wffs to which it applies.

Section 1.1 Review

Techniques

- Construct truth tables for compound wffs.
- Recognize tautologies and contradictions.

Main Ideas

Wffs are symbolic representations of statements.

Truth values for compound wffs depend on the truth values of their components and the types of connectives used.

Tautologies are "intrinsically true" wffs—true for all truth values.

Exercises 1.1

Answers to starred items are given in the back of the book.

★1. Which of the following are statements?
   a. The moon is made of green cheese.
   b. He is certainly a tall man.
   c. Two is a prime number.
   d. Will the game be over soon?
   e. Next year interest rates will rise.
   f. Next year interest rates will fall.
   g. $x^2 - 4 = 0$

2. Given the truth values A true, B false, and C true, what is the truth value of each of the following wffs?
   a. $A \wedge (B \vee C)$
   b. $(A \wedge B) \vee C$
   c. $(A \wedge B)' \vee C$
   d. $A' \vee (B' \wedge C)'$

3. What is the truth value of each of the following statements?
   a. 8 is even or 6 is odd.
   b. 8 is even and 6 is odd.
   c. 8 is odd or 6 is odd.
   d. 8 is odd and 6 is odd.
   e. If 8 is odd, then 6 is odd.
   f. If 8 is even, then 6 is odd.
   g. If 8 is odd, then 6 is even.
   h. If 8 is odd and 6 is even, then $8 < 6$.

★4. Find the antecedent and consequent in each of the following statements.

    a. Healthy plant growth follows from sufficient water.

    b. Increased availability of information is a necessary condition for further technological advances.

    c. Errors will be introduced only if there is a modification of the program.

    d. Fuel savings implies good insulation or storm windows throughout.

5. Several forms of negation are given for each of the following statements. Which are correct?

    a. The answer is either 2 or 3.

        1. Neither 2 nor 3 is the answer.

        2. The answer is not 2 or not 3.

        3. The answer is not 2 and it is not 3.

    b. Cucumbers are green and seedy.

        1. Cucumbers are not green and not seedy.

        2. Cucumbers are not green or not seedy.

        3. Cucumbers are green and not seedy.

    c. $2 < 7$ and 3 is odd.

        1. $2 > 7$ and 3 is even.

        2. $2 \geq 7$ and 3 is even.

        3. $2 \geq 7$ or 3 is odd.

        4. $2 \geq 7$ or 3 is even.

6. Write the negation of each wff.

    ★**a.** If the food is good, then the service is excellent.

    ★**b.** Either the food is good or the service is excellent.

    c. Either the food is good and the service is excellent, or else the price is high.

    d. Neither the food is good nor the service excellent.

    e. If the price is high, then the food is good and the service is excellent.

7. Let A, B, and C be the following statements:

    A Roses are red.

    B Violets are blue.

    C Sugar is sweet.

    Translate the following compound statements into symbolic notation.

    a. Roses are red and violets are blue.

    b. Roses are red, and either violets are blue or sugar is sweet.

    c. Whenever violets are blue, roses are red and sugar is sweet.

    d. Roses are red only if violets aren't blue or sugar is sour.

    e. Roses are red and, if sugar is sour, then either violets aren't blue or sugar is sweet.

8. Use A, B, and C as defined in Exercise 7 to translate the following statements into English.

   a. $B \lor C'$
   b. $B' \lor (A \to C)$
   c. $(C \land A') \leftrightarrow B$
   d. $C \land (A' \leftrightarrow B)$
   e. $(B \land C')' \to A$
   f. $A \lor (B \land C')$
   g. $(A \lor B) \land C'$

★9. Using letters for the component statements, translate the following compound statements into symbolic notation.

   a. If prices go up, then housing will be plentiful and expensive; but if housing is not expensive, then it will still be plentiful.
   b. Either going to bed or going swimming is a sufficient condition for changing clothes; however, changing clothes does not mean going swimming.
   c. Either it will rain or it will snow but not both.
   d. If Janet wins or if she loses, she will be tired.
   e. Either Janet will win or, if she loses, she will be tired.

10. Using letters for the component statements, translate the following compound statements into symbolic notation.

   a. If the horse is fresh, then the knight will win.
   b. The knight will win only if the horse is fresh and the armor is strong.
   c. A fresh horse is a necessary condition for the knight to win.
   d. The knight will win if and only if the armor is strong.
   e. A sufficient condition for the knight to win is that the armor is strong or the horse is fresh.

11. Construct truth tables for the following wffs. Note any tautologies or contradictions.

   ★**a.** $(A \to B) \leftrightarrow A' \lor B$
   ★**b.** $(A \land B) \lor C \to A \land (B \lor C)$
   c. $A \land (A' \lor B')'$
   d. $A \land B \to A'$
   e. $(A \to B) \to [(A \lor C) \to (B \lor C)]$
   f. $A \to (B \to A)$
   g. $A \land B \leftrightarrow B' \lor A'$
   h. $(A \lor B') \land (A \land B)'$
   i. $[(A \lor B) \land C'] \to A' \lor C$

★12. A memory chip from a microcomputer has 24 bistable (ON–OFF) memory elements. What is the total number of ON–OFF configurations?

13. Consider the following pseudocode.

   repeat
       $i = 1$
       read a value for x
       if $((\, x < 5.0)$ and $(2x < 10.7))$ or $\left(\sqrt{5x} > 5.1 \right)$ **then**
         write the value of x
       end if
       increase i by 1
     until $i > 5$

   The input values for x are 1.0, 5.1, 2.4, 7.2, and 5.3. What are the output values?

14. Verify the equivalences in the list on page 10 by constructing truth tables. (We have already verified 1a, 4b, and 5a.)

15. Verify by constructing truth tables that the following wffs are tautologies.
   ⋆**a.** $A \vee A'$
   b. $(A')' \leftrightarrow A$
   ⋆**c.** $A \wedge B \rightarrow B$
   d. $A \rightarrow A \vee B$
   e. $(A \vee B)' \leftrightarrow A' \wedge B'$ (De Morgan's Law)
   f. $(A \wedge B)' \leftrightarrow A' \vee B'$ (De Morgan's Law)

16. Suppose that A, B, and C represent conditions that will be true or false when a certain computer program is executed. Suppose further that you want the program to carry out a certain task only when A or B is true (but not both) and C is false. Using A, B, and C and the connectives AND, OR, and NOT, write a statement that will be true only under these conditions.

17. Rewrite the following statement form with a simplified conditional expression, where the function $odd(n)$ returns true if n is odd.

   if not$((Value1 < Value2)$ **or** $odd(Number))$
   or $(\mathbf{not}(Value1 < Value2)$ **and** $odd(Number))$ **then**
       statement1
   else
       statement2
   end if

18. **a.** Verify that $A \rightarrow B$ is equivalent to $A' \vee B$.
   b. Using part (a) and other equivalences, write the negation of the statement "If Sam passed his bar exam, then he will get the job."

19. Use algorithm *TautologyTest* to prove that the following are tautologies.
   ⋆**a.** $[B' \wedge (A \rightarrow B)] \rightarrow A'$
   b. $[(A \rightarrow B) \wedge A] \rightarrow B$
   c. $(A \vee B) \wedge A' \rightarrow B$
   d. $(A \wedge B) \wedge B' \rightarrow A$

20. In each case, construct compound wffs P and Q so that the given statement is a tautology.

   a. $P \wedge Q$
   b. $P \rightarrow P'$
   c. $P \wedge (Q \rightarrow P')$

21. From the truth table for $A \vee B$, the value of $A \vee B$ is true if A is true, if B is true, or if both are true. This use of the word "or," where the result is true if both components are true, is called the *inclusive or*. It is the inclusive or that is understood in the sentence "We may have rain or drizzle tomorrow." Another use of the word "or" in the English language is the *exclusive or*, sometimes written *XOR*, in which the result is false when both components are true. The exclusive or is understood in the sentence "At the intersection, you should turn north or south." Exclusive or is symbolized by $A \oplus B$.

   a. Write the truth table for the exclusive or.
   b. Show that $A \oplus B \leftrightarrow (A \leftrightarrow B)'$ is a tautology.

22. Every compound statement is equivalent to a statement using only the connectives of conjunction and negation. To see this, we need to find equivalent wffs for $A \vee B$ and for $A \rightarrow B$ that use only $\wedge$ and $'$. These new statements can replace, respectively, any occurrences of $A \vee B$ and $A \rightarrow B$. (The connective $\leftrightarrow$ was defined in terms of other connectives, so we already know that it can be replaced by a statement using these other connectives.)

   a. Show that $A \vee B$ is equivalent to $(A' \wedge B')'$.
   b. Show that $A \rightarrow B$ is equivalent to $(A \wedge B')'$.

23. Show that every compound wff is equivalent to a wff using only the connectives of

   a. $\vee$ and $'$
   b. $\rightarrow$ and $'$
   (*Hint:* See Exercise 22.)

24. Prove that there are compound statements that are not equivalent to any statement using only the connectives $\rightarrow$ and $\vee$.

★25. The binary connective $|$ is defined by the following truth table:

A	B	$A|B$
T	T	F
T	F	T
F	T	T
F	F	T

   Show that every compound wff is equivalent to a wff using only the connective $|$. (*Hint:* Use Exercise 22 and find equivalent statements for $A \wedge B$ and A' in terms of $|$.)

26. The binary connective $\downarrow$ is defined by the following truth table:

A	B	$A{\downarrow}B$
T	T	F
T	F	F
F	T	F
F	F	T

Show that every compound statement is equivalent to a statement using only the connective $\downarrow$. (*Hint:* See Exercise 25.)

In Exercises 27–29, you are traveling in a certain country where every inhabitant is either a truthteller who always tells the truth or a liar who always lies.[2]

★**27.** You meet two of the inhabitants of this country, Percival and Llewellyn. Percival says, "At least one of us is a liar." Is Percival a liar or a truthteller? What about Llewellyn? Explain your answer.

28. Traveling on, you meet Merlin and Meredith. Merlin says, "If I am a truthteller, then Meredith is a truthteller." Is Merlin a liar or a truthteller? What about Meredith? Explain your answer.

29. Finally, you meet Rothwold and Grymlin. Rothwold says, "Either I am a liar or Grymlin is a truthteller." Is Rothwold a liar or a truthteller? What about Grymlin? Explain your answer.

Section 1.2   Propositional Logic

The argument of the defense attorney at the beginning of this chapter made a number of (supposedly true) statements and then asked the jury to draw a specific conclusion based on those statements. In Section 1.1, we used the notation of formal logic to represent statements in symbolic form as wffs; because statements are sometimes called *propositions*, these wffs are also called **propositional wffs**. Now we want to use tools from formal logic to see how to reach logical conclusions based on given statements. The formal system that uses propositional wffs is called **propositional logic, statement logic**, or **propositional calculus**. (The word calculus is used here in the more general sense of "calculation" or "reasoning," not "differentiating" or "integrating.")

Valid Arguments

An argument can be represented in symbolic form as

$$P_1 \wedge P_2 \wedge P_3 \wedge \cdots \wedge P_n \rightarrow Q$$

where $P_1, P_2, \dots, P_n$ are the given statements, called the **hypotheses** of the argument, and Q is the **conclusion** of the argument. As usual, the P's and the Q represent wffs,

[2]*For more puzzles about "knights" and "knaves," see* What Is the Name of This Book? *by the logician—and magician—Raymond Smullyan (Prentice-Hall, 1978).*

not merely statement letters. When should this be considered a *valid argument*? This question can be stated in several equivalent ways:

- When can Q be *logically deduced from* $P_1, \dots, P_n$?
- When is Q a *logical conclusion from* $P_1, \dots, P_n$?
- When does $P_1, \dots, P_n$ *logically imply* Q?
- When does Q *follow logically from* $P_1, \dots, P_n$?

and so forth.

An informal answer is that Q is a logical conclusion from $P_1, \dots, P_n$ whenever the truth of $P_1, \dots, P_n$ implies the truth of Q. In other words, when the implication

$$P_1 \wedge P_2 \wedge P_3 \wedge \cdots \wedge P_n \to Q$$

is true. (Of course, this implication is true if any of the hypotheses is false, but in an argument we usually care about what happens when all the hypotheses are true.) Furthermore, this implication should be true based on the relationship of the conclusion to the hypotheses, not on any incidental knowledge we may happen to have about Q.

EXAMPLE 9   Consider the following argument:

George Washington was the first president of the United States. Thomas Jefferson wrote the Declaration of Independence. Therefore, every day has 24 hours.

This argument has the two hypotheses

1. George Washington was the first president of the United States.
2. Thomas Jefferson wrote the Declaration of Independence.

and the conclusion

Every day has 24 hours.

Even though each of the individual hypotheses, as well as the conclusion, is a true statement, we would *not* consider this argument valid. The conclusion is merely an isolated true fact, not at all related to or "following from" the hypotheses.   ●

A valid argument should therefore be true based entirely on its internal structure; it should be "intrinsically true." Therefore we make the following formal definition.

Definition: Valid Argument

The propositional wff

$$P_1 \wedge P_2 \wedge P_3 \wedge \cdots \wedge P_n \to Q$$

is a **valid argument** when it is a tautology.

The argument in Example 8 would be symbolized as

$$A \wedge B \to C$$

which is clearly not a tautology.

To test whether $P_1 \wedge P_2 \wedge P_3 \wedge \cdots \wedge P_n \to Q$ is a tautology, we could build a truth table or use algorithm *TautologyTest*. Instead, we will turn to formal logic, which uses a system of **derivation rules** that manipulate wffs in a truth-preserving manner. You begin with the hypotheses $P_1, \dots, P_n$ (assumed true) and attempt to apply the manipulation rules in such a way as to end up with the conclusion Q (which must then also be true because truth is preserved under the rules).

Definition: Proof Sequence

A **proof sequence** is a sequence of wffs in which each wff is either a hypothesis or the result of applying one of the formal system's derivation rules to earlier wffs in the sequence.

Using formal logic to prove that Q is a valid conclusion from $P_1, \dots, P_n$, we must produce a proof sequence of the form

P_1	(hypothesis)
P_2	(hypothesis)
$\vdots$	
P_n	(hypothesis)
wff_1	(obtained by applying a derivation rule to earlier wffs)
wff_2	(obtained by applying a derivation rule to earlier wffs)
$\vdots$	
Q	(obtained by applying a derivation rule to earlier wffs)

The derivation rules for a formal system must be carefully chosen. If they are too powerful, then they won't be truth-preserving and we'll be able to deduce anything at all from a given set of hypotheses. If they are too weak, there will be logical conclusions that we won't be able to prove from given hypotheses. We want a formal logic system that is **correct** (*only* valid arguments should be provable) and **complete** (*every* valid argument should be provable). In addition, the derivation rules should be kept to a minimum in order to make the formal system manageable. We would like the system to have the smallest set of rules that still allows it to be complete.

Derivation Rules for Propositional Logic

The derivation rules for propositional logic fall into two categories, equivalence rules and inference rules. Equivalence rules allow individual wffs to be rewritten, while inference rules allow new wffs to be derived from previous wffs in the proof sequence.

Equivalence rules state that certain pairs of wffs R and S are equivalent. Remember from Section 1.1 that $R \Leftrightarrow S$ means that $R \leftrightarrow S$ is a tautology and that S can be substituted for R in any wff with no change to its truth values. Equivalence rules are therefore truth-preserving; a true wff remains true if such a substitution is done within it.

Table 1.12 lists the equivalence rules we will use in our formal system for propositional logic. (Additional rules could be formulated based on other tautologies, but we are trying to keep our rule set to a minimum.) Each is given a name to make it easier to identify its use in a proof sequence. We saw the commutative and associative rules, as well as De Morgan's laws, in Section 1.1. There they were given for statement letters only, here they are given for any wffs P, Q, R, but they are still tautologies.

	Equivalence Rules	
Expression	Equivalent to	Name/Abbreviation for Rule
$P \lor Q$ $P \land Q$	$Q \lor P$ $Q \land P$	Commutative—comm
$(P \lor Q) \lor R$ $(P \land Q) \land R$	$P \lor (Q \lor R)$ $P \land (Q \land R)$	Associative—ass
$(P \lor Q)'$ $(P \land Q)'$	$P' \land Q'$ $P' \lor Q'$	De Morgan's Laws—De Morgan
$P \to Q$	$P' \lor Q$	Implication—imp
P	$(P')'$	Double negation—dn

Table 1.12

PRACTICE 9   Prove the Implication rule, that is, prove that

$$(P \to Q) \leftrightarrow (P' \lor Q)$$

is a tautology.

EXAMPLE 10   Suppose that one hypothesis of a propositional argument can be symbolized as

$$(A' \lor B') \lor C$$

Then a proof sequence for the argument could begin with the following steps:

1. $(A' \lor B') \lor C$ hyp (hypothesis)
2. $(A \land B)' \lor C$ 1, De Morgan
3. $(A \land B) \to C$ 2, imp

The justification given for each step is not a required part of the proof sequence, but it does confirm that the step is a legitimate one. Step 1 is a hypothesis. Step 2 is derived from step 1 by applying one of De Morgan's Laws. Step 3 is derived from step 2 by using the Implication rule that $P \to Q$ is equivalent to $P' \lor Q$, where P is the wff $A \land B$, and Q is the wff C.

The equivalence rules allow substitution in either direction. That is, in Example 10 we replaced $A' \lor B'$ with $(A \land B)'$, but in some other proof sequence, using the same rule, we might replace $(A \land B)'$ with $A' \lor B'$.

Inference rules say that if one or more wffs that match the first part of the rule pattern are already part of the proof sequence, we can add to the proof sequence a new wff that matches the last part of the rule pattern. Table 1.13 shows the propositional inference rules we will use, again along with their identifying names.

		Inference Rules	
From	Can Derive	Name/Abbreviation for Rule	
$P, P \rightarrow Q$	Q	Modus ponens—mp	
$P \rightarrow Q, Q'$	P'	Modus tollens—mt	
P, Q	$P \wedge Q$	Conjunction—con	
$P \wedge Q$	P, Q	Simplification—sim	
P	$P \vee Q$	Addition—add	

Table 1.13

Unlike equivalence rules, inference rules do not work in both directions. We cannot "reverse" the Addition rule in Table 1.13; from $P \vee Q$, we cannot infer either P or Q.

EXAMPLE 11

Suppose that $A \rightarrow (B \wedge C)$ and A are two hypotheses of an argument. A proof sequence for the argument could begin with the following steps:

1. $A \rightarrow (B \wedge C)$ hyp
2. A hyp
3. $B \wedge C$ 1, 2, mp

The justification at step 3 is that steps 1 and 2 exactly match the pattern required for modus ponens, where P is A and Q is $B \wedge C$. Modus ponens ("method of assertion" in Latin) says that Q can be derived from P and $P \rightarrow Q$.   ●

PRACTICE 10

Give a next step and a justification for a proof sequence that begins

1. $(A \wedge B') \rightarrow C$ hyp
2. C' hyp   ●

The inference rules are also truth-preserving. For example, suppose that P and $P \rightarrow Q$ are both true wffs in a proof sequence. Then Q is deducible from these two wffs by modus ponens. If P and $P \rightarrow Q$ are both true, then—by the truth table for implication—Q is also true.

The derivation rules represent recipes or patterns for transforming wffs. A rule can be applied only when the wffs exactly match the pattern.

EXAMPLE 12   Suppose that $(A \rightarrow B) \vee C$ and A are two hypotheses of an argument. A proof sequence for the argument could begin with the following steps:

1. $(A \rightarrow B) \vee C$ hyp
2. A hyp

REMINDER:
To use a derivation rule, wffs must exactly match the rule pattern.

Unlike Example 11, however, nothing further can be done. Modus ponens requires the presence of wffs matching the pattern P and $P \rightarrow Q$. In $P \rightarrow Q$, the main connective is an implication. The wff $(A \rightarrow B) \vee C$ has disjunction, not implication, as its main connective. Modus ponens does not apply. ●

Now we are ready to work our way through a complete proof of an argument.

EXAMPLE 13   Using propositional logic, prove that the argument

$$A \wedge (B \rightarrow C) \wedge [(A \wedge B) \rightarrow (D \vee C')] \wedge B \rightarrow D$$

is valid.

We must produce a proof sequence that begins with the hypotheses and ends with the conclusion. There are four hypotheses, so this gives us lots of "ammunition" to use in the proof. The beginning of the proof is easy enough because it just involves listing the hypotheses:

1. A hyp
2. $B \rightarrow C$ hyp
3. $(A \wedge B) \rightarrow (D \vee C')$ hyp
4. B hyp

Our final goal is to arrive at D, the conclusion. But without even looking ahead, there are a couple of fairly obvious steps we can take that may or may not be helpful.

5. C 2, 4, mp
6. $A \wedge B$ 1, 4, con
7. $D \vee C'$ 3, 6, mp

At least at this point we have introduced D, but it's not by itself. Note that from step 5 we have C, which we haven't made use of. If only we had $C \rightarrow D$, we'd be home free. Ah, look at the form of step 7; it's a disjunction, and the Implication rule says that we can transform a disjunction of a certain form into an implication. The disjunction must have a negated wff on the left. We can do that:

8. $C' \vee D$ 7, comm
9. $C \rightarrow D$ 8, imp

and so

10. D 5, 9, mp ●

As in Example 13, proof sequences involve a certain amount of rewriting just because you can and a certain amount of keeping an eye on the desired goal and what it would take to get there. Although not as mechanical as constructing a truth table, the strict rules of the game nevertheless provide a more or less mechanical way to construct the proof sequence. There are only a certain number of legitimate things that can be done at any one point in the sequence. If one choice seems to lead down a blind alley, go back and take another. Also, there may be more than one correct proof sequence; as a relatively trivial instance, steps 6 and 7 could have been done before step 5 in Example 13.

PRACTICE 11   Using propositional logic, prove the validity of the argument

$$[(A \lor B') \to C] \land (C \to D) \land A \to D$$

●

> **Derivation Hints**
>
> 1. Modus ponens is probably the most intuitive inference rule. Think often about trying to use it.
> 2. Wffs of the form $(P \land Q)'$ or $(P \lor Q)'$ are seldom helpful in a proof sequence. Try using De Morgan's Laws to convert them into $P' \lor Q'$ and $P' \land Q'$, respectively, which breaks out the individual components.
> 3. Wffs of the form $P \lor Q$ are also seldom helpful in a proof sequence because they do not imply either P or Q. Try using Double negation to convert $P \lor Q$ to $(P')' \lor Q$, and then using Implication to convert to $P' \to Q$.

Deduction Method and Other Rules

Suppose the argument we seek to prove has the form

$$P_1 \land P_2 \land P_3 \land \cdots \land P_n \to (R \to S)$$

where the conclusion is itself an implication. Instead of using $P_1, \dots, P_n$ as the hypotheses and deriving $R \to S$, the *deduction method* lets us add R as an additional hypothesis and then derive S. In other words, we can instead prove

$$P_1 \land P_2 \land P_3 \land \cdots \land P_n \land R \to S$$

This is to our advantage because it gives us one more hypothesis, i.e., additional ammunition for the proof, and it simplifies the desired conclusion.

The deduction method approach agrees with our understanding of implication, but Exercise 36 at the end of this section provides a formal justification.

EXAMPLE 14   Use propositional logic to prove

$$[A \to (A \to B)] \to (A \to B)$$

Using the deduction method, we get two hypotheses instead of one, and we want to derive B.

1. $A \to (A \to B)$	hyp
2. A	hyp
3. $A \to B$	1, 2, mp
4. B	2, 3, mp

●

PRACTICE 12   Use propositional logic to prove

$$(A \to B) \land (B \to C) \to (A \to C)$$

●

The formal system we have described is correct and complete. Every argument we can prove is a tautology (the system is correct), and every implication that is a tautology is provable (the system is complete). We can easily argue for correctness because each of the derivation rules is truth-preserving. Completeness would be more difficult to prove, and we will not do so.

Correctness and completeness say that the set of derivation rules we have used is exactly right—not too strong, not too weak. Nonetheless, many formal systems for propositional logic use additional truth-preserving inference rules. We can prove these additional rules using our original rule set. Once such a rule is proved, it can be used as justification in a proof sequence because, if required, the single step invoking this rule could be replaced with the proof sequence for the rule. Nothing more can be proved by the addition of these rules, but the proof sequences might be shorter. (See Exercises 1.2 for a list of additional rules.)

EXAMPLE 15   The rule of Hypothetical Syllogism (hs) is

> From $P \rightarrow Q$ and $Q \rightarrow R$, one can derive $P \rightarrow R$.

This rule is making the claim that

$$(P \rightarrow Q) \wedge (Q \rightarrow R) \rightarrow (P \rightarrow R)$$

is a valid argument. The proof sequence for this argument looks just like that for Practice 12. Because it is a legitimate derivation rule, Hypothetical Syllogism can be used to justify a step in a proof sequence.   ●

EXAMPLE 16   Use propositional logic to prove

$$(A' \vee B) \wedge (B \rightarrow C) \rightarrow (A \rightarrow C)$$

The following proof sequence will do.

1. $A' \vee B$ hyp
2. $B \rightarrow C$ hyp
3. $A \rightarrow B$ 1, imp
4. $A \rightarrow C$ 2, 3, hs

Without use of the new rule, we could still have produced a proof sequence by essentially proving the new rule as part of this proof:

1. $A' \vee B$ hyp
2. $B \rightarrow C$ hyp
3. $A \rightarrow B$ 1, imp
4. A hyp
5. B 3, 4, mp
6. C 2, 5, mp

Additional rules thus can shorten proof sequences but at the expense of having to remember additional rules!   ●

PRACTICE 13   Prove

$$(A \rightarrow B) \wedge (C' \vee A) \wedge C \rightarrow B$$ ●

Verbal Arguments

An argument in English (an attorney's trial summary, an advertisement, or a political speech) that consists of simple statements can be tested for validity by a two-step process.

1. Symbolize the argument using propositional wffs.
2. Prove that the argument is valid by constructing a proof sequence for it using the derivation rules for propositional logic.

EXAMPLE 17

Consider the argument "If interest rates drop, the housing market will improve. Either the federal discount rate will drop or the housing market will not improve. Interest rates will drop. Therefore the federal discount rate will drop." Using

I Interest rates drop.
H The housing market will improve.
F The federal discount rate will drop.

the argument is

$$(I \rightarrow H) \wedge (F \vee H') \wedge I \rightarrow F$$

A proof sequence to establish validity is

1.	$I \rightarrow H$	hyp
2.	$F \vee H'$	hyp
3.	I	hyp
4.	$H' \vee F$	2, comm
5.	$H \rightarrow F$	4, imp
6.	$I \rightarrow F$	1, 5, hs
7.	F	3, 6, mp

EXAMPLE 18

Is the following argument valid? "My client is left-handed, but if the diary is not missing, then my client is not left-handed; therefore, the diary is missing." There are only two simple statements involved here, so we symbolize them as follows:

L My client is left-handed.
D The diary is missing.

The argument is then

$$L \wedge (D' \rightarrow L') \rightarrow D$$

The validity of the argument is established by the following proof sequence.

1.	L	hyp
2.	$D' \rightarrow L'$	hyp
3.	$(D')' \vee L'$	2, imp
4.	$D \vee L'$	3. dn
5.	$L' \vee D$	4, comm
6.	$L \rightarrow D$	5, imp
7.	D	1, 6, mp

The argument says that *if* the hypotheses are true, then the conclusion will be true. The validity of the argument is a function only of its logical form and has nothing to do with the actual truth of any of its components. We still have no idea about whether the diary is really missing. Furthermore, the argument "Skooses are pink, but if Gingoos does not like perskees, then skooses are not pink; therefore Gingoos does like perskees," which has the same logical form, is also valid, even though it does not make sense.

PRACTICE 14   Use propositional logic to prove that the following argument is valid. Use statement letters S, R, and B: "If security is a problem, then regulation will be increased. If security is not a problem, then business on the Web will grow. Therefore if regulation is not increased, then business on the Web will grow."

Formal logic is not necessary to prove the validity of propositional arguments. A valid argument is represented by a tautology, and truth tables provide a mechanical test for whether a wff is a tautology. So, what was the point of all of this? In the next section we will see that propositional wffs are not sufficient to represent everything we would like to say, and we will devise new wffs called *predicate wffs*. There is no mechanical test for the predicate wff analogue of tautology, and in the absence of such a test, we will have to rely on formal logic to justify arguments. We have developed formal logic for propositional arguments as a sort of dry run for the predicate case.

In addition, the sort of reasoning we have used in propositional logic carries over into everyday life. It is the foundation for logical thinking in computer science, mathematics, the courtroom, the marketplace, the laboratory. Although we have approached logic as a mechanical system of applying rules, enough practice should ingrain this way of thinking so that you no longer need to consult tables of rules, but can draw logical conclusions and recognize invalid arguments on your own.

Section 1.2 Review

Techniques

- Apply derivation rules for propositional logic.
- Use propositional logic to prove the validity of a verbal argument.

Main Ideas

A valid argument can be represented by a wff of the form $P_1 \wedge P_2 \wedge P_3 \wedge \cdots \wedge P_n \rightarrow Q$ that is a tautology.

A proof sequence in a formal logic system is a sequence of wffs that are either hypotheses or derived from earlier wffs in the sequence by the derivation rules of the system.

The propositional logic system is complete and correct; valid arguments and only valid arguments are provable.

Exercises 1.2

★1. Justify each step in the proof sequence of

$$[A \to (B \lor C)] \land B' \land C' \to A'$$

1. $A \to (B \lor C)$
2. B'
3. C'
4. $B' \land C'$
5. $(B \lor C)'$
6. A'

2. Justify each step in the proof sequence of

$$A' \land B \land [B \to (A \lor C)] \to C$$

1. A'
2. B
3. $B \to (A \lor C)$
4. $A \lor C$
5. $(A')' \lor C'$
6. $A' \to C$
7. C

In Exercises 3–11, use propositional logic to prove that the argument is valid.

3. $A' \land (B \to A) \to B'$

★4. $(A \to B) \land [A \to (B \to C)] \to (A \to C)$

5. $[(C \to D) \to C] \to [(C \to D) \to D]$

★6. $A' \land (A \lor B) \to B$

7. $[A \to (B \to C)] \land (A \lor D') \land B \to (D \to C)$

8. $(A' \to B') \land B \land (A \to C) \to C$

9. $(A \to B) \land [B \to (C \to D)] \land [A \to (B \to C)] \to (A \to D)$

10. $[A \to (B \to C)] \to [B \to (A \to C)]$

11. $(A \land B) \to (A \to B')'$

Use propositional logic to prove the validity of the arguments in Exercises 12–20. These will become additional derivation rules for propositional logic, summarized in Table 1.14.

★12. $(P \lor Q) \land P' \to Q$

13. $(P \to Q) \to (Q' \to P')$

14. $(Q' \to P') \to (P \to Q)$

15. $P \rightarrow P \wedge P$

16. $P \vee P \rightarrow P$ (*Hint:* Instead of assuming the hypothesis, begin with a version of Exercise 15; also make use of Exercise 14.)

17. $[(P \wedge Q) \rightarrow R] \rightarrow [P \rightarrow (Q \rightarrow R)]$

★18. $P \wedge P' \rightarrow Q$

19. $P \wedge (Q \vee R) \rightarrow (P \wedge Q) \vee (P \wedge R)$ (*Hint:* First rewrite the conclusion.)

20. $P \vee (Q \wedge R) \rightarrow (P \vee Q) \wedge (P \vee R)$ (*Hint:* Prove both $P \vee (Q \wedge R) \rightarrow (P \vee Q)$ and $P \vee (Q \wedge R) \rightarrow (P \vee R)$; for each proof, first rewrite the conclusion.)

More Inference Rules		
From	Can Derive	Name/Abbreviation for Rule
$P \rightarrow Q, Q \rightarrow R$	$P \rightarrow R$ [Example 15]	Hypothetical syllogism—hs
$P \vee Q, P'$	Q [Exercise 12]	Disjunctive syllogism—ds
$P \rightarrow Q$	$Q' \rightarrow P'$ [Exercise 13]	Contraposition—cont
$Q' \rightarrow P'$	$P \rightarrow Q$ [Exercise 14]	Contraposition—cont
P	$P \wedge P$ [Exercise 15]	Self-reference—self
$P \vee P$	P [Exercise 16]	Self-reference—self
$(P \wedge Q) \rightarrow R$	$P \rightarrow (Q \rightarrow R)$ [Exercise 17]	Exportation—exp
P, P'	Q [Exercise 18]	Inconsistency—inc
$P \wedge (Q \vee R)$	$(P \wedge Q) \vee (P \wedge R)$ [Exercise 19]	Distributive—dist
$P \vee (Q \wedge R)$	$(P \vee Q) \wedge (P \vee R)$ [Exercise 20]	Distributive—dist

Table 1.14

For Exercises 21–28, use propositional logic to prove the arguments valid; you may use any of the rules in Table 1.14 or any previously proved exercise.

21. $A' \rightarrow (A \rightarrow B)$

★22. $(P \rightarrow Q) \wedge (P' \rightarrow Q) \rightarrow Q$

23. $(A' \rightarrow B') \wedge (A \rightarrow C) \rightarrow (B \rightarrow C)$

24. $(A' \rightarrow B) \wedge (B \rightarrow C) \wedge (C \rightarrow D) \rightarrow (A' \rightarrow D)$

25. $(A \vee B) \wedge (A \rightarrow C) \wedge (B \rightarrow C) \rightarrow C$

26. $(Y \to Z') \wedge (X' \to Y) \wedge [Y \to (X \to W)] \wedge (Y \to Z) \to (Y \to W)$

★27. $(A \wedge B)' \wedge (C' \wedge A)' \wedge (C \wedge B')' \to A'$

28. $(P \vee (Q \wedge R)) \wedge (R' \vee S) \wedge (S \to T') \to (T \to P)$

Using propositional logic, including the rules in Table 1.14, prove that each argument in Exercises 29–35 is valid. Use the statement letters shown.

29. If the program is efficient, it executes quickly: Either the program is efficient, or it has a bug. However, the program does not execute quickly. Therefore it has a bug. E, Q, B

30. If Jane is more popular, then she will be elected. If Jane is more popular, then Craig will resign. Therefore if Jane is more popular, she will be elected and Craig will resign. J, E, C

31. The crop is good, but there is not enough water. If there is a lot of rain or not a lot of sun, then there is enough water. Therefore the crop is good and there is a lot of sun. C, W, R, S

32. If the ad is successful, then the sales volume will go up. Either the ad is successful or the store will close. The sales volume will not go up. Therefore the store will close. A, S, C

★33. Russia was a superior power, and either France was not strong or Napoleon made an error. Napoleon did not make an error, but if the army did not fail, then France was strong. Hence the army failed and Russia was a superior power. R, F, N, A

34. It is not the case that if electric rates go up, then usage will go down, nor is it true that either new power plants will be built or bills will not be late. Therefore usage will not go down and bills will be late. R, U, P, B

35. If Jose took the jewelry or Mrs. Krasov lied, then a crime was committed. Mr. Krasov was not in town. If a crime was committed, then Mr. Krasov was in town. Therefore Jose did not take the jewelry. J, L, C, T

★36. **a.** Use a truth table to verify that $A \to (B \to C) \leftrightarrow (A \wedge B) \to C$ is a tautology.
   b. Prove that $A \to (B \to C) \Leftrightarrow (A \wedge B) \to C$ by using a series of equivalences.
   c. Explain how this equivalence justifies the deduction method that says: to prove $P_1 \wedge P_2 \wedge \cdots \wedge P_n \to (R \to S)$, deduce S from $P_1, P_2, \ldots, P_n$, and R.

37. The argument of the defense attorney at the beginning of this chapter was

   If my client is guilty, then the knife was in the drawer. Either the knife was not in the drawer or Jason Pritchard saw the knife. If the knife was not there on October 10, it follows that Jason Pritchard didn't see the knife. Furthermore, if the knife was there on October 10, then the knife was in the drawer and also the hammer was in the barn. But we all know that the hammer was not in the barn. Therefore, ladies and gentlemen of the jury, my client is innocent.

   Use propositional logic to prove that this is a valid argument.

Section 1.3   Quantifiers, Predicates, and Validity

Quantifiers and Predicates

Propositional wffs have rather limited expressive power. For example, we would consider the sentence "For every x, $x > 0$" to be a true statement about the positive integers, yet it cannot be adequately symbolized using only statement letters, parentheses, and logical connectives. It contains two new features, a *quantifier* and a *predicate*. Quantifiers are phrases such as "for every" or "for each" or "for some" that tell in some sense *how many* objects have a certain property. The **universal quantifier** is symbolized by an upside down A, $\forall$, and is read "for all," "for every," "for each," or "for any." Thus the example sentence can be symbolized by

$$(\forall x)(x > 0)$$

A quantifier and its named variable are always placed in parentheses. The second set of parentheses shows that the quantifier acts upon the enclosed expression, which in this case is "$x > 0$."

The phrase "$x > 0$" describes a property of the variable x, that of being positive. A property is also called a **predicate;** the notation $P(x)$ is used to represent some unspecified predicate or property that x may have. Thus, our original sentence is an example of the more general form

$$(\forall x)P(x)$$

The truth value of the expression $(\forall x)(x > 0)$ depends on the domain of objects in which we are "interpreting" this expression, that is, the collection of objects from which x may be chosen. This collection of objects is called the *domain of interpretation.* We have already agreed that if the domain of interpretation consists of the positive integers, the expression has the truth value true because every possible value for x has the required property of being greater than zero. If the domain of interpretation consists of all the integers, the expression has the truth value false, because not every x has the required property. We impose the condition that the domain of interpretation contain at least one object so that we are not talking about a trivial case.

An interpretation of the expression $(\forall x)P(x)$ would consist of not only the collection of objects from which x could take its value but also the particular property that $P(x)$ represents in this domain. Thus an interpretation for $(\forall x)P(x)$ could be the following: The domain consists of all the books in your local library, and $P(x)$ is the property that x has a red cover. In this interpretation, $(\forall x)P(x)$ says that every book in your local library has a red cover. The truth value of this expression, in this interpretation, is undoubtedly false.

PRACTICE 15   What is the truth value of the expression $(\forall x)P(x)$ in each of the following interpretations?

    a. $P(x)$ is the property that x is yellow, and the domain of interpretation is the collection of all buttercups.

b. $P(x)$ is the property that x is yellow, and the domain of interpretation is the collection of all flowers.

c. $P(x)$ is the property that x is a plant, and the domain of interpretation is the collection of all flowers.

d. $P(x)$ is the property that x is either positive or negative, and the domain of interpretation consists of the integers.

The **existential quantifier** is symbolized by a backwards E, $\exists$, and is read "there exists one," "for at least one," or "for some." Thus the expression

$$(\exists x)(x > 0)$$

is read "there exists an x such that x is greater than zero."

Again, the truth value of this expression depends on the interpretation. If the domain of interpretation contains a positive number, the expression has the value true; otherwise, it has the value false. The truth value of $(\exists x)P(x)$, if the domain consists of all the books in your local library and $P(x)$ is the property that x has a red cover, is true if there is at least one book in the library with a red cover.

PRACTICE 16

REMINDER:
All, every, each, any—use $\forall$

some, one, at least one— use $\exists$

a. Construct an interpretation (i.e., give the domain and the meaning of $P(x)$) in which $(\forall x)P(x)$ has the value true.

b. Construct an interpretation in which $(\forall x)P(x)$ has the value false.

c. Can you find one interpretation in which both $(\forall x)P(x)$ is true and $(\exists x)P(x)$ is false?

d. Can you find one interpretation in which both $(\forall x)P(x)$ is false and $(\exists x)P(x)$ is true?

The predicates we have seen so far, involving properties of a single variable, are **unary predicates.** Predicates can be **binary,** involving properties of two variables, **ternary,** involving properties of three variables, or, more generally, **n-ary,** involving properties of n variables.

EXAMPLE 19

The expression $(\forall x)(\exists y)Q(x, y)$ is read "for every x there exists a y such that $Q(x, y)$." Note how there are two quantifiers for the two variables of the binary property. In the interpretation where the domain consists of the integers and $Q(x, y)$ is the property that $x < y$, this just says that for any integer, there is a larger integer. The truth value of the expression is true. In the same interpretation, the expression $(\exists y)(\forall x)Q(x, y)$ says that there is a single integer y that is larger than any integer x. The truth value here is false.

Example 19 illustrates that the order in which the quantifiers appear is important. In expressions such as $(\forall x)P(x)$ or $(\exists x)P(x)$, x is a *dummy variable;* that is, the truth values of the expressions remain the same in a given interpretation if they are written, say, as $(\forall y)P(y)$ or $(\exists z)P(z)$, respectively. Similarly, the truth value of $(\forall x)(\exists y)Q(x, y)$ is the same as that of $(\forall z)(\exists w)Q(z, w)$ in any interpretation. However, $(\forall x)(\exists x)Q(x, x)$ says something quite different. In the interpretation of Example 19, for instance, $(\forall x)(\exists x)Q(x, x)$ says that for every integer x, there is an integer x such that $x < x$. This statement is false, even though $(\forall x)(\exists y)Q(x, y)$ was true in this interpretation. We cannot collapse separate variables together into one without changing the nature of the expression we obtain.

Constants are also allowed in expressions. A constant symbol (a, b, c, 0, 1, 2, etc.) is interpreted as some specific object in the domain. This specification is part of the interpretation. For example, the expression $(\forall x)Q(x, a)$ is false in the interpretation where the domain consists of the integers, $Q(x, y)$ is the property $x < y$, and a is assigned the value 7; it is not the case that every integer is less than 7.

Now we can sum up what is required in an interpretation.

Definition: Interpretation

An **interpretation** for an expression involving predicates consists of the following:

a. A collection of objects, called the **domain** of the interpretation, which must include at least one object

b. An assignment of a property of the objects in the domain to each predicate in the expression

c. An assignment of a particular object in the domain to each constant symbol in the expression

Expressions can be built by combining predicates with quantifiers, grouping symbols (parentheses or brackets), and the logical connectives of Section 1.1. As before, an expression must obey rules of syntax in order to be considered a well-formed formula. Well-formed formulas containing predicates and quantifiers are called **predicate wffs** to distinguish them from propositional wffs, which contain only statement letters and logical connectives.

The expression $P(x)(\forall x) \wedge \exists y$ is not a well-formed formula. Examples of predicate wffs are

$$P(x) \vee Q(y) \tag{1}$$

$$(\forall x)[P(x) \rightarrow Q(x)] \tag{2}$$

$$(\forall x)((\exists y)[P(x, y) \wedge Q(x, y)] \rightarrow R(x)) \tag{3}$$

and

$$(\exists x)S(x) \vee (\forall y)T(y) \tag{4}$$

"Grouping symbols" such as parentheses and brackets identify the **scope** of a quantifier, the section of the wff to which the quantifier applies. (This is analogous to the scope of an identifier in a computer program as the section of the program in which that identifier has meaning.) There are no quantifiers in wff (1). In (2), the scope of the quantifier $(\forall x)$ is $P(x) \rightarrow Q(x)$. In (3), the scope of $(\exists y)$ is $P(x, y) \wedge Q(x, y)$, while the scope of $(\forall x)$ is the entire expression in parentheses following it. In (4), the scope of $(\exists x)$ is $S(x)$ and the scope of $(\forall y)$ is $T(y)$.

If a variable occurs somewhere in a wff where it is not part of a quantifier and is not within the scope of a quantifier involving that variable, it is called a **free variable.** For example, y is a free variable in

$$(\forall x)[Q(x, y) \rightarrow (\exists y)R(x, y)]$$

because of the first occurrence of y, which is neither the variable of a quantifier nor within the scope of a quantifier using y. A wff with free variables may not have a truth value at all in a given interpretation. For example, in the interpretation where the domain is all of the integers, the predicate $P(x)$ means "$x > 0$," and 5 means (of course) the integer 5, the wff

$$P(y) \wedge P(5)$$

has no truth value because we don't know which element of the domain y refers to. Some elements of the domain are positive and others are not. The wff

$$P(y) \vee P(5)$$

is true in this interpretation even though we don't know what y refers to because $P(5)$ is true. In both of these wffs y is a free variable.

EXAMPLE 20   In the wff

$$(\forall x)(\exists y)[S(x, y) \wedge L(y, a)]$$

the scope of $(\exists y)$ is all of $S(x, y) \wedge L(y, a)$. The scope of $(\forall x)$ is $(\exists y)[S(x, y) \wedge L(y, a)]$; parentheses or brackets can be eliminated when the scope is clear. Consider the interpretation where the domain consists of all the cities in the United States, $S(x, y)$ is the property "x and y are in the same state," $L(y, z)$ is the property "y's name begins with the same letter as z's name," and a is assigned the value Albuquerque. So the interpretation of the entire wff is that for any city x there is a city y in the same state that begins with the letter A. The wff is true in this interpretation. (At least it is true if every state has a city beginning with the letter "A.")   ●

PRACTICE 17   What is the truth value of the wff

$$(\exists x)(A(x) \wedge (\forall y)[B(x, y) \rightarrow C(y)])$$

in the interpretation where the domain consists of all integers, $A(x)$ is "$x > 0$," $B(x, y)$ is "$x > y$," and $C(y)$ is "$y \leq 0$"? Construct another interpretation with the same domain in which the statement has the opposite truth value.   ●

Translation

Many English language statements can be expressed as predicate wffs. For example, "Every parrot is ugly" is really saying "For any thing, if it is a parrot, then it is ugly."

Letting $P(x)$ denote "x is a parrot" and $U(x)$ denote "x is ugly," the statement can be symbolized as

$$(\forall x)[P(x) \rightarrow U(x)]$$

Other English language variations that take the same symbolic form are "All parrots are ugly" and "Each parrot is ugly." Notice that the quantifier is the universal quantifier and the logical connective is implication; $\forall$ and $\rightarrow$ almost always belong together. The wff $(\forall x)[P(x) \land U(x)]$ is an incorrect translation because it says that everything in the domain—understood here to be the whole world—is an ugly parrot. This says something much stronger than the original English statement.

Similarly, "There is an ugly parrot" is really saying "There exists something that is both a parrot and ugly." In symbolic form,

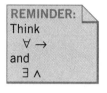

REMINDER:
Think
$\forall \rightarrow$
and
$\exists \land$

$$(\exists x)[P(x) \land U(x)]$$

Variations are "Some parrots are ugly" and "There are ugly parrots." Here the quantifier is the existential quantifier and the logical connective is conjunction; $\exists$ and $\land$ almost always belong together. The wff $(\exists x)[P(x) \rightarrow U(x)]$ is an incorrect translation. This wff is true as long as there is anything, call it x, in the domain (the whole world) that is not a parrot, because then $P(x)$ is false and the implication is true. Indeed, this wff is true if there are no parrots in the world at all!

To translate an English statement into a wff, it may help to first write some intermediate English language statement and then symbolize that statement. We did this with the parrot examples above.

The word "only" seems particularly troublesome in translations because its placement in a sentence can completely change the meaning. For example, the English statements

John loves only Mary.
Only John loves Mary
John only loves Mary.

say three entirely different things. They can be rewritten as

If John loves any thing, then that thing is Mary.
If any thing loves Mary, then that thing is John.
If John does any thing to Mary, then that thing is love.

EXAMPLE 21   Given the predicate symbols

$D(x)$ is "x is a dog."
$R(x)$ is "x is a rabbit."
$C(x, y)$ is "x chases y."

Table 1.15 shows examples of an English statement, an intermediate English statement, and a wff translation. Note that in wff 2, the connective associated with $\exists$ is $\land$

and the connective associated with $\forall$ is $\rightarrow$. In wff 3, the main connective (because of the order of precedence of connectives) is $\rightarrow$, which goes along with the universal quantifiers.

English Statement	Intermediate Statement	Wff
1. All dogs chase all rabbits.	For any thing, if it is a dog, then for any other thing, if that thing is a rabbit, then the dog chases it.	$(\forall x)[D(x) \rightarrow (\forall y)(R(y) \rightarrow C(x,y))]$
2. Some dogs chase all rabbits.	There is some thing that is a dog and, for any other thing, if that thing is a rabbit, then the dog chases it.	$(\exists x)[D(x) \wedge (\forall y)(R(y) \rightarrow C(x,y))]$
3. Only dogs chase rabbits.	For any two things, if one is a rabbit and the other chases it, then the other is a dog.	$(\forall y)(\forall x)[R(y) \wedge C(x,y) \rightarrow D(x)]$

Table 1.15

PRACTICE 18   Using the predicate symbols $S(x)$ for "x is a student," $I(x)$ for "x is intelligent," and $M(x)$ for "x likes music," write wffs that express the following statements. (The domain is the collection of all people.)

a. All students are intelligent.
b. Some intelligent students like music.
c. Everyone who likes music is a stupid student.
d. Only intelligent students like music.

Negating statements with quantifiers, as in negating compound statements, requires care. The negation of the statement "Everything is beautiful" is "It is false that everything is beautiful" or "Something is nonbeautiful." Symbolically,

$[(\forall x)A(x)]'$ is equivalent to $(\exists x)[A(x)]'$

Note that "Everything is nonbeautiful," or $(\forall x)[A(x)]'$, says something *stronger* than the negation of the original statement.

The negation of "Something is beautiful" is "Nothing is beautiful" or "Everything fails to be beautiful." Symbolically,

$[(\exists x)A(x)]'$ is equivalent to $(\forall x)[A(x)]'$

In English, the statement "Everything is not beautiful" would often be misinterpreted as "Not everything is beautiful" or "There is something nonbeautiful." However, this misinterpretation, symbolized by $(\exists x)[A(x)]'$, is *not as strong* as the negation of the original statement.

PRACTICE 19   Which of the following expresses the negation of "Everybody loves somebody sometime"?

    a. Everybody hates somebody sometime.
    b. Somebody loves everybody all the time.
    c. Everybody hates everybody all the time.
    d. Somebody hates everybody all the time.

Validity

The truth value of a propositional wff depends on the truth values assigned to the statement letters. The truth value of a predicate wff depends on the interpretation. Choosing an interpretation for a predicate wff is thus analogous to choosing truth values in a propositional wff. However, there are an infinite number of possible interpretations for a predicate wff and only 2^n possible rows in the truth table for a propositional wff with n statement letters.

A tautology is a propositional wff that is true for all rows of the truth table. The analogue to tautology for predicate wffs is *validity*—a predicate wff is **valid** if it is true in all possible interpretations. The validity of a wff must be derived from the form of the wff itself, since validity is independent of any particular interpretation; a valid wff is "intrinsically true."

An algorithm exists to decide whether a propositional wff is a tautology—construct the truth table and examine all possible truth assignments. How can we go about deciding validity for predicate wffs? We clearly cannot look at all possible interpretations, because there are an infinite number of them. As it turns out, no algorithm to decide validity exists. (This does not mean simply that no algorithm has yet been found—it means that it has been proved that there is no such algorithm.) We must simply use reasoning to determine whether the form of a wff makes the wff true in all interpretations. Of course, if we can find a single interpretation in which the wff has the truth value false or has no truth value at all, then the wff is not valid.

Table 1.16 compares propositional and predicate wffs.

	Propositional Wffs	Predicate Wffs
Truth values	True or false, depending on truth value assignments to statement letters	True, false, or perhaps (if the wff has a free variable) neither, depending on interpretation
"Intrinsic truth"	Tautology—true for all truth value assignments	Valid wff—true for all interpretations
Methodology	Algorithm (truth table) to determine whether wff is a tautology	No algorithm to determine whether wff is valid

Table 1.16

Now let's try our hand at determining validity.

EXAMPLE 22   **a.** The wff

$$(\forall x)P(x) \rightarrow (\exists x)P(x)$$

is valid. In any interpretation, if every element of the domain has a certain property, then there exists an element of the domain that has that property. (Remember that the domain of any interpretation must have at least one object in it.) Therefore, whenever the antecedent is true, so is the consequent, and the implication is therefore true.

b. The wff

$$(\forall x)P(x) \rightarrow P(a)$$

is valid because in any interpretation, a is a particular member of the domain and therefore has the property that is shared by all members of the domain.

c. The wff

$$(\forall x)[P(x) \wedge Q(x)] \leftrightarrow (\forall x)P(x) \wedge (\forall x)Q(x)$$

is valid. If both P and Q are true for all the elements of the domain, then P is true for all elements and Q is true for all elements, and vice versa.

d. The wff

$$P(x) \rightarrow [Q(x) \rightarrow P(x)]$$

is valid, even though it contains a free variable. To see this, consider any interpretation, and let x be any member of the domain. Then x either does or does not have property P. If x does not have property P, then $P(x)$ is false; because $P(x)$ is the antecedent of the main implication, this implication is true. If x does have property P, then $P(x)$ is true; regardless of the truth value of $Q(x)$, the implication $Q(x) \rightarrow P(x)$ is true, and so the main implication is also true.   ●

EXAMPLE 23   The wff

$$(\exists x)P(x) \rightarrow (\forall x)P(x)$$

is not valid. For example, in the interpretation where the domain consists of the integers and $P(x)$ means that x is even, it is true that there exists an integer that is even, but it is false that every integer is even. The antecedent of the implication is true and the consequent is false, so the value of the implication is false.   ●

We do not necessarily have to go to a mathematical context to construct an interpretation in which a wff is false, but it is frequently easier to do so because the relationships among objects are relatively clear.

PRACTICE 20   Is the wff

$$(\forall x)[P(x) \vee Q(x)] \rightarrow (\forall x)P(x) \vee (\forall x)Q(x)$$

valid or invalid? Explain.   ●

Section 1.3 Review

Techniques

● Determine the truth value of a predicate wff in a given interpretation.

● Translate English language statements into predicate wffs, and vice versa.

● Recognize a valid wff and explain why it is valid.

● Recognize a nonvalid wff and construct an interpretation in which it is false or has no truth value.

Main Ideas

The truth value of predicate wffs depends on the interpretation considered.

Valid predicate wffs are "intrinsically true"—true in all interpretations.

Exercises 1.3

1. What is the truth value of each of the following wffs in the interpretation where the domain consists of the integers, $O(x)$ is "x is odd," $L(x)$ is "$x < 10$," and $G(x)$ is "$x > 9$"?

   a. $(\exists x)O(x)$ **b.** $(\forall x)[L(x) \rightarrow O(x)]$
   c. $(\exists x)[L(X) \wedge G(x)]$ **d.** $(\forall x)[L(x) \vee G(x)]$

2. What is the truth value of each of the following wffs in the interpretation where the domain consists of the integers?

   ★**a.** $(\forall x)(\exists y)(x + y = x)$ ★**b.** $(\exists y)(\forall x)(x + y = x)$
   ★**c.** $(\forall x)(\exists y)(x + y = 0)$ ★**d.** $(\exists y)(\forall x)(x + y = 0)$
   e. $(\forall x)(\forall y)(x < y \vee y < x)$ **f.** $(\forall x)[x < 0 \rightarrow (\exists y)(y > 0 \wedge x + y = 0)]$
   g. $(\exists x)(\exists y)(x^2 = y)$ **h.** $(\forall x)(x^2 > 0)$

3. Give the truth value of each of the following wffs in the interpretation where the domain consists of the states of the United States, $Q(x, y)$ is "x is north of y," $P(x)$ is "x starts with the letter M," and a is "Mississippi."

   a. $(\forall x)P(x)$ **b.** $(\forall x)(\forall y)(\forall z)[Q(x, y) \wedge Q(y, z) \rightarrow Q(x, z)]$
   c. $(\exists y)(\exists x)Q(y, x)$ **d.** $(\forall x)(\exists y)[P(y) \wedge Q(x, y)]$
   e. $(\exists y)Q(a, y)$

4. For each wff, find an interpretation in which it is true and one in which it is false.

   ★**a.** $(\forall x)([A(x) \vee B(x)] \wedge [A(x) \wedge B(x)]')$
   b. $(\forall x)(\forall y)[P(x, y) \rightarrow P(y, x)]$
   c. $(\forall x)[P(x) \rightarrow (\exists y)Q(x, y)]$
   d. $(\exists x)[A(x) \wedge (\forall y)B(x, y)]$
   e. $[(\forall x)A(x) \rightarrow (\forall x)B(x)] \rightarrow (\forall x)[A(x) \rightarrow B(x)]$

5. Identify the scope of each of the quantifiers in the following wffs and indicate any free variables.

   a. $(\forall x)[P(x) \rightarrow Q(y)]$ **b.** $(\exists x)[A(x) \land (\forall y)B(y)]$
   c. $(\exists x)[(\forall y)P(x, y) \land Q(x, y)]$ **d.** $(\exists x)(\exists y)[A(x, y) \land B(y, z) \rightarrow A(a, z)]$

6. Using the predicate symbols shown and appropriate quantifiers, write each English language statement as a predicate wff. (The domain is the whole world.)

   $D(x)$ is "x is a day." M is "Monday."
   $S(x)$ is "x is sunny." T is "Tuesday."
   $R(x)$ is "x is rainy."

   ★**a.** All days are sunny.
   ★**b.** Some days are not rainy.
   ★**c.** Every day that is sunny is not rainy.
   d. Some days are sunny and rainy.
   e. No day is both sunny and rainy.
   f. It is always a sunny day only if it is a rainy day.
   g. No day is sunny.
   h. Monday was sunny; therefore every day will be sunny.
   i. It rained both Monday and Tuesday.
   j. If some day is rainy, then every day will be sunny.

7. Using the predicate symbols shown and appropriate quantifiers, write each English language statement as a predicate wff. (The domain is the whole world.)

   $P(x)$ is "x is a person."
   $T(x)$ is "x is a time."
   $F(x, y)$ is "x is fooled at y."

   a. You can fool some of the people all of the time.
   b. You can fool all of the people some of the time.
   c. You can't fool all of the people all of the time.

8. Using the predicate symbols shown and appropriate quantifiers, write each English language statement as a predicate wff. (The domain is the whole world.)

   $J(x)$ is "x is a judge." $C(x)$ is "x is a chemist."
   $L(x)$ is "x is a lawyer." $A(x, y)$ is "x admires y."
   $W(x)$ is "x is a woman."

   a. There are some women lawyers who are chemists.
   ★**b.** No woman is both a lawyer and a chemist.
   ★**c.** Some lawyers admire only judges.
   d. All judges admire only judges.
   e. Only judges admire judges.
   f. All women lawyers admire some judge.
   g. Some women admire no lawyer.

9. Using the predicate symbols shown and appropriate quantifiers, write each English language statement as a predicate wff. (The domain is the whole world.)

$C(x)$ is "x is a Corvette." $P(x)$ is "x is a Porsche."
$F(x)$ is "x is a Ferrari." $S(x, y)$ is "x is slower than y."

★**a.** Nothing is both a Corvette and a Ferrari.
★**b.** Some Porsches are slower than only Ferraris.
  c. Only Corvettes are slower than Porsches.
  d. All Ferraris are slower than some Corvette.
  e. Some Porsches are slower than no Corvette.
  f. If there is a Corvette that is slower than a Ferrari, then all Corvettes are slower than all Ferraris.

10. Using the predicate symbols shown and appropriate quantifiers, write each English language statement as a predicate wff. (The domain is the whole world.)

$B(x)$ is "x is a bee."
$F(x)$ is "x is a flower."
$L(x, y)$ is "x loves y."

a. All bees love all flowers.   **b.** Some bees love all flowers.
c. All bees love some flowers.   **d.** Every bee hates only flowers.
e. Only bees love flowers.   **f.** Every bee loves only flowers.
g. No bee loves only flowers.   **h.** Some bees love some flowers.
i. Some bees love only flowers.   **j.** Every bee hates some flowers.
k. Every bee hates all flowers.   **l.** No bee hates all flowers.

11. Using the predicate symbols shown and appropriate quantifiers, write each English language statement as a predicate wff. (The domain is the whole world.)

$S(x)$ is "x is a spy novel." $L(x)$ is "x is long."
$M(x)$ is "x is a mystery." $B(x, y)$ is "x is better than y."

a. All spy novels are long.
b. Not every mystery is a spy novel.
c. Only mysteries are long.
d. Some spy novels are mysteries.
e. Spy novels are better than mysteries.
f. Some mysteries are better than all spy novels.
g. Only spy novels are better than mysteries.

12. Give English language translations of the following wffs if

 L(x, y) is "x loves y."   j is "John."
 H(x) is "x is handsome."   k is "Kathy."
 M(x) is "x is a man."   W(x) is "x is a woman."
 P(x) is "x is pretty."

★a. $H(j) \wedge L(k, j)$
★b. $(\forall x)[M(x) \rightarrow H(x)]$
 c. $(\forall x)(W(x) \rightarrow (\forall y)[L(x, y) \rightarrow M(y) \wedge H(y)])$
 d. $(\exists x)[M(x) \wedge H(x) \wedge L(x, k)]$
 e. $(\exists x)(W(x) \wedge P(x) \wedge (\forall y)[L(x, y) \rightarrow H(y) \wedge M(y)])$
 f. $(\forall x)[W(x) \wedge P(x) \rightarrow L(j, x)]$

13. Several forms of negation are given for each of the following statements. Which is correct?

★a. Some people like mathematics.
    1. Some people dislike mathematics.
    2. Everybody dislikes mathematics.
    3. Everybody likes mathematics.
 b. Everyone loves ice cream.
    1. No one loves ice cream.
    2. Everyone dislikes ice cream.
    3. Someone doesn't love ice cream.
 c. All people are tall and thin.
    1. Someone is short and fat.
    2. No one is tall and thin.
    3. Someone is short or fat.
 d. Some pictures are old or faded.
    1. Every picture is neither old nor faded.
    2. Some pictures are not old or faded.
    3. All pictures are not old or not faded.

14. Write the negation of each of the following.

 a. Only students eat pizza.
 b. Every student eats pizza
 c. Some students eat only pizza.

15. Explain why each wff is valid.

 a. $(\forall x)(\forall y)A(x, y) \leftrightarrow (\forall y)(\forall x)A(x, y)$
 b. $(\exists x)(\exists y)A(x, y) \leftrightarrow (\exists y)(\exists x)A(x, y)$
★c. $(\exists x)(\forall y)P(x, y) \rightarrow (\forall y)(\exists x)P(x, y)$
 d. $A(a) \rightarrow (\exists x)A(x)$
 e. $(\forall x)[A(x) \rightarrow B(x)] \rightarrow [(\forall x)A(x) \rightarrow (\forall x)B(x)]$

16. Give interpretations to prove that each of the following wffs is not valid:

★**a.** $(\exists x)A(x) \wedge (\exists x)B(x) \rightarrow (\exists x)[A(x) \wedge B(x)]$
   b. $(\forall x)(\exists y)P(x, y) \rightarrow (\exists x)(\forall y)P(x, y)$
   c. $(\forall x)[P(x) \rightarrow Q(x)] \rightarrow [(\exists x)P(x) \rightarrow (\forall x)Q(x)]$
   d. $(\forall x)[A(x)]' \leftrightarrow [(\forall x)A(x)]'$

17. Decide whether each of the following wffs is valid or invalid. Justify your answer.

   a. $(\exists x)A(x) \leftrightarrow ((\forall x)[A(x)]')'$
   b. $(\forall x)P(x) \vee (\exists x)Q(x) \rightarrow (\forall x)[P(x) \vee Q(x)]$
   c. $(\forall x)A(x) \rightarrow ((\exists x)[A(x)]')'$
   d. $(\forall x)[P(x) \vee Q(x)] \rightarrow (\forall x)P(x) \vee (\exists y)Q(y)$

Section 1.4   Predicate Logic

We can imagine arguments of the form

$$P_1 \wedge P_2 \wedge P_3 \wedge \cdots \wedge P_n \rightarrow Q$$

where the wffs are built from predicates and quantifiers as well as logical connectives and grouping symbols. For a **valid argument**, Q must follow logically from $P_1, \ldots,$ P_n based solely on the internal structure of the argument, not on the truth or falsity of Q in any particular interpretation. In other words, the wff

$$P_1 \wedge P_2 \wedge P_3 \wedge \cdots \wedge P_n \rightarrow Q$$

must be valid—true in all possible interpretations. No equivalent of the truth table exists to easily prove validity, so we turn to a formal logic system called **predicate logic.** We again use a system of derivation rules to build a proof sequence leading from the hypotheses to the conclusion. The rules should once more be truth-preserving, so that if in some interpretation all the hypotheses are true, then the conclusion will also be true in that interpretation. The system will then be correct (*only* valid arguments will be provable). We also want the system to be complete (*every* valid argument should be provable), yet at the same time the rule set should be minimal.

Derivation Rules for Predicate Logic

The equivalence rules and inference rules of propositional logic are still part of predicate logic. An argument of the form

$$P \wedge (P \rightarrow Q) \rightarrow Q$$

is still valid by modus ponens, even if the wffs involved are predicate wffs.

EXAMPLE 24   Use predicate logic to prove the validity of the argument

$$(\forall x)R(x) \wedge [(\forall x)R(x) \rightarrow (\forall x)S(x)] \rightarrow (\forall x)S(x)$$

A proof sequence is

1. $(\forall x)R(x)$ hyp
2. $(\forall x)R(x) \rightarrow (\forall x)S(x)$ hyp
3. $(\forall x)S(x)$ 1, 2, mp

However, there are many arguments with predicate wffs that are not tautologies but are still valid because of their structure and the meaning of the universal and existential quantifiers (see Example 22). The overall approach to proving these arguments is to strip off the quantifiers, manipulate the unquantified wffs, and then put the quantifiers back in. The new rules of inference provide mechanisms to strip off and insert quantifiers. Hence there are four new rules—one each to strip off the universal and existential quantifier, respectively, and one each to insert the universal and existential quantifier, respectively. The four rules are given in Table 1.17; their details will be explained shortly. In Table 1.17, the notation $P(x)$ does *not* imply that P is a unary predicate with x as its only variable; it simply means that x is one of the variables in the predicate P. Thus $P(x)$ might actually be something like $(\exists y)(\forall z)Q(x, y, z)$.

Inference Rules			
From	Can Derive	Name/Abbreviation for Rule	Restrictions on Use
$(\forall x)P(x)$	$P(t)$ where t is a variable or constant symbol	Universal Instantiation—ui	If t is a variable, it must not fall within the scope of a quantifier for t.
$(\exists x)P(x)$	$P(t)$ where t is a variable or constant symbol not previously used in proof sequence	Existential Instantiation—ei	Must be the first rule used that introduces t.
$P(x)$	$(\forall x)P(x)$	Universal Generalization—ug	$P(x)$ has not been deduced from any hypotheses in which x is a free variable nor has $P(x)$ been deduced by ei from any wff in which x is a free variable.
$P(x)$ or $P(a)$ where a is a constant symbol	$(\exists x)P(x)$	Existential Generalization—eg	To go from $P(a)$ to $(\exists x)P(x)$, x must not appear in $P(a)$.

Table 1.17

Now let's examine these rules more closely, particularly the necessity for their restrictions.

Universal Instantiation

This rule says that from $(\forall x)P(x)$ we can derive $P(x)$, $P(y)$, $P(z)$, $P(a)$, etc., thus stripping off a universal quantifier. The justification is that if P is true for every element of the domain, we can name such an element by an arbitrary variable name like x, y, or z, or we can specify a particular constant in the domain, and P is still true for all of these things.

EXAMPLE 25

Universal Instantiation can be used to prove one of the classical "syllogisms" of the Greek philosopher and scientist Aristotle, who lived from 384–322 B.C. and who first developed a system of formal logic.

The argument has the form "All humans are mortal. Socrates is human. Therefore Socrates is mortal." Using the notation

$H(x)$ is "x is human."
s is a constant symbol (Socrates).
$M(x)$ is "x is mortal."

the argument is

$$(\forall x)[H(x) \rightarrow M(x)] \wedge H(s) \rightarrow M(s)$$

and a proof sequence is

1. $(\forall x)(H(x) \rightarrow M(x))$ hyp
2. $H(s)$ hyp
3. $H(s) \rightarrow M(s)$ 1, ui
4. $M(s)$ 2, 3, mp

In step 3, a constant symbol has been substituted for x throughout the scope of the universal quantifier, as allowed by Universal Instantiation.   ●

Without the restriction on Universal Instantiation, a hypothesis of the form $(\forall x)(\exists y)P(x, y)$ could lead to the wff $(\exists y)P(y, y)$; here y has been substituted for x within the scope of a quantifier on y. This would be invalid. For example, in the domain of the integers, if $P(x, y)$ means "$y > x$," then $(\forall x)(\exists y)P(x, y)$ is true (for every integer there is a bigger integer) but $(\exists y)P(y, y)$ is false (no integer has the property that it is bigger than itself).

PRACTICE 21

Prove the argument $(\forall x)[P(x) \rightarrow R(x)] \wedge [R(y)]' \rightarrow [P(y)]'$.   ●

Existential Instantiation

This rule allows us to strip off an existential quantifier. It says that from $(\exists x)P(x)$ we can derive $P(x)$, $P(y)$, $P(z)$, $P(a)$, etc., provided that these are essentially new names or constant symbols. The justification is that if P is true for some element of the domain, we can give that element a specific name, but we cannot assume anything else about it.

EXAMPLE 26   The following would be legitimate steps in a proof sequence:

1. $(\forall x)[P(x) \rightarrow Q(x)]$ hyp
2. $(\exists y)P(y)$ hyp
3. $P(a)$ 2, ei
4. $P(a) \rightarrow Q(a)$ 1, ui
5. $Q(a)$ 3, 4, mp

REMINDER:
Use Existential
Instantiation
early in the
proof sequence.

In step 3, the specific element with property P was given the name a. In step 4, ui was then used to say that an implication that is universally true in the domain is certainly true for this a. Steps 3 and 4 *cannot be reversed*. If ui is first used on hypothesis 1 to name a constant a, there is then no reason to assume that this particular a is the one that is guaranteed by hypothesis 2 to have property P.   ●

The effect of the restriction on Existential Instantiation is that you should look at all your hypotheses and, if you plan to use ei on any of them, do it first.

Universal Generalization

Universal Generalization allows a universal quantifier to be inserted. This must be done pretty carefully, however. If we know that $P(x)$ is true and that the x is absolutely arbitrary, i.e., that x could be any element of the domain, then we can conclude $(\forall x)P(x)$. But if x is supposed to represent some specific element of the domain that has property P, then we can't generalize that every element of the domain has property P.

EXAMPLE 27   Use predicate logic to prove

$$(\forall x)[P(x) \rightarrow Q(x)] \wedge (\forall x)P(x) \rightarrow (\forall x)Q(x)$$

Here is a proof sequence:

1. $(\forall x)[P(x) \rightarrow Q(x)]$ hyp
2. $(\forall x)P(x)$ hyp
3. $P(x) \rightarrow Q(x)$ 1, ui
4. $P(x)$ 2, ui Note that there is no restriction on ui about reusing
               a name.
5. $Q(x)$ 3, 4, mp
6. $(\forall x)Q(x)$ 5, ug

The use of Universal Generalization at step 6 is legitimate because x was not a free variable in any hypothesis nor was ei used anywhere in the proof. The variable x in steps 3 and 4 is just an arbitrary name, representative of any element in the domain.   ●

There are two restrictions on Universal Generalization. Without the first restriction, the sequence

1. $P(x)$ hyp
2. $(\forall x)P(x)$ 1, incorrect ug; x was free in the hypothesis.

would be a proof of the wff $P(x) \rightarrow (\forall x)P(x)$, but this is not a valid wff. Element x of the domain may have property P, but that does not mean that every element of the domain has property P. In the hypothesis, x is naming some fixed if unspecified element of the domain. For instance, in the interpretation where the domain consists of automobiles and $P(x)$ means "x is yellow," some particular car may be yellow but it is certainly not true that all cars are yellow.

Without the second restriction, the sequence

1. $(\forall x)(\exists y)Q(x, y)$ hyp
2. $(\exists y)Q(x, y)$ 1, ui
3. $Q(x, t)$ 2, ei
4. $(\forall x)Q(x, t)$ 3, incorrect ug; $Q(x, t)$ was deduced by ei from the wff in step 2, in which x is free.

would be a proof of the wff $(\forall x)(\exists y)Q(x, y) \rightarrow (\forall x)Q(x, t)$. This is also not a valid wff. For instance, in the interpretation where the domain consists of the integers and $Q(x, y)$ means that $x + y = 0$, then it is the case that for every integer x there is an integer y (the negative of x) such that $x + y = 0$. However, if t is a particular fixed element in the domain, then it will not be true that adding that same integer t to every x will always produce zero.

PRACTICE 22   Prove the argument $(\forall x)[P(x) \wedge Q(x)] \rightarrow (\forall x)[Q(x) \wedge P(x)]$.   ●

Existential Generalization

The last rule allows insertion of an existential quantifier. From $P(x)$ or $P(a)$ we can derive $(\exists x)P(x)$; something has been named as having property P, so we can say that there exists something that has property P.

EXAMPLE 28   Prove the argument $(\forall x)P(x) \rightarrow (\exists x)P(x)$.
Here is a proof sequence:

1. $(\forall x)P(x)$ hyp
2. $P(x)$ 1, ui
3. $(\exists x)P(x)$ 2, eg   ●

Without the restriction on Existential Generalization, from $P(a, y)$ one could derive $(\exists y)P(y, y)$; here the quantified variable y, which replaced the constant symbol a, already appeared in the wff to which Existential Generalization was applied. But the argument $P(a, y) \rightarrow (\exists y)P(y, y)$ is not valid. In the domain of integers, if $P(x, y)$ means "$y > x$" and a stands for 0, then if $y > 0$, this does not mean that there is an integer y that is greater than itself.

More Work with Rules

As is the case with propositional logic rules, predicate logic rules can be applied only when the exact pattern of the rule is matched (and, of course, when no restrictions on use of the rule are violated). In particular, notice that the instantiation rules strip off a quantifier from the front of an entire wff that is in the scope of that quantifier. Both of the following would be illegal uses of Existential Instantiation:

1. $(\exists x)P(x) \vee (\exists x)Q(x)$ hyp
2. $P(a) \vee Q(a)$ 1, incorrect ei. The scope of the first existential quantifier in step 1 does not extend to the whole rest of the wff.

1. $(\forall x)(\exists y)Q(x, y)$ hyp
2. $(\forall x)Q(x, a)$ 1, incorrect ei. The existential quantifier in step 1 is not at the front.

Similarly, the rules to insert a quantifier put that quantifier in the front of a wff that is then entirely within its scope.

Even though we have added only four new derivation rules, the rule set is complete and correct. We can prove every valid argument and only valid arguments using these rules. Application of the rules, as in the case of propositional logic, is somewhat mechanical because there are only a limited number of options at each step. Again, the general plan of attack is usually as follows:

- Strip off the quantifiers.
- Work with the separate wffs.
- Insert quantifiers as necessary.

EXAMPLE 29   Using predicate logic, prove the argument

$$(\forall x)[P(x) \wedge Q(x)] \rightarrow (\forall x)P(x) \wedge (\forall x)Q(x)$$

In Example 22(c) we noted that this wff is valid, so if all valid arguments are provable, we should be able to find a proof sequence. As usual, the hypothesis gives us a starting point.

1. $(\forall x)[P(x) \wedge Q(x)]$ hyp

Stripping off the universal quantifier that appears in step 1 will yield access to $P(x) \wedge Q(x)$, which can then be separated. The universal quantifier can then be inserted separately on each of those two wffs using Universal Generalization. The conclusion $(\forall x)P(x) \wedge (\forall x)Q(x)$ will follow. A proof sequence is

1. $(\forall x)[P(x) \wedge Q(x)]$ hyp
2. $P(x) \wedge Q(x)$ 1, ui
3. $P(x)$ 2, sim
4. $Q(x)$ 2, sim
5. $(\forall x)P(x)$ 3, ug
6. $(\forall x)Q(x)$ 4, ug
7. $(\forall x)P(x) \wedge (\forall x)Q(x)$ 5, 6, con

Neither restriction on Universal Generalization has been violated because x is not free in the hypothesis and Existential Instantiation has not been used.   ●

PRACTICE 23   Using predicate logic, prove the following argument. (*Hint:* The deduction method still applies.)

$$(\forall y)[P(x) \rightarrow Q(x, y)] \rightarrow [P(x) \rightarrow (\forall y)Q(x, y)]$$ ●

As an extension to the deduction method, we can insert a "temporary" hypothesis into a proof. If some wff *T* is introduced into a proof sequence as a temporary hypothesis, and eventually a wff *W* is deduced from *T* and other hypotheses, then the wff $T \rightarrow W$ has been deduced from the other hypotheses and can be inserted in the proof sequence.

EXAMPLE 30   The argument

$$[P(x) \rightarrow (\forall y)Q(x, y)] \rightarrow (\forall y)[P(x) \rightarrow Q(x, y)]$$

is valid. In the following proof sequence, $P(x)$ is introduced at step 2 as a temporary hypothesis, which allows us to deduce $Q(x, y)$ at step 4. The indented steps show that these wffs depend on the temporary hypothesis. At step 5, the temporary hypothesis is "discharged," as the dependency of $Q(x, y)$ on the temporary hypothesis is explicitly acknowledged as an implication. Of course, the entire wff at step 5, $P(x) \rightarrow Q(x, y)$, still depends on the hypothesis of step 1. At step 6, neither restriction on Universal Generalization is violated because *y* is not a free variable in step 1 (the only hypothesis at this point) and Existential Instantiation is not used in the proof.

1.	$P(x) \rightarrow (\forall y)Q(x, y)$	hyp
2.	$P(x)$	temporary hyp
3.	$\quad(\forall y)Q(x, y)$	1, 2, mp
4.	$\quad Q(x, y)$	3, ui
5.	$P(x) \rightarrow Q(x, y)$	temp. hyp discharged
6.	$(\forall y)[P(x) \rightarrow Q(x, y)]$	5, ug

Notice how the temporary hypothesis gives us enough ammunition to make something happen. Without this technique, it would be difficult to know what to do after step 1.   ●

Practice 23 and Example 30 show that the wff

$$(\forall y)[P(x) \rightarrow Q(x, y)] \leftrightarrow [P(x) \rightarrow (\forall y)Q(x, y)]$$

is valid. It says that the universal quantifier can be "passed over" subwffs that do not contain the quantified variable; in this case, $(\forall y)$ is passed over $P(x)$. A similar result holds for the existential quantifier. As a result, there may be two or more equivalent ways of expressing English language sentences as predicate wffs, as in Exercises 6 through 11 of Section 1.3.

PRACTICE 24   Prove the argument

$$(\forall x)[(B(x) \lor C(x)) \rightarrow A(x)] \rightarrow (\forall x)[B(x) \rightarrow A(x)]$$ ●

In Section 1.3 we observed that, based on our understanding of negation and the meaning of the quantifiers, $[(\forall x)A(x)]'$ is equivalent to $(\exists x)[A(x)]'$. We should be able to formally prove

$$[(\exists x)A(x)]' \leftrightarrow (\forall x)[A(x)]'$$

EXAMPLE 31

Prove that

$$[(\exists x)A(x)]' \leftrightarrow (\forall x)[A(x)]'$$

We must prove the implication in each direction.

a. $[(\exists x)A(x)]' \rightarrow (\forall x)[A(x)]'$

The hypothesis alone gives us little to work with, so we introduce a (somewhat surprising) temporary hypothesis. A proof sequence is

1. $[(\exists x)A(x)]'$ hyp
2. $A(x)$ temporary hyp
3. $(\exists x)A(x)$ 2, eg
4. $A(x) \rightarrow (\exists x)A(x)$ temporary hyp discharged
5. $[A(x)]'$ 1, 4, mt
6. $(\forall x)[A(x)]'$ 5, ug

b. $(\forall x)[A(x)]' \rightarrow [(\exists x)A(x)]'$

This proof also requires a temporary hypothesis. It is even more surprising than case (a) because we assume the exact opposite of the conclusion we are trying to reach.

1. $(\forall x)[A(x)]'$ hyp
2. $(\exists x)A(x)$ temporary hyp
3. $A(a)$ 2, ei
4. $[A(a)]'$ 1, ui
5. $[(\forall x)[A(x)]']'$ 3, 4, inc
6. $(\exists x)A(x) \rightarrow [(\forall x)[A(x)]']'$ temporary hyp discharged
7. $[((\forall x)[A(x)]')']'$ 1, dn
8. $[(\exists x)A(x)]'$ 6, 7, mt

The proofs of Example 31 are rather difficult because they require considerably more imagination than most. As a result, however, we do have the following equivalence, to which we've given a name:

$$[(\exists x)A(x)]' \leftrightarrow (\forall x)[A(x)]' \quad \text{(Negation—neg)}$$

This might be useful in a proof sequence. As an extension of the equivalence rules, whenever $P \leftrightarrow Q$ is valid, Q can be substituted for P within an expression in a proof sequence. The argument in Practice 22, $(\forall x)[P(x) \wedge Q(x)] \rightarrow (\forall x)[Q(x) \wedge P(x)]$, can be proved in two steps simply by substituting $Q(x) \wedge P(x)$ for $P(x) \wedge Q(x)$ in the hypothesis, since these wffs are equivalent.

EXAMPLE 32   Is the wff

$$(\forall x)[P(x) \lor Q(x)] \rightarrow (\exists x)P(x) \lor (\forall x)Q(x)$$

a valid argument? Prove or disprove.

Let's first consider whether the wff seems valid. If so, we should try to find a proof sequence for it; if not, we should try to find an interpretation in which it is not true. This wff says that if every element of the domain has either property P or property Q, then at least one element must have property P or else all elements have property Q. This seems very reasonable, so we'll try to find a proof.

First we'll use an equivalence to rewrite the conclusion in a more useful form. Changing the $\lor$ to an implication will allow use of the deduction method. Thus we want to prove

$$(\forall x)[P(x) \lor Q(x)] \rightarrow [\ [(\exists x)P(x)]' \rightarrow (\forall x)Q(x)]$$

A proof sequence is

1.	$(\forall x)[P(x) \lor Q(x)]$	hyp
2.	$[(\exists x)P(x)]'$	hyp
3.	$(\forall x)[P(x)]'$	2, neg
4.	$[P(x)]'$	3, ui
5.	$P(x) \lor Q(x)$	1, ui
6.	$[[P(x)]']' \lor Q(x)$	5, dn
7.	$[P(x)]' \rightarrow Q(x)$	6, imp
8.	$Q(x)$	4, 7, mp
9.	$(\forall x)Q(x)$	8, ug

●

EXAMPLE 33   Is the wff

$$(\exists x)P(x) \land (\exists x)Q(x) \rightarrow (\exists x)[P(x) \land Q(x)]$$

a valid argument? Prove or disprove.

If something in a domain has property P and something has property Q, that does not mean that some one thing has both property P and Q. For example, in the domain of integers, if $P(x)$ means "x is even" and $Q(x)$ means "x is odd," then the hypotheses are true, but the conclusion is false because there is no single integer that is both even and odd. One interpretation in which the wff is false is enough to disprove it.

It is useful, however, to see where a potential proof sequence goes wrong. We begin with the two hypotheses and then remove one of the existential quantifiers.

1.	$(\exists x)P(x)$	hyp
2.	$(\exists x)Q(x)$	hyp
3.	$P(a)$	1, ei

Now here's the problem. The next step would be to remove the existential quantifier from the wff at step 2, but, according to the rules for ei, we have to name the object that has property Q by some different name, not a. So we could eventually get to a wff in the proof sequence that looks like

$$P(a) \land Q(b)$$

but this does us no good. Existential generalization could not be used to replace both constant symbols with a single variable. At best, we could arrive at

$$(\exists y)(\exists x)[P(x) \land Q(y)]$$

which is not what we want.

PRACTICE 25   Is the wff

$$(\exists x)R(x) \land [(\exists x)[R(x) \land S(x)]]' \rightarrow (\exists x)[S(x)]'$$

a valid argument? Prove or disprove.

Verbal Arguments

To prove the validity of a verbal argument, we proceed much as before. We cast the argument in symbolic form and show that the conclusion can be deduced from the hypotheses. If the argument involves predicate wffs, then the derivation rules of predicate logic are available.

EXAMPLE 34   Show that the following argument is valid: "Every microcomputer has a serial interface port. Some microcomputers have a parallel port. Therefore some microcomputers have both a serial and a parallel port." Using

$M(x)$ is "x is a microcomputer."
$S(x)$ is "x has a serial port."
$P(x)$ is "x has a parallel port."

the argument is

$$(\forall x)[M(x) \rightarrow S(x)] \land (\exists x)[M(x) \land P(x)] \rightarrow (\exists x)[M(x) \land S(x) \land P(x)]$$

Note that if we attempt to symbolize this argument in propositional logic, we get $A \land B \rightarrow C$, which is not a valid argument. Propositional logic is simply not expressive enough to capture the interrelationships among the parts of this argument that serve to make it valid.

A proof sequence is

1.	$(\forall x)[M(x) \rightarrow S(x)]$	hyp
2.	$(\exists x)[M(x) \land P(x)]$	hyp
3.	$M(a) \land P(a)$	2, ei
4.	$M(a) \rightarrow S(a)$	1, ui
5.	$M(a)$	3, sim
6.	$S(a)$	4, 5, mp
7.	$M(a) \land P(a) \land S(a)$	3, 6, con
8.	$M(a) \land S(a) \land P(a)$	7, comm
9.	$(\exists x)[M(x) \land S(x) \land P(x)]$	8, eg

Once again, it is the form of the argument that matters, not the content.

PRACTICE 26   Show that the following argument is valid: "All rock music is loud music. Some rock music exists; therefore some loud music exists." Use predicates $R(x)$ and $L(x)$.   ●

Conclusion

We've now finished our study of formal logic. What has been accomplished? The goal of formal logic, often called *symbolic logic,* is to make arguments as meaningless as possible! The symbolic notation of propositional and predicate logic allows us to symbolize arguments. An argument cast in symbolic notation removes any possibility that we will be swayed by our opinions or our external knowledge about the topic of the argument, and we can concentrate solely on its structure to determine its logical validity. Furthermore, the inference rules allow the proof of an argument's validity to be produced by symbol manipulation. Again, it requires no external knowledge, only a careful adherence to the forms and restrictions of the rules. In theory, then, producing a proof sequence should be almost mechanical.

Nonetheless, you may still feel that it is difficult to produce a proof sequence. Practice does make the process easier, because after a while, you become familiar with the various forms an argument might take and you recognize which rules you should try to apply. At any rate, you should at least find it easy at this point to *check* that a *proposed* proof sequence is logically correct.

Again, one objective of practice with this mechanical process of applying rules is that it will ultimately transform into a habit of logical thinking in everyday life. But aside from logical thinking in its pure sense, the notions of formal rules of inference have two very direct applications to computer science. An entire system of programming, and some programming languages, are based on applying rules of inference. We will see such a language in Section 1.5. Similarly, rules of inference can be applied to formally prove program correctness, leading to increased confidence that code is error-free. We'll look at some of the inference rules for program correctness in Section 1.6.

Section 1.4 Review

Techniques

- Apply derivation rules for predicate logic.
- Use predicate logic to prove the validity of a verbal argument.

Main Idea

The predicate logic system is correct and complete; valid arguments and only valid arguments are provable.

Exercises 1.4

1. Justify each step in the following proof sequence of

$$(\exists x)[P(x) \to Q(x)] \to [(\forall x)P(x) \to (\exists x)Q(x)]$$

1. $(\exists x)[P(x) \to Q(x)]$
2. $P(a) \to Q(a)$
3. $(\forall x)P(x)$
4. $P(a)$
5. $Q(a)$
6. $(\exists x)Q(x)$

★2. Consider the wff $(\forall x)[(\exists y)P(x, y) \wedge (\exists y)Q(x, y)] \to (\forall x)(\exists y)[P(x, y) \wedge Q(x, y)]$.

   a. Find an interpretation to prove that this wff is not valid.
   b. Find the flaw in the following "proof" of this wff.

1. $(\forall x)[(\exists y)P(x, y) \wedge (\exists y)Q(x, y)]$ hyp
2. $(\forall x)[P(x, a) \wedge Q(x, a)]$ 1, ei
3. $(\forall x)(\exists y)[P(x, a) \wedge Q(x, a)]$ 2, eg

3. Consider the wff $(\forall y)(\exists x)Q(x, y) \to (\exists x)(\forall y)Q(x, y)$.

   a. Find an interpretation to prove that this wff is not valid.
   b. Find the flaw in the following "proof" of this wff.

1. $(\forall y)(\exists x)Q(x, y)$ hyp
2. $(\exists x)Q(x, y)$ 1, ui
3. $Q(a, y)$ 2, ei
4. $(\forall y)Q(a, y)$ 3, ug
5. $(\exists x)(\forall y)Q(x, y)$ 4, eg

In Exercises 4–8, prove that each wff is a valid argument.

★4. $(\forall x)P(x) \to (\forall x)[P(x) \vee Q(x)]$

5. $(\forall x)P(x) \wedge (\exists x)Q(x) \to (\exists x)[P(x) \wedge Q(x)]$

6. $(\exists x)(\exists y)P(x, y) \to (\exists y)(\exists x)P(x, y)$

7. $(\forall x)(\forall y)Q(x, y) \to (\forall y)(\forall x)Q(x, y)$

8. $(\forall x)P(x) \wedge (\exists x)[P(x)]' \to (\exists x)Q(x)$

In Exercises 9–18, either prove that the wff is a valid argument or give an interpretation in which it is false.

★9. $(\exists x)[A(x) \wedge B(x)] \to (\exists x)A(x) \wedge (\exists x)B(x)$

10. $(\exists x)[R(x) \vee S(x)] \to (\exists x)R(x) \vee (\exists x)S(x)$

11. $(\forall x)[P(x) \rightarrow Q(x)] \rightarrow [(\forall x)P(x) \rightarrow (\forall x)Q(x)]$

12. $[(\forall x)P(x) \rightarrow (\forall x)Q(x)] \rightarrow (\forall x)[P(x) \rightarrow Q(x)]$

★13. $(\exists x)(\forall y Q(x, y) \rightarrow (\forall y)(\exists x)Q(x, y)$

14. $(\forall x)P(x) \vee (\exists x)Q(x) \rightarrow (\forall x)[P(x) \vee Q(x)]$

15. $(\forall x)[A(x) \rightarrow B(x)] \rightarrow [(\exists x)A(x) \rightarrow (\exists x)B(x)]$

16. $(\forall y)[Q(x, y) \rightarrow P(x)] \rightarrow [(\exists y)Q(x, y) \rightarrow P(x)]$

★17. $[P(x) \rightarrow (\exists y)Q(x, y)] \rightarrow (\exists y)[P(x) \rightarrow Q(x, y)]$

18. $(\exists x)[P(x) \rightarrow Q(x)] \wedge (\forall y)[Q(y) \rightarrow R(y)] \wedge (\forall x)P(x) \rightarrow (\exists x)R(x)$

Using predicate logic, prove that each argument in Exercises 19–25 is valid. Use the predicate symbols shown.

19. Some plants are flowers. All flowers smell sweet. Therefore, some plants smell sweet. $P(x), F(x), S(x)$

★20. Every crocodile is bigger than every alligator. Sam is a crocodile. But there is a snake, and Sam isn't bigger than that snake. Therefore, something is not an alligator. $C(x), A(x), B(x, y), s, S(x)$

21. There is an astronomer who is not nearsighted. Everyone who wears glasses is nearsighted. Furthermore, everyone either wears glasses or wears contact lenses. Therefore, some astronomer wears contact lenses. $A(x), N(x), G(x), C(x)$

★22. Every member of the board comes from industry or government. Everyone from government who has a law degree is in favor of the motion. John is not from industry, but he does have a law degree. Therefore, if John is a member of the board, he is in favor of the motion. $M(x), I(x), G(x), L(x), F(x), j$

23. There is some movie star who is richer than everyone. Anyone who is richer than anyone else pays more taxes than anyone else does. Therefore, there is a movie star who pays more taxes than anyone. $M(x), R(x, y), T(x, y)$

24. Every computer science student works harder than somebody, and everyone who works harder than any other person gets less sleep than that person. Maria is a computer science student. Therefore, Maria gets less sleep than someone else. $C(x), W(x, y), S(x, y), m$

25. Every ambassador speaks only to diplomats, and some ambassador speaks to someone. Therefore, there is a diplomat. $A(x), S(x, y), D(x)$

26. Prove that

$$[(\forall x)P(x)]' \rightarrow (\exists x)[P(x)]'$$

is valid. (*Hint:* Instead of a proof sequence, use Example 31 and substitute equivalent expressions.)

27. The equivalence of Exercise 26 says that if it is false that every element of the domain has property P, then some element of the domain fails to have property P, and vice versa. The element that fails to have property P is called a counterexample to the assertion that every element has property P. Thus a counterexample to the assertion

$$(\forall x)(x \text{ is odd})$$

in the domain of integers is the number 10, an even integer. (Of course, there are lots of other counterexamples to this assertion.) Find counterexamples in the domain of integers to the following assertions. (An integer $x > 1$ is prime if the only factors of x are 1 and x.)

a. $(\forall x)(x \text{ is negative})$ b. $(\forall x)(x \text{ is the sum of even integers})$
c. $(\forall x)(x \text{ is prime} \rightarrow x \text{ is odd})$ d. $(\forall x)(x \text{ prime} \rightarrow (-1)^x = -1)$
e. $(\forall x)(x \text{ prime} \rightarrow 2^x - 1 \text{ is prime})$

Section 1.5   Logic Programming

The programming languages with which you are probably familiar, such as C++, Java, or Pascal, are known as **procedural languages.** Much of the content of a program written in a procedural language consists of instructions to carry out the algorithm the programmer believes will solve the problem at hand. The programmer, therefore, is telling the computer how to solve the problem in a step-by-step fashion.

Some programming languages, rather than being procedural, are **declarative languages** or **descriptive languages.** A declarative language is based on predicate logic; such a language comes equipped with its own rules of inference. A program written in a declarative language consists only of statements—actually predicate wffs—that are declared as hypotheses. Execution of a declarative program allows the user to pose queries, asking for information about possible conclusions that can be derived from the hypotheses. After obtaining the user's query, the language turns on its "inference engine" and applies its rules of inference to the hypotheses to see which conclusions fit the user's query. The program, remember, contains only the hypotheses, not any explicit instructions as to what steps to perform in what order. The inference engine of the language acts behind the scenes, so to speak, to construct a proof sequence. It is the mechanical nature of applying inference rules that makes this "automated theorem proving" possible.

Prolog

The programming language Prolog, which stands for PROgramming in LOGic, is a declarative programming language. The set of declarations that constitutes a Prolog program is also known as a **Prolog database.** Items in a Prolog database take on one of two forms, known in Prolog as *facts* and *rules*. (Prolog rules, however, are just another kind of fact and should not be confused with a rule of inference.)

Prolog facts allow predicates to be defined by stating which items in some domain of interpretation satisfy the predicates. As an example, suppose we wish to create a Prolog program that describes food chains in a given ecological region. We might begin with a binary predicate *eat*. We then describe the predicate by giving the pairs of elements in the domain that make *eat* true. Thus we might have the facts

eat(bear, fish)
eat(bear, fox)
eat(deer, grass)

in our database. (The exact details of Prolog statements vary from one Prolog implementation to another, so we are only giving the spirit of the language here by using a Prolog-like pseudocode.) Here "bear," "fish," "fox," "deer," and "grass" are constants because they represent specific elements in the domain. Because the domain itself is never specified except by describing predicates, at this point we may take the domain to consist of "bear," "fish," "fox," "deer," and "grass." It is up to the user to maintain a consistent understanding and use of the predicates in a Prolog program. Thus

eat(bear, fish)

can be used either to represent the fact that bears eat fish or the fact that fish eat bears! We impose the convention that *eat*(x, y) means "x eats y."

We could add descriptions of two unary predicates, *animal* and *plant,* to the database by adding the facts

animal(bear)
animal(fish)
animal(fox)
animal(deer)
plant(grass)

Armed with this Prolog program (database), we can pose some simple queries.

EXAMPLE 35   The query

is(*animal*(bear))

merely asks if the fact *animal*(bear) is in the database. Because this fact is in the database, Prolog would respond to the query by answering yes. (This is a one-step proof sequence—no rules of inference are required). Further dialogue with Prolog could include

is(*eat*(deer, grass))

yes

is(*eat*(bear, rabbit))

no

Queries may include variables, as shown in the next example.

EXAMPLE 36   The query

>   **which**(*x*: *eat*(bear, *x*))

produces

>   fish
>   fox

as a response. Prolog has answered the query by searching the database for all facts that match the pattern *eat*(bear, *x*), where *x* is a variable. The answer "fish" is given first because the rules are searched in order from top to bottom.   ●

Queries may contain the logical connectives **and, or,** and **not.**

PRACTICE 27   Given the database

>   *eat*(bear, fish)
>   *eat*(bear, fox)
>   *eat*(deer, grass)
>   *animal*(bear)
>   *animal*(fish)
>   *animal*(fox)
>   *animal*(deer)
>   *plant*(grass)

what will be Prolog's response to the query

>   **which**(*x*: *eat*(*x*, *y*) **and** *plant*(*y*))   ●

The second type of item in a Prolog database is a **Prolog rule.** A rule is a description of a predicate by means of an implication. For example, we might use a rule to define a predicate of *prey:*

>   *prey*(*x*) **if** *eat*(*y*, *x*) **and** *animal*(*x*)

This says that *x* is a prey if it is an animal that is eaten. If we add this rule to our database, then in response to the query

>   **which**(*x*: *prey*(*x*))

we would get

>   fish
>   fox

Horn Clauses and Resolution

How do Prolog facts and rules relate to more formal predicate logic? We can describe the facts in our database by the wffs

$E(b, fi)$
$E(b, fo)$
$E(d, g)$
$A(b)$
$A(fi)$
$A(fo)$
$A(d)$
$P(g)$

and the rule by the wff

$E(y, x) \wedge A(x) \rightarrow Pr(x)$

Universal quantifiers are not explicitly part of the rule as it appears in a Prolog program, but Prolog treats the rule as being universally quantified,

$(\forall y)(\forall x)[E(y, x) \wedge A(x) \rightarrow Pr(x)]$

and repeatedly uses Universal Instantiation to strip off the universal quantifiers and allow the variables to assume in turn each value of the domain.

Both facts and rules are examples of Horn clauses. A **Horn clause** is a wff composed of predicates or the negations of predicates (with either variables or constants as arguments) joined by disjunctions, where at most one predicate is unnegated. Thus the fact

$E(d, g)$

is an example of a Horn clause because it consists of a single unnegated predicate. The wff

$[E(y, x)]' \vee [A(x)]' \vee Pr(x)$

is an example of a Horn clause because it consists of three predicates joined by disjunction where only $Pr(x)$ is unnegated. By De Morgan's law, it is equivalent to

$[E(y, x) \wedge A(x)]' \vee Pr(x)$

which in turn is equivalent to

$E(y, x) \wedge A(x) \rightarrow Pr(x)$

and therefore represents the rule in our Prolog program.

The rule of inference used by Prolog is called **resolution.** Two Horn clauses in a Prolog database are resolved into a new Horn clause if one contains an unnegated predicate that matches a negated predicate in the other clause. The new clause eliminates the matching term and is then available to use in answering the query. For example,

$A(a)$
$[A(a)]' \vee B(b)$

resolves to $B(b)$. This says that from

$A(a), [A(a)]' \vee B(b)$

REMINDER:
Prolog's resolution rule looks for a term and its negation to infer one Horn clause from two.

which is equivalent to

$$A(a), A(a) \rightarrow B(b)$$

Prolog infers

$$B(b)$$

which is just an application of modus ponens. Therefore Prolog's rule of inference includes modus ponens as a special case.

In applying the resolution rule, variables are considered to "match" any constant symbol. (This is the repeated application of Universal Instantiation.) In any resulting new clause, the variables are replaced with their associated constants in a consistent manner. Thus in response to the query "which x is a prey," Prolog searches the database for a rule with the desired predicate $Pr(x)$ as the consequent. It finds

$$[E(y, x)]' \lor [A(x)]' \lor Pr(x)$$

It then proceeds through the database looking for other clauses that can be resolved with this clause. The first such clause is the fact $E(b, fi)$. These two clauses resolve into

$$[A(fi)]' \lor Pr(fi)$$

(Note that the constant fi has replaced x everywhere.) Using this new clause, it can be resolved with the fact $A(fi)$ to conclude $Pr(fi)$. Having reached all conclusions possible from resolution with the fact $E(b, fi)$, Prolog backtracks to search for another clause to resolve with the rule clause; this time around it would find $E(b, fo)$.

As a more complex example of resolution, suppose we add the rule

$$hunted(x) \textbf{ if } prey(x)$$

to the database. This rule in symbolic form is

$$[Pr(x)] \rightarrow H(x)$$

or, as a Horn clause,

$$[Pr(x)]' \lor H(x)$$

It resolves with the rule defining prey

$$[E(y, x)]' \lor [A(x)]' \lor Pr(x)$$

to give the new rule

$$[E(y, x)]' \lor [A(x)]' \lor H(x)$$

The query

$$\textbf{which}(x: hunted(x))$$

will use this new rule to conclude

fish
fox

EXAMPLE 37   Suppose that a Prolog database contains the following entries:

eat(bear, fish)
eat(fish, little-fish)
eat(little-fish, algae)
eat(raccoon, fish)
eat(bear, raccoon)
eat(bear, fox)
eat(fox, rabbit)
eat(rabbit, grass)
eat(bear, deer)
eat(deer, grass)
eat(wildcat, deer)

animal(bear)
animal(fish)
animal(little-fish)
animal(raccoon)
animal(fox)
animal(rabbit)
animal(deer)
animal(wildcat)
plant(grass)
plant(algae)

prey(x) if *eat*(y, x) and *animal*(x)

Then the following dialog with Prolog could take place:

is(*animal*(rabbit))

yes

is(*eat*(wildcat, grass))

no

which(x: *eat*(x, fish))

bear
raccoon

which(x, y: *eat*(x, y) **and** *plant*(y))

little-fish algae
rabbit grass
deer grass

which(*x*: *prey*(*x*))

fish
little-fish
fish
raccoon
fox
rabbit
deer
deer

Note that fish is listed twice as satisfying the last query because fish are eaten by bear (fact 1) and by raccoon (fact 3). Similarly, deer are eaten by both bear and wildcat.   ●

PRACTICE 28

    a. Formulate a Prolog rule that defines the predicate *predator*.
    b. Adding this rule to the database of Example 37, what would be the response to the query

    which(*x*: *predator*(*x*))   ●

Recursion

Prolog rules are implications. Their antecedents may depend on facts, as in

    prey(*x*) **if** *eat*(*y*, *x*) **and** *animal*(*x*)

or on other rules, as in

    hunted(*x*) **if** *prey*(*x*)

The antecedent of a rule may also depend on that rule itself, in which case the rule is defined in terms of itself. A definition in which the item being defined is itself part of the definition is called a **recursive definition.**

As an example, suppose we wish to use the ecology database of Example 37 to study food chains. We can then define a binary relation *in-food-chain*(*x*, *y*), meaning "*y* is in *x*'s food chain." This, in turn, means one of two things:

    1. *x* eats *y* directly

or

    2. *x* eats something that eats something that eats something ... that eats *y*.

Case 2 can be rewritten as follows:

    2′. *x* eats *z* and *y* is in *z*'s food chain.

Case 1 is simple to test from our existing facts, but without (2′), *in-food-chain* means nothing different from *eat*. On the other hand, (2′) without (1) sends us down an infinite path of something eating something eating something and so on, with nothing telling us when to stop. Recursive definitions always need a stopping point that consists of specific information.

The Prolog rule for *in-food-chain* incorporates (1) and (2'):

in-food-chain(x, y) **if** *eat*(x, y)
in-food-chain(x, y) **if** *eat*(x, z) **and** *in-food-chain*(z, y)

It is a recursive rule because it defines the predicate *in-food-chain* in terms of *in-food-chain*. A recursive rule is necessary when the predicate being described is passed on from one object to the next. The predicate *in-food-chain* has this property:

in-food-chain(x, y) ∧ *in-food-chain*(y, z) → *in-food-chain*(x, z)

EXAMPLE 38   After the in-food-chain rule is added to the database of Example 26, the following query is made:

which(*y: in-food-chain*(bear, y))

The response follows (numbers are added for reference purposes):

1. fish
2. raccoon
3. fox
4. deer
5. little-fish
6. algae
7. fish
8. little-fish
9. algae
10. rabbit
11. grass
12. grass

Prolog applies the simple case of

in-food-chain(bear, y) **if** *eat*(bear, y)

first, obtaining answers 1 through 4 directly from the facts *eat*(bear, fish), *eat*(bear, raccoon), and so on. Moving to the recursive case,

in-food-chain(bear, y) **if** *eat*(bear, z) **and** *in-food-chain*(z, y)

a match of *eat*(bear, z) occurs with z equal to "fish." Prolog then looks for all solutions to the relation *in-food-chain*(fish, y). Using first the simple case of *in-food-chain*, a match occurs with the fact *eat*(fish, little-fish). This results in response 5, little-fish. There are no other facts of the form *eat*(fish, y), so the next thing to try is the recursive case of *in-food-chain*(fish, y):

in-food-chain(fish, y) **if** *eat*(fish, z) **and** *in-food-chain*(z, y)

A match of *eat*(fish, z) occurs with z equal to "little-fish." Prolog then looks for all solutions to the relation *in-food-chain*(little-fish, y). Using the simple case of *in-food-chain*, a match occurs with the fact *eat*(little-fish, algae). This results in response 6, algae. There are no other facts of the form *eat*(little-fish, y), so the next thing to try is the recursive case of *in-food-chain*(little-fish, y):

in-food-chain(little-fish, y) **if** *eat*(little-fish, z) **and** *in-food-chain*(z, y)

A match of *eat*(little-fish, *z*) occurs with *z* equal to "algae." Prolog then looks for all solutions to the relation *in-food-chain*(algae, *y*). A search of the entire database reveals no facts of the form *eat*(algae, *y*) (or *eat*(algae, *z*)), so neither the simple case nor the recursive case of *in-food-chain*(algae, *y*) can be pursued further.

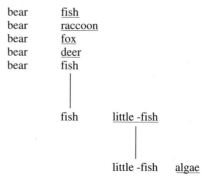

Figure 1.2

Figure 1.2 shows the situation at this point. Prolog has reached a dead end with *in-food-chain*(algae, *y*) and will backtrack up the path. Because there are no other facts of the form *eat*(little-fish, *z*), the search for solutions to *in-food-chain*(little-fish, *y*) terminates. Then, because there are no other facts of the form *eat*(fish, *z*), the search for solutions to *in-food-chain*(fish, *y*) terminates. Backing up still further, there is another match of *eat*(bear, *z*) with *z* equal to "raccoon" that will generate another search path.   ●

In Example 38, once Prolog began to investigate *in-food-chain*(fish, *y*), all query answers that could be obtained from exploring this path (responses 5 and 6) were generated before other answers (responses 7–12). Exploring as far as possible down a given path and then backtracking up that path before exploring other paths is called a **depth-first search** strategy.

PRACTICE 29   Trace the execution of the Prolog program of Example 38 and explain why responses 7–12 occur.   ●

Expert Systems

Many interesting applications programs have been developed, in Prolog and similar logic programming languages, that gather a database of facts and rules about some domain and then use this database to draw conclusions. Such programs are known as **expert systems, knowledge-based systems,** or **rule-based systems.** The database in an expert system attempts to capture the knowledge ("elicit the expertise") of a human expert in a particular field, including both the facts known to the expert and the expert's reasoning path in reaching conclusions from those facts. The completed expert system not only simulates the human expert's actions but can be questioned to reveal why it made certain choices and not others.

Expert systems have been built that simulate a medical specialist's diagnosis from a patient's symptoms, a factory manager's decisions regarding valve control in a chemical plant based on sensor readings, the decisions of a fashion buyer for a retail store based on market research, the choices made by a consultant specifying a computer system configuration based on customer needs, and many more. The challenging part of building an expert system lies in extracting all pertinent facts and rules from the human expert.

Section 1.5 Review

Techniques

- Formulate Prolog-like facts and rules.
- Formulate Prolog-like queries.
- Determine the answer(s) to a query using a Prolog database.

Main Idea

A declarative language incorporates predicate wffs and rules of inference to draw conclusions from hypotheses. The elements of such a language are based on predicate logic rather than instructions that carry out an algorithm.

Exercises 1.5

Exercises 1–6 refer to the database of Example 37; find the results of the query in each case.

1. **is**(*eat*(bear, little-fish))

2. **is**(*eat*(fox, rabbit))

★3. **which**(x: *eat*(raccoon, x))

4. **which**(x: *eat*(x, grass))

5. **which**(x: *eat*(bear, x) **and** *eat*(x, rabbit))

6. **which**(x: *prey*(x) **and** **not**(*eat*(fox, x)))

★7. Formulate a Prolog rule that defines "herbivore" to add to the database of Example 37.

8. If the rule of Exercise 7 is included in the database of Example 37, what is the response to the query

   which(x: *herbivore*(x))

9. A Prolog database contains the following, where *boss*(*x*, *y*) means "*x* is *y*'s boss" and *supervisor*(*x*, *y*) means "*x* is *y*'s supervisor":

boss(Mike, Joan)
boss(Judith, Mike)
boss(Anita, Judith)
boss(Judith, Kim)
boss(Kim, Enrique)
boss(Anita, Sam)
boss(Enrique, Jefferson)
boss(Mike, Hamal)

supervisor(*x*, *y*) **if** *boss*(*x*, *y*)
supervisor(*x*, *y*) **if** *boss*(*x*, *z*) **and** *supervisor*(*z*, *y*)

Find the results of the following queries:

a. which(*x*: *boss*(*x*, Sam))
b. which(*x*: *boss*(Judith, *x*))
c. which(*x*: *supervisor*(Anita, *x*))

10. Declare a Prolog database that gives information about states and capital cities. Some cities are big, others small. Some states are eastern, others western.

a. Write a query to find all the small capital cities.
b. Write a query to find all the states with small capital cities.
c. Write a query to find all the eastern states with big capital cities.
d. Formulate a rule to define cosmopolitan cities as big capitals of western states.
e. Write a query to find all the cosmopolitan cities.

★11. Suppose a Prolog database exists that gives information about authors and the books they have written. Books are classified as either fiction, biography, or reference.

a. Write a query to ask whether Mark Twain wrote *Hound of the Baskervilles*.
b. Write a query to find all books written by Faulkner.
c. Formulate a rule to define nonfiction authors.
d. Write a query to find all nonfiction authors.

12. Suppose a Prolog database exists that gives information about a family. Predicates of *male, female,* and *parent-of* are included.

a. Formulate a rule to define *father-of*.
b. Formulate a rule to define *daughter-of*.
★**c.** Formulate a recursive rule to define *ancestor-of*.

13. Suppose a Prolog database exists that gives information about the parts in an automobile engine. Predicates of *big, small,* and *part-of* are included.

a. Write a query to find all small items that are part of other items.
b. Write a query to find all big items that have small subitems.
c. Formulate a recursive rule to define *component-of*.

Section 1.6   Proof of Correctness

As our society becomes ever more dependent on computers, it is more and more important that the programs computers run are reliable and error-free. **Program verification** attempts to ensure that a computer program is correct. "Correctness" has a narrower definition here than in everyday usage. A program is **correct** if it behaves in accordance with its specifications. However, this does not necessarily mean that the program solves the problem that it was intended to solve; the program's specifications may be at odds with or not address all aspects of a client's requirements. **Program validation**, which we won't discuss further, attempts to ensure that the program indeed meets the client's original requirements. In a large program development project, "program V & V" or "Software Quality Assurance" is considered so important that a group of people separate from the programmers is often designated to carry out these tasks.

Program verification may be approached both through program testing and through proof of correctness. **Program testing** seeks to show that particular input values produce acceptable output values. Program testing is a major part of any software development effort, but it is well-known folklore that "testing can prove the presence of errors but never their absence." If a test run under a certain set of conditions with a certain set of input data reveals a "bug" in the code, then that bug can be corrected. But except for rather simple programs, multiple tests that reveal no bugs do not guarantee that the code is bug-free, that there is not some error lurking in the code waiting to strike under the right circumstances.

As a complement to testing, computer scientists have developed a more mathematical approach to "prove" that a program is correct. **Proof of correctness** uses the techniques of a formal logic system to prove that if the input variables satisfy certain specified predicates or properties, the output variables produced by executing the program satisfy other specified properties.

To distinguish between proof of correctness and program testing, consider a program to compute the length c of the hypotenuse of a right triangle, given positive values a and b for the lengths of the legs. Proving the program correct would establish that whenever a and b satisfy the predicates $a > 0$ and $b > 0$, then after the program is executed, the predicate $a^2 + b^2 = c^2$ is satisfied. Testing such a program would require taking various specific values for a and b, computing the resulting c, and checking that $a^2 + b^2$ equals c^2 in each case. However, only representative values for a and b can be tested, not all possible values.

Again, testing and proof of correctness are complementary aspects of program verification. All programs undergo program testing; they may or may not undergo proof of correctness as well. Proof of correctness generally is applied only to small and critical sections of code rather than to the entire program.

Assertions

Describing proof of correctness more formally, let us denote by X an arbitrary collection of input values to some program or program segment P. The actions of P transform X into a corresponding group of output values Y; the notation $Y = P(X)$ suggests that the Y values depend on the X values through the actions of program P.

A predicate $Q(X)$ describes conditions that the input values are supposed to satisfy. For example, if a program is supposed to find the square root of a positive number, then X consists of one input value, x, and $Q(x)$ might be "$x > 0$." A predicate R describes conditions that the output values are supposed to satisfy. These conditions will often involve the input values as well, so R has the form $R(X, Y)$ or $R[X, P(X)]$. In our square root case, if y is the single output value, then y is supposed to be the square root of x, so $R(x, y)$ would be "$y^2 = x$." Program P is correct if the implication

$$(\forall X)(Q(X) \to R[X, P(X)]) \tag{1}$$

is valid. In other words, whenever Q is true about the input values, R should be true about the input and output values. For the square root case, (1) is

$$(\forall x)(x > 0 \to [P(x)]^2 = x)$$

The implication (1) is standard predicate wff notation, but the traditional program correctness notation for (1) is

$$\{Q\}P\{R\} \tag{2}$$

$\{Q\}P\{R\}$ is called a **Hoare triple,** named for the British computer scientist Anthony Hoare. Condition Q is called the **precondition** for program P, and condition R is the **postcondition.** In the Hoare notation, the universal quantifier does not explicitly appear; it is understood.

Rather than simply having an initial predicate and a final predicate, a program or program segment is broken down into individual statements s_i, with predicates inserted between statements as well as at the beginning and end. These predicates are also called **assertions** because they assert what is supposed to be true about the program variables at that point in the program. Thus we have

$$\{Q\}$$
$$s_0$$
$$\{R_1\}$$
$$s_1$$
$$\{R_2\}$$
$$\vdots$$
$$s_{n-1}$$
$$\{R\}$$

where $Q, R_1, R_2, \ldots , R_n = R$ are assertions. The intermediate assertions are often obtained by working backward from the output assertion R.

P is provably correct if each of the following implications holds:

$$\{Q\}s_0\{R_1\}$$
$$\{R_1\}s_1\{R_2\}$$
$$\{R_2\}s_2\{R_3\}$$
$$\vdots$$
$$\{R_{n-1}\}s_{n-1}\{R\}$$

A proof of correctness for P consists of producing this sequence of valid implications, that is, producing a proof sequence of predicate wffs. Some new rules of inference can be used, based on the nature of the program statement s_i.

Assignment Rule

Suppose that statement s_i is an assignment statement of the form $x = e$; that is, the variable x takes on the value of e, where e is some expression. The Hoare triple to prove correctness of this one statement has the form

$$\{R_i\}\ x = e\ \{R_{i+1}\}$$

For this to be valid, the assertions R_i and R_{i+1} must be related in a particular way.

EXAMPLE 39

Consider the following assignment statement together with the given precondition and postcondition:

$$\{x - 1 > 0\}$$
$$x = x - 1$$
$$\{x > 0\}$$

For every x, if $x - 1 > 0$ before the statement is executed (note that this says that $x > 1$), then after the value of x is reduced by 1, it will be the case that $x > 0$. Therefore

$$\{x - 1 > 0\}\ x = x - 1\ \{x > 0\}$$

is valid.
●

In Example 39, we just reasoned our way through the validity of the wff represented by the Hoare triple. The point of predicate logic is to allow us to determine validity in a more mechanical fashion by the application of rules of inference. (After all, we don't want to just "reason our way through" the entire program to convince ourselves of its correctness; the programmer already did that when the program was written!)

The appropriate rule of inference for assignment statements is the **assignment rule**, given in Table 1.18. It says that if the precondition and postcondition are appropriately related, the Hoare triple can be inserted at any time in a proof sequence without having to be inferred from something earlier in the proof sequence. This makes the Hoare triple for an assignment statement akin to a hypothesis in our previous proofs.

From	Can Derive	Name of Rule	Restrictions on Use
	$\{R_i\}s_i\{R_{i+1}\}$	Assignment	1. s_i has the form $x = e$. 2. R_i is R_{i+1} with e substituted everywhere for x.

Table 1.18

EXAMPLE 40   For the case of Example 39,

$$\{x - 1 > 0\}$$
$$x = x - 1$$
$$\{x > 0\}$$

the triple

$$\{x - 1 > 0\} \; x = x - 1 \; \{x > 0\}$$

is valid by the assignment rule. The postcondition is

$$x > 0$$

Substituting $x - 1$ for x throughout the postcondition results in

$$x - 1 > 0 \quad \text{or} \quad x > 1$$

which is the precondition. Here we didn't have to think at all; we just checked that the assignment rule had been followed.   ●

PRACTICE 30   According to the assignment rule, what should be the precondition in the following program segment?

$$\{\text{precondition}\}$$
$$x = x - 2$$
$$\{x = y\}$$ ●

REMINDER:
To use the assignment rule, work from the bottom to the top.

Because the assignment rule tells us what a precondition should look like based on what a postcondition looks like, a proof of correctness often begins with the final desired postcondition and works its way back up through what the earlier assertions should look like according to the assignment rule. Once it has been determined what the topmost assertion must be, a check is done to see that this assertion is really true.

EXAMPLE 41   Verify the correctness of the following program segment to exchange the values of x and y:

temp $= x$
$x = y$
$y =$ temp

At the beginning of this program segment, x and y have certain values. Thus we may express the actual precondition as $x = a$ and $y = b$. The desired postcondition is then $x = b$ and $y = a$. Using the assignment rule, we can work backward from the postcondition to find the earlier assertions (read the following from the bottom to the top).

$$\{y = b, x = a\}$$
$$\text{temp} = x$$
$$\{y = b, \text{temp} = a\}$$
$$x = y$$
$$\{x = b, \text{temp} = a\}$$
$$y = \text{temp}$$
$$\{x = b, y = a\}$$

The first assertion agrees with the precondition; the assignment rule, applied repeatedly, assures us that the program segment is correct. ●

PRACTICE 31   Verify the correctness of the following program segment with the precondition and postcondition shown:

$\{x = 3\}$
    $y = 4$
    $z = x + y$
$\{z = 7\}$ ●

Sometimes the necessary precondition is trivially true, as shown in the next example.

EXAMPLE 42   Verify the correctness of the following program segment to compute $y = x - 4$.

$y = x$
$y = y - 4$

Here the desired postcondition is $y = x - 4$. Using the assignment rule to work backward from the postcondition, we get (again, read bottom to top)

$\{x - 4 = x - 4\}$
        $y = x$
$\{y - 4 = x - 4\}$
        $y = y - 4$
$\{y = x - 4\}$

The precondition is always true; therefore, by the assignment rule, each successive assertion, including the postcondition, is true. ●

Conditional Rule

A **conditional statement** is a program statement of the form

if condition B **then**
    P_1
else
    P_2
end if

When this statement is executed, a condition B that is either true or false is evaluated. If B is true, program segment P_1 is executed, but if B is false, program segment P_2 is executed.

A **conditional rule of inference,** shown in Table 1.19, determines when a Hoare triple

$\{Q\}s_i\{R\}$

can be inserted in a proof sequence if s_i is a conditional statement. The Hoare triple is inferred from two other Hoare triples. One of these says that if Q is true and B is true and program segment P_1 is executed, then R holds; the other says that if Q is true and B is false and program segment P_2 is executed, then R holds. This simply says that each branch of the conditional statement must be proved correct.

From	Can Derive	Name of Rule	Restrictions on Use

$\{Q \wedge B\}\ P_1\ \{R\},$
 $\{Q \wedge B'\}\ P_2\ \{R\}$
	$\{Q\}s_i\{R\ \}$	Conditional	s_i has the form
			 if condition B **then**
			 P_1
			 else
			 P_2
			 end if

Table 1.19

EXAMPLE 43   Verify the correctness of the following program segment with the precondition and postcondition shown.

$\{n = 5\}$
  if $n >= 10$ **then**
    $y = 100$
  else
    $y = n + 1$
  end if
$\{y = 6\}$

Here the precondition is $n = 5$, and the condition B to be evaluated is $n >= 10$. In order to apply the conditional rule, we must first prove that

$\{n = 5 \text{ and } n \geq 10\}\ y = 100\ \{y = 6\}$

holds. Remember that this stands for an implication, which will be true because its antecedent, $n = 5$ and $n \geq 10$, is false. We must also show that

$\{n = 5 \text{ and } n < 10\}\ y = n + 1\ \{y = 6\}$

holds. Working back from the postcondition, using the assignment rule, we get

$\{n + 1 = 6 \text{ or } n = 5\}$
  $y = n + 1$
$\{y = 6\}$

Thus

$\{n = 5\}\ y = n + 1\ \{y = 6\}$

is true by the assignment rule and therefore

$\{n = 5 \text{ and } n < 10\}\ y = n + 1\ \{y = 6\}$

is also true because the condition $n < 10$ adds nothing new to the assertion. The conditional rule allows us to conclude that the program segment is correct.   ●

PRACTICE 32   Verify the correctness of the following program segment with the precondition and postcondition shown.

$\{x = 4\}$
    if $x < 5$ **then**
        $y = x - 1$
    else
        $y = 7$
    end if
$\{y = 3\}$ ●

EXAMPLE 44   Verify the correctness of the following program segment to compute max(x, y), the maximum of two distinct values x and y.

$\{x \neq y\}$
    if $x >= y$ **then**
        max $= x$
    else
        max $= y$
    end if

The desired postcondition reflects the definition of the maximum, $(x > y$ and max $= x)$ or $(x < y$ and max $= y)$. The two implications to prove are

$$\{x \neq y \text{ and } x \geq y\} \text{ max} = x \; \{(x > y \text{ and max} = x) \text{ or } (x < y \text{ and max} = y)\}$$

and

$$\{x \neq y \text{ and } x < y\} \text{ max} = y \; \{(x > y \text{ and max} = x) \text{ or } (x < y \text{ and max} = y)\}$$

Using the assignment rule on the first case (substituting x for max in the postcondition) would give the precondition

$$(x > y \wedge x = x) \vee (x < y \wedge x = y)$$

Since the second disjunct is always false, this is equivalent to

$$(x > y \wedge x = x)$$

which in turn is equivalent to

$$x > y \quad \text{or} \quad x \neq y \text{ and } x \geq y$$

The second implication is proved similarly.   ●

In Chapter 2, we will see how to verify correctness for a loop statement, where a section of code can be repeated many times.

As we have seen, proof of correctness involves a lot of detailed work. It is a difficult tool to apply to large programs that already exist. It is generally easier to prove correctness while the program is being developed. Indeed, the list of assertions from beginning to end specify the intended behavior of the program and can be used early in its design. In addition, the assertions serve as valuable documentation after the program is complete.

Section 1.6 Review

Techniques

- Verify the correctness of a program segment that includes assignment statements.
- Verify the correctness of a program segment that includes conditional statements.

Main Idea

A formal system of rules of inference can be used to prove the correctness of program segments.

Exercises 1.6

Note: In Exercises 2, 3, 5, and 8, the $*$ denotes multiplication.

★1. According to the assignment rule, what is the precondition in the following program segment?

$\{$precondition$\}$
$\quad x = x + 1$
$\{x = y - 1\}$

2. According to the assignment rule, what is the precondition in the following program segment?

$\{$precondition$\}$
$\quad x = 2 * x$
$\{x > y\}$

3. Verify the correctness of the following program segment with the precondition and postcondition shown.

$\{x = 1\}$
$\quad y = x + 3$
$\quad y = 2 * y$
$\{y = 8\}$

4. Verify the correctness of the following program segment with the precondition and postcondition shown.

$\{x > 0\}$
$\quad y = x + 2$
$\quad z = y + 1$
$\{z > 3\}$

★5. Verify the correctness of the following program segment to compute $y = x(x - 1)$.

$\quad y = x - 1$
$\quad y = x * y$

6. Verify the correctness of the following program segment to compute $y = 2x + 1$.

   $y = x$
   $y = y + y$
   $y = y + 1$

★7. Verify the correctness of the following program segment with the precondition and postcondition shown.

   $\{y = 0\}$
      if $y < 5$ **then**
         $y = y + 1$
      else
         $y = 5$
      end if
   $\{y = 1\}$

8. Verify the correctness of the following program segment with the precondition and postcondition shown.

   $\{x = 7\}$
      if $x <= 0$ **then**
         $y = x$
      else
         $y = 2 * x$
      end if
   $\{y = 14\}$

9. Verify the correctness of the following program segment to compute $\min(x, y)$, the minimum of two distinct values x and y.

   $\{x \neq y\}$
      if $x <= y$ **then**
         $\min = x$
      else
         $\min = y$
      end if

10. Verify the correctness of the following program segment to compute $|x|$, the absolute value of x, for a nonzero number x.

   $\{x \neq 0\}$
      if $x >= 0$ **then**
         $abs = x$
      else
         $abs = -x$
      end if

11. Verify the correctness of the following program segment with the assertions shown.

$\{z = 3\}$
    $x = z + 1$
    $y = x + 2$
$\{y = 6\}$
    if $y > 0$ **then**
        $z = y + 1$
    else
        $z = 2 * y$
    end if
$\{z = 7\}$

Chapter 1 Review

Terminology

algorithm (p. 13)
antecedent (p. 4)
assertion (p. 70)
assignment rule (p. 71)
binary connective (p. 5)
binary predicate (p. 34)
complete formal system (p. 22)
conclusion (p. 20)
conditional rule of inference (p. 73)
conditional statement (p. 73)
conjunct (p. 3)
conjunction (p. 3)
consequent (p. 4)
contradiction (p. 9)
correct formal system (p. 22)
correct program (p. 69)
De Morgan's laws (p. 11)
declarative language (p. 58)
depth-first search (p. 66)
derivation rule (p. 22)
descriptive language (p. 58)
disjunct (p. 4)
disjunction (p. 4)

domain (p. 35)
dual of an equivalence (p. 10)
equivalence (p. 4)
equivalence rule (p. 22)
equivalent wffs (p. 9)
existential generalization (p. 46)
existential instantiation (p. 46)
existential quantifier (p. 34)
expert system (p. 66)
free variable (p. 36)
Hoare triple (p. 70)
Horn clause (p. 61)
hypothesis (p. 20)
implication (p. 4)
inference rule (p. 24)
interpretation (p. 35)
knowledge-based system (p. 66)
logical connective (p. 3)
main connective (p. 7)
n-ary predicate (p. 34)
negation (p. 5)
postcondition (p. 70)
precondition (p. 70)
predicate (p. 33)

predicate logic (p. 45)
predicate wff (p. 35)
procedural language (p. 58)
program testing (p. 69)
program validation (p. 69)
program verification (p. 69)
Prolog database (p. 58)
Prolog fact (p. 59)
Prolog rule (p. 60)
proof of correctness (p. 69)
proof sequence (p. 22)
proposition (p. 3)
propositional calculus (p. 20)
propositional logic (p. 20)
propositional wff (p. 20)
pseudocode (p. 13)
recursive definition (p. 64)
resolution (p. 61)
rule-based system (p. 66)
scope (p. 35)
statement (p. 3)
statement letter (p. 3)

statement logic (p. 20)	universal generalization	valid argument
tautology (p. 9)	(p. 46)	(pp. 21, 45)
ternary predicate (p. 34)	universal instantiation	valid wff (p. 39)
unary connective (p. 5)	(p. 46)	well-formed formula
unary predicate (p. 34)	universal quantifier (p. 33)	(wff) (p. 7)

Self-Test

Answer the following true–false questions without looking back in the chapter.

Section 1.1

1. A contradiction is any propositional wff that is not a tautology.
2. The disjunction of any propositional wff with a tautology has the truth value true.
3. Algorithm *TautologyTest* determines whether any propositional wff is a tautology.
4. Equivalent propositional wffs have the same truth values for every truth value assignment to the components.
5. One of De Morgan's laws states that the negation of a disjunction is the disjunction of the negations (of the disjuncts).

Section 1.2

6. An equivalence rule allows one wff to be substituted for another in a proof sequence.
7. If a propositional wff can be derived using modus ponens, then its negation can be derived using modus tollens.
8. Propositional logic is complete because every tautology is provable.
9. A valid argument is one in which the conclusion is always true.
10. The deduction method applies when the conclusion is an implication.

Section 1.3

11. A predicate wff that begins with a universal quantifier is universally true, that is, true in all interpretations.
12. In the predicate wff $(\forall x)P(x, y)$, y is a free variable.
13. An existential quantifier is usually found with the conjunction connective.
14. The domain of an interpretation consists of the values for which the predicate wff defined on that interpretation is true.
15. A valid predicate wff has no interpretation in which it is false.

Section 1.4

16. The inference rules of predicate logic allow existential and universal quantifiers to be added or removed during a proof sequence.
17. Existential instantiation should be used only after universal instantiation.
18. $P(x) \wedge (\exists x)Q(x)$ can be deduced from $(\forall x)[P(x) \wedge (\exists y)Q(y)]$ using universal instantiation.
19. Every provable wff of propositional logic is also provable in predicate logic.
20. A predicate wff that is not valid cannot be proved using predicate logic.

Section 1.5

21. A Prolog rule describes a predicate.

22. Horn clauses are wffs consisting of single negated predicates.

23. Modus ponens is a special case of Prolog resolution.

24. A Prolog recursive rule is a rule of inference that is used more than once.

25. A Prolog inference engine applies its rule of inference without guidance from either the programmer or the user.

Section 1.6

26. A provably correct program always gives the right answers to a given problem.

27. If an assertion after an assignment statement is $y > 4$, then the precondition must be $y \geq 4$.

28. Proof of correctness involves careful development of test data sets.

29. Using the conditional rule of inference in proof of correctness involves proving that two different Hoare triples are valid.

30. The assertions used in proof of correctness can also be used as a program design aid before the program is written, and as program documentation.

On the Computer

For Exercises 1–5, write a computer program that produces the desired output from the given input.

1. *Input:* Truth values for two statement letters A and B
   Output: Corresponding truth values (appropriately labeled, of course) for

   $$A \wedge B, A \vee B, A \rightarrow B, A \leftrightarrow B, A'$$

2. *Input:* Truth values for two statement letters A and B
   Output: Corresponding truth values for the wffs

   $$A \rightarrow B' \quad \text{and} \quad B' \wedge [A \vee (A \wedge B)]$$

3. *Input:* Truth values for three statement letters A, B, and C
   Output: Corresponding truth values for the wffs

   $$A \vee (B \wedge C') \rightarrow B' \quad \text{and} \quad A \vee C' \leftrightarrow (A \vee C)'$$

4. *Input:* Truth values for three statement letters A, B, and C, and a representation of a simple propositional wff. Special symbols can be used for the logical connectives, and postfix notation can be used; for example,

   $$A\ B \wedge C \vee \quad \text{for} \quad (A \wedge B) \vee C$$

or

$$A'\ B\ \wedge \qquad \text{for} \qquad A' \wedge B$$

Output: Corresponding truth value of the wff

5. *Input:* Representation of a simple propositional wff as in the previous exercise
   Output: Decision on whether the wff is a tautology

6. If you have a version of Prolog available, enter the Prolog database of Example 37 and perform the queries there. Also add the recursive rule for *in-food-chain* and perform the query

   which(y: *in-food-chain*(bear, y))

2

Proofs, Recursion, and Analysis of Algorithms

Chapter Objectives

After studying this chapter, you will be able to:

- Attack the proofs of conjectures using the techniques of direct proof, proof by contraposition, and proof by contradiction.
- Recognize when a proof by induction is appropriate and carry out such a proof using either the first or second principle of induction.
- Mathematically prove the correctness of programs that use loop statements.
- Understand recursive definitions of sequences, collections of objects, and operations on objects.
- Write recursive definitions for certain sequences, collections of objects, and operations on objects.
- Understand how recursive algorithms execute.
- Write recursive algorithms to generate sequences defined recursively.
- Find closed-form solutions for certain recurrence relations found in analysis of algorithms.

You are serving on the city council's Board of Land Management, which is considering a proposal by a private contractor to manage a chemical disposal site. The material to be stored at the site degrades to inert matter at the rate of 5% per year. The contractor claims that, at this rate of stabilization, only about one-third of the original active material will remain at the end of 20 years.

Question: Is the contractor's estimate correct?

It is possible to check this estimate by doing some brute-force calculations: If there is this much initially, then there will be that much next year, and then so much the following year, and so on through the 20 years. But a quick and elegant solution can be obtained by solving a recurrence relation; recurrence relations are discussed in Section 2.4.

First, however, we'll consider how to prove "real-world" arguments as opposed to the formal arguments of Chapter 1. It is helpful to have an arsenal of techniques for attacking a proof. Direct proof, proof by contraposition, and proof by contradiction are examined in Section 2.1. Section 2.2 concentrates on mathematical induction, a proof technique with particularly wide application in computer science. In Section 2.3 we will see how, using induction, proof of correctness can be extended to cover looping statements.

Section 2.4 discusses recursion, which is closely related to mathematical induction and is important in expressing many definitions and even algorithms. Some sequences defined recursively can also be defined by a formula; finding such a formula involves solving a recurrence relation, and induction is used to verify that the formula is correct. The use of recurrence relations to determine the amount of work a particular algorithm must do is explored in Section 2.5.

Section 2.1   Proof Techniques

Theorems and Informal Proofs

The formal arguments of Chapter 1 have the form $P \rightarrow Q$, where P and Q may represent compound statements. The point there was to prove that an argument is valid—true in all interpretations by nature of its internal form or structure, not because of its content or the meaning of its component parts. However, we often want to prove arguments that are not universally true, just true within some context. Meaning becomes important because we are discussing a particular subject—graph algorithms or Boolean algebra or compiler theory or whatever—and we want to prove that if P is true in this context, then so is Q. If we can do this, then $P \rightarrow Q$ becomes a *theorem* about that subject. To prove a theorem about subject XXX, we can introduce facts about XXX into the proof; these facts act like additional hypotheses.

It may not be easy to recognize which subject-specific facts will be helpful or to arrange a sequence of steps that will logically lead from P to Q. Unfortunately, there is no formula for constructing proofs and no practical general algorithm or computer program for proving theorems. Experience is helpful, not only because you get better with practice, but also because a proof that works for one theorem can sometimes be modified to work for a new but similar theorem.

Theorems are often stated and proved in a somewhat less formal way than the propositional and predicate arguments of Chapter 1. For example, a theorem may express the fact that every object in the domain of interpretation (the subject matter under discussion) having property P also has property Q. The formal statement of the theorem would be $(\forall x)[P(x) \rightarrow Q(x)]$. But the theorem would be informally stated as $P(x) \rightarrow Q(x)$. If we can prove $P(x) \rightarrow Q(x)$ where x is treated as an arbitrary element of the domain, Universal Generalization would then give $(\forall x)[P(x) \rightarrow Q(x)]$.

As another example, we may know that all objects in the domain have some property; that is, something of the form $(\forall x)P(x)$ can be considered as a subject-specific fact. An informal proof might proceed by saying "Let x be any element of the domain. Then x has property P." (Formally, we are making use of Universal Instantiation to get $P(x)$ from $(\forall x)P(x)$.)

Similarly, proofs are usually not written a step at a time with formal justifications for each step. Instead, the important steps and their rationale are outlined in narrative form. Such a narrative, however, can be translated into a formal proof if required. In fact, the value of a formal proof is that it serves as a sort of insurance—if a narrative proof *cannot* be translated into a formal proof, it should be viewed with great suspicion.

To Prove or Not to Prove

A textbook will often say, "Prove the following theorem," and the reader will know that the theorem is true; furthermore, it is probably stated in its most polished form. But suppose you are doing research in some subject. You observe a number of cases in which whenever P is true, Q is also true. On the basis of these experiences, you may formulate a conjecture: $P \rightarrow Q$. The more cases you find where Q follows from P, the more confident you are in your conjecture. This process illustrates **inductive reasoning,** drawing a conclusion based on experience.

No matter how reasonable the conjecture sounds, however, you will not be satisfied until you have applied **deductive reasoning** to it as well. In this process, you try to verify the truth or falsity of your conjecture. You produce a proof of $P \rightarrow Q$ (thus making it a theorem), or else you find a **counterexample** that disproves the conjecture, a case in which P is true but Q is false. (We were using deductive reasoning in predicate logic when we either proved that a wff was valid or found an interpretation in which the wff was false.)

REMINDER:
One counter-example is enough to disprove a conjecture.

If you are simply presented with a conjecture, it may be difficult to decide which of the two approaches you should try—to prove the conjecture or to disprove it! A single counterexample to a conjecture is sufficient to disprove it. Of course, merely hunting for a counterexample and being unsuccessful does not constitute a proof that the conjecture is true.

EXAMPLE 1   For a positive integer n, **n factorial** is defined as $n(n-1)(n-2) \cdots 1$, and is denoted by $n!$. Prove or disprove the conjecture "For every positive integer n, $n! \leq n^2$."

Let's begin by testing some cases:

n	$n!$	n^2	$n! \leq n^2$
1	1	1	yes
2	2	4	yes
3	6	9	yes

So far, this conjecture seems to be looking good. But for the next case,

n	$n!$	n^2	$n! \leq n^2$
4	24	16	no

we have found a counterexample. The fact that the conjecture is true for $n = 1$, 2, and 3 does nothing to prove the conjecture, but the single case $n = 4$ is enough to disprove it. ●

PRACTICE 1   Provide counterexamples to the following conjectures:
  a. All animals living in the ocean are fish.
  b. Every integer less than 10 is bigger than 5. ●

If a counterexample is not forthcoming, what techniques can we use to try to prove a conjecture? For the rest of this section, we'll examine various methods of attacking a proof.

Exhaustive Proof

While "disproof by counterexample" always works, "proof by example" seldom does. The one exception to this situation occurs when the conjecture is an assertion about a finite collection. In this case, the conjecture can be proved true by showing that it is true for each member of the collection. **Proof by exhaustion** means that all possible cases have been exhausted, although it often means that the person doing the proof is exhausted as well!

EXAMPLE 2   Prove the conjecture "If an integer between 1 and 20 is divisible by 6, then it is also divisible by 3." ("Divisible by 6" means "evenly divisible by 6," that is, the number is an integral multiple of 6.)
  Because there is only a finite number of cases, the conjecture can be proved by simply showing it to be true for all the integers between 1 and 20. Table 2.1 is the proof. ●

Number	Divisible by 6	Divisible by 3
1	no	
2	no	
3	no	
4	no	
5	no	
6	yes: 6 = 1×6	yes: 6 = 2×3
7	no	
8	no	
9	no	
10	no	
11	no	
12	yes: 12 = 2×6	yes: 12 = 4×3
13	no	
14	no	
15	no	
16	no	
17	no	
18	yes: 18 = 3×6	yes: 18 = 6×3
19	no	
20	no	

Table 2.1

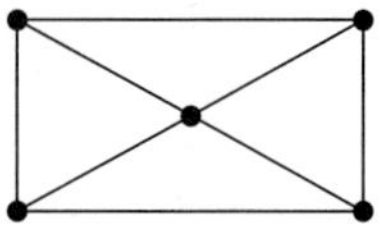

Figure 2.1

EXAMPLE 3

Prove the conjecture "It is not possible to trace all the lines in Figure 2.1 without lifting your pencil and without retracing any lines."

There is only a finite number of different ways to trace the lines in the figure. By careful bookkeeping, each of the possibilities can be attempted, and each will fail. In Chapter 6, we will learn a much less tedious way to solve this problem than proof by exhaustion. ●

PRACTICE 2

a. Prove the conjecture "For any positive integer less than or equal to 5, the square of the integer is less than or equal to the sum of 10 plus 5 times the integer."
b. Disprove the conjecture "For any positive integer, the square of the integer is less than or equal to the sum of 10 plus 5 times the integer." ●

Direct Proof

In general (where exhaustive proof won't work), how can you prove that $P \rightarrow Q$ is true? The obvious approach is the **direct proof**—assume the hypothesis P and deduce the conclusion Q. A formal proof would require a proof sequence leading from P to Q.

EXAMPLE 4   Consider the conjecture

$(\forall x)(\forall y)(x$ is an even integer $\wedge$ y is an even integer $\rightarrow$ the product xy is an even integer)

A complete formal proof sequence might look like the following:

1. $(\forall x)[x$ is even integer $\rightarrow (\exists k)(k$ an integer $\wedge x = 2k)]$ number fact (definition of even integer)

2. x is even integer $\rightarrow (\exists k)(k$ an integer $\wedge x = 2k)$ 1, ui
3. y is even integer $\rightarrow (\exists k)(k$ an integer $\wedge y = 2k)$ 1, ui
4. x is an even integer $\wedge$ y is an even integer   temporary hyp
5. x is an even integer   4, sim
6. $(\exists k)(k$ is an integer $\wedge x = 2k)$ 2, 5, mp
7. m is an integer $\wedge x = 2m$ 6, ei
8. y is an even integer   4, sim
9. $(\exists k)(k$ an integer $\wedge y = 2k)$ 3, 8, mp
10. n is an integer and $y = 2n$ 9, ei
11. $x = 2m$ 7, sim
12. $y = 2n$ 10, sim
13. $xy = (2m)(2n)$ 11, 12, substitution of equals
14. $xy = 2(2mn)$ 13, multiplication fact
15. m is an integer   7, sim
16. n is an integer   10, sim
17. $2mn$ is an integer   15, 16, number fact
18. $xy = 2(2mn) \wedge 2mn$ is an integer   14, 17, con
19. $(\exists k)(k$ an integer $\wedge xy = 2k)$ 18, eg
20. $(\forall x)((\exists k)(k$ an integer $\wedge x = 2k) \rightarrow x$ is even integer)   number fact (definition of even integer)

21. $(\exists k)(k$ an integer $\wedge xy = 2k) \rightarrow xy$ is even integer   20, ui
22. xy is an even integer   19, 21, mp
23. x is an even integer $\wedge$ y is an even integer $\rightarrow xy$ is an even integer   temp. hyp discharged
24. $(\forall x)(\forall y)(x$ is an even integer $\wedge$ y is an even integer $\rightarrow$ the product xy is an even integer)   23, ug twice   ●

We'll never do such a proof again, and you won't have to either! A much more informal proof would be perfectly acceptable in most circumstances.

EXAMPLE 5   Following is an informal direct proof that the product of two even integers is even. Let $x = 2m$ and $y = 2n$, where m and n are integers. Then $xy = (2m)(2n) = 2(2mn)$, where $2mn$ is an integer. Thus xy has the form $2k$, where k is an integer, and xy is therefore even.   ●

The proof in Example 5 does not explicitly state the hypothesis (that x and y are even), and it makes implicit use of the definition of an even integer. Even in informal proofs, however, it is important to identify the hypothesis and the conclusion, not just what they are in words but what they really mean, by applying appropriate definitions. If we do not clearly understand what we have (the hypothesis) or what we want (the conclusion), we cannot hope to build a bridge from one to the other. That's why it is important to know definitions.

PRACTICE 3   Give a direct proof (informal) of the theorem "If an integer is divisible by 6, then twice that integer is divisible by 4."   ●

Contraposition

If you have tried diligently but failed to produce a direct proof of your conjecture $P \rightarrow Q$, and you still feel that the conjecture is true, you might try some variants on the direct proof technique. If you can prove the theorem $Q' \rightarrow P'$, you can conclude $P \rightarrow Q$ by making use of the tautology $(Q' \rightarrow P') \rightarrow (P \rightarrow Q)$. $Q' \rightarrow P'$ is the **contrapositive** of $P \rightarrow Q$, and the technique of proving $P \rightarrow Q$ by doing a direct proof of $Q' \rightarrow P'$ is called **proof by contraposition.** (The contraposition rule of inference in propositional logic, Table 1.14, says that $P \rightarrow Q$ could be derived from $Q' \rightarrow P'$.)

EXAMPLE 6   Prove that if the square of an integer is odd, then the integer must be odd.

The conjecture is n^2 odd $\rightarrow n$ odd. We do a proof by contraposition, and prove n even $\rightarrow n^2$ even. Let n be even. Then $n^2 = n(n)$ is even by Example 5.   ●

EXAMPLE 7   Prove that if $n + 1$ separate passwords are issued to n students, then some student gets ≥ 2 passwords.

The contrapositive is "If every student gets < 2 passwords, then $n + 1$ passwords were not issued." Suppose every student has < 2 passwords; then every one of the n students has at most 1 password. The total number of passwords issued is at most n, not $n + 1$.   ●

Example 7 is an illustration of the Pigeonhole Principle, which we will see in Chapter 3.

PRACTICE 4   Write the contrapositive of each statement in Practice 5 of Chapter 1.   ●

Practice 7 of Chapter 1 showed that the wffs $A \rightarrow B$ and $B \rightarrow A$ are not equivalent. $B \rightarrow A$ is the **converse** of $A \rightarrow B$. If an implication is true, its converse may be true or false. Therefore, you cannot prove $P \rightarrow Q$ by looking at $Q \rightarrow P$.

EXAMPLE 8   The implication "If $a > 5$, then $a > 2$" is true, but its converse, "If $a > 2$, then $a > 5$," is false.   ●

PRACTICE 5   Write the converse of each statement in Practice 5 of Chapter 1.   ●

REMINDER:
"If and only
if" requires two
proofs, one in
each direction.

Theorems are often stated in the form "P if and only if Q," meaning P if Q and P only if Q, or $Q \rightarrow P$ and $P \rightarrow Q$. To prove such a theorem, you must prove both an implication and its converse. Again, the truth of one does not imply the truth of the other.

EXAMPLE 9   Prove that the product xy is odd if and only if both x and y are odd integers.

We first prove that if x and y are odd, so is xy. A direct proof will work. Suppose that both x and y are odd. Then $x = 2n + 1$ and $y = 2m + 1$, where m and n are integers. Then $xy = (2n + 1)(2m + 1) = 4nm + 2m + 2n + 1 = 2(2nm + m + n) + 1$. This has the form $2k + 1$, where k is an integer, so xy is odd.
Next we prove that if xy is odd, both x and y must be odd, or

xy odd $\rightarrow x$ odd and y odd

A direct proof would begin with the hypothesis that xy is odd, which leaves us little more to say. A proof by contraposition works well because we'll get more useful information as hypotheses. So we will prove

$(x$ odd and y odd$)' \rightarrow (xy$ odd$)'$

By De Morgan's law $(A \wedge B)' \Leftrightarrow A' \vee B'$, we see that this can be written as

x even or y even $\rightarrow xy$ even   (1)

The hypothesis "x even or y even" breaks down into three cases. We consider each case in turn.

1. x even, y odd: Here $x = 2m$, $y = 2n + 1$, and then $xy = (2m)(2n + 1) = 2(2mn + m)$, which is even.
2. x odd, y even: This works just like case 1.
3. x even, y even: Then xy is even by Example 5.

This completes the proof of (1) and thus of the theorem.   ●

The second part of the proof of Example 9 uses **proof by cases,** a form of exhaustive proof. It involves identifying all the possible cases consistent with the given information and then proving each case separately.

Contradiction

In addition to direct proof and proof by contraposition, you might use the technique of **proof by contradiction.** (Proof by contradiction is sometimes called *indirect proof,* but this term more properly means any argument that is not a direct proof.) As we did in Chapter 1, we will let 0 stand for any contradiction, that is, any wff whose truth value is always false. ($A \wedge A'$ would be such a wff.) Once more, suppose you are trying to prove $P \to Q$. By constructing a truth table, we see that

$$(P \wedge Q' \to 0) \to (P \to Q)$$

is a tautology, so to prove the theorem $P \to Q$, it is sufficient to prove $P \wedge Q' \to 0$. Therefore, in a proof by contradiction you assume that both the hypothesis and the negation of the conclusion are true and then try to deduce some contradiction from these assumptions.

EXAMPLE 10

Let's use proof by contradiction on the statement "If a number added to itself gives itself, then the number is 0." Let x represent any number. The hypothesis is $x + x = x$ and the conclusion is $x = 0$. To do a proof by contradiction, assume $x + x = x$ and $x \neq 0$. Then $2x = x$ and $x \neq 0$. Because $x \neq 0$, we can divide both sides of the equation $2x = x$ by x and arrive at the contradiction $2 = 1$. Hence, $(x + x = x) \to (x = 0)$.   ●

REMINDER:
To prove that something is not true, try proof by contradiction

Example 10 notwithstanding, a proof by contradiction most immediately comes to mind when you want to prove that something is *not* true. It's hard to prove that something *is not true*; it's much easier to *assume it is true* and obtain a contradiction.

EXAMPLE 11

A well-known proof by contradiction shows that $\sqrt{2}$ is not a rational number. Recall that a **rational number** is one that can be written in the form p/q where p and q are integers, $q \neq 0$, and p and q have no common factors (other than ± 1).

Let us assume that $\sqrt{2}$ is rational. Then $\sqrt{2} = p/q$, and $2 = p^2/q^2$, or $2q^2 = p^2$. Then 2 divides p^2, so—since 2 is itself indivisible—2 must divide p. This means that 2 is a factor of p, hence 4 is a factor of p^2, and the equation $2q^2 = p^2$ can be written as $2q^2 = 4x$, or $q^2 = 2x$. We see from this equation that 2 divides q^2; hence 2 divides q. At this point, 2 is a factor of q and a factor of p, which contradicts the statement that p and q have no common factors. Therefore $\sqrt{2}$ is not rational.   ●

The proof of Example 11 involves more than just algebraic manipulations. It is often necessary to use lots of words in a proof.

PRACTICE 6

Prove by contradiction that the product of odd integers is not even. (We did a direct proof of an equivalent statement in Example 9.)   ●

Proof by contradiction can be a valuable technique, but it is easy to think we have done a proof by contradiction when we really haven't. For example, suppose we assume $P \wedge Q'$ and are able to deduce Q without using the assumption Q'. Then we assert $Q \wedge Q'$ as a contradiction. What really happened here is a direct proof of

$P \to Q$, and the proof should be rewritten in this form. Thus in Example 9, we could assume $x + x = x$ and $x \neq 0$, as before. Then we could argue that from $x + x = x$ we get $2x = x$ and, after subtracting x from both sides, $x = 0$. We then have $x = 0$ and $x \neq 0$, a contradiction. However, in this argument we never made use of the assumption $x \neq 0$; we actually proved directly that $x + x = x$ implies x = 0.

Another misleading claim of proof by contradiction occurs when we assume $P \wedge Q'$ and are able to deduce P' without using the assumption P. Then we assert $P \wedge P'$ as a contradiction. What really happened here is a direct proof of $Q' \to P'$, and we have constructed a proof by contraposition, not a proof by contradiction. In both this case and the previous one, it is not that the proofs are wrong, just that they are not proofs by contradiction.

Table 2.2 summarizes useful proof techniques we have discussed so far.

Proof Technique	Approach to Prove $P \to Q$	Remarks
Exhaustive Proof	Demonstrate $P \to Q$ for all possible cases.	May only be used to prove a finite number of cases.
Direct Proof	Assume P, deduce Q.	The standard approach—usually the thing to try.
Proof by Contraposition	Assume Q', deduce P'.	Use this if Q' as a hypothesis seems to give more ammunition than P would.
Proof by Contradiction	Assume $P \wedge Q'$, deduce a contradiction.	Use this when Q says something is not true.

Table 2.2

Serendipity

Serendipity means a fortuitous happening, or good luck. While this isn't really a general proof technique, some of the most interesting proofs come from clever observations that we can admire, even if we would never have thought of them ourselves. We'll look at two such proofs, just for fun.

EXAMPLE 12   A tennis tournament has 342 players. A single match involves 2 players. The winner of a match will play the winner of a match in the next round, while losers are eliminated from the tournament. The 2 players who have won all previous rounds play in the final game, and the winner wins the tournament. Prove that the total number of matches to be played is 341.

The hard way to prove this result is to compute $342/2 = 171$ to get the number of matches in the first round, resulting in 171 winners to go on to the second round. For the second round, $171/2 = 85$ plus 1 left over; there are 85 matches and 85 winners, plus the one left over, to go on to the third round. The third round has $86/2 = 43$ matches, and so forth. The total number of matches is the sum of $171 + 85 + 43 + \cdots$.

The clever observation is to note that each match results in exactly 1 loser, so there must be the same number of matches as losers in the tournament. Because there is only 1 winner, there are 341 losers, and therefore 341 matches.   ●

EXAMPLE 13   A standard 64-square checkerboard is arranged in 8 rows of 8 squares each. Adjacent squares are alternating colors of red and black. A set of 32 1 × 2 tiles, each covering 2 squares, will cover the board completely (4 tiles per row, 8 rows). Prove that if the squares at diagonally opposite corners of the checkerboard are removed, the remaining board cannot be covered with 31 tiles.

The hard way to prove this result is to try all possibilities with 31 tiles and see that they all fail. The clever observation is to note that opposing corners are the same color, so the checkerboard with the corners removed has two less squares of one color than of the other. Each tile covers one square of each color, so any set of tiles must cover an equal number of squares of each color and cannot cover the board with the corners removed.   ●

Section 2.1 Review

Techniques

- Look for a counterexample.
- Construct direct proofs, proofs by contraposition, and proofs by contradiction.

Main Ideas

Inductive reasoning is used to formulate a conjecture based on experience. Deductive reasoning is used either to refute a conjecture by finding a counterexample or to prove a conjecture.

In proving a conjecture about some subject, facts about that subject can be used.

Under the right circumstances, proof by contraposition or contradiction may work better than a direct proof.

Exercises 2.1

The following definitions may be helpful in working some of these exercises.

- A *perfect square* is an integer n such that $n = k^2$ for some integer k.
- A *prime number* is an integer $n > 1$ such that n is not divisible by any integers other than 1 and n.
- For two numbers x and y, $x < y$ means $y - x > 0$.
- The *absolute value* of a number x, $|x|$, is x if $x \geq 0$ and is $-x$ if $x < 0$.

★1. Write the converse and the contrapositive of each statement in Exercise 4 of Section 1.1.

2. Provide counterexamples to the following statements.

   a. Every geometric figure with four right angles is a square.
   b. If a real number is not positive, then it must be negative.
   c. All people with red hair have green eyes or are tall.
   d. All people with red hair have green eyes and are tall.

For Exercises 3–28, prove the given statement.

★3. If $n = 25$, 100, or 169, then n is a perfect square and is a sum of two perfect squares.

4. If n is an even integer, $4 \leq n \leq 12$, then n is a sum of two prime numbers.

5. For any positive integer n less than or equal to 3, $n! < 2^n$.

6. For $2 \leq n \leq 4$, $n^2 \geq 2^n$.

7. The sum of even integers is even (do a direct proof).

8. The sum of even integers is even (do a proof by contradiction).

★9. The sum of two odd integers is even.

10. The sum of an even integer and an odd integer is odd.

11. The product of any two consecutive integers is even.

12. The sum of an integer and its square is even.

★13. The square of an even number is divisible by 4.

14. For every integer n, the number

$$3(n^2 + 2n + 3) - 2n^2$$

is a perfect square.

15. If a number x is positive, so is $x + 1$ (do a proof by contraposition).

16. The number n is an odd integer if and only if $3n + 5$ is an even integer.

★17. For x and y positive numbers, $x < y$ if and only if $x^2 < y^2$.

18. If $x^2 + 2x - 3 = 0$, then $x \neq 2$.

19. If x is an even prime number, then $x = 2$.

★20. If two integers are each divisible by some integer n, then their sum is divisible by n.

21. If the product of two integers is not divisible by an integer n, then neither integer is divisible by n.

22. The sum of three consecutive integers is divisible by 3.

★23. The square of an odd integer equals $8k + 1$ for some integer k.

24. The difference of two consecutive cubes is odd.

25. The sum of the squares of two odd integers cannot be a perfect square. (*Hint:* Use Exercise 23.)

★26. The product of the squares of two integers is a perfect square.

27. For any two numbers x and y, $|xy| = |x||y|$.

28. For any two numbers x and y, $|x + y| \leq |x| + |y|$.

29. The value A is the average of the n numbers $x_1, x_2, \ldots, x_n$. Prove that at least one of $x_1, x_2, \ldots, x_n$ is greater than or equal to A.

30. Suppose you were to use the steps of Example 11 to attempt to prove that $\sqrt{4}$ is not a rational number. At what point would the proof not be valid?

31. Prove that $\sqrt{3}$ is not a rational number.

32. Prove that $\sqrt{5}$ is not a rational number.

33. Prove that $\sqrt[3]{2}$ is not a rational number.

For Exercises 34–46, prove or disprove the given statement.

★34. The product of any three consecutive integers is even.

35. The sum of any three consecutive integers is even.

36. The product of an integer and its square is even.

★37. The sum of an integer and its cube is even.

38. Any positive integer can be written as the sum of the squares of two integers.

39. For a positive integer x, $x + \dfrac{1}{x} \geq 2$.

40. For every prime number n, $n + 4$ is prime.

41. For every positive integer n, $2^n + 1$ is prime.

42. For n an even integer, $n > 2$, $2^n - 1$ is not prime.

43. The product of two rational numbers is rational.

★44. The sum of two rational numbers is rational.

45. The product of two irrational numbers is irrational.

46. The sum of a rational number and an irrational number is irrational.

For Exercises 47–49, use the accompanying figure and the following facts from geometry:

- The interior angles of a triangle sum to 180°.
- Vertical angles (opposite angles formed when two lines intersect) are the same size.
- A straight angle is 180°.
- A right angle is 90°.

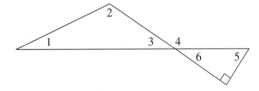

47. Prove that the measure of angle 4 is the sum of the measures of angles 1 and 2.

★48. Prove that the measure of angle 5 plus the measure of angle 3 is 90°.

49. If angle 1 and angle 5 are the same size, then angle 2 is a right angle.

50. Prove that the sum of the integers from 1 through 100 is 5050. (*Hint:* Instead of actually adding all the numbers, try to make the same clever observation that the German mathematician Karl Frederick Gauss [1777–1855] made as a school-child: Group the numbers into pairs, using 1 and 100, 2 and 99, etc.)

Section 2.2   Induction

First Principle of Induction

There is one final proof technique especially useful in computer science. To illustrate how the technique works, imagine that you are climbing an infinitely high ladder. How do you know whether you will be able to reach an arbitrarily high rung? Suppose we make the following two assertions about your climbing abilities:

1. You can reach the first rung.
2. Once you get to a rung, you can always climb to the next one up. (Notice that this assertion is an implication.)

If both statement 1 and the implication of statement 2 are true, then by statement 1 you can get to the first rung and therefore by statement 2 you can get to the second; by statement 2 again, you can get to the third rung; by statement 2 again you can get to the fourth; and so on. You can climb as high as you wish. Both assertions here are necessary. If only statement 1 is true, you have no guarantee of getting beyond the first rung, and if only statement 2 is true, you may never be able to get started. Let's assume the rungs of the ladder are numbered by positive integers—1, 2, 3, and so on.

Now think of a specific property a number might have. Instead of "reaching an arbitrarily high rung," we can talk about an arbitrary, positive integer having that property. We will use the shorthand notation $P(n)$ to mean that the positive integer n has the property P. How can we use the ladder-climbing technique to prove that for all positive integers n, we have $P(n)$? The two assertions we need to prove are

1. $P(1)$ (1 has property P.)
2. For any positive integer k,   (If any number has property P, so does the
   $P(k) \rightarrow P(k + 1)$.   next number.)

If we can prove both assertions 1 and 2, then $P(n)$ holds for any positive integer n, just as you could climb to an arbitrary rung on the ladder.

The foundation for arguments of this type is the first principle of mathematical induction.

First Principle of Mathematical Induction

1. $P(1)$ is true
2. $(\forall k)[P(k) \text{ true} \rightarrow P(k + 1) \text{ true}]$ $\Big\}$ $\rightarrow P(n)$ true for all positive integers n

---

REMINDER:
To prove
something true
for all n ≥
some value,
think induction.

The first principle of mathematical induction is an implication. The conclusion is a statement of the form "$P(n)$ is true for all positive integers n." Therefore, whenever we want to prove that something is true for every positive integer n, it is a good bet that mathematical induction is an appropriate proof technique to use.

In order to know that the conclusion of this implication is true, we show that the two hypotheses, statements 1 and 2, are true. To prove statement 1, we need only show that property P holds for the number 1, usually a trivial task. Statement 2 is also an implication that must hold for all k. To prove this implication, we assume for an arbitrary positive integer k that $P(k)$ is true and show, based on this assumption, that $P(k + 1)$ is true. You should convince yourself that assuming that property P holds for the number k is not the same as assuming what we ultimately want to prove (a frequent source of confusion when one first encounters proofs of this kind). It is merely the way to proceed with a direct proof that the implication $P(k) \rightarrow P(k + 1)$ is true.

In doing a proof by induction, establishing the truth of statement 1, $P(1)$, is called the **basis**, or **basis step**, for the inductive proof. Establishing the truth of $P(k) \rightarrow P(k + 1)$ is called the **inductive step**. When we assume $P(k)$ to be true in order to prove the inductive step, $P(k)$ is called the **inductive assumption**, or **inductive hypothesis**.

All of the proof methods we have talked about in this chapter are techniques for deductive reasoning—ways to prove a conjecture that perhaps was formulated by inductive reasoning. Mathematical induction is also a *deductive* technique, not a method for inductive reasoning (don't get confused by the terminology here). For the other proof techniques, we can begin with a hypothesis and string facts together until we more or less stumble on a conclusion. In fact, even if our conjecture is slightly incorrect, we might see what the correct conclusion is in the course of doing the proof. In mathematical induction, however, we must know right at the outset the exact form of the property $P(n)$ that we are trying to establish. Mathematical induction, therefore, is not an exploratory proof technique—it can only confirm a correct conjecture.

Proofs by Mathematical Induction

Suppose that the ancestral progenitor Smith married and had two children. Let's call these two children generation 1. Now suppose each of those two children had two children; then in generation 2, there were four offspring. This continued from generation unto generation. The Smith family tree therefore looks like Figure 2.2. (This looks exactly like Figure 1.1b, where we looked at the possible T–F values for n statement letters.)

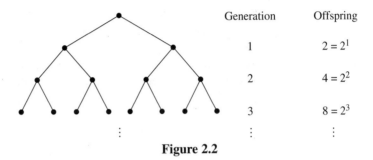

Generation	Offspring
1	$2 = 2^1$
2	$4 = 2^2$
3	$8 = 2^3$
$\vdots$	$\vdots$

Figure 2.2

It appears that generation n contains 2^n offspring. More formally, if we let $P(n)$ denote the number of offspring at generation n, then we guess that

$$P(n) = 2^n$$

We can use induction to *prove* that our guess for $P(n)$ is correct.

The basis step is to establish $P(1)$, which is the equation

$$P(1) = 2^1 = 2$$

This is true because we are told that Smith had two children. We now assume that our guess is correct for an arbitrary generation k, $k \geq 1$, that is, we assume

$$P(k) = 2^k$$

and try to show that

$$P(k + 1) = 2^{k+1}$$

In this family, each offspring has two children; thus the number of offspring at generation $k + 1$ will be twice the number at generation k, or $P(k + 1) = 2P(k)$. By the inductive assumption, $P(k) = 2^k$, so

$$P(k + 1) = 2P(k) = 2(2^k) = 2^{k+1}$$

so indeed

$$P(k + 1) = 2^{k+1}$$

This completes our proof. Now that we have set our mind at ease about the Smith clan, we can apply the inductive proof technique to less obvious problems.

EXAMPLE 14   Prove that the equation

$$1 + 3 + 5 + \cdots + (2n - 1) = n^2 \tag{1}$$

is true for any positive integer n. Here the property $P(n)$ is that equation (1) above is true. The left side of this equation is the sum of all the odd integers from 1 to $2n - 1$. Although we can verify the truth of this equation for any particular value of n by substituting that value for n, we cannot substitute all possible positive integer values. Thus a proof by exhaustion does not work. A proof by mathematical induction is appropriate.

The basis step is to establish $P(1)$, which is equation (1) when n has the value 1, or

$$P(1): \quad 1 = 1^2$$

This is certainly true. For the inductive hypothesis, we assume $P(k)$ for an arbitrary positive integer k, which is equation (1) when n has the value k, or

$$P(k): \quad 1 + 3 + 5 + \cdots + (2k - 1) = k^2 \tag{2}$$

(Note that $P(k)$ is *not* the equation $(2k - 1) = k^2$, which is true only for $k = 1$.) Using the inductive hypothesis, we want to show $P(k + 1)$, which is equation (1) when n has the value $k + 1$, or

$$P(k + 1): \quad 1 + 3 + 5 + \cdots + [2(k + 1) - 1] \stackrel{?}{=} (k + 1)^2 \tag{3}$$

(The question mark over the equals sign is to remind us that this is the fact we want to prove, as opposed to something we already know.)

The key to an inductive proof is to find a way to relate what we want to show—$P(k + 1)$, equation (3)—to what we have assumed—$P(k)$, equation (2). The left side of $P(k + 1)$ can be rewritten to show the next-to-last term:

$$1 + 3 + 5 + \cdots + (2k - 1) + [2(k + 1) - 1]$$

This expression contains the left side of equation (2) as a subexpression. Because we have assumed $P(k)$ to be true, we can substitute the right side of equation (2) for this subexpression. Thus,

$$
\begin{aligned}
1 + 3 + 5 + \cdots + [2(k + 1) - 1] \\
&= 1 + 3 + 5 + \cdots + (2k - 1) + [2(k + 1) - 1] \\
&= k^2 + [2(k + 1) - 1] \\
&= k^2 + [2k + 2 - 1] \\
&= k^2 + 2k + 1 \\
&= (k + 1)^2
\end{aligned}
$$

Therefore,

$$1 + 3 + 5 + \cdots + [2(k + 1) - 1] = (k + 1)^2$$

which verifies $P(k + 1)$ and proves that equation (1) is true for any positive integer n.   ●

Table 2.3 summarizes the three steps necessary for a proof using the first principle of induction.

To prove by first principle of induction	
Step 1	Prove base case.
Step 2	Assume $P(k)$.
Step 3	Prove $P(k + 1)$.

Table 2.3

EXAMPLE 15

Prove that

$$1 + 2 + 2^2 + \cdots + 2^n = 2^{n+1} - 1$$

for any $n \geq 1$.

Again, induction is appropriate. $P(1)$ is the equation

$$1 + 2 = 2^{1+1} - 1 \qquad \text{or} \qquad 3 = 2^2 - 1$$

REMINDER:
To prove
$P(k) \rightarrow$
$P(k + 1)$
you have to
discover the
$P(k)$ case
within the
$P(k + 1)$ case.

which is true. We take $P(k)$

$$1 + 2 + 2^2 + \cdots + 2^k = 2^{k+1} - 1$$

as the inductive hypothesis and try to establish $P(k + 1)$:

$$1 + 2 + 2^2 + \cdots + 2^{k+1} \stackrel{?}{=} 2^{k+1+1}$$

Again, rewriting the sum on the left side of $P(k + 1)$ reveals how the inductive assumption can be used:

$$
\begin{aligned}
1 + 2 + 2^2 + \cdots + 2^{k+1} \\
= 1 + 2 + 2^2 + \cdots + 2^k + 2^{k+1} \\
= 2^{k+1} - 1 + 2^{k+1} \qquad \text{(from the inductive assumption } P(k)) \\
= 2(2^{k+1}) - 1 \\
= 2^{k+1+1} - 1
\end{aligned}
$$

Therefore,

$$1 + 2 + 2^2 + \cdots + 2^{k+1} = 2^{k+1+1} - 1$$

which verifies $P(k + 1)$ and completes the proof.

PRACTICE 7

Prove that for any positive integer n,

$$1 + 2 + 3 + \cdots + n = \frac{n(n+1)}{2}$$

Not all proofs by induction involve formulas with sums. Other algebraic identities about the positive integers can be proved by induction, as well as nonalgebraic assertions like the number of offspring in generation n of the Smith family.

EXAMPLE 16   Prove that for any positive integer n, $2^n > n$.

$P(1)$ is the assertion $2^1 > 1$, which is surely true. Now we assume $P(k)$, $2^k > k$, and try to conclude $P(k + 1)$, $2^{k+1} > k + 1$. Beginning with the left side of $P(k + 1)$, we note that $2^{k+1} = 2^k \cdot 2$. Using the inductive assumption $2^k > k$ and multiplying both sides of this inequality by 2, we get $2^k \cdot 2 > k \cdot 2$. We complete the argument

$$2^{k+1} = 2^k \cdot 2 > k \cdot 2 = k + k \geq k + 1$$

or

$$2^{k+1} > k + 1$$

EXAMPLE 17   Prove that for any positive integer n, the number $2^{2n} - 1$ is divisible by 3.

The basis step is to show $P(1)$, that $2^{2(1)} - 1 = 4 - 1 = 3$ is divisible by 3. Clearly this is true.

We assume that $2^{2k} - 1$ is divisible by 3, which means that $2^{2k} - 1 = 3m$ for some integer m, or $2^{2k} = 3m + 1$. We want to show that $2^{2(k+1)} - 1$ is divisible by 3.

$$
\begin{aligned}
2^{2(k+1)} - 1 &= 2^{2k+2} - 1 \\
&= 2^2 \cdot 2^{2k} - 1 \\
&= 2^2(3m + 1) - 1 \quad \text{(by the inductive hypothesis)} \\
&= 12m + 4 - 1 \\
&= 12m + 3 \\
&= 3(4m + 1) \quad \text{where } 4m + 1 \text{ is an integer}
\end{aligned}
$$

Thus $2^{2(k+1)} - 1$ is divisible by 3.

For the first step of the induction process, it may be appropriate to begin at 0 or at 2 or 3 instead of at 1. The same principle applies, no matter where you first hop on the ladder.

EXAMPLE 18   Prove that $n^2 > 3n$ for $n > 4$.

Here we should use induction and begin with a basis step of $P(4)$. (Testing values of $n = 1, 2,$ and 3 shows that the inequality does not hold for these values.) $P(4)$ is the inequality $4^2 > 3(4)$, or $16 > 12$, which is true. The inductive hypothesis is that $k^2 > 3k$ and that $k \geq 4$, and we want to show that $(k + 1)^2 > 3(k + 1)$.

$$
\begin{aligned}
(k + 1)^2 &= k^2 + 2k + 1 \\
&> 3k + 2k + 1 \quad \text{(by the inductive hypothesis)} \\
&\geq 3k + 8 + 1 \quad \text{(since } k \geq 4) \\
&> 3k + 3 \\
&= 3(k + 1)
\end{aligned}
$$

PRACTICE 8   Prove that $2^{n+1} < 3^n$ for all $n > 1$.   ●

Misleading claims of proof by induction are also possible. When we prove the truth of $P(k + 1)$ without relying on the truth of $P(k)$, we have done a direct proof of $P(k + 1)$ where $k + 1$ is arbitrary. The proof is not invalid, but it should be rewritten to show that it is a direct proof of $P(n)$ for any n, not a proof by induction.

An inductive proof may be called for when its application is not as obvious as in the above examples. This usually arises when there is some quantity in the statement to be proved that can take on arbitrary nonnegative integer values.

EXAMPLE 19   A programming language might be designed with the following convention regarding multiplication: A single factor requires no parentheses, but the product "a times b" must be written as $(a)b$. So the product

$$a \cdot b \cdot c \cdot d \cdot e \cdot f \cdot g$$

could be written in this language as

$$((((((a)b)c)d)e)f)g$$

or as, for example,

$$((a)b)(((c)d)(e)f)g$$

depending on the order in which the products are formed. The result is the same in either case.

We want to show that any product of factors can be written with an even number of parentheses. The proof is by induction on the number of factors. For a single factor, there are 0 parentheses, an even number. Assume that for any product of k factors there is an even number of parentheses. Now consider a product P of $k + 1$ factors. P can be thought of as r times s where r has k factors and s is a single factor. By the inductive hypothesis, r has an even number of parentheses. Then we write r times s as $(r)s$. This adds 2 more parentheses to the even number of parentheses in r, giving P an even number of parentheses.   ●

EXAMPLE 20   A "tiling" problem gives a nice illustration of induction in a geometric setting. An *angle iron* is an L-shaped piece that can cover three squares on a checkerboard (Figure 2.3a). The problem is to show that for any positive integer n, a $2^n \times 2^n$ checkerboard with one square removed can be tiled—completely covered—by angle irons.

The base case is $n = 1$, which gives a 2×2 checkerboard. Figure 2.3b shows the solution to this case if the upper right corner is removed. Removing any of the other three corners works the same way. Assume that any $2^k \times 2^k$ checkerboard with one square removed can be tiled using angle irons. Now consider a checkerboard with dimensions $2^{k+1} \times 2^{k+1}$. We need to show that it can be tiled when one square is removed. To relate the $k + 1$ case to the inductive hypothesis, divide the $2^{k+1} \times 2^{k+1}$ checkerboard into four quarters. Each quarter will be a $2^k \times 2^k$ checkerboard, and one will have a missing square (Figure 2.3c). By the inductive hypothesis, this checkerboard can be tiled. Remove a corner from each of the other three checkerboards, as in

Figure 2.3d. By the inductive hypothesis, the three boards with the holes removed can be tiled, and one angle iron can tile the three holes. Hence the original $2^{k+1} \times 2^{k+1}$ board with its one hole can be tiled. ●

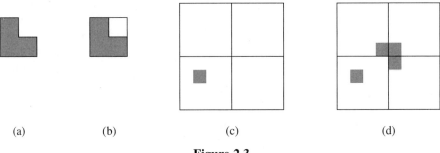

(a)   (b)   (c)   (d)

Figure 2.3

Second Principle of Induction

In addition to the first principle of induction, which we have been using,

1. $P(1)$ is true
2. $(\forall k)[P(k) \text{ true} \to P(k + 1) \text{ true}]$ $\Bigr\}\to P(n)$ true for all positive integers n

there is a second principle of induction.

Second Principle of Mathematical Induction

1′. $P(1)$ is true
2′. $(\forall k)[P(r)$ true for all r,
    $1 \le r \le k \to P(k + 1) \text{ true}]$ $\Bigr\}\to P(n)$ true for all positive integers n

These two induction principles differ in statements 2 and 2′. In statement 2, we must be able to prove for an arbitrary positive integer k that $P(k + 1)$ is true based only on the assumption that $P(k)$ is true. In statement 2′, we can assume that $P(r)$ is true for all integers r between 1 and an arbitrary positive integer k in order to prove that $P(k + 1)$ is true. This seems to give us a great deal more "ammunition," so we might sometimes be able to prove the implication in 2′ when we cannot prove the implication in 2.

What allows us to deduce $(\forall n)P(n)$ in either case? We shall see that the two induction principles themselves, that is, the two methods of proof, are equivalent. In other words, if we accept the first principle of induction as valid, then the second principle of induction is valid, and conversely. In order to prove the equivalence of the two induction principles, we'll introduce another principle, which seems so obvious as to be unarguable.

Principle of Well-Ordering

Every collection of positive integers that contains any members at all has a smallest member.

We shall see that the following implications are true:

second principle of induction → first principle of induction
first principle of induction → well-ordering
well-ordering → second principle of induction

As a consequence, all three principles are equivalent, and accepting any one of them as true means accepting the other two as well.

To prove that the second principle of induction implies the first principle of induction, suppose we accept the second principle as valid reasoning. We then want to show that the first principle is valid, that is, that we can conclude $P(n)$ for all n from statements 1 and 2. If statement 1 is true, so is statement 1'. If statement 2 is true, then so is statement 2', because we can say that we concluded $P(k + 1)$ from $P(r)$ for all r between 1 and k, even though we used only the single condition $P(k)$. (More precisely, statement 2' requires that we prove $P(1) \wedge P(2) \wedge \cdots \wedge P(k) \rightarrow P(k + 1)$, but $P(1) \wedge P(2) \wedge \cdots \wedge P(k) \rightarrow P(k)$, and from statement 2, $P(k) \rightarrow P(k + 1)$, so $P(1) \wedge P(2) \wedge \cdots \wedge P(k) \rightarrow P(k + 1)$.) By the second principle of induction, we conclude $P(n)$ for all n. The proofs that the first principle of induction implies well-ordering and that well-ordering implies the second principle of induction are left as exercises in Section 3.1.

In order to distinguish between a proof by the first principle of induction and a proof by the second principle of induction, let's look at a rather picturesque example that can be proved both ways.

EXAMPLE 21   Prove that a straight fence with n fence posts has $n - 1$ sections for any $n \geq 1$ (see Figure 2.4a).

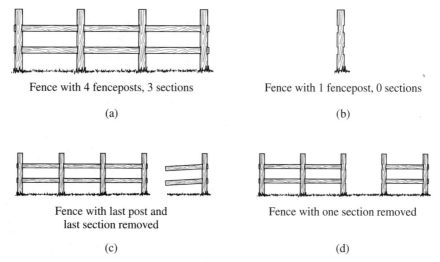

Fence with 4 fenceposts, 3 sections

(a)

Fence with 1 fencepost, 0 sections

(b)

Fence with last post and
last section removed

(c)

Fence with one section removed

(d)

Figure 2.4

Let $P(n)$ be the statement that a fence with n fence posts has $n - 1$ sections, and prove $P(n)$ true for all $n \geq 1$.

We'll start with the first principle of induction. For the basis step, $P(1)$ says that a fence with only 1 fence post has 0 sections, which is clearly true (see Figure 2.4b). Assume that $P(k)$ is true:

a fence with k fence posts has $k - 1$ sections

and try to prove $P(k + 1)$:

(?) a fence with $k + 1$ fence posts has k sections

Given a fence with $k + 1$ fence posts, how can we relate that to a fence with k fence posts so that we can make use of the inductive hypothesis? We can chop off the last post and the last section (Figure 2.4c). The remaining fence has k fence posts and, by the inductive hypothesis, $k - 1$ sections. Therefore the original fence had k sections.

Now we'll prove the same result using the second principle of induction. The basis step is the same as before. For the inductive hypothesis, we assume

for all r, $1 \leq r \leq k$, a fence with r fence posts has $r - 1$ sections

and try to prove $P(k + 1)$:

(?) a fence with $k + 1$ fence posts has k sections

For a fence with $k + 1$ fence posts, split the fence into two parts by removing one section (Figure 2.4d). The two parts of the fence have r_1 and r_2 fence posts, where $1 \leq r_1 \leq k$, $1 \leq r_2 \leq k$, and $r_1 + r_2 = k + 1$. By the inductive hypothesis, the two parts have, respectively, $r_1 - 1$ and $r_2 - 1$ sections, so that the original fence has

$(r_1 - 1) + (r_2 - 1) + 1$ sections

(The extra 1 is for the one that we removed.) Simple arithmetic then yields

$r_1 + r_2 - 1 = (k + 1) - 1 = k$ sections

This proves that a fence with $k + 1$ fence posts has k sections, which verifies $P(k + 1)$ and completes the proof using the second principle of induction. ●

Example 21 allowed for either form of inductive proof because we could either reduce the fence at one end or split it at an arbitrary point. The problem of Example 19 is similar.

EXAMPLE 22   We again want to show that any product of factors can be written in this programming language with an even number of parentheses, this time using the second principle of induction. The base case is the same as in Example 19: A single factor has 0 parentheses, an even number. Assume that any product of r factors, $1 \leq r \leq k$, can be written with an even number of parentheses. Then consider a product P with $k + 1$ factors. P can be written as $(S)T$, a product of two factors S and T, where S has r_1 factors and T has r_2 factors. Then $1 \leq r_1 \leq k$ and $1 \leq r_2 \leq k$, with $r_1 + r_2 = k + 1$. By the inductive hypothesis, S and T each have an even number of parentheses, and therefore so does $(S)T = P$. ●

Most problems do not work equally well with either form of induction; the fence post and the programming language problem were somewhat artificial. Generally, complete induction is called for when the problem "splits" most naturally in the middle instead of growing from the end.

EXAMPLE 23

Prove that for every $n \geq 2$, n is a prime number or a product of prime numbers.

> **REMINDER:**
> Use second priciple of induction when the $k + 1$ case depends on results further back than k.

We will postpone the decision of whether to use the first or the second principle of induction; the basis step is the same in each case and need not start with 1. Obviously here we should start with 2. $P(2)$ is the statement that 2 is a prime number or a product of primes. Because 2 is a prime number, $P(2)$ is true. Jumping ahead, for either principle we will be considering the number $k + 1$. If $k + 1$ is prime, we are done. If $k + 1$ is not prime, then it is a composite number and can be written as $k + 1 = ab$. Here $k + 1$ has been split into two factors. Maybe neither of these factors has the value k, so an assumption only about $P(k)$ isn't enough. Hence, we'll use the second principle of induction.

So let's start again. We assume that for all r, $2 \leq r \leq k$, $P(r)$ is true—r is prime or the product of primes. Now consider the number $k + 1$. If $k + 1$ is prime, we are done. If $k + 1$ is not prime, then it is a composite number and can be written as $k + 1 = ab$, where $1 < a < k + 1$ and $1 < b < k + 1$. (This is a nontrivial factorization, so neither factor can be 1 or $k + 1$.) Therefore $2 \leq a \leq k$ and $2 \leq b \leq k$. The inductive hypothesis applies to both a and b, so a and b are either prime or the product of primes. Thus, $k + 1$ is the product of prime numbers. This verifies $P(k + 1)$ and completes the proof by the second principle of induction.   ●

EXAMPLE 24

Prove that any amount of postage greater than or equal to 8 cents can be built using only 3-cent and 5-cent stamps.

Here we let $P(n)$ be the statement that only 3-cent and 5-cent stamps are needed to build n cents worth of postage, and prove that $P(n)$ is true for all $n \geq 8$. The basis step is to establish $P(8)$, which is done by the equation

$$8 = 3 + 5$$

For reasons that will be clear momentarily, we'll also establish two additional cases, $P(9)$ and $P(10)$, by the equations

$$9 = 3 + 3 + 3$$
$$10 = 5 + 5$$

Now we assume that $P(r)$ is true for any r, $8 \leq r \leq k$, and consider $P(k + 1)$. We may assume that $k + 1$ is at least 11, since we have already proved $P(r)$ true for $r = 8, 9$, and 10. If $k + 1 \geq 11$, then $(k + 1) - 3 = k - 2 \geq 8$, and by the inductive hypothesis, $P(k - 2)$ is true. Therefore $k - 2$ can be written as a sum of 3s and 5s, and adding an additional 3 gives us $k + 1$ as a sum of 3s and 5s. This verifies that $P(k + 1)$ is true and completes the proof.   ●

PRACTICE 9

a. Why are the additional cases $P(9)$ and $P(10)$ proved separately in Example 24?
b. Why can't the first principle of induction be used in the proof of Example 24?   ●

Section 2.2 Review

Techniques

- Use the first principle of induction in proofs.
- Use the second principle of induction in proofs.

Main Ideas

Mathematical induction is a technique to prove properties of positive integers.

An inductive proof need not begin with 1.

Induction can be used to prove statements about quantities whose values are arbitrary nonnegative integers.

The first and second principles of induction each prove the same conclusion, but one approach may be easier to use than the other in a given situation.

Exercises 2.2

In Exercises 1–20, use mathematical induction to prove that the statements are true for every positive integer n.

★1. $2 + 6 + 10 + \cdots + (4n - 2) = 2n^2$

2. $2 + 4 + 6 + \cdots + 2n = n(n + 1)$

★3. $1 + 5 + 9 + \cdots + (4n - 3) = n(2n - 1)$

4. $1 + 3 + 6 + \cdots + \dfrac{n(n + 1)}{2} = \dfrac{n(n + 1)(n + 2)}{6}$

★5. $4 + 10 + 16 + \cdots + (6n - 2) = n(3n + 1)$

6. $5 + 10 + 15 + \cdots + 5n = \dfrac{5n(n + 1)}{2}$

7. $1^2 + 2^2 + \cdots + n^2 = \dfrac{n(n + 1)(2n + 1)}{6}$

8. $1^3 + 2^3 + \cdots + n^3 = \dfrac{n^2(n + 1)^2}{4}$

★9. $1^2 + 3^2 + \cdots + (2n - 1)^2 = \dfrac{n(2n - 1)(2n + 1)}{3}$

10. $1^4 + 2^4 + \cdots + n^4 = \dfrac{n(n + 1)(2n + 1)(3n^2 + 3n - 1)}{30}$

11. $1 \cdot 3 + 2 \cdot 4 + 3 \cdot 5 + \cdots + n(n + 2) = \dfrac{n(n + 1)(2n + 7)}{6}$

12. $1 + a + a^2 + \cdots + a^{n-1} = \dfrac{a^n - 1}{a - 1}$ for $a \neq 0$, $a \neq 1$

★13. $\dfrac{1}{1 \cdot 2} + \dfrac{1}{2 \cdot 3} + \dfrac{1}{3 \cdot 4} + \cdots + \dfrac{1}{n(n+1)} = \dfrac{n}{n+1}$

14. $\dfrac{1}{1 \cdot 3} + \dfrac{1}{3 \cdot 5} + \dfrac{1}{5 \cdot 7} + \cdots + \dfrac{1}{(2n-1)(2n+1)} = \dfrac{n}{2n+1}$

★15. $1^2 - 2^2 + 3^2 - 4^2 + \cdots + (-1)^{n+1}n^2 = \dfrac{(-1)^{n+1}(n)(n+1)}{2}$

16. $2 + 6 + 18 + \cdots + 2 \cdot 3^{n-1} = 3^n - 1$

17. $2^2 + 4^2 + \cdots + (2n)^2 = \dfrac{2n(n+1)(2n+1)}{3}$

18. $1 \cdot 2 + 2 \cdot 3 + 3 \cdot 4 + \cdots + n(n+1) = \dfrac{n(n+1)(n+2)}{3}$

19. $\dfrac{1}{1 \cdot 4} + \dfrac{1}{4 \cdot 7} + \dfrac{1}{7 \cdot 10} + \cdots + \dfrac{1}{(3n-2)(3n+1)} = \dfrac{1}{3n+1)}$

20. $1 \cdot 1! + 2 \cdot 2! + 3 \cdot 3! + \cdots + n \cdot n! = (n+1)! - 1$ where $n!$ is the product of the positive integers from 1 to n.

★21. A *geometric progression* (*geometric sequence*) is a sequence of terms where there is an initial term a and each succeeding term is obtained by multiplying the previous term by a *common ratio r*. Prove the formula for the sum of the first n terms of a geometric sequence ($n \geq 1$):

$$a + ar + ar^2 + \cdots + ar^n = \dfrac{a - ar^n}{1 - r}$$

22. An *arithmetic progression* (*arithmetic sequence*) is a sequence of terms where there is an initial term a and each succeeding term is obtained by adding a *common difference d* to the previous term. Prove the formula for the sum of the first n terms of an arithmetic sequence ($n \geq 1$):

$$a + (a + d) + (a + 2d) + \cdots + [a + (n-1)d] = \dfrac{n}{2}[2a + (n-1)d]$$

23. Prove that

$$(-2)^0 + (-2)^1 + (-2)^2 + \cdots + (-2)^n = \dfrac{1 - 2^{n+1}}{3}$$

for every positive odd integer n.

24. Prove that $n^2 \geq 2n + 3$ for $n \geq 3$.

★25. Prove that $n^2 > n + 1$ for $n \geq 2$.

26. Prove that $n^2 > 5n + 10$ for $n > 6$.

27. Prove that $2^n > n^2$ for $n \geq 5$.

28. Prove that $n! > n^2$ for $n \geq 4$, where $n!$ is the product of the positive integers from 1 to n.

★29. Prove that $2^n < n!$ for $n \geq 4$.

30. Prove that $2^{n-1} \leq n!$ for $n \geq 1$.

31. Prove that $n! < n^n$ for $n \geq 2$.

32. Prove that $(1 + x)^n > 1 + x^n$ for $n > 1$, $x > 0$.

33. Prove that $\left(\dfrac{a}{b}\right)^{n+1} < \left(\dfrac{a}{b}\right)^n$ for $n \geq 1$ and $0 < a < b$.

★34. Prove that $1 + 2 + \cdots + n < n^2$ for $n > 1$.

35. **a.** Try to use induction to prove that

$$1 + \frac{1}{2} + \frac{1}{4} + \cdots + \frac{1}{2^n} < 2 \text{ for } n \geq 1$$

What goes wrong?

b. Prove that

$$1 + \frac{1}{2} + \frac{1}{4} + \cdots + \frac{1}{2^n} = 2 - \frac{1}{2^n} \text{ for } n \geq 1$$

thus showing that

$$1 + \frac{1}{2} + \frac{1}{4} + \cdots + \frac{1}{2^n} < 2 \text{ for } n \geq 1$$

For Exercises 36–40, prove that the statements are true for every positive integer.

★36. $2^{3n} - 1$ is divisible by 7.

37. $3^{2n} + 7$ is divisible by 8.

38. $7^n - 2^n$ is divisible by 5.

39. $13^n - 6^n$ is divisible by 7.

★40. $2^n + (-1)^{n+1}$ is divisible by 3.

41. $2^{5n+1} + 5^{n+2}$ is divisible by 27.

42. $3^{4n+2} + 5^{2n+1}$ is divisible by 14.

43. $7^{2n} + 16n - 1$ is divisible by 64.

★44. $10^n + 3 \cdot 4^{n+2} + 5$ is divisible by 9.

45. $n^3 - n$ is divisible by 3.

46. $n^3 + 2n$ is divisible by 3.

47. $x^n - 1$ is divisible by $x - 1$ for $x \neq 1$.

★**48.** Prove *DeMoivre's theorem:*

$$(\cos\theta + i\sin\theta)^n = \cos n\theta + i\sin n\theta$$

for all $n \geq 1$. *Hint:* Recall the addition formulas from trigonometry:

$$\cos(\alpha + \beta) = \cos\alpha\cos\beta - \sin\alpha\sin\beta$$
$$\sin(\alpha + \beta) = \sin\alpha\cos\beta + \cos\alpha\sin\beta$$

49. Prove that

$$\sin\theta + \sin 3\theta + \cdots + \sin(2n-1)\theta = \frac{\sin^2 n\theta}{\sin\theta}$$

for all $n \geq 1$ and all θ for which $\sin\theta \neq 0$.

★**50.** Use induction to prove that the product of any three consecutive positive integers is divisible by 3.

51. Suppose that exponentiation is defined by the equation

$$x^j \cdot x = x^{j+1}$$

for any $j \geq 1$. Use induction to prove that $x^n \cdot x^m = x^{n+m}$, for $n \geq 1$, $m \geq 1$. (*Hint:* Do induction on m for a fixed, arbitrary value of n.)

52. According to Example 20, it is possible to use angle irons to tile a 4×4 checkerboard with the upper right corner removed. Sketch such a tiling.

53. Example 20 does not cover the case of checkerboards that are not sized by powers of 2. Determine whether it is always possible to tile a 3×3 checkerboard.

54. Consider n infinitely long straight lines, none of which are parallel and no three of which have a common point of intersection. Show that for $n \geq 1$, the lines divide the plane into $(n^2 + n + 2)/2$ separate regions.

55. A string of 0s and 1s is to be processed and converted to an even-parity string by adding a parity bit to the end of the string. The parity bit is initially 0. When a 0 character is processed, the parity bit remains unchanged. When a 1 character is processed, the parity bit is switched from 0 to 1 or from 1 to 0. Prove that the number of 1s in the final string, that is, including the parity bit, is always even. (*Hint:* Consider various cases.)

★**56.** What is wrong with the following "proof" by mathematical induction? We will prove that for any positive integer n, n is equal to 1 more than n. Assume that $P(k)$ is true.

$$k = k + 1$$

Adding 1 to both sides of this equation, we get

$$k + 1 = k + 2$$

Thus,

$$P(k + 1) \text{ is true}$$

57. What is wrong with the following "proof" by mathematical induction? We will prove that all computers are built by the same manufacturer. In particular, we will prove that in any collection of n computers where n is a positive integer, all of the computers are built by the same manufacturer. We first prove $P(1)$, a trivial process, because in any collection consisting of one computer, there is only one manufacturer. Now we assume $P(k)$; that is, in any collection of k computers, all the computers were built by the same manufacturer. To prove $P(k + 1)$, we consider any collection of $k + 1$ computers. Pull one of these $k + 1$ computers (call it HAL) out of the collection. By our assumption, the remaining k computers all have the same manufacturer. Let HAL change places with one of these k computers. In the new group of k computers, all have the same manufacturer. Thus, HAL's manufacturer is the same one that produced all the other computers, and all $k + 1$ computers have the same manufacturer.

58. An obscure tribe has only three words in its language, *moon, noon,* and *soon.* New words are composed by juxtaposing these words in any order, as in *soon-noonmoonnoon.* Any such juxtaposition is a legal word.

   a. Use the first principle of induction (on the number of subwords in the word) to prove that any word in this language has an even number of *o*'s.
   b. Use the second principle of induction (on the number of subwords in the word) to prove that any word in this language has an even number of *o*'s.

★59. Consider propositional wffs that contain only the connectives ∧, ∨, and → (no negation) and where wffs must be parenthesized when joined by a logical connective. Count each statement letter, connective, or parenthesis as one symbol. For example, $((A) \land (B)) \lor ((C) \land (D))$ is such a wff, with 19 symbols. Prove that any such wff has an odd number of symbols.

60. A *simple closed polygon* consists of n points in the plane joined in pairs by n line segments; each point is the endpoint of exactly two line segments (see accompanying figure for examples). Prove that the sum of the interior angles of an n-sided simple closed polygon is $(n - 2)180°$ for all $n \geq 3$. (*Hint:* For the $k + 1$ case, split the polygon into two pieces.)

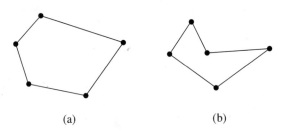

(a)   (b)

★61. Prove that any amount of postage greater than or equal to 2 cents can be built using only 2-cent and 3-cent stamps.

62. Prove that any amount of postage greater than or equal to 12 cents can be built using only 4-cent and 5-cent stamps.

63. Prove that any amount of postage greater than or equal to 14 cents can be built using only 3-cent and 8-cent stamps.

★64. Prove that any amount of postage greater than or equal to 64 cents can be built using only 5-cent and 17-cent stamps.

65. In any group of k people, $k \geq 1$, each person is to shake hands with every other person. Find a formula for the number of handshakes, and prove the formula using induction.

Section 2.3 More on Proof of Correctness

In Section 1.6, we explained the use of a formal logic system to prove mathematically the correctness of a program. Assertions or predicates involving the program variables are inserted at the beginning, at the end, and at intermediate points between the program statements. Then proving the correctness of any particular program statement s_i involves proving that the implication represented by the Hoare triple

$$\{Q\}\ s_i\ \{R\} \tag{1}$$

is true. Here Q and R are assertions known, respectively, as the precondition and postcondition for the statement. The program is provably correct if all such implications for the statements in the program are true.

In Chapter 1, we discussed rules of inference that give conditions under which implication (1) is true when s_i is an assignment statement and when s_i is a conditional statement. Now we will use a rule of inference that gives conditions under which implication (1) is true when s_i is a loop statement. We have deferred consideration of loop statements until now because mathematical induction is used in applying this rule of inference.

Loop Rule

Suppose that s_i is a loop statement in the form

while condition B **do**
   P
end while

where B is a condition that is either true or false and P is a program segment. When this statement is executed, condition B is evaluated. If B is true, program segment P is executed and then B is evaluated again. If B is still true, program segment P is executed again, then B is evaluated again, and so forth. If condition B ever evaluates to false, the loop terminates.

The form of implication (1) that can be used when s_i is a loop statement imposes (like the assignment rule did) a relationship between the precondition and postcondition. The precondition Q holds before the loop is entered; strangely enough, one requirement is that Q must continue to hold after the loop terminates (which means that we should look for a Q that we want to be true when the loop terminates). In addition, B'—the condition for loop termination—must be true then as well. Thus (1) will have the form

$$\{Q\}\ s_i\ \{Q \land B'\} \tag{2}$$

EXAMPLE 25   Consider the following pseudocode function, which is supposed to return the value $x * y$ for nonnegative integers x and y.

Product(nonnegative integer x; nonnegative integer y)

Local variables:

integers i, j

$\quad i = 0$

$\quad j = 0$

$\quad$ **while** $i \ne x$ **do**

$\quad\quad j = j + y$

$\quad\quad i = i + 1$

$\quad$ **end while**

$\quad$ //j now has the value $x * y$

$\quad$ return j

end function Product

This function contains a loop; the condition B for continued loop execution is $i \ne x$. The condition B' for loop termination is $i = x$. When the loop terminates, it is claimed in the comment that j has the value $x * y$. Given that $i = x$ when the loop terminates, the assertion that $j = i * y$ would also have to be true. Thus, upon loop termination, if the function does what is claimed, the assertion

$$j = i * y \wedge i = x$$

is true. This matches the form $Q \wedge B'$ if we take the assertion

$$j = i * y$$

as Q. In order to match the form of (2), the assertion $j = i * y$ would have to be true before the loop statement. This is indeed the case because right before the loop statement, $i = j = 0$.

It would seem that for this example we have a candidate assertion Q for implication (2), but we do not yet have the rule of inference that allows us to say when (2) is a true implication. (Remember that we discovered our Q by "wishful thinking" about the correct operation of the function code.)   ●

Assertion Q must be true before the loop is entered. If implication (2) is to hold, Q must remain true after the loop terminates. Because it may not be known exactly when the loop will terminate, Q must remain true after each iteration through the loop, which will include the final iteration. Q represents a predicate, or relation, among the values of the program variables. If this relation holds among the values of the program variables before a loop iteration executes and holds among the values after the iteration executes, then the *relation* among these variables is unaffected by the action of the loop iteration, even though the values themselves may be changed. Such a relation is called a **loop invariant**.

The **loop rule of inference** allows the truth of (2) to be inferred from an implication stating that Q is a loop invariant, that is, that if Q is true and condition B is true, so that another loop iteration is executed, then Q remains true after that iteration. The rule is formally stated in Table 2.4.

From	Can Derive	Name of Rule	Restrictions on Use

$\{Q \wedge B\}\ P\ \{Q\}$	$\{Q\}\ s_i\ \{Q \wedge B'\}$	loop	s_i has the form
			while condition B **do**
			P
			end while

Table 2.4

To use this rule of inference, we must find a *useful* loop invariant Q—one that asserts what we want and expect to have happen—and then prove the implication

$$\{Q \wedge B\}\ P\ \{Q\}$$

Here is where induction comes into play. We denote by $Q(n)$ the statement that a proposed loop invariant Q is true after n iterations of the loop. Because we do not necessarily know how many iterations the loop may execute (that is, how long condition B remains true), we want to show that $Q(n)$ is true for all $n \geq 0$. (The value of $n = 0$ corresponds to the assertion upon entering the loop, after zero loop iterations.)

EXAMPLE 26   Consider again the pseudocode function of Example 25. In that example, we guessed that Q is the relation

$$j = i * y$$

In order to use the loop rule of inference, we must prove that Q is a loop invariant.

The quantities x and y remain unchanged throughout the function, but values of i and j change within the loop. We let i_n and j_n denote the values of i and j, respectively, after n iterations of the loop. Then $Q(n)$ is the statement $j_n = i_n * y$.

We prove by induction that $Q(n)$ holds for all $n \geq 0$. $Q(0)$ is the statement

$$j_0 = i_0 * y$$

which, as we noted in Example 25, is true, because after zero iterations of the loop, when we first get to the loop statement, both i and j have been assigned the value 0. (Formally, the assignment rule could be used to prove that these conditions on i and j hold at this point.)

Assume $Q(k)$: $j_k = i_k * y$

Show $Q(k + 1)$: $j_{k+1} = i_{k+1} * y$

Between the time j and i have the values j_k and i_k and the time they have the values j_{k+1} and i_{k+1}, one iteration of the loop takes place. In that iteration, j is changed by adding y to the previous value, and i is changed by adding 1. Thus,

$$j_{k+1} = j_k + y \tag{3}$$
$$i_{k+1} = i_k + 1 \tag{4}$$

Then

$$j_{k+1} = j_k + y \qquad \text{(by (3))}$$
$$= i_k * y + y \qquad \text{(by the inductive hypothesis)}$$
$$= (i_k + 1)y$$
$$= i_{k+1} * y \qquad \text{(by (4))}$$

We have proved that Q is a loop invariant.

The loop rule of inference allows us to infer that after the loop statement is exited, the condition $Q \wedge B'$ holds, which in this case becomes

$$j = i * y \wedge i = x$$

Therefore at this point the statement

$$j = x * y$$

is true, which is exactly what the function is intended to compute.   ●

Example 26 illustrates that loop invariants say something stronger about the program than we actually want to show; what we want to show is the special case of the loop invariant on termination of the loop. Finding the appropriate loop invariant requires working backward from the desired conclusion, as in Example 25.

We did not, in fact, prove that the loop in this example actually does terminate. What we proved was **partial correctness**—the program produces the correct answer, given that execution does terminate. Since x is a nonnegative integer and i is an integer that starts at 0 and is then incremented by 1 at each pass through the loop, we know that eventually $i = x$ will become true.

PRACTICE 10   Show that the following function returns the value $x + y$ for nonnegative integers x and y by proving the loop invariant Q: $j = x + i$ and evaluating Q when the loop terminates.

Sum(nonnegative integer x; nonnegative integer y)

Local variables:
integers i, j

```
i = 0
j = x
while i ≠ y do
    j = j + 1
    i = i + 1
end while
//j now has the value x + y
return j
end function Sum
```
    ●

The two functions of Example 25 and Practice 10 are somewhat unrealistic; after all, if we wanted to compute $x * y$ or $x + y$, we could no doubt do it with a single program statement. However, the same techniques apply to more meaningful computations, such as the Euclidean algorithm.

Euclidean Algorithm

The **Euclidean algorithm** was devised by the Greek mathematician Euclid over 2300 years ago, making it one of the oldest known algorithms. This algorithm finds the greatest common divisor of two nonnegative integers a and b, where not both a and b are zero. The **greatest common divisor** of a and b, denoted by $\gcd(a, b)$, is the largest integer that divides (evenly into) both a and b. For example, $\gcd(12, 18)$ is 6, $\gcd(420, 66) = 6$, and $\gcd(18, 0) = 18$ (this last because any number except 0 divides 0).

The Euclidean algorithm works by a succession of divisions. To find $\gcd(a, b)$, assuming that $a \geq b$, you first divide a by b, getting a quotient and a remainder. More formally, at this point $a = q_1 b + r_1$, where $0 \leq r_1 < b$. Next you divide the divisor, b, by the remainder, r_1, getting $b = q_2 r_1 + r_2$, where $0 \leq r_2 < r_1$. Again divide the divisor, r_1, by the remainder, r_2, getting $r_1 = q_3 r_2 + r_3$, where $0 \leq r_3 < r_2$. Clearly, there is a looping process going on. The process terminates when the remainder is 0, at which point the greatest common divisor is the last divisor used.

EXAMPLE 27   To find $\gcd(420, 66)$ the following divisions are performed:

$$
\begin{array}{cccc}
6 & 2 & 1 & 3 \\
66)\overline{420} & 24)\overline{66} & 18)\overline{24} & 6)\overline{18} \\
\underline{396} & \underline{48} & \underline{18} & \underline{18} \\
24 & 18 & 6 & 0
\end{array}
$$

The answer is 6, the divisor used when the remainder became 0.   ●

A pseudocode version of the algorithm follows, given in the form of a function to return $\gcd(a, b)$.

ALGORITHM   **Euclidean Algorithm**

GCD(nonnegative integer a; nonnegative integer b)
//$a \geq b$, not both a and b are zero

Local variables:
integers i, j

    $i = a$
    $j = b$
    while $j \neq 0$ **do**
        compute $i = qj + r, 0 \leq r < j$
        $i = j$
        $j = r$
    end while

    //i now has the value $\gcd(a, b)$
    return i;
end function GCD

We intend to prove the correctness of this function, but we will need one additional fact first, namely

$$(\forall \text{ integers } a, b, q, r)[(a = qb + r) \rightarrow (\gcd(a, b) = \gcd (b, r))] \tag{5}$$

To prove (5), assume that $a = qb + r$, and suppose that c divides both a and b, so that $a = q_1c$ and $b = q_2c$. Then

$$r = a - qb = q_1c - qq_2c = c(q_1 - qq_2)$$

so that c divides r as well. Therefore anything that divides a and b also divides b and r. Now suppose d divides both b and r so that $b = q_3d$ and $r = q_4d$. Then

$$a = qb + r = qq_3d + q_4d = d(qq_3 + q_4)$$

so that d divides a as well. Therefore anything that divides b and r also divides a and b. Because (a, b) and (b, r) have identical divisors, they must have the same greatest common divisor.

EXAMPLE 28   Prove the correctness of the Euclidean algorithm.

Using function GCD, we will prove the loop invariant Q: $\gcd(i, j) = \gcd(a, b)$ and evaluate Q when the loop terminates. We use induction to prove $Q(n)$: $\gcd(i_n, j_n) = \gcd(a, b)$ for all $n \geq 0$. $Q(0)$ is the statement

$$\gcd(i_0, j_0) = \gcd(a, b)$$

which is true because when we first get to the loop statement, i and j have the values a and b, respectively.

Assume $Q(k)$: $\gcd(i_k, j_k) = \gcd(a, b)$
Show $Q(k + 1)$: $\gcd(i_{k+1}, j_{k+1}) = \gcd(a, b)$

By the assignment statements within the loop body, we know that

$$i_{k+1} = j_k$$
$$j_{k+1} = r_k$$

Then

$$\gcd(i_{k+1}, j_{k+1}) = \gcd(j_k, r_k)$$
$$= \gcd(i_k, j_k) \quad \text{by (5)}$$
$$= \gcd(a, b) \quad \text{by the inductive hypothesis}$$

Q is therefore a loop invariant. At loop termination, $\gcd(i, j) = \gcd(a, b)$ and $j = 0$, so $\gcd(i, 0) = \gcd(a, b)$. But $\gcd(i, 0)$ is i, so $i = \gcd(a, b)$. Therefore function GCD is correct.   ●

Section 2.3 Review

Techniques

- Verify the correctness of a program segment that includes a loop statement.
- Compute gcd(a, b) using Euclid's algorithm.

Main Ideas

A loop invariant, proved by induction on the number of loop iterations, can be used to prove correctness of a program loop.

The classic Euclidean algorithm for finding the greatest common divisor of two nonnegative integers is provably correct.

Exercises 2.3

In Exercises 1–4, prove that the pseudocode program segment is correct by proving the loop invariant Q and evaluating Q at loop termination.

1. Function to return the value of x^2 for $x \geq 1$

   Square(positive integer x)

   Local variables:
   integers i, j

       $i = 1$
       $j = 1$
       while $i \neq x$ **do**
           $j = j + 2i + 1$
           $i = i + 1$
       end while
       //j now has the value x^2
       return j
     end function Square

       $Q: j = i^2$

★2. Function to return the value of $x!$ for $x \geq 1$

   Factorial(positive integer x)

   Local variables:
   integers i, j

       $i = 2$
       $j = 1$
       while $i \neq x + 1$ **do**
           $j = j * i$
           $i = i + 1$
       end while

//j now has the value $x!$
return j
end function Factorial

$Q: j = (i - 1)!$

3. Function to return the value of x^y for $x, y \geq 1$

   Power(positive integer x; positive integer y)

   Local variables:
   integers i, j

   $i = 1$
   $j = x$
   while $i \neq y$ **do**
      $j = j * x$
      $i = i + 1$
   end while

   //j now has the value x^y
   return j
   end function Power

   $Q: j = x^i$

4. Function to compute and write out quotient q and remainder r when x is divided by y, $x \geq 0$, $y \geq 1$

   Divide(nonnegative integer x; positive integer y)

   Local variables:
   nonnegative integers q, r

   $q = 0$
   $r = x$
   while $r \geq y$ **do**
      $q = q + 1$
      $r = r - y$
   end while

   //q and r are now the quotient and remainder
   write("The quotient is" q "and the remainder is" r)
   end function Divide

   $Q: x = q * y + r$

For Exercises 5–8, use the Euclidean algorithm to find the greatest common divisor of the given numbers.

5. (2420, 70)

★6. (735, 90)

7. (1326, 252)

8. (1018215, 2695)

In Exercises 9–13, prove that the program segment is correct by finding and proving the appropriate loop invariant Q and evaluating Q at loop termination.

9. Function to return the value $x - y$ for $x, y \geq 0$

   Difference(nonnegative integer x; nonnegative integer y)

   Local variables:
   integers i, j

       $i = 0$
       $j = x$
       while $i \neq y$ **do**
           $j = j - 1$
           $i = i + 1$
       end while

       //j now has the value $x - y$
       return j
   end function Difference

★10. Function to return the value $x * y^n$ for $n \geq 0$

   Computation(integer x; integer y; nonnegative integer n)

   Local variables:
   integers i, j

       $i = 0$
       $j = x$
       while $i \neq n$ **do**
           $j = j * y$
           $i = i + 1$
       end while

       //j now has the value $x * y^n$
       return j
   end function Computation

11. Function to return the value 2^n for $n \geq 1$

   TwosPower(positive integer n)

   Local variables:
   integers i, j

       $i = 1$
       $j = 2$
       while $i \neq n$ **do**
           $j = j * 2$
           $i = i + 1$
       end while

       //j now has the value 2^n
       return j
   end function TwosPower

★12. Function to return the value $x * n!$ for $n \geq 1$

AnotherOne(integer x; positive integer n)

Local variables:

integers i, j

$\quad i = 1$

$\quad j = x$

$\quad$ **while** $i \neq n$ **do**

$\quad\quad i = j * (i + 1)$

$\quad\quad i - i + 1$

$\quad$ **end while**

$\quad$ //j now has the value $x * n!$

$\quad$ return j

end function AnotherOne

13. Function to return the value of the polynomial

$$a_n x^n + a_{n-1} x^{n-1} + \cdots + a_1 x + a_0$$

at a given value of x

Polly(real a_n; ... ; real a_0; real x)

Local variables:

integers i, j

$\quad i = n$

$\quad j = a$

$\quad$ **while** $i \neq 0$ **do**

$\quad\quad j = j * x + a_{i-1}$

$\quad\quad i := i - 1$

$\quad$ **end while**

$\quad$ //j now has value of the polynomial evaluation

$\quad$ return j

end function Polly

Section 2.4   Recursion and Recurrence Relations

Recursive Definitions

A definition in which the item being defined appears as part of the definition is called an **inductive definition** or a **recursive definition**. At first this seems like nonsense — how can we define something in terms of itself? This works because there are two parts to a recursive definition:

1. A basis, where some simple cases of the item being defined are explicitly given
2. An inductive or recursive step, where new cases of the item being defined are given in terms of previous cases

Part 1 gives us a place to start by providing some simple, concrete cases; part 2 allows us to construct new cases from these simple ones and then to construct still other cases from these new ones, and so forth. (The analogy with proofs by mathematical induction accounts for the name "inductive definition." In a proof by induction, there is a basis step, namely, to show that $P(1)$—or P at some other initial value—holds, and there is an inductive step where the truth of $P(k + 1)$ is deduced from the truth of P at previous values.)

Recursion is an important idea that can be used to define sequences of objects, more general collections of objects, and operations on objects. (The Prolog predicate *in-food-chain* of Section 1.5 was defined recursively.) Even algorithms can be recursive.

Recursively Defined Sequences

A **sequence** S is a list of objects that are enumerated in some order; there is a first such object, then a second, and so on. $S(k)$ denotes the kth object in the sequence. A sequence is defined recursively by explicitly naming the first value (or the first few values) in the sequence and then defining later values in the sequence in terms of earlier values.

EXAMPLE 29   The sequence S is defined recursively by

1. $S(1) = 2$
2. $S(n) = 2S(n - 1)$ for $n \geq 2$

By statement 1, $S(1)$, the first object in S, is 2. Then by statement 2, the second object in S is $S(2) = 2S(1) = 2(2) = 4$. By statement 2 again, $S(3) = 2S(2) = 2(4) = 8$. Continuing in this fashion, we can see that S is the sequence

$$2, 4, 8, 16, 32, ...$$

A rule like that of statement 2 in Example 29, which defines a sequence value in terms of one or more earlier values, is called a **recurrence relation.**

PRACTICE 11   The sequence T is defined recursively as follows:

1. $T(1) = 1$
2. $T(n) = T(n - 1) + 3$ for $n \geq 2$

Write the first five values in the sequence T.

EXAMPLE 30   The famous **Fibonacci sequence** of numbers, introduced in the thirteenth century by an Italian merchant and mathematician, is defined recursively by

$$F(1) = 1$$
$$F(2) = 1$$
$$F(n) = F(n - 2) + F(n - 1) \text{ for } n > 2$$

Here the first two values of the sequence are given, and the recurrence relation defines the nth value in terms of the two preceding values. It's best to think of the recurrence relation in its most general form, which says that F at any value—except 1 and 2—is the sum of F at the two previous values. ●

PRACTICE 12   Write the first eight values of the Fibonacci sequence. ●

EXAMPLE 31   Prove that in the Fibonacci sequence

$$F(n + 4) = 3F(n + 2) - F(n) \text{ for all } n \geq 1$$

Because we want to prove something true for all $n \geq 1$, it is natural to think of a proof by induction. And because the value of $F(n)$ depends on both $F(n - 1)$ and $F(n - 2)$, the second principle of induction should be used. For the basis step of the inductive proof, we'll prove two cases, $n = 1$ and $n = 2$. For $n = 1$, we get

$$F(5) = 3F(3) - F(1)$$

or (using values computed in Practice 12)

$$5 = 3(2) - 1$$

which is true. For $n = 2$,

$$F(6) = 3F(4) - F(2)$$

or

$$8 = 3(3) - 1$$

which is also true. Assume that for all r, $1 \leq r \leq k$,

$$F(r + 4) = 3F(r + 2) - F(r)$$

Now show the case for $k + 1$, where $k + 1 \geq 3$. Thus we want to show

$$F(k + 1 + 4) \stackrel{?}{=} 3F(k + 1 + 2) - F(k + 1)$$

or

$$F(k + 5) \stackrel{?}{=} 3F(k + 3) - F(k + 1)$$

From the recurrence relation for the Fibonacci sequence, we have

$$F(k + 5) = F(k + 3) + F(k + 4) \qquad (F \text{ at any value is the sum of } F \text{ at the two previous values})$$

and by the inductive hypothesis, with $r = k - 1$ and $r = k$, respectively,

$$F(k + 3) = 3F(k + 1) - F(k - 1)$$

and

$$F(k + 4) = 3F(k + 2) - F(k)$$

Therefore

$$F(k + 5) = F(k + 3) + F(k + 4)$$
$$= [3F(k + 1) - F(k - 1)] + [3F(k + 2) - F(k)]$$
$$= 3[F(k + 1) + F(k + 2)] - [F(k - 1) + F(k)]$$
$$= 3F(k + 3) - F(k + 1) \qquad \text{(using the recurrence relation again)}$$

This completes the inductive proof.

EXAMPLE 32   The formula

$$F(n + 4) = 3F(n + 2) - F(n) \text{ for all } n \geq 1$$

of Example 31 can also be proved without induction, using just the recurrence relation from the definition of Fibonacci numbers. The recurrence relation

$$F(n + 2) = F(n) + F(n + 1)$$

can be rewritten as

$$F(n + 1) = F(n + 2) - F(n) \tag{1}$$

Then

$$F(n + 4) = F(n + 3) + F(n + 2)$$
$$= F(n + 2) + F(n + 1) + F(n + 2) \qquad \text{rewriting } F(n + 3)$$
$$= F(n + 2) + [F(n + 2) - F(n)] + F(n + 2) \qquad \text{rewriting } F(n + 1)$$
$$\qquad\qquad\qquad\qquad\qquad\qquad\qquad\qquad\qquad \text{using (1)}$$
$$= 3F(n + 2) - F(n)$$

PRACTICE 13   In the inductive proof of Example 31, why is it necessary to prove $n = 2$ as a special case?

Recursively Defined Sets

The objects in a sequence are ordered—there is a first object, a second object, and so on. A set of objects is a collection of objects on which no ordering is imposed. Some sets can be defined recursively.

EXAMPLE 33   In Section 1.1 we noted that certain strings of statement letters, logical connectives, and parentheses, such as $(A \wedge B)' \vee C$, are considered legitimate, while other strings, such as $\wedge \wedge A''B$, are not legitimate. The syntax for arranging such symbols constitutes the definition of the set of propositional well-formed formulas, and it is a recursive definition.

1. Any statement letter is a wff.
2. If P and Q are wffs, so are $(P \wedge Q)$, $(P \vee Q)$, $(P \rightarrow Q)$, (P'), and $(P \leftrightarrow Q)$.

We often omit parentheses when doing so causes no confusion; thus we write $(P \lor Q)$ as $P \lor Q$, or (P') as P'. By beginning with statement letters and repeatedly using rule 2, any propositional wff can be built. For example, A, B, and C are all wffs by rule 1. By rule 2,

$(A \land B)$ and (C')

are both wffs. By rule 2 again,

$((A \land B) \to (C'))$

is a wff. Applying rule 2 yet again, we get the wff

$(((A \land B) \to (C'))')$

Eliminating some parentheses, we can write this wff as

$((A \land B) \to C')'$

PRACTICE 14   Show how to build the wff $((A \lor (B') \to C)$ from the definition in Example 33.

PRACTICE 15   A recursive definition for the set of people who are ancestors of James could have the following basis:

James's parents are ancestors of James.

Give the inductive step.

Strings of symbols drawn from a finite "alphabet" set are objects that are commonly encountered in computer science. Computers store data as **binary strings**, strings from the alphabet consisting of 0s and 1s; compilers view program statements as strings of *tokens,* such as key words and identifiers. The collection of all finite-length strings of symbols from an alphabet, usually called strings *over* an alphabet, can be defined recursively (see Example 34). Many sets of strings with special properties also have recursive definitions.

EXAMPLE 34   The set of all (finite-length) strings of symbols over a finite alphabet A is denoted by $A*$. The recursive definition of $A*$ is

1. The **empty string** λ (the string with no symbols) belongs to $A*$.
2. Any single member of A belongs to $A*$.
3. If x and y are strings in $A*$, so is xy, the **concatenation** of strings x and y.

Parts 1 and 2 constitute the basis, and part 3 is the recursive step of this definition. Note that for any string x, $x\lambda = \lambda x = x$.

PRACTICE 16   If $x = 1011$ and $y = 001$, write the strings xy, yx, and $yx\lambda x$.

PRACTICE 17   Give a recursive definition for the set of all binary strings that are **palindromes**, strings that read the same forwards and backwards.

EXAMPLE 35   Suppose that in a certain programming language, identifiers can be alphanumeric strings of arbitrary length but must begin with a letter. A recursive definition for the set of such strings is

1. A single letter is an identifier.
2. If A is an identifier, so is the concatenation of A and any letter or digit.

A more symbolic notation for describing sets of strings that are recursively defined is called **Backus Naur form**, or **BNF**, originally developed to define the programming language ALGOL. In BNF notation, items that are defined in terms of other items are enclosed in angle brackets, while specific items that are not further broken down do not appear in brackets. The vertical line $|$ denotes a choice, with the same meaning as the English word *or*. The BNF definition of an identifier is

<identifier> ::= <letter> $|$ <identifier> <letter > $|$ <identifier> <digit>

<letter> ::= $a \,|\, b \,|\, c \,|\, \cdots \,|\, z$

<digit> ::= $1 \,|\, 2 \,|\, \cdots \,|\, 9$

Thus the identifier *me2* is built from the definition by a sequence of choices such as

<identifier>	can be	<identifier> <digit>
	which can be	<identifier>2
	which can be	<identifier> <letter>2
	which can be	<identifier>e2
	which can be	<letter>e2
	which can be	*me2*

Recursively Defined Operations

Certain operations performed on objects can be defined recursively, as in Examples 36 and 37.

EXAMPLE 36   A recursive definition of the exponentiation operation a^n on a nonzero real number a, where n is a nonnegative integer, is

1. $a^0 = 1$
2. $a^n = (a^{n-1})a$ for $n \geq 1$

EXAMPLE 37   A recursive definition for multiplication of two positive integers m and n is

1. $m(1) = m$
2. $m(n) = m(n - 1) + m$ for $n \geq 2$

PRACTICE 18   Let x be a string over some alphabet. Give a recursive definition for the operation x^n (concatenation of x with itself n times) for $n \geq 1$.

In Section 1.1, we defined the operation of logical disjunction on two statement letters. This can serve as the basis step for a recursive definition of the disjunction of n statement letters, $n \geq 2$:

1. $A_1 \vee A_2$ defined as in Section 1.1
2. $A_1 \vee \cdots \vee A_n = (A_1 \vee \cdots \vee A_{n-1}) \vee A_n$ for $n > 2$ \hfill (2)

Using this definition, we can generalize the associative property of disjunction (tautological equivalence 2a) to say that in a disjunction of n statement letters, grouping by parentheses is unnecessary because all such groupings are equivalent to the general expression for the disjunction of n statement letters. In symbolic form, for any n with $n \geq 3$ and any p with $1 \leq p \leq n - 1$,

$$(A_1 \vee \cdots \vee A_p) \vee (A_{p+1} \vee \cdots \vee A_n) \Leftrightarrow A_1 \vee \cdots \vee A_n$$

This equivalence can be proved by induction on n. For $n = 3$,

$$A_1 \vee (A_2 \vee A_3) \Leftrightarrow (A_1 \vee A_2) \vee A_3 \quad \text{(by equivalence 2a)}$$
$$= A_1 \vee A_2 \vee A_3 \quad \text{(by equation (2))}$$

Assume that for $n = k$ and $1 \leq p \leq k - 1$,

$$(A_1 \vee \cdots \vee A_p) \vee (A_{p+1} \vee \cdots \vee A_k) \Leftrightarrow A_1 \vee \cdots \vee A_k$$

Then for $n = k + 1$ and $1 \leq p \leq k$,

$$(A_1 \vee \cdots \vee A_p) \vee (A_{p+1} \vee \cdots \vee A_{k+1})$$
$$= (A_1 \vee \cdots \vee A_p) \vee [(A_{p+1} \vee \cdots \vee A_k) \vee A_{k+1}] \quad \text{(by equation (2))}$$
$$\Leftrightarrow [(A_1 \vee \cdots \vee A_p) \vee (A_{p+1} \vee \cdots \vee A_k)] \vee A_{k+1} \quad \text{(by equivalence 2a)}$$
$$\Leftrightarrow (A_1 \vee \cdots \vee A_k) \vee A_{k+1} \quad \text{(by inductive hypothesis)}$$
$$= A_1 \vee \cdots \vee A_{k+1} \quad \text{(by equation (2))}$$

Recursively Defined Algorithms

Example 29 gives a recursive definition for a sequence S. Suppose we want to write a computer program to evaluate $S(n)$ for some positive integer n. We can use either of two approaches. If we want to find $S(12)$, for example, we can begin with $S(1) = 2$ and then compute $S(2)$, $S(3)$, and so on, much as we did in Example 29, until we finally get to $S(12)$. This approach no doubt involves iterating through some sort of loop. A pseudocode function S that uses this iterative algorithm follows. The basis, where $n = 1$, is handled in the first clause of the **if** statement; the value 2 is returned. The **else** clause, for $n > 1$, does some initializing and then goes into the **while** loop that computes larger values of the sequence until the correct upper limit is reached. You can trace the execution of this algorithm for a few values of n to convince yourself that it works.

ALGORITHM

S(integer *n*)
//function that iteratively computes the value *S*(*n*)
//for the sequence *S* of Example 29

Local variables:
integer *i*   //loop index
CurrentValue   //current value of function *S*

   if *n* = 1 **then**
      return 2
   else
      i = 2
      CurrentValue = 2
      while *i* <= *n* **do**
         CurrentValue = 2∗CurrentValue
         i = *i* + 1
      end while

      //CurrentValue now has the value *S*(*n*)
      return CurrentValue
   end if
end function *S*

The second approach to computing $S(n)$ uses the recursive definition of S directly. Following is a version of the *recursive algorithm*, written again as a pseudocode function.

ALGORITHM

S(integer *n*)
//function that recursively computes the value *S*(*n*)
//for the sequence *S* of Example 29

   if *n* = 1 **then**
      return 2
   else
      return 2∗*S*(*n* − 1)
   end if
end function *S*

The body of this function consists of a single **if-then-else** statement. To understand how the function works, let's trace the execution to compute the value of $S(3)$. The function is first invoked with an input value of $n = 3$. Because n is not 1, execution is directed to the **else** clause. At this point, activity on computing $S(3)$ must be suspended until the value of $S(2)$ is known. Any known information relevant to the computation of $S(3)$ is stored within computer memory on a stack, to be retrieved when the computation can be completed. (A stack is a collection of data where any new item goes on top of the stack, and only the item on top of the stack at any given time can be accessed or removed from the stack. A stack is thus a LIFO—last in, first out—structure.) The function is invoked again with an input value of $n = 2$. Again, the **else** clause is executed, and computation of $S(2)$ is suspended, with relevant information stored on the stack, while the function is invoked again with $n = 1$ as input.

This time the first clause of the **if** statement applies, and the functional value, 2, can be computed directly. This final invocation of the function is now complete, and its value of 2 is returned to the second-to-last invocation, which can now remove any information relevant to the $n = 2$ case from the stack, compute $S(2)$, and return the result to the previous (initial) invocation. Finally, this original invocation of S is able to empty the stack and complete its calculation, returning the value of $S(3)$.

What are the relative advantages of iterative and recursive algorithms for doing the same task? In this example, the recursive version is certainly shorter because it does not have to manage a loop computation. Describing the execution of the recursive version makes it sound more complex than the iterative version, but all steps are carried out automatically. One need not be aware of what is happening internally except to note that a long series of recursive invocations can use a lot of memory by storing information relevant to previous invocations on the stack. If too much memory is consumed, a "stack overflow" can result. Besides using more memory, recursive algorithms can require many more computations and can run more slowly than nonrecursive ones (see Exercise 7 in On the Computer at the end of this chapter).

Nonetheless, recursion provides a natural way to think about many problems, some of which would have very complex nonrecursive solutions. The problem of computing values for a sequence that has itself been defined recursively is well-suited to a recursive solution. Many programming languages support recursion.

PRACTICE 19   Write the body of a recursive function to compute $T(n)$ for the sequence T defined in Practice 11.   ●

EXAMPLE 38   In Example 37, a recursive definition was given for multiplying two positive integers m and n. A recursive pseudocode function for multiplication based on this definition follows.   ●

ALGORITHM

Product(integer m; integer n)
//Function that recursively computes the product of m and n
    if $n = 1$ **then**
        return m;
    else
        return *Product*$(m, n - 1) + m$
    end if
end function *Product*

A recursive algorithm invokes itself with "smaller" input values. Suppose a problem can be solved by solving smaller versions of the same problem, and the smaller versions eventually become trivial cases that are easily handled. Then a recursive algorithm can be useful, even if the original problem was not stated recursively.

To convince ourselves that a given recursive algorithm works, we don't have to start with a particular input and go down through smaller and smaller cases to the trivial case and then back up again. We did this when discussing the computation of $S(3)$, but that was just to illustrate the mechanics of a recursive computation. Instead, we can verify the trivial case (like proving the base case in an induction proof) and verify that if the algorithm works correctly when invoked on smaller input values, then it indeed solves the problem for the original input values (this is similar to proving $P(k + 1)$ from the assumption $P(k)$ in an inductive proof).

EXAMPLE 39

One of the most common tasks in data processing is to sort a list L of n items into increasing or decreasing numerical or alphabetical order. (The list might consist of customer names, for example, and in sorted order "Valdez, Juanita" should come after "Tucker, Joseph.") The **selection sort** algorithm—a simple but not particularly efficient sorting algorithm—is described in pseudocode in the accompanying box.

This function sorts the first j items in L into increasing order; when the function is initially invoked, j has the value n (thus, the first invocation ultimately sorts the entire list). The recursive part of the algorithm lies within the **else** clause; the algorithm examines the section of the list under consideration and finds the location i such that $L(i)$ is the maximum value. It then exchanges $L(i)$ and $L(j)$, after which the maximum value occurs at position j, the last position in the part of the list being considered. $L(j)$ is now correct and should never change again, so this process is repeated on the list $L(1)$ through $L(j - 1)$. If this part of the list is sorted correctly, then the entire list will be sorted correctly. Whenever j has the value 1, the part of the list being considered consists of only one entry, which must be in the right place. The entire list is sorted at that point.

ALGORITHM   SelectionSort

SelectionSort(list L; integer j)
//recursively sorts the items from 1 to j in list L into increasing order
    if $j = 1$ **then**
        sort is complete, write out the sorted list
    else
        find the index i of the maximum item in L between 1 and j
        exchange $L(i)$ and $L(j)$
        SelectionSort($L, j - 1$)
    end if
end function SelectionSort

EXAMPLE 40   Now that we have sorted our list, another common task is to search the list for a particular item. (Is Juanita Valdez already a customer?) An efficient search technique for
a sorted list is the recursive **binary search algorithm**, which is described in
pseudocode in the accompanying box.

ALGORITHM   BinarySearch

BinarySearch(list L; integer i; integer j; itemtype x)
//searches sorted list L from $L(i)$ to $L(j)$ for item x
   if $i > j$ **then**
      write("not found")
   else
      find the index k of the middle item in the list $L(i)$–$L(j)$
      if $x =$ middle item **then**
         write("found")
      else
         if $x <$ middle item **then**
            BinarySearch($L, i, k - 1, x$)
         else
            BinarySearch($L, k + 1, j, x$)
         end if
      end if
   end if
end function BinarySearch

This algorithm searches the section of list L between $L(i)$ and $L(j)$ for item x; initially
i and j have the values 1 and n, respectively. The first clause of the major **if** statement
is the basis step that says x cannot be found in an empty list, one where the first index
exceeds the last index. In the major **else** clause, the middle item in a section of the list
must be found. (If the section contains an odd number of items, there is indeed a middle item; if the section contains an even number of items, it is sufficient to take as the
"middle" item the one at the end of the first half of the list section.) Comparing x with
the middle item either locates x or indicates which half of the list to search next. ●

EXAMPLE 41   Let's apply the binary search algorithm to the list

    3, 7, 8, 10, 14, 18, 22, 34

where the target item x is the number 25. The initial list is not empty, so the middle
item is located and determined to have the value 10. Then x is compared with the
middle item. Because $x > 10$, the search is invoked on the second half of the list,
namely, the items

    14, 18, 22, 34

Again, this list is nonempty, and the middle item is 18. Because $x > 18$, the second half of this list is searched, namely, the items

22, 34

In this nonempty list, the middle item is 22. Because $x > 22$, the search continues on the second half of the list, namely,

34

This is a one-element list, with the middle item being the only item. Because $x < 34$, a search is begun on the "first half" of the list; but the first half is empty. The algorithm terminates at this point with the information that x is not in the list.

This execution requires four comparisons in all; x is compared, in turn, to 10, 18, 22, and 34. ●

PRACTICE 20   In a binary search of the list in Example 41, name the elements against which x is compared if x has the value 8. ●

We have now seen a number of recursive definitions. Table 2.5 summarizes their features.

Recursive Definitions	
What Is Being Defined	Characteristics
Recursive Sequence	The first one or two values in the sequence are known; later items in the sequence are defined in terms of earlier items.
Recursive Set	A few specific items are known to be in the set; other items in the set are built from combinations of items already in the set.
Recursive Operation	A "small" case of the operation gives a specific value; other cases of the operation are defined in terms of smaller cases.
Recursive Algorithm	For the smallest values of the arguments, the algorithm behavior is known; for larger values of the arguments, the algorithm invokes itself with smaller argument values.

Table 2.5

Solving Recurrence Relations

We developed two algorithms, one iterative and one recursive, to compute a value $S(n)$ for the sequence S of Example 29. However, there is a still easier way to compute $S(n)$. Recall that

$$S(1) = 2 \tag{1}$$
$$S(n) = 2S(n - 1) \text{ for } n \geq 2 \tag{2}$$

Because

$$S(1) = 2 = 2^1$$
$$S(2) = 4 = 2^2$$
$$S(3) = 8 = 2^3$$
$$S(4) = 16 = 2^4$$

and so on, we can see that

$$S(n) = 2^n \qquad (3)$$

Using equation (3), we can plug in a value for n and compute $S(n)$ without having to compute—either explicitly, or, through recursion, implicitly—all the lower values of S first. An equation such as (3), where we can substitute a value and get the output value back directly, is called a **closed-form solution** to the recurrence relation (2) subject to the basis step (1). Finding a closed-form solution is called **solving** the recurrence relation. Clearly, it is nice to find closed-form solutions whenever possible.

One technique for solving recurrence relations is an "expand, guess, and verify" approach that repeatedly uses the recurrence relation to expand the expression for the nth term until the general pattern can be guessed. Finally the guess is verified by mathematical induction.

EXAMPLE 42   Consider again the basis step and recurrence relation for the sequence S of Example 29:

$$S(1) = 2 \qquad (4)$$
$$S(n) = 2S(n - 1) \text{ for } n \geq 2 \qquad (5)$$

Let's pretend we don't already know the closed-form solution and use the expand, guess, and verify approach to find it. Beginning with $S(n)$, we expand by using the recurrence relation repeatedly. Keep in mind that the recurrence relation is a recipe that says S at any value can be replaced by two times S at the previous value. We apply this recipe to S at the values $n, n - 1, n - 2$, and so on:

$$S(n) = 2S(n - 1)$$
$$= 2[2S(n - 2)] = 2^2S(n - 2)$$
$$= 2^2[2S(n - 3)] = 2^3S(n - 3)$$

By looking at the developing pattern, we guess that after k such expansions, the equation has the form

$$S(n) = 2^kS(n - k)$$

This expansion of S-values in terms of lower S-values must stop when $n - k = 1$, that is, when $k = n - 1$. At that point,

$$S(n) = 2^{n-1}S[n - (n - 1)]$$
$$= 2^{n-1}S(1) = 2^{n-1}(2) = 2^n$$

which expresses the closed-form solution.

We are not yet done, however, because we guessed at the general pattern. We now confirm our closed-form solution by induction on the value of n. The statement we want to prove is therefore $S(n) = 2^n$ for $n \geq 1$.

For the basis step, $S(1) = 2^1$. This is true by equation (4). We assume that $S(k) = 2^k$. Then

$$S(k + 1) = 2S(k) \quad \text{(by equation (5))}$$
$$= 2(2^k) \quad \text{(by the inductive hypothesis)}$$
$$= 2^{k+1}$$

This proves that our closed-form solution is correct.   ●

PRACTICE 21   Find a closed-form solution for the recurrence relation, subject to the basis step, for sequence T.

1. $T(1) = 1$
2. $T(n) = T(n - 1) + 3$ for $n \geq 2$

(*Hint:* Expand, guess, and verify.)   ●

Methods for solving recurrence relations are somewhat like those used for solving differential equations. In particular, both recurrence relations and differential equations are classified into various types, many of which have solution formulas that are known.

A recurrence relation for a sequence $S(n)$ is **linear** if the earlier values of S appearing in the definition occur only to the first power. The most general linear recurrence relation has the form

$$S(n) = f_1(n)S(n - 1) + f_2(n)S(n - 2) + \cdots + f_k(n)S(n - k) + g(n)$$

where the f_i's and g can be expressions involving n. The recurrence relation has **constant coefficients** if the f_i's are all constants. It is **first-order** if the nth term depends only on term $n - 1$. Linear first-order recurrence relations with constant coefficients therefore have the form

$$S(n) = cS(n - 1) + g(n) \tag{6}$$

Finally, a recurrence relation is **homogeneous** if $g(n) = 0$ for all n.

We will find the solution formula for equation (6), the general linear first-order recurrence relation with constant coefficients, subject to the basis that $S(1)$ is known. We will use the expand, guess, and verify approach. The work here is a generalization of what was done in Example 42. Repeatedly applying equation (6) and simplifying, we get

$$S(n) = cS(n - 1) + g(n)$$
$$= c[cS(n - 2) + g(n - 1)] + g(n)$$
$$= c^2 S(n - 2) + cg(n - 1) + g(n)$$
$$= c^2 [cS(n - 3) + g(n - 2)] + cg(n - 1) + g(n)$$
$$= c^3 S(n - 3) + c^2 g(n - 2) + cg(n - 1) + g(n)$$
$$\vdots$$

After k expansions, the general form appears to be

$$S(n) = c^k S(n - k) + c^{k-1} g(n - (k - 1)) + \cdots + cg(n - 1) + g(n)$$

If the sequence has a base value at 1, then the expansion terminates when $n - k = 1$ or $k = n - 1$, at which point

$$S(n) = c^{n-1} S(1) + c^{n-2} g(2) + \cdots + cg(n - 1) + g(n)$$
$$= c^{n-1} S(1) + c^{n-2} g(2) + \cdots + c^1 g(n - 1) + c^0 g(n) \tag{7}$$

We can use **summation notation** to write part of this expression more compactly. The uppercase Greek letter sigma, Σ, stands for summation. The notation

$$\sum_{i=p}^{q} (\text{expression})$$

says to substitute into the expression successive values of i, the **index of summation**, from the lower limit p to the upper limit q, and then sum the results. (See Appendix A for further discussion of summation notation.) Thus, for example,

$$\sum_{i=1}^{n} (2i - 1) = 1 + 3 + 5 + \cdots + (2n - 1)$$

In Example 14, Section 2.2, we proved by induction that the value of this summation is n^2.

In summation notation, equation (7) becomes

$$S(n) = c^{n-1} S(1) + \sum_{i=2}^{n} c^{n-i} g(i)$$

Induction can be used, much as was done in Example 42, to verify that this formula is the solution to recurrence relation (6) (see Exercise 82).

Therefore, the solution to the recurrence relation (6) is

$$S(n) = c^{n-1} S(1) + \sum_{i=2}^{n} c^{n-i} g(i) \tag{8}$$

This is not yet a closed-form solution, however, because we must find an expression for the summation. Usually it is either trivial to find the sum or we found its value in Section 2.2 using mathematical induction.

The work we've done here found a general solution once and for all for any recurrence relation of the form shown in (6); this work *need not be repeated*. All that is necessary is to match your problem to equation (6) in order to find the value for c and the formula for $g(n)$ and then plug these results into the expression in (8).

EXAMPLE 43   The sequence $S(n)$ of Example 42,

$$S(1) = 2$$
$$S(n) = 2S(n - 1) \text{ for } n \geq 2$$

is a linear, first-order, homogeneous recurrence relation with constant coefficients. In other words, it matches equation (6) with $c = 2$ and $g(n) = 0$. From formula (8), the closed-form solution is

$$S(n) = 2^{n-1}(2) + \sum_{i=2}^{n} 0 = 2^n$$

which agrees with our previous result.

PRACTICE 22   Rework Practice 21 using equation (8).

You now have a choice of two alternative ways to solve a linear, first-order recurrence relation with constant coefficients. Table 2.6 summarizes these approaches.

To Solve Recurrence Relations of the Form $S(n) = cS(n-1) + g(n)$ Subject to Basis $S(1)$	
Method	Steps
Expand, guess, verify	1. Repeatedly use the recurrence relation until you can guess a pattern. 2. Decide what that pattern will be when $n - k = 1$. 3. Verify the resulting formula by induction.
Solution formula	1. Match your recurrence relation to the form $S(n) = cS(n-1) + g(n)$ to find c and $g(n)$. 2. Use c, $g(n)$, and $S(1)$ in the formula $$S(n) = c^{n-1}S(1) + \sum_{i=2}^{n} c^{n-i}g(i)$$ 3. Evaluate the resulting summation to get the final expression.

Table 2.6

EXAMPLE 44   Find a closed-form solution to the recurrence relation

$$S(n) = 2S(n-1) + 3 \text{ for } n \geq 2$$

subject to the basis step

$$S(1) = 4$$

We'll use the solution formula method. Comparing our recurrence relation

$$S(n) = 2S(n-1) + 3$$

with the general form $S(n) = cS(n-1) + g(n)$, we see that

$$c = 2 \qquad g(n) = 3$$

The fact that $g(n) = 3$ says that g has a constant value of 3 no matter what the value of n. Substituting into the general solution form $S(n) = c^{n-1}S(1) + \sum_{i=2}^{n} c^{n-i}g(i)$, we get

$$S(n) = 2^{n-1}(4) + \sum_{i=2}^{n} 2^{n-i}(3)$$

$$= 2^{n-1}(2^2) + 3\sum_{i=2}^{n} 2^{n-i}$$

$$= 2^{n+1} + 3[2^{n-2} + 2^{n-3} + \cdots + 2^1 + 2^0] \qquad \text{(from Example 15)}$$

$$= 2^{n+1} + 3[2^{n-1} - 1]$$

So the value of $S(5)$, for example, is $2^6 + 3(2^4 - 1) = 64 + 3(15) = 109$.

Alternatively, by expand-guess-verify, we expand

$$S(n) = 2S(n - 1) + 3$$

$$= 2[2S(n - 2) + 3] + 3 = 2^2 S(n - 2) + 2 \cdot 3 + 3$$

$$= 2^2[2S(n - 3) + 3] + 2 \cdot 3 + 3 = 2^3 S(n - 3) + 2^2 \cdot 3 + 2 \cdot 3 + 3$$

$$\vdots$$

> **REMINDER:**
> When expanding, be sure to pick up all the pieces of the recurrence relation recipe, like the + 3 in this example.

The general pattern seems to be

$$S(n) = 2^k S(n - k) + 2^{k-1} \cdot 3 + 2^{k-2} \cdot 3 + \cdots + 2^2 \cdot 3 + 2 \cdot 3 + 3$$

which, when $n - k = 1$ or $k = n - 1$, becomes

$$S(n) = 2^{n-1}S(1) + 2^{n-2} \cdot 3 + 2^{n-3} \cdot 3 + \cdots + 2^2 \cdot 3 + 2 \cdot 3 + 3$$

$$= 2^{n-1}(4) + 3[2^{n-2} + 2^{n-3} + \cdots + 2^2 + 2 + 1]$$

$$= 2^{n+1} + 3[2^{n-1} - 1] \qquad \text{(from Example 15)}$$

Finally, we must prove by induction that $S(n) = 2^{n+1} + 3[2^{n-1} - 1]$.

Base case: $n = 1$: $S(1) = 4 = 2^2 + 3[2^0 - 1]$, true

Assume $S(k) = 2^{k+1} + 3[2^{k-1} - 1]$

Show $S(k + 1) = 2^{k+2} + 3[2^k - 1]$

$$S(k + 1) = 2S(k) + 3 \qquad\qquad \text{by the recurrence relation}$$

$$= 2(2^{k+1} + 3[2^{k-1} - 1]) + 3 \qquad \text{by the inductive hypothesis}$$

$$= 2^{k+2} + 3 \cdot 2^k - 6 + 3 \qquad\qquad \text{multiplying out}$$

$$= 2^{k+2} + 3[2^k - 1]$$

Section 2.4 Review

Techniques

- Generate values in a sequence defined recursively.
- Prove properties of the Fibonacci sequence.
- Recognize objects in a recursively defined collection of objects.

- Give recursive definitions for particular sets of objects.
- Give recursive definitions for certain operations on objects.
- Write recursive algorithms to generate sequences defined recursively.
- Solve linear, first-order recurrence relations with constant coefficients by using a solution formula.
- Solve recurrence relations by the expand, guess, and verify technique.

Main Ideas

Recursive definitions can be given for sequences of objects, sets of objects, and operations on objects where basis information is known and new information depends on already known information.

Recursive algorithms provide a natural way to solve certain problems by invoking the same task on a smaller version of the problem.

Certain recurrence relations have closed-form solutions.

Exercises 2.4

For Exercises 1–10, write the first five values in the sequence.

★1. $S(1) = 10$
    $S(n) = S(n - 1) + 10$ for $n \geq 2$

2. $A(1) = 2$
   $$A(n) = \frac{1}{A(n - 1)} \text{ for } n \geq 2$$

3. $B(1) = 1$
   $B(n) = B(n - 1) + n^2$ for $n \geq 2$

★4. $S(1) = 1$
    $$S(n) = S(n - 1) + \frac{1}{n} \text{ for } n \geq 2$$

5. $T(1) = 1$
   $T(n) = nT(n - 1)$ for $n \geq 2$

6. $P(1) = 1$
   $P(n) = n^2P(n - 1) + (n - 1)$ for $n \geq 2$

★7. $M(1) = 2$
    $M(2) = 2$
    $M(n) = 2M(n - 1) + M(n - 2)$ for $n > 2$

8. $D(1) = 3$
   $D(2) = 5$
   $D(n) = (n - 1)D(n - 1) + (n - 2)D(n - 2)$ for $n > 2$

9. $W(1) = 2$
   $W(2) = 3$
   $W(n) = W(n - 1)W(n - 2)$ for $n > 2$

10. $T(1) = 1$
    $T(2) = 2$
    $T(3) = 3$
    $T(n) = T(n - 1) + 2T(n - 2) + 3T(n - 3)$ for $n > 3$

In Exercises 11–13, prove the given property of the Fibonacci numbers directly from the definition.

★11. $F(n + 1) + F(n - 2) = 2F(n)$ for $n \geq 3$

12. $F(n) = 5F(n - 4) + 3F(n - 5)$ for $n \geq 6$

13. $[F(n + 1)]^2 = [F(n)]^2 + F(n - 1)F(n + 2)$ for $n \geq 2$

14. $F(n + 3) = 2F(n + 1) + F(n)$ for $n \geq 1$

15. $F(n + 6) = 4F(n + 3) + F(n)$ for $n \geq 1$

In Exercises 16–19, prove the given property of the Fibonacci numbers for all $n \geq 1$. (*Hint:* The first principle of induction will work.)

★16. $F(1) + F(2) + \cdots + F(n) = F(n + 2) - 1$

17. $F(2) + F(4) + \cdots + F(2n) = F(2n + 1) - 1$

18. $F(1) + F(3) + \cdots + F(2n - 1) = F(2n)$

19. $[F(1)]^2 + [F(2)]^2 + \cdots + [F(n)]^2 = F(n)F(n + 1)$

In Exercises 20–23, prove the given property of the Fibonacci numbers using the second principle of induction.

★20. Exercise 14

21. Exercise 15

22. $F(n) < 2^n$ for $n \geq 1$

23. $F(n) > \left(\dfrac{3}{2}\right)^{n-1}$ for $n \geq 6$

24. The values p and q are defined as follows:

$$p = \frac{1 + \sqrt{5}}{2} \quad \text{and} \quad q = \frac{1 - \sqrt{5}}{2}$$

   a. Prove that $1 + p = p^2$ and $1 + q = q^2$.
   b. Prove that

$$F(n) = \frac{p^n - q^n}{p - q}$$

c. Use part (b) to prove that

$$F(n) = \frac{\sqrt{5}}{5}\left(\frac{1+\sqrt{5}}{2}\right)^n - \frac{\sqrt{5}}{5}\left(\frac{1-\sqrt{5}}{2}\right)^n$$

is a closed-form solution for the Fibonacci sequence.

For Exercises 25–28, decide whether the sequences described are subsequences of the Fibonacci sequence, that is, their members are some or all of the members, in the right order, of the Fibonacci sequence.[1]

25. The sequence $A(n)$, where $A(n) = (n-1)2^{n-2} + 1$, $n \geq 1$. The first four values are 1, 2, 5, 13, which—so far—form a subsequence of the Fibonacci sequence.

★26. The sequence $B(n)$, where $B(n) = 1 +$ (the sum of the first n terms of the Fibonacci sequence), $n \geq 1$. The first four values are 2, 3, 5, 8, which—so far—form a subsequence of the Fibonacci sequence.

27. The sequence $C(n)$, where $C(n)$ is the number of ways in which n coins can be arranged in horizontal rows with all the coins in each row touching and every coin above the bottom row touching two coins in the row below it, $n \geq 1$. The first five values are 1, 1, 2, 3, 5, which—so far—form a subsequence of the Fibonacci sequence.

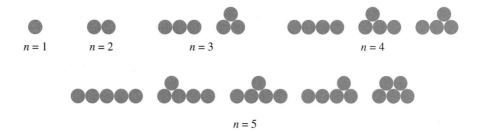

28. The sequence $D(n)$, where $D(n)$ describes the number of ways to paint the floors on an n-story building where each floor is painted yellow or blue and no two adjacent floors can be blue (although adjacent floors can be yellow), $n \geq 1$. The first four values are 2, 3, 5, 8, which—so far—form a subsequence of the Fibonacci sequence. For example, $D(3) = 3$ because a 3-story building can be painted

Y	Y	Y	B	B
Y	Y	B	Y	Y
Y	B	Y	B	Y

(*Hint:* Think about a recursive expression for $D(n + 1)$.)

[1]*Exercises 25–28 are taken from "Mathematical Recreations" by Ian Stewart,* Scientific American, *May 1995.*

29. The original problem posed by Fibonacci concerned pairs of rabbits. Two rabbits do not breed until they are two months old. After that, each pair of rabbits produces a new pair each month. No rabbits ever die. Let $R(n)$ denote the number of rabbit pairs at the end of n months if you start with a single rabbit pair. Show that $R(n)$ is the Fibonacci sequence.

★30. Write a recursive definition for a geometric progression with initial term a and common ratio r (see Exercise 21, Section 2.2).

31. Write a recursive definition for an arithmetic progression with initial term a and common difference d (see Exercise 22, Section 2.2).

★32. In an experiment, a certain colony of bacteria initially has a population of 50,000. A reading is taken every 2 hours, and at the end of every 2-hour interval, there are 3 times as many bacteria as before.

    a. Write a recursive definition for $A(n)$, the number of bacteria present at the beginning of the nth time period.

    b. At the beginning of which interval are there 1,350,000 bacteria present?

33. An amount of $500 is invested in an account paying 10% interest compounded annually.

    a. Write a recursive definition for $P(n)$, the amount in the account at the beginning of the nth year.

    b. After how many years will the account balance exceed $700?

34. A collection T of numbers is defined recursively by

    1. 2 belongs to T.
    2. If X belongs to T, so does $X + 3$ and $2 * X$.

    Which of the following belong to T?

    a. 6   **b.** 7   **c.** 19   **d.** 12

35. A collection M of numbers is defined recursively by

    1. 2 and 3 belong to M.
    2. If X and Y belong to M, so does $X * Y$.

    Which of the following belong to M?

    a. 6   **b.** 9   **c.** 16   **d.** 21   **e.** 26   **f.** 54   **g.** 72   **h.** 218

★36. A collection S of strings of characters is defined recursively by

    1. a and b belong to S.
    2. If X belongs to S, so does Xb.

    Which of the following belong to S?

    a. a **b.** ab **c.** aba **d.** $aaab$ **e.** $bbbbb$

37. A collection W of strings of symbols is defined recursively by
    1. a, b, and c belong to W.
    2. If X belongs to W, so does $a(X)c$.

    Which of the following belong to W?

    a. $a(b)c$ **b.** $a(a(b)c)c$ **c.** $a(abc)c$ **d.** $a(a(a(a)c)c)c$ **e.** $a(aacc)c$

★38. Give a recursive definition for the set of all unary predicate wffs in x.

39. Give a recursive definition for the set of all well-formed formulas of integer arithmetic, involving integers together with the arithmetic operations of $+$, $-$, $*$, and $/$.

40. Give a recursive definition for the set of all strings of well-balanced parentheses.

41. Give a recursive definition for the set of all binary strings containing an odd number of 0s.

★42. Give a recursive definition for x^R, the reverse of the string x.

43. Give a recursive definition for $|x|$, the length of the string x.

★44. Use BNF notation to define the set of positive integers.

45. Use BNF notation to define the set of decimal numbers, which consist of an optional sign ($+$ or $-$), followed by one or more digits, followed by a decimal point, followed by zero or more digits.

★46. Give a recursive definition for the factorial operation $n!$ for $n \geq 1$.

47. Give a recursive definition for the addition of two nonnegative integers m and n.

48. **a.** Write a recursive definition for the operation of taking the maximum of n integers $a_1, \ldots, a_n$, $n \geq 2$.
    b. Write a recursive definition for the operation of taking the minimum of n integers $a_1, \ldots, a_n$, $n \geq 2$.

49. **a.** Give a recursive definition for the conjunction of n statement letters in propositional logic, $n \geq 2$.
    b. Write a generalization of the associative property of conjunction (tautological equivalence 2b of Section 1.1) and use induction to prove it.

★50. Let A and $B_1, B_2, \ldots, B_n$ be statement letters. Prove the finite extension of the distributive equivalences of propositional logic:

$$A \lor (B_1 \land B_2 \land \cdots \land B_n) \Leftrightarrow (A \lor B_1) \land (A \lor B_2) \land \cdots \land (A \lor B_n)$$

and

$$A \land (B_1 \lor B_2 \lor \cdots \lor B_n) \Leftrightarrow (A \land B_1) \lor (A \land B_2) \lor \cdots \lor (A \land B_n)$$

for $n \geq 2$.

51. Let B_1, B_2, B_n be statement letters. Prove the finite extension of De Morgan's laws:

$$(B_1 \lor B_2 \lor \cdots \lor B_n)' \Leftrightarrow B_1' \land B_2' \land \cdots \land B_n'$$

and

$$(B_1 \land B_2 \land \cdots \land B_n)' \Leftrightarrow B_1' \lor B_2' \lor \cdots \lor B_n'$$

for $n \geq 2$.

In Exercises 52–57, write the body of a recursive function to compute $S(n)$ for the given sequence S.

52. 1, 3, 9, 27, 51, ...

53. 2, 1, 1/2, 1/4, 1/8, ...

★54. 1, 2, 4, 7, 11, 16, 22, ...

55. 2, 4, 16, 256, ...

56. $a, b, a + b, a + 2b, 2a + 3b, 3a + 5b, ...$

★57. $p, p - q, p + q, p - 2q, p + 2q, p - 3q, ...$

58. What value is returned by the following recursive function Mystery for an input value of n?

Mystery (integer n)

   if $n = 1$ **then**

      return 1

   else

      return Mystery$(n - 1) + 1$

   end if

end function Mystery

59. The following recursive function is initially invoked with an i-value of 1. L is a list (array) of 10 integers. What does the function do?

g(list L; integer i; integer x)

   if $i > 10$ **then**

      return 0

   else

      if $L(i) = x$ **then**

         $g = 10$

      else

         return $g(L, i + 1, x)$

      end if

   end if

end function g

★60. Informally describe a recursive algorithm to reverse the entries in a list of items.

61. Informally describe a recursive algorithm to compute the sum of the digits of a positive integer.

62. Informally describe a recursive algorithm to compute the greatest common divisor of two positive integers a and b where $a \geq b$. (*Hint:* The solution is based on the Euclidean algorithm, discussed in Section 2.3. In particular, make use of expression (5) on page 116.)

★63. Simulate the execution of algorithm *SelectionSort* on the following list L; write the list after every exchange that changes the list.

4, 10, −6, 2, 5

64. Simulate the execution of algorithm *SelectionSort* on the following list L; write the list after every exchange that changes the list.

9, 0, 2, 6, 4

65. The binary search algorithm is used with the following list; x has the value "Chicago." Name the elements against which x is compared.

Boston, Charlotte, Indianapolis, New Orleans, Philadelphia, San Antonio, Yakima

66. The binary search algorithm is used with the following list; x has the value "flour." Name the elements against which x is compared.

butter, chocolate, eggs, flour, shortening, sugar

67. Do a proof of correctness for the iterative function given in this section to compute $S(n)$ of Example 29, where $S(n) = 2^n$.

In Exercises 68–73, solve the recurrence relation subject to the basis step.

★68. $S(1) = 5$
$S(n) = S(n - 1) + 5$ for $n \geq 2$

69. $F(1) = 2$
$F(n) = 2F(n - 1) + 2^n$ for $n \geq 2$

70. $T(1) = 1$
$T(n) = 2T(n - 1) + 1$ for $n \geq 2$
(*Hint:* See Example 15.)

★71. $A(1) = 1$
$A(n) = A(n - 1) + n$ for $n \geq 2$
(*Hint:* See Practice 7.)

72. $S(1) = 1$
$S(n) = S(n - 1) + 2n - 1$ for $n \geq 2$
(*Hint:* See Example 14.)

73. $P(1) = 2$
$P(n) = 2P(n - 1) + n2^n$ for $n \geq 2$
(*Hint:* See Practice 7.)

For Exercises 74 and 75, solve the recurrence relation subject to the basis step by using the expand, guess, and verify approach.

★74. $F(1) = 1$
$F(n) = nF(n - 1)$ for $n \geq 2$

75. $S(1) = 1$
$S(n) = nS(n - 1) + n!$

76. A colony of bats is counted every two months. The first four counts are 1200, 1800, 2700, and 4050. If this growth rate continues, what will the 12th count be? (*Hint:* Write and solve a recurrence relation.)

★77. At the beginning of this chapter the contractor claimed:

The material to be stored at the chemical disposal site degrades to inert matter at the rate of 5% per year. Therefore only about one-third of the original active material will remain at the end of 20 years.

Write and solve a recurrence relation to check the contractor's claim; note that the end of 20 years is the beginning of the 21st year.

78. In an account that pays 8% annually, $1000 is deposited. At the end of each year, an additional $100 is deposited into the account. What is the account worth at the end of 7 years (that is, at the beginning of the 8th year)? (*Hint:* Write and solve a recurrence relation. Also consult Exercise 21 of Section 2.2 for the formula for the sum of a geometric sequence.)

79. The shellfish population in a bay is estimated to have a count of about 1,000,000. Studies show that pollution reduces this population by about 2% per year, while other hazards are judged to reduce the population by about 10,000 per year. After 9 years, that is, at the beginning of year 10, what is the approximate shellfish population? (*Hint:* Write and solve a recurrence relation. Also consult Exercise 21 of Section 2.2 for the formula for the sum of a geometric sequence.)

★80. Early members of the Pythagorean Society defined *figurate numbers* to be the number of dots in certain geometrical configurations. The first few *triangular numbers* are 1, 3, 6, and 10:

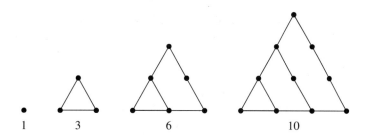

Find a formula for the *n*th triangular number. (*Hint:* See Practice 7.)

81. The first few *pentagonal numbers* (see Exercise 80) are 1, 5, 12, and 22:

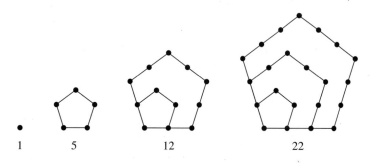

Find a formula for the *n*th pentagonal number. (*Hint:* See Exercise 22 of Section 2.2 for the formula for the sum of an arithmetic sequence.)

82. Use induction to verify that equation (8) of this section is the solution to the recurrence relation (6) subject to the basis condition that $S(1)$ is known.

Section 2.5   Analysis of Algorithms

Often more than one algorithm exists to perform the same task. When comparing algorithms, several criteria can be used to judge which is the "best" algorithm. We might ask, for example, which is easiest to understand or which runs most efficiently. One way to judge the efficiency of an algorithm is to estimate the number of operations that it must perform. We count only the operations that are basic to the task at hand, not "housekeeping" operations that make just a small contribution to the total work required.

For example, suppose the task is to search a sorted list of n words or numbers for a particular item x. Since any possible algorithm seems to require comparing elements from the list with x until a match is found, the basic operation to count is comparisons.

The **sequential search algorithm** simply compares x with each entry in the list in turn until either x is found or the list is exhausted. A pseudocode description of the sequential search algorithm is given in the accompanying box. This algorithm does the maximum amount of work when x is the last item in the list or when x does not appear in the list at all. In either case, all elements are compared to x, so n comparisons are done.

ALGORITHM   SequentialSearch

SequentialSearch(list L; integer n; itemtype x)
//searches a list L of n items for item x

Local variable:
integer i //marks position in the list
   $i = 1$
   while $L(i) \neq x$ and $i < n$ **do**
      $i = i + 1$
   end while
   if $L(i) = x$ **then**
      write("Found")
   else
      write("Not found")
   end if
end function SequentialSearch

Of course, x may be the very first item in the list, in which case only one comparison is made; then again, x may be in the middle of the list, requiring roughly $n/2$ comparisons. Clearly, there are many possibilities, and it would be helpful to obtain some measure of the average amount of work done. Such a measure would require some way to describe the average list being searched and the average relationship of a target item x to that list. But average behavior is generally very difficult to determine, not only for this algorithm but for most algorithms. To compare the efficiency of algorithms, therefore, we will often content ourselves with the worst-case count of the number of operations required.

The study of the efficiency of algorithms, that is, the number of operations they perform, is called **analysis of algorithms**. Various techniques for algorithm analysis have been developed. Sometimes a rather straightforward analysis can be done simply by inspecting the algorithm.

EXAMPLE 45

Following is an algorithm to write out the sum of the m quiz grades for each student on a grade roster after dropping each student's lowest grade. The outer loop goes through each of the n students; the inner loop goes through the quizzes for the current student. The algorithm assigns values to variables and performs comparisons (to find the lowest quiz grade for each student). It also does some adding and subtracting, and we will count the work contributed by these arithmetic operations.

> **for** $i = 1$ to n **do**
>   low = roster(i).quiz(1)
>   sum = roster(i).quiz(1)
>
>   **for** $j = 2$ to m **do**
>   sum = sum + roster(i).quiz(j)   //A
>   **if** roster(i).quiz(j) < low **then**
>   low = roster(i).quiz(j)
>   **end if**
>   **end for**
>   sum = sum − low;   //S
>   write("Total for student ", i, " is ", sum)
> **end for**

Subtraction occurs at the line marked //S, which is executed once for each pass through the outer loop (once for each student), a total of n times. Addition, however, occurs at the line marked //A. This is done within the inner loop, which executes $m - 1$ times for each student, that is, for each of the n passes of the outer loop. The total number of additions is therefore $n(m - 1)$. The total number of arithmetic operations is $n + n(m - 1)$. The same number of arithmetic operations is always done; there is no best, worst, or average case.

In this section we will analyze algorithms that are defined recursively. Because much of the activity of a recursive algorithm takes place "out of sight" in the many invocations that can occur, an analysis using the direct counting technique of Example 45 won't work. Analysis of recursive algorithms often involves solving a recurrence relation.

Analysis Using Recurrence Relations (Binary Search)

As an example, consider another algorithm (besides sequential search) for searching a sorted list, the binary search algorithm of Section 2.4. How many comparisons are required by the binary search algorithm in the worst case? One comparison is done at the middle value, and then the process is repeated on half the list. If the original list is n elements long, then half the list is at worst $n/2$ elements long. (In Example 41, for instance, when 10 is the middle value, the right "half" of the list has 4 elements, but the left "half" has only 3.) If we are to keep cutting the list in half, it is convenient to consider only the case where we get an integer value each time we cut in half, so we will assume that $n = 2^m$ for some $m \geq 0$. If $C(n)$ stands for the maximum number of comparisons required for an n-element list, then we see that

$$C(n) = 1 + C\left(\frac{n}{2}\right)$$

This reflects one comparison at the middle value plus however many comparisons are needed for half the list. We don't actually know what the number of comparisons is for half the list, just as we don't know what it is for the total list, but we have the notation to express this value symbolically. We've now written a recurrence relation. The basis step is

$$C(1) = 1$$

because only one comparison is required to search a one-element list.

Because this is not a first-order recurrence relation, to solve it we will start from scratch with the expand, guess, and verify approach. Also, the solution will involve the logarithm function; for a review of the logarithm function and its properties, see Appendix B.

EXAMPLE 46   Solve the recurrence relation

$$C(n) = 1 + C\left(\frac{n}{2}\right) \text{ for } n \geq 2, n = 2^m$$

subject to the basis step

$$C(1) = 1$$

Expanding, we get

$$C(n) = 1 + C\left(\frac{n}{2}\right)$$
$$= 1 + \left(1 + C\left(\frac{n}{4}\right)\right)$$
$$= 1 + 1 + \left(1 + C\left(\frac{n}{8}\right)\right)$$

and the general term seems to be

$$C(n) = k + C\left(\frac{n}{2^k}\right)$$

The process stops when $2^k = n$, or $k = \log_2 n$. (We'll omit the base 2 notation from now on—$\log n$ will mean $\log_2 n$.) Then

$$C(n) = \log n + C(1) = 1 + \log n$$

Now we will use induction to show that $C(n) = 1 + \log n$ for all $n \geq 1$, $n = 2^m$. This is a somewhat different form of induction, because the only values of interest are powers of 2. We still take 1 as the basis step for the induction, but then we prove that if our statement is true for a value k, it is true for $2k$. The statement will then be true for 1, 2, 4, 8, ... , that is, for all nonnegative integer powers of 2, which is just what we want.

$C(1) = 1 + \log 1 = 1 + 0 = 1$, true

Assume that $C(k) = 1 + \log k$. Then

$$\begin{aligned} C(2k) &= 1 + C(k) && \text{(by the recurrence relation)} \\ &= 1 + 1 + \log k && \text{(by the inductive hypothesis)} \\ &= 1 + \log 2 + \log k && (\log 2 = 1) \\ &= 1 + \log 2k && \text{(property of logarithms)} \end{aligned}$$

This completes the inductive proof. ●

By Example 46, the maximum number of comparisons required to do a binary search on an n-element ordered list, with $n = 2^m$, is $1 + \log n$. In Example 41, n was 8, and four comparisons $(1 + \log 8)$ were required in the worst case (x not in the list). A sequential search would require eight comparisons. Because

$$1 + \log n < n \text{ for } n = 2^m, n \geq 4$$

binary search is almost always more efficient than sequential search. However, the sequential search algorithm does have one big advantage—if the list being searched is *unsorted*, the sequential search algorithm works, but the binary search algorithm does not. If we first sort the list and then use the binary search algorithm, we must then consider the number of operations involved in sorting the list. Exercises 8–15 at the end of this section ask you to count the operations required to sort a list by several different algorithms.

The binary search algorithm is recursive, which means that the algorithm invokes itself with smaller input values. In this case, the smaller version is significantly smaller—the input is about half the size of that for the original problem. This is clear from the form of the recurrence relation, where $C(n)$ depends not on $C(n - 1)$ but on $C(n/2)$. Algorithms with recurrence relations of this form, where the problem is decomposed into significantly smaller subproblems, are sometimes called **divide-and-conquer** algorithms.

The binary search recurrence relation is a special case of the general form

$$S(n) = cS\left(\frac{n}{2}\right) + g(n) \text{ for } n \geq 2, n = 2^m \tag{1}$$

where c is a constant and g can be an expression involving n. We'd like to find a closed-form solution for (1) subject to the basis that $S(1)$ is known. Then we could solve any recurrence relation that matches (1) by just substituting into the solution formula, just as we did in the last section for first-order recurrence relations. Notice that (1) is not a first-order recurrence relation because the value at n does not depend on the value at $n - 1$. We could use the expand, guess, and verify approach to find the solution, but instead we will do some transformations on (1) to convert it to a first-order recurrence relation with constant coefficients and then use what we already know.

Equation (1) assumes that $n = 2^m$ with $n \geq 2$. From this it follows that $m = \log n$ and $m \geq 1$. Substituting 2^m for n in equation (1) results in

$$S(2^m) = cS(2^{m-1}) + g(2^m) \tag{2}$$

Now, letting $T(m)$ represent $S(2^m)$ in equation (2), we get

$$T(m) = cT(m - 1) + g(2^m) \text{ for } m \geq 1 \tag{3}$$

Equation (3) is a linear, first-order equation with constant coefficients; from equation (8) of Section 2.4, we obtain the solution

$$T(m) = c^{m-1}T(1) + \sum_{i=2}^{m} c^{m-i}g(2^i) \tag{4}$$

subject to the basis condition that $T(1)$ is known. Because equation (3) holds for $m = 1$, we know that

$$T(1) = cT(0) + g(2)$$

Making this substitution in (4) results in

$$T(m) = c^m T(0) + \sum_{i=1}^{m} c^{m-i}g(2^i) \tag{5}$$

Now reversing the substitution $T(m) = S(2^m)$, (5) becomes

REMINDER:

In the summation part of the general solution formula, c is raised to the $(\log n) - i$ power, not $(\log n) - 1$.

$$S(2^m) = c^m S(2^0) + \sum_{i=1}^{m} c^{m-i}g(2^i)$$

Finally, letting $2^m = n$, or $m = \log n$, we get

$$S(n) = c^{\log n}S(1) + \sum_{i=1}^{\log n} c^{(\log n)-i}g(2^i) \tag{6}$$

Equation (6) is thus the solution for the recurrence relation (1). As before, to use this general solution you need only match your recurrence relation to (1) to determine c and $g(n)$, then substitute into equation (6). If you can evaluate the summation, the result will be a closed-form solution.

EXAMPLE 47   The recurrence relation for the binary search algorithm

$$C(1) = 1$$

$$C(n) = 1 + C\left(\frac{n}{2}\right) \text{ for } n \geq 2, n = 2^m$$

matches equation (1) above, with $c = 1$ and $g(n) = 1$. The solution, according to formula (6), is

$$C(n) = 1^{\log n} C(1) + \sum_{i=1}^{\log n} 1^{(\log n)-i}(1)$$

$$= 1 + (\log n)(1) = 1 + \log n$$

which agrees with our previous result.

PRACTICE 23   Show that the solution to the recurrence relation

$$S(1) = 1$$

$$S(n) = 2S\left(\frac{n}{2}\right) + 1 \text{ for } n \geq 2, n = 2^m$$

is $2n - 1$. (*Hint:* See Example 15 and note that $2^{\log n} = n$.)

Upper Bound (Euclidean Algorithm)

The Euclidean algorithm, as presented in Section 2.3, uses a while loop to do successive divisions in order to find gcd(a, b) for nonnegative integers a and b, $a \geq b$. To analyze the Euclidean algorithm, we must first decide on the operation we are counting. Because the Euclidean algorithm does repeated divisions, we'll take the division operation as our unit of work. Given a and b, assume that $b \leq a = n$ so that n is a measure of the size of the input values. We want to find $E(n)$, where this denotes the amount of work (the number of divisions) required to find gcd(a, b) in the worst case.

A recursive version of the Euclidean algorithm can also be written (see Exercise 62, Section 2.4); the key to the recursive version is to recognize that gcd(a, b) involves finding gcd(b, r), where r is the remainder upon dividing a by b. We've just seen a case where the operations of a recursive algorithm (binary search) could be expressed neatly as a recurrence relation where the input size gets halved after each operation. A recurrence relation would express $E(n)$ in terms of E at smaller values. But what are these smaller values? To find gcd(a, b) we find gcd(b, r), so it is clear that the input size is getting smaller, but in what way? Consider Example 27, where to find gcd(420, 66) the following divisions were performed:

$$
\begin{array}{cccc}
6 & 2 & 1 & 3 \\
66\overline{)420} & 24\overline{)66} & 18\overline{)24} & 6\overline{)18} \\
\underline{396} & \underline{48} & \underline{18} & \underline{18} \\
24 & 18 & 6 & 0
\end{array}
$$

Here the successive values being divided are 420, 66, 24, 18. The change from 420 to 66 is much larger than cutting in half, while the change from 24 to 18 is less.

In fact, we won't find a recurrence relation or an exact expression for $E(n)$. But we will at least find an *upper bound* for $E(n)$. An **upper bound** is a ceiling on the amount of work an algorithm does; the algorithm can require *no more steps* than the upper bound, but it may not require that many.

To find this upper bound, we will show that if $i > j$ and i is divided by j with remainder r, then $r < i/2$. There are two cases:

1. If $j \leq i/2$, then $r < i/2$ since $r < j$.
2. If $j > i/2$, then $i = 1 * j + (i - j)$; in other words, the quotient is 1 and r is $i - j$, which is $< i/2$.

In the Euclidean algorithm, the remainder r at any step becomes the dividend (the number being divided) two steps later. So the successive dividends are at least halved every two divisions. The value n can be halved $\log n$ times; therefore at most $2 \log n$ divisions are done. Thus

$$E(n) \leq 2 \log n \tag{7}$$

The value of $2 \log n$ for $n = 420$ is almost 18, whereas it took only 4 divisions to find gcd(420, 66). Evidently this upper bound estimate is rather loose, like saying that every student in this class is under 12 feet in height. An improved (that is, lower) upper bound is derived in Exercises 16–18 at the end of this section.

Section 2.5 Review

Technique

- Solve recurrence relations arising from divide-and-conquer algorithms by using a solution formula.

Main Ideas

Analysis of an algorithm estimates the number of basic operations that the algorithm performs.

Analysis of recursive algorithms often leads to recurrence relations.

Lacking an exact expression for the number of operations an algorithm performs, it may be possible to find an upper bound.

Exercises 2.5

1. For the algorithm of Example 45, count the total number of assignments and comparisons done in the best case (least work) and the worst case (most work); describe each of these cases.

★2. Describe a recursive version of the sequential search algorithm to search a list L of n items for item x.

3. Using the recursive version from Exercise 2, write a recurrence relation for the number of comparisons of x against list elements done by the sequential search algorithm in the worst case, and solve this recurrence relation. (As before, the answer should be n.)

In Exercises 4–7, solve the recurrence relation subject to the basis step. (*Hint:* See Example 15, and note that $2^{\log n} = n$.)

4. $T(1) = 3$

$$T(n) = T\left(\frac{n}{2}\right) + n \text{ for } n \geq 2, n = 2^m$$

★5. $P(1) = 1$

$$P(n) = 2P\left(\frac{n}{2}\right) + 3$$

6. $S(1) = 1$

$$S(n) = 2S\left(\frac{n}{2}\right) + n$$

7. $P(1) = 1$

$$P(n) = 2P\left(\frac{n}{2}\right) + n^2$$

Exercises 8–10 refer to algorithm *SelectionSort* of Section 2.4.

★8. In one part of algorithm *SelectionSort*, the index of the maximum item in a list must be found. This requires comparisons between list elements. In an n-element (unsorted) list, how many such comparisons are needed in the worst case to find the maximum element? How many such comparisons are needed in the average case?

9. Defining the basic operation as the comparison of list elements and ignoring the amount of work required to exchange list elements, write a recurrence relation for the amount of work done by selection sort on an n-element list. (*Hint:* Use the result from Exercise 8.)

10. Solve the recurrence relation of Exercise 9.

Exercises 11–15 relate to a sorting algorithm called *MergeSort*, which is described as follows: A one-element list is already sorted; no further work is required. Otherwise, split the list in half, sort each half, and then merge the two halves back into one sorted list.

★11. The merge part of algorithm *MergeSort* requires comparing elements from each of two sorted lists to see which goes next into the combined, sorted list. When one list runs out of elements, the remaining elements from the other list can be added without further comparisons. Given the following pairs of lists, perform a merge and count the number of comparisons to merge the two lists into one.

    a. 6, 8, 9 and 1, 4, 5   **b.** 1, 5, 8 and 2, 3, 4

    c. 0, 2, 3, 4, 7, 10 and 1, 8, 9

12. Under what circumstances will the maximum number of comparisons take place while merging two sorted lists? If the lengths of the lists are r and s, what is the maximum number of comparisons?

★13. Write a recurrence relation for the number of comparisons between list elements done by algorithm *MergeSort* in the worst case. Assume that $n = 2^m$.

14. Solve the recurrence relation of Exercise 13.

15. Compare the worst-case behavior of algorithms *SelectionSort* and *MergeSort* for $n = 4, 8, 16$, and 32 (use a calculator).

Exercises 16–19 concern a better upper bound for the number of divisions required by the Euclidean algorithm in finding gcd(a, b). Assume that a and b are nonnegative integers with $a > b$. (If $a = b$, then only 1 division is required, so this is a minimal amount of work; we want an upper bound on the maximal amount of work.)

16. Suppose that m divisions are required to find gcd(a, b). Prove by induction that for $m \geq 1$, $a \geq F(m + 2)$ and $b \geq F(m + 1)$, where $F(n)$ is the Fibonacci sequence. (*Hint:* To find gcd(a, b), after the first division the algorithm computes gcd(b, r).)

★17. Suppose that m divisions are required to find gcd(a, b), with $m \geq 4$, and that $a = n$. Prove that

$$\left(\frac{3}{2}\right)^{m+1} < F(m + 2) \leq n$$

(*Hint:* Use the result of Exercise 16 here and Exercise 23 of Section 2.4.)

18. Suppose that m divisions are required to find $\gcd(a, b)$, with $m \geq 4$, and that $a = n$. Prove that

$$m < (\log_{1.5} n) - 1$$

(*Hint:* Use the result of Exercise 17.)

19. **a.** Compute $\gcd(89, 55)$ and count the number of divisions required.
    b. Compute the upper bound on the number of divisions required for $\gcd(89, 55)$ using equation (7).
    c. Compute the upper bound on the number of divisions required for $\gcd(89, 55)$ using the result of Exercise 18.

Chapter 2 Review

Terminology

analysis of algorithms (p. 147)

Backus-Naur form (BNF) (p. 125)

basis step (p. 96)

binary search algorithm (p. 130)

binary string (p. 124)

closed-form solution (p. 132)

concatenation (p. 124)

constant coefficient recurrence relation (p. 133)

contrapositive (p. 88)

converse (p. 89)

counterexample (p. 84)

deductive reasoning (p. 84)

direct proof (p. 87)

divide-and-conquer algorithm (p. 149)

empty string (p. 124)

Euclidean algorithm (p. 115)

Fibonacci sequence (p. 121)

first principle of mathematical induction (p. 96)

first-order recurrence relation (p. 133)

greatest common divisor (p. 115)

homogeneous recurrence relation (p. 133)

index of summation (p. 134)

inductive assumption (p. 96)

inductive definition (p. 120)

inductive hypothesis (p. 96)

inductive reasoning (p. 84)

inductive step (p. 96)

linear recurrence relation (p. 133)

loop invariant (p. 112)

loop rule of inference (p. 112)

n factorial (p. 84)

palindrome (p. 124)

partial correctness (p. 114)

proof by cases (p. 89)

proof by contradiction (p. 90)

proof by contraposition (p. 88)

proof by exhaustion (p. 85)

rational number (p. 90)

recurrence relation (p. 121)

recursive definition (p. 120)

second principle of mathematical induction (p. 102)

selection sort algorithm (p. 129)

sequence (p. 121)

sequential search algorithm (p. 146)

solving a recurrence relation (p. 132)

summation notation (p. 134)

upper bound (p. 152)

well-ordering principle (p. 102)

Self-Test   Answer the following true–false questions without looking back in the chapter.

Section 2.1

1. A conjecture can never be proved merely by proving a finite number of cases.

2. A proof by contradiction of $P \rightarrow Q$ begins by assuming both P and Q'.

3. In the statement of the theorem "twice an odd integer is even," an existential quantifier is understood.

4. To prove the conjecture "If Laramie is the capital, then Wyoming is the state," it is sufficient to prove "If Wyoming is the state, then Laramie is the capital."

5. To prove "A if and only if B" requires a proof of $A \rightarrow B$ and a proof of $B \rightarrow A$.

Section 2.2

6. Induction is an appropriate proof technique for proving a statement about all the positive integers.

7. The basis step of an inductive proof requires proving a property true for $n = 1$.

8. If the truth of $P(k + 1)$ depends on the truth of other previous values besides $P(k)$, then the second principle of induction should be used.

9. The key to a proof by the first principle of induction is to see how the truth of P at the value $k + 1$ depends on the truth of P at value k.

10. The inductive hypothesis in an inductive proof of the statement

$$1^3 + 2^2 + \cdots + n^3 = \frac{n^2(n + 1)^2}{4}$$

is

$$k^3 = \frac{k^2(k + 1)^2}{4}$$

Section 2.3

11. A loop invariant remains true until the loop is exited, at which point it becomes false.

12. Partial correctness of a loop statement in a program means that the loop behaves correctly for some input values but not for others.

13. The second principle of induction is used to prove loop invariants because the loop can be executed an arbitrary number of times.

14. If a loop statement has the form

    while (condition B)
    P
    end while

    then the loop invariant Q will be B'.

15. When computing the gcd(42, 30) by the Euclidean algorithm, the computation of dividing 30 by 12 is carried out.

Section 2.4

16. A sequence defined by

    $$S(1) = 7$$
    $$S(n) = 3S(n - 1) + 2 \text{ for n} \geq 2$$

    contains the number 215.

17. Recursive algorithms are valuable primarily because they run more efficiently than iterative algorithms.

18. In applying the binary search algorithm to the list

    2, 5, 7, 10, 14, 20

    where $x = 8$ is the target item, x is never compared to 5.

19. A closed-form solution to a recurrence relation is obtained by applying mathematical induction to the recurrence relation.

20. $S(n) = 2S(n - 1) + 3S(n - 2) + 5n$ is a linear, first-order recurrence relation with constant coefficients.

Section 2.5

21. Analysis of an algorithm generally finds the amount of work done in the worst case because it is too difficult to analyze an average case.

22. Divide-and-conquer algorithms lead to recurrence relations that are not first-order.

23. Binary search is more efficient than sequential search on a sorted list of more than three elements.

24. If sequential search were rewritten as a recursive algorithm, it would be a divide-and-conquer algorithm.

25. An upper bound for the Euclidean algorithm gives a ceiling on the number of divisions required to find gcd(a, b).

On the Computer

For Exercises 1–10, write a computer program that produces the desired output from the given input.

1. *Input:* Number n of terms in a geometric progression (see Exercise 21, Section 2.2), the initial term a, and the common ratio r
   Output: Sum of the first n terms using

   a. iteration
   b. formula of Exercise 21, Section 2.2

2. *Input:* Number n of terms in an arithmetic progression (see Exercise 22, Section 2.2), the initial term a, and the common difference d
   Output: Sum of the first n terms using

   a. iteration
   b. formula of Exercise 22, Section 2.2

3. *Input:* Number n
   Output: Sum of the first n cubes using

   a. iteration, using only multiplication and addition; output the number of multiplications and additions used
   b. formula of Exercise 8, Section 2.2, using only multiplication, addition, and division; output the number of multiplications, additions, and divisions used

4. *Input:* None
   Output: Table showing every integer n, $8 \le n \le 100$, as the sum of 3s and 5s (see Example 24)

5. *Input:* Binary string
   Output: Message indicating whether the input string is a palindrome (see Practice 17)
   Algorithm: Use recursion.

6. *Input:* String of characters x and a positive integer n
   Output: Concatenation of n copies of x
   Algorithm: Use recursion.
   (Some programming languages provide built-in string manipulation capabilities, such as concatenation.)

7. *Input:* Positive integer n
   Output: nth value in the Fibonacci sequence using

   a. iteration
   b. recursion

   Now insert a counter in each version to indicate the total number of addition operations done. Run each version for various values of n and, on a single graph, plot the number of additions as a function of n for each version.

8. *Input:* Two positive integers m and n
   Output: gcd(m, n) using

   a. the iterative version of the Euclidean algorithm
   b. a recursive version of the Euclidean algorithm

9. *Input:* Unsorted list of 10 integers
   Output: Input list sorted in increasing order
   Algorithm: Use the recursive selection sort of Example 39.

10. *Input:* Sorted list of 10 integers and an integer x
    Output: Message indicating whether x is in the list
    Algorithm: Use the binary search algorithm of Example 40.

11. The formula $3^n < n!$ is true for all $n \geq N$. Write a program to determine N and then prove the result by induction.

12. The formula $2^n > n^3$ is true for all $n \geq N$. Write a program to determine N and then prove the result by induction.

13. The value $(1 + \sqrt{5})/2$, known as the *golden ratio*, is related to the Fibonacci sequence by

$$\lim_{n \to \infty} \frac{F(n + 1)}{F(n)} = \frac{1 + \sqrt{5}}{2}$$

Verify this limit by computing $F(n + 1)/F(n)$ for $n = 10, 15, 25, 50,$ and 100 and comparing the result with the golden ratio.

14. Compare the work done by sequential search and binary search on an ordered list of n entries by computing n and $1 + \log n$ for values of n from 1 to 100. Present the results in graphical form.

Sets and Combinatorics

Chapter Objectives

After studying this chapter, you will be able to:

- Use the notation of set theory.
- Find the power set of a finite set.
- Find the union, intersection, difference, complement, and Cartesian product of sets.
- Identify binary and unary operations on a set.
- Prove set identities.
- Recognize that not all sets are countable.
- Apply the Multiplication Principle and the Addition Principle to solve counting problems.
- Use decision trees to solve counting problems.
- Use the Principle of Inclusion and Exclusion to find the number of elements in the union of sets.
- Use the Pigeonhole Principle to decide when certain common events must occur.
- Use the formulas for permutations and combinations of r objects, with and without repetition, from a set of n distinct objects.
- Find the number of distinct permutations of n objects that are not all distinct.
- Use the binomial theorem to expand $(a + b)^n$.

You survey the 87 subscribers to your newsletter in preparation for the release of your new software product. The results of your survey reveal that 68 have a Windows-based system available to them, 34 have a Unix system available, and 30 have access to a Mac. In addition, 19 have access to both Windows and Unix systems, 11 have access to both Unix systems and Macs, and 23 can use both Macs and Windows.

Question: how many of your subscribers have access to all three types of systems?

This is an example of a counting problem; you want to count the number of elements in a certain collection or set—the set of all subscribers with access to all three systems. A formula that easily solves this counting problem will be developed in Section 3.3.

Set theory is one of the cornerstones of mathematics. Many concepts in mathematics and computer science can be conveniently expressed in the language of sets. Operations can be performed on sets to generate new sets. Although most sets of interest to computer scientists are finite or countable, there are sets with so many members that they cannot be enumerated. Set theory is discussed in Section 3.1.

It is often of interest to count the number of elements in a finite set. This may not be a trivial task. Section 3.2 provides some ground rules for counting the number of elements in a set consisting of the outcomes of an event. Counting the elements in such a set can be made manageable by breaking the event down into a sequence of subevents or into disjoint subevents that have no outcomes in common. Some specialized counting principles appear in Section 3.3, and Section 3.4 provides formulas for counting the number of ways to arrange objects in a set and to select objects from a set.

Section 3.5 discusses the binomial theorem, an algebraic result that can also be viewed as a consequence of the counting formulas.

Section 3.1   Sets

Definitions are important in any science because they contribute to precise communication. However, if we look up a word in the dictionary, the definition is expressed using other words, which are defined using still other words, and so on. Thus, we have to have a starting point for definitions where the meaning is taken to be understood; our starting point in this discussion will be the idea of a set, a term that we shall not formally define. Instead, we shall simply use the intuitive idea that a set is a collection of objects. Usually all of the objects in a set share some common property (aside from that of belonging to the same set!); any object that has the property is a member of the set, and any object that does not have the property is not a member. (This is consistent with our use of the word *set* in Section 2.4, where we talked about the set of propositional well-formed formulas, the set of all strings of symbols from a finite alphabet, and the set of identifiers in some programming language.)

Notation

We use capital letters to denote sets and the symbol $\in$ to denote membership in a set. Thus $a \in A$ means that object a is a member, or element, of set A, and $b \notin A$ means that object b is not an element of set A. Braces are used to indicate a set.

EXAMPLE 1   If $A = \{\text{violet, chartreuse, burnt umber}\}$, then chartreuse $\in A$ and magenta $\notin A$. ●

No ordering is imposed on the elements in a set, therefore {violet, chartreuse, burnt umber} is the same as {chartreuse, burnt umber, violet}. Also, each element of a set is listed only once; it is redundant to list it again.

Two sets are **equal** if they contain the same elements. (In a definition, "if" really means "if and only if"; thus two sets are equal if and only if they contain the same elements.) Using predicate logic notation,

$A = B$ means $(\forall x)[(x \in A \rightarrow x \in B) \wedge (x \in B \rightarrow x \in A)]$

In describing a particular set, we have to identify its elements. For a finite set (one with n elements for some nonnegative integer n), we might do this by simply listing all the elements, as in set A of Example 1. Although it is impossible to list all elements of an infinite set, for some infinite sets we can indicate a pattern for listing elements indefinitely. Thus, we might write {2, 4, 6, ... } to express the set S of all positive even integers. (Although this is a common practice, the danger exists that the reader will not see the pattern that the writer has in mind.) S can also be defined recursively by giving an explicit member of S and then describing other members of S in terms of already known members. For example:

1. $2 \in S$
2. If $n \in S$, then $(n + 2) \in S$

But the clearest way to describe this particular set S is to describe the characterizing property of the set elements in words and write

$S = \{x \mid x \text{ is a positive even integer}\}$

read as "the set of all x such that x is a positive even integer."

The various ways we might try to describe a set are thus:

- List (or partially list) its elements.
- Use recursion to describe how to generate the set elements.
- Describe a property P that characterizes the set elements.

We shall see later in this section that there are sets for which the first approach won't work; often the second approach is difficult to use. The third method is usually the best choice.

The notation for a set whose elements are characterized as having property P is $\{x \mid P(x)\}$. Property P here is a *unary predicate*; this term was introduced in Chapter 1. In fact, the formal logic notation of Chapter 1 again comes to the rescue to clarify what we mean by a characterizing property of a set's elements:

$S = \{x \mid P(x)\}$ means $(\forall x)[(x \in S \rightarrow P(x)) \wedge (P(x) \rightarrow x \in S)]$

In words, every element of S has property P and everything that has property P is an element of S.

PRACTICE 1   Describe each of the following sets by listing its elements.

a. $\{x \mid x \text{ is an integer and } 3 < x \leq 7\}$
b. $\{x \mid x \text{ is a month with exactly 30 days)}$
c. $\{x \mid x \text{ is the capital of the United States}\}$

PRACTICE 2   Describe each of the following sets by giving a characterizing property.

  a. $\{1, 4, 9, 16\}$
  b. {the butcher, the baker, the candlestick maker}
  c. $\{2, 3, 5, 7, 11, 13, 17, ...\}$ ●

It is convenient to name certain standard sets so that we can refer to them easily. We shall use

$\mathbb{N}$ = set of all nonnegative integers (note that $0 \in \mathbb{N}$)
$\mathbb{Z}$ = set of all integers
$\mathbb{Q}$ = set of all rational numbers
$\mathbb{R}$ = set of all real numbers
$\mathbb{C}$ = set of all complex numbers

Sometimes we will also want to talk about the set with no elements (the **empty set**, or **null set**), denoted by $\varnothing$ or $\{\ \}$. For example, if $S = \{x \mid x \in \mathbb{N} \text{ and } x < 0\}$, then $S = \varnothing$.

Now suppose that a set A is described as

$$A = \{x \mid (\exists y)(y \in \{0, 1, 2\} \text{ and } x = y^3)\}$$

Because y is not a free variable here, this is still of the form $A = \{x \mid P(x)\}$. The members of A can be found by letting y assume each of the values 0, 1, and 2 and then taking the cube of each such value. Therefore $A = \{0, 1, 8\}$. If we follow the same process with the set B,

$$B = \{x \mid x \in \mathbb{N} \text{ and } (\exists y)(y \in \mathbb{N} \text{ and } x \leq y))$$

then choosing $y = 0$ gives us $x = 0$; choosing $y = 1$ gives $x = 0$ or 1; choosing $y = 2$ gives $x = 0$, 1, or 2; and so on. In other words, B consists of all nonnegative integers that are less than or equal to some nonnegative integer, which means that $B = \mathbb{N}$. But for the set C,

$$C = \{x \mid x \in \mathbb{N} \text{ and } (\forall y)(y \in \mathbb{N} \rightarrow x \leq y)\}$$

we get $C = \{0\}$ because 0 is the only nonnegative integer that is less than or equal to every nonnegative integer.

PRACTICE 3   Describe each set.

  a. $A = \{x \mid x \in \mathbb{N} \text{ and } (\forall y)(y \in \{2, 3, 4, 5\} \rightarrow x \geq y)\}$
  b. $B = \{x \mid (\exists y)(\exists z)(y \in \{1, 2\} \text{ and } z \in \{2, 3\} \text{ and } x = y + z)\}$ ●

Relationships between Sets

For $A = \{2, 3, 5, 12\}$ and $B = \{2, 3, 4, 5, 9, 12\}$, every member of A is also a member of B. When this happens, A is said to be a *subset* of B.

PRACTICE 4   Complete the definition: A is a **subset** of B if

  $(\forall x)(x \in A \rightarrow \underline{\quad\quad}).$ ●

If A is a subset of B, we denote this by $A \subseteq B$. If $A \subseteq B$ but $A \neq B$ (there is at least one element of B that is not an element of A), then A is a **proper subset** of B, denoted by $A \subset B$.

PRACTICE 5   Use formal logic notation to define $A \subset B$.   ●

EXAMPLE 2   Let

$A = \{1, 7, 9, 15\}$
$B = \{7, 9\}$
$C = \{7, 9, 15, 20\}$

Then the following statements (among others) are all true:

$B \subseteq C$	$15 \in C$
$B \subseteq A$	$\{7, 9\} \subseteq B$
$B \subset A$	$\{7\} \subset A$
$A \nsubseteq C$	$\varnothing \subseteq C$

The last statement ($\varnothing \subseteq C$) is true because the statement $(\forall x)(x \in \varnothing \rightarrow x \in C)$ is true since $x \in \varnothing$ is always false.   ●

PRACTICE 6   Let

$A = \{x \mid x \in \mathbb{N} \text{ and } x \geq 5\}$
$B = \{10, 12, 16, 20\}$
$C = \{x \mid (\exists y)(y \in \mathbb{N} \text{ and } x = 2y)\}$

Which of the following statements are true?

a. $B \subseteq C$
b. $B \subset A$
c. $A \subseteq C$
d. $26 \in C$
e. $\{11, 12, 13\} \subseteq A$
f. $\{11, 12, 13\} \subset C$
g. $\{12\} \in B$
h. $\{12\} \subseteq B$
i. $\{x \mid x \in \mathbb{N} \text{ and } x < 20\} \nsubseteq B$
j. $5 \subseteq A$
k. $\{\varnothing\} \subseteq B$
l. $\varnothing \notin A$ ●

Suppose that $B = \{x \mid P(x)\}$ and that $A \subseteq B$. Because every element of A is also an element of B, and P is a property characterizing all elements of B, then every element in A also has property $P(x)$. The elements of A "inherit" property P. In fact, to prove that $A \subseteq B$, we pick an arbitrary $x \in A$ and show that $P(x)$ holds. If A is a proper subset of B, A's elements will usually have some additional characterizing property not shared by all elements of B. (This is the same notion of "inheritance" that prevails when a child type or subtype or derived type is defined in an object-oriented programming language. The child type inherits all of the properties and operations from the parent type with the addition of specialized local properties or operations as needed.)

EXAMPLE 3   Let

$$B = \{x \mid x \text{ is a multiple of } 4\}$$

and let

$$A = \{x \mid x \text{ is a multiple of } 8\}$$

Then we have $A \subseteq B$. To prove this, let $x \in A$; note that x is a completely arbitrary member of A. We must show that x satisfies the characterizing property of B; this means we must show that x is a multiple of 4. Because we have $x \in A$, x satisfies the characterizing property of A; that is, x is a multiple of 8 and thus we can write $x = m \cdot 8$ for some integer m. This equation can be written as $x = m \cdot 2 \cdot 4$ or $x = k \cdot 4$, where $k = 2m$, so k is an integer. This shows that x is a multiple of 4, and therefore $x \in B$.

There are numbers (like 12) that are multiples of 4 but not multiples of 8; so $A \subset B$. Another way to describe A is

$$A = \{x \mid x = k \cdot 4 \text{ and } k \text{ is an even number}\}$$

In this form it is clear that A's elements have inherited the characterizing property of B—being a multiple of 4—but that there is an additional restriction that makes A less general than B.

PRACTICE 7   Let

$$A = \{x \mid x \in \mathbb{R} \text{ and } x^2 - 4x + 3 = 0\}$$
$$B = \{x \mid x \in \mathbb{N} \text{ and } 1 \le x \le 4\}$$

Prove that $A \subset B$.

We know that A and B are equal sets if they have the same elements. We can restate this equality in terms of subsets: $A = B$ if and only if $A \subseteq B$ and $B \subseteq A$. Proving set inclusion in both directions is the usual way to establish the equality of two sets.

EXAMPLE 4   We will prove that $\{x \mid x \in \mathbb{N} \text{ and } x^2 < 15\} = \{x \mid x \in \mathbb{N} \text{ and } 2x < 7\}$.

Let $A = \{x \mid x \in \mathbb{N} \text{ and } x^2 < 15\}$ and $B = \{x \mid x \in \mathbb{N} \text{ and } 2x < 7\}$. To show that $A = B$, we show $A \subseteq B$ and $B \subseteq A$. For $A \subseteq B$, we must choose an arbitrary member of A, that is, anything satisfying the characterizing property of A, and show that it also satisfies the characterizing property of B. Let $x \in A$. Then x is a nonnegative integer satisfying the inequality $x^2 < 15$. The nonnegative integers with squares less than 15 are 0, 1, 2, and 3, so these are the members of A. The double of each of these non-negative integers is a number less than 7. Hence, each member of A is a member of B, and $A \subseteq B$.

Now we show $B \subseteq A$. Any member of B is a nonnegative integer whose double is less than 7. These numbers are 0, 1, 2, and 3, each of which has a square less than 15, so $B \subseteq A$.

Sets of Sets

For a set S, we can form a new set whose elements are all of the subsets of S. This new set is called the **power set** of S, $\wp(S)$.

EXAMPLE 5   For $S = \{0, 1\}$, $\wp(S) = \{\varnothing, \{0\}, \{1\}, \{0, 1\}\}$. Note that the members of the power set of a set are themselves sets.   ●

For any set S, $\wp(S)$ will always have at least $\varnothing$ and S itself as members, since $\varnothing \subseteq S$ and $S \subseteq S$ are always true.

PRACTICE 8   For $A = \{1, 2, 3\}$, what is $\wp(A)$?   ●

In Practice 8, A has three elements and $\wp(A)$ has eight elements. Try finding $\wp(S)$ for other sets S until you can guess the answer to the following practice problem.

PRACTICE 9   If S has n elements, then $\wp(S)$ has _____ elements. (Does your answer work for $n = 0$, too?)   ●

> **REMINDER:**
> To find $\wp(S)$, start with $\varnothing$, then add sets taking 1 element from S at a time, then 2 elements at a time, then 3 at a time, and so forth.

There are several ways we can show that for a set S with n elements, $\wp(S)$ will have 2^n elements. The following proof uses induction. For the basis step of the induction, we let $n = 0$. The only set with 0 elements is $\varnothing$. The only subset of $\varnothing$ is $\varnothing$, so $\wp(\varnothing) = \{\varnothing\}$, a set with $1 = 2^0$ elements. We assume that for any set with k elements, the power set has 2^k elements.

Now let S have $k + 1$ elements and put one of these elements, call it x, aside. The remaining set has k elements, so by our inductive assumption, its power set has 2^k elements. Each of these elements is also a member of $\wp(S)$. The only members of $\wp(S)$ not counted by this procedure are those including element x. All the subsets including x can be found by taking all those subsets not including x (of which there are 2^k) and throwing in the x; thus, there will be 2^k subsets including x. Altogether, there are 2^k subsets without x and 2^k subsets with x, or $2^k + 2^k = 2 \cdot 2^k = 2^{k+1}$ subsets. Therefore, $\wp(S)$ has 2^{k+1} elements.

Analogy with the truth tables of Section 1.1 is another way to show that $\wp(S)$ has 2^k elements for a set S with n elements. There we had n statement letters and showed that there were 2^k true–false combinations among these letters. But we can also think of each true–false combination as representing a particular subset, with T indicating membership and F indicating nonmembership in that subset. (For example, the row of the truth table with all statement letters F corresponds to the empty set.) Thus, the number of true–false combinations among n statement letters equals the number of subsets of a set with n elements; both are 2^n.

Binary and Unary Operations

By itself a set is not very interesting until we do something with its elements. For example, we can perform several arithmetic operations on elements of the set $\mathbb{Z}$. We might subtract two integers, or we might take the negative of an integer. Subtraction acts on two integers; it is a *binary* operation on $\mathbb{Z}$. Negation acts on one integer; it is a *unary* operation on $\mathbb{Z}$.

To see exactly what is involved in a binary operation, let's look at subtraction more closely. For any two integers x and y, $x - y$ will produce an answer, and only one answer, and that answer will always be an integer. Finally, subtraction is performed on an **ordered pair** of numbers. For example, $7 - 5$ does not produce the same result as $5 - 7$. An ordered pair is denoted by (x, y), where x is the first component of the ordered pair and y is the second component. Order is important in an ordered pair; thus, the sets $\{1, 2\}$ and $\{2, 1\}$ are equal, but the ordered pairs $(1, 2)$ and $(2, 1)$ are not. You are probably familiar with ordered pairs used as coordinates to locate a point in the plane. The point $(1, 2)$ is different from the point $(2, 1)$. Two ordered pairs (x, y) and (u, v) are equal only when it is the case that $x = u$ and $y = v$.

PRACTICE 10   Given that $(2x - y, x + y) = (7, -1)$, solve for x and y.   ●

PRACTICE 11   Let $S = \{3, 4\}$. List all the ordered pairs (x, y) of elements of S.   ●

We shall generalize the properties of subtraction on the integers to define a binary operation $\circ$ on a set S. The symbol $\circ$ is merely a placeholder; in any specific discussion, it will be replaced by the appropriate operation symbol, such as a subtraction sign.

Definition: Binary Operation
$\circ$ is a **binary operation** on a set S if for every ordered pair (x, y) of elements of S, $x \circ y$ exists, is unique, and is a member of S.

That the value $x \circ y$ always exists and is unique is described by saying that the binary operation $\circ$ is **well-defined**. The property that $x \circ y$ always belongs to S is described by saying that S is **closed** under the operation $\circ$.

EXAMPLE 6   Addition, subtraction, and multiplication are all binary operations on $\mathbb{Z}$. For example, when we perform addition on the ordered pair of integers (x, y), $x + y$ exists and is a unique integer.   ●

EXAMPLE 7   The logical operations of conjunction, disjunction, implication, and equivalence are binary operations on the set of propositional wffs. If P and Q are propositional wffs, then $P \wedge Q, P \vee Q, P \to Q$, and $P \leftrightarrow Q$ are unique propositional wffs.   ●

A candidate $\circ$ for an operation can fail to be a binary operation on a set S in any of three ways: (1) There are elements $x, y \in S$ for which $x \circ y$ does not exist; (2) there are elements $x, y \in S$ for which $x \circ y$ gives more than one result; or (3) there are elements $x, y \in S$ for which $x \circ y$ does not belong to S.

EXAMPLE 8   Division is not a binary operation on $\mathbb{Z}$ because $x \div 0$ does not exist.   ●

EXAMPLE 9   Define $x \circ y$ on $\mathbb{N}$ by

$$x \circ y = \begin{cases} 1 \text{ if } x \geq 5 \\ 0 \text{ if } x \leq 5 \end{cases}$$

Then, by the first part of the definition for $\circ$, $5 \circ 1 = 1$, but by its second part, $5 \circ 1 = 0$. Thus, $\circ$ is not well-defined on $\mathbb{N}$.   ●

EXAMPLE 10   Subtraction is not a binary operation on $\mathbb{N}$ because $\mathbb{N}$ is not closed under subtraction. (For example, $1 - 10 \notin \mathbb{N}$.)   ●

For # to be a **unary operation** on a set S, it must be true that for any $x \in S$, $x^\#$ is well-defined and S is closed under #; in other words, for any $x \in S$, $x^\#$ exists, is unique, and is a member of S. We do not have a unary operation if any of these conditions is not met.

EXAMPLE 11   Let $x^\#$ be defined by $x^\# = -x$, so that $x^\#$ is the negative of x. Then # is a unary operation on $\mathbb{Z}$ but not on $\mathbb{N}$ because $\mathbb{N}$ is not closed under #.   ●

EXAMPLE 12   The logical connective of negation is a unary operation on the set of propositional wffs. If P is a propositional wff, then P' is a unique propositional wff.   ●

From these examples it is clear that whether $\circ$ (or #) is a binary (or unary) operation can depend not only on its definition but also on the set involved.

PRACTICE 12   Which of the following are neither binary nor unary operations on the given sets? Why not?

a. $x \circ y = x \div y$; S = set of all positive integers
b. $x \circ y = x \div y$; S = set of all positive rational numbers
c. $x \circ y = x^y$; $S = \mathbb{R}$
d. $x \circ y = $ maximum of x and y; S = $\mathbb{N}$
e. $x^\# = \sqrt{x}$; S = set of all positive real numbers
f. $x^\# = $ solution to equation $(x^\#)^2 = x$; $S = \mathbb{C}$ ●

So far, all our binary operations have been defined by means of a description or an equation. Suppose S is a finite set, $S = \{x_1, x_2, \ldots, x_n\}$. Then a binary operation $\circ$ on S can be defined by an array, or table, where element i, j (ith row and jth column) denotes $x_i \circ x_j$.

EXAMPLE 13   Let $S = \{2, 5, 9\}$, and let $\circ$ be defined by the array

$\circ$	2	5	9
2	2	2	9
5	5	9	2
9	5	5	9

Thus, $2 \circ 5 = 2$ and $9 \circ 2 = 5$. Inspecting the table, we see that $\circ$ is a binary operation on S.   ●

Operations on Sets

Most of the operations we have seen operate on numbers, but we can also operate on sets. Given an arbitrary set S, we can define some binary and unary operations on the set $\wp(S)$. S in this case is called the **universal set** or the **universe of discourse**. The universal set defines the context of the objects being discussed. If $S = \mathbb{Z}$, for example, then all subsets will contain only integers.

A binary operation on $\wp(S)$ must act on any two subsets of S to produce a unique subset of S. There are at least two natural ways in which this can happen.

EXAMPLE 14 Let S be the set of all students at Silicon U. Then the members of $\wp(S)$ are sets of students. Let A be the set of computer science majors, and let B be the set of business majors. Both A and B belong to $\wp(S)$. A new set of students can be defined as consisting of everybody who is majoring in either computer science or business (or both); this set is called the *union* of A and B. Another new set can be defined as consisting of everybody who is majoring in both computer science and business. This set (which might be empty) is called the *intersection* of A and B. ●

Definitions: Union and Intersection of Sets

Let A, $B \in \wp(S)$. The **union** of A and B, denoted by $A \cup B$, is $\{x \mid x \in A$ or $x \in B\}$. The **intersection** of A and B, denoted by $A \cap B$, is $\{x \mid x \in A$ and $x \in B\}$.

EXAMPLE 15 Let $A = \{1, 3, 5, 7, 9\}$ and $B = \{3, 5, 6, 10, 11\}$. Here we may consider A and B as members of $\wp(\mathbb{N})$. Then $A \cup B = \{1, 3, 5, 6, 7, 9, 10, 11\}$ and $A \cap B = \{3, 5\}$. Both $A \cup B$ and $A \cap B$ are members of $\wp(\mathbb{N})$. ●

PRACTICE 13 Let A, $B \in \wp(S)$ for any set S. Is it always the case that $A \cap B \subseteq A \cup B$? ●

We can use *Venn diagrams* (named for the nineteenth-century British mathematician John Venn) to visualize the binary operations of union and intersection. The shaded areas in Figures 3.1 and 3.2 illustrate the set that results from performing the binary operation on the two given sets.

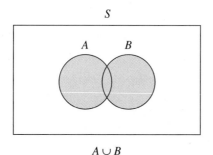

$A \cup B$

Figure 3.1

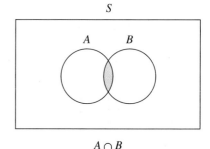

$A \cap B$

Figure 3.2

We will define one unary operation on $\wp(S)$.

Definition: Complement of a Set

For a set $A \in \wp(S)$, the **complement** of A, A', is $\{x \mid x \in S \text{ and } x \notin A\}$.

PRACTICE 14   Illustrate A' in a Venn diagram.

Another binary operation on sets A and B in $\wp(S)$ is **set difference**: $A - B = \{x \mid x \in A \text{ and } x \notin B\}$. This operation can be rewritten as $A - B = \{x \mid x \in A \text{ and } x \in B'\}$ and, finally, as $A - B = A \cap B'$.

PRACTICE 15   Illustrate $A - B$ in a Venn diagram.

Two sets A and B such that $A \cap B = \varnothing$ are said to be **disjoint**. Thus, $A - B$ and $B - A$, for example, are disjoint sets.

EXAMPLE 16   Let

$$A = \{x \mid x \text{ is an even nonnegative integer}\}$$
$$B = \{x \mid (\exists y)(y \in \mathbb{N} \text{ and } x = 2y + 1)\}$$
$$C = \{x \mid (\exists y)(y \in \mathbb{N} \text{ and } x = 4y)\}$$

be subsets of $\mathbb{N}$. Because B represents the set of nonnegative odd integers, A and B are disjoint sets. Also, every nonnegative integer is either even or odd, so $A \cup B = \mathbb{N}$. These two facts also tell us that $A' = B$. Every multiple of 4 is an even number, so C is a subset of A, from which it follows that $A \cup C = A$. C is in fact a proper subset of A, and $A - C = \{x \mid (\exists y)(y \in \mathbb{N} \text{ and } x = 4y + 2)\}$.

PRACTICE 16   Let

$$A = \{1, 2, 3, 5, 10\}$$
$$B = \{2, 4, 7, 8, 9\}$$
$$C = \{5, 8, 10\}$$

be subsets of $S = \{1, 2, 3, 4, 5, 6, 7, 8, 9, 10\}$. Find

  a. $A \cup B$
  b. $A - C$
  c. $B' \cap (A \cup C)$

We shall define one final operation using elements of $\wp(S)$.

Definition: Cartesian Product

Let A and B be subsets of S. The **Cartesian product (cross product)** of A and B, denoted by $A \times B$, is defined by

$$A \times B = \{(x, y) \mid x \in A \text{ and } y \in B\}$$

Thus, the Cartesian product of two sets A and B is the set of all ordered pairs whose first component comes from A and whose second comes from B. The cross product is not a binary operation on $\wp(S)$. Although it acts on an ordered pair of members of $\wp(S)$ and gives a unique result, the resulting set is not, in general, a subset of S. The elements are not members of S but ordered pairs of members of S. So the resulting set is not a member of $\wp(S)$. The closure property for a binary operation fails to hold.

Because we will often be interested in the cross product of a set with itself, we will abbreviate $A \times A$ as A^2; in general, we use A^n to mean the set of all ordered n-tuples $(x_1, x_2, \dots, x_n)$ of elements of A.

PRACTICE 17 Let $A = \{1, 2\}$ and $B = \{3, 4\}$.

    a. Find $A \times B$.
    b. Find $B \times A$.
    c. Find A^2.
    d. Find A^3.   ●

Set Identities

There are many set equalities involving the operations of union, intersection, difference, and complementation that are true for all subsets of a given set S. Because they are independent of the particular subsets used, these equalities are called set identities. Some basic set identities follow. The names and forms of these identities are very similar to the tautological equivalences of Section 1.1 (check back and compare). We shall see in Chapter 7 that this similarity is not a coincidence.

Basic Set Identities

1a. $A \cup B = B \cup A$	1b. $A \cap B = B \cap A$	(commutative properties)
2a. $(A \cup B) \cup C =$	2b. $(A \cap B) \cap C =$	(associative properties)
$A \cup (B \cup C)$	$A \cap (B \cap C)$	
3a. $A \cup (B \cap C) =$	3b. $A \cap (B \cup C) =$	(distributive properties)
$(A \cup B) \cap (A \cup C)$	$(A \cap B) \cup (A \cap C)$	
4a. $A \cup \emptyset = A$	4b. $A \cap S = A$	(identity properties)
5a. $A \cup A' = S$	5b. $A \cap A' = \emptyset$	(complement properties)

(Note that 2a allows us to write $A \cup B \cup C$ with no need for parentheses; 2b allows us to write $A \cap B \cap C$.)

EXAMPLE 17 Let's prove identity 3a. We might draw Venn diagrams for each side of the equation and see that they look the same. However, identity 3a is supposed to hold for all subsets A, B, and C, and whatever one picture we draw cannot be completely general. Thus, if we draw A and B disjoint, that's a special case, but if we draw A and B not disjoint, that doesn't take care of the case where A and B are disjoint. To do a proof by Venn diagrams requires a picture for each possible case, and the more sets involved (A, B, and C in this problem), the more cases there are. To avoid drawing a picture for each case, let's prove set equality by proving set inclusion in each direction. Thus, we want to prove

$$A \cup (B \cap C) \subseteq (A \cup B) \cap (A \cup C)$$

and also

$$(A \cup B) \cap (A \cup C) \subseteq A \cup (B \cap C)$$

To show that $A \cup (B \cap C) \subseteq (A \cup B) \cap (A \cup C)$, we let x be an arbitrary member of $A \cup (B \cap C)$. Then we can proceed as follows:

$$x \in A \cup (B \cap C) \rightarrow x \in A \text{ or } x \in (B \cap C)$$
$$\rightarrow x \in A \text{ or } (x \in B \text{ and } x \in C)$$
$$\rightarrow (x \in A \text{ or } x \in B) \text{ and } (x \in A \text{ or } x \in C)$$
$$\rightarrow x \in (A \cup B) \text{ and } x \in (A \cup C)$$
$$\rightarrow x \in (A \cup B) \cap (A \cup C)$$

To show that $(A \cup B) \cap (A \cup C) \subseteq A \cup (B \cap C)$, we reverse the above argument. ●

PRACTICE 18   Prove identity 4a.   ●

Once we have proved the set identities in this list, we can use them to prove other set identities. Just as the tautological equivalences of propositional logic represent recipes or patterns for transforming wffs, the set identities represent patterns for transforming set expressions. And, as with tautologies, the set identity can be applied only when the set expression exactly matches the pattern.

EXAMPLE 18   We can use the basic set identities to prove

$$[A \cup (B \cap C)] \cap ([A' \cup (B \cap C)] \cap (B \cap C)') = \varnothing$$

> **REMINDER:**
> You must match the pattern of a set identity in order to use it. In the set identities, A, B, and C can represent any sets.

for A, B, and C, any subsets of S. In the following proof, the number to the right is that of the basic set identity used to validate each step. The first step uses identity 2b because the expression $[A \cup (B \cap C)] \cap ([A' \cup (B \cap C)] \cap (B \cap C)')$ matches the right side of 2b, $A \cap (B \cap C)$, where A is $[A \cup (B \cap C)]$, B is $[A' \cup (B \cap C)]$, and C is $(B \cap C)'$.

$$[A \cup (B \cap C)] \cap ([A' \cup (B \cap C)] \cap (B \cap C)')$$
$$= ([A \cup (B \cap C)] \cap [A' \cup (B \cap C)]) \cap (B \cap C)' \quad \text{(2b)}$$
$$= ([(B \cap C) \cup A] \cap [(B \cap C) \cup A']) \cap (B \cap C)' \quad \text{(1a twice)}$$
$$= [(B \cap C) \cup (A \cap A')] \cap (B \cap C)' \quad \text{(3a)}$$
$$= [(B \cap C) \cup \varnothing] \cap (B \cap C)' \quad \text{(5b)}$$
$$= (B \cap C) \cap (B \cap C)' \quad \text{(4a)}$$
$$= \varnothing \quad \text{(5b)}$$ ●

The **dual** for each set identity in our list also appears in the list. The dual is obtained by interchanging $\cup$ and $\cap$ and interchanging S and $\varnothing$. The dual of the identity in Example 18 is

$$[A \cap (B \cup C)] \cup ([A' \cap (B \cup C)] \cup (B \cup C)') = S$$

which we could prove true by replacing each basic set identity used in the proof of Example 18 with its dual. Because this method always works, any time we have proved a set identity by using the basic identities, we have also proved its dual.

PRACTICE 19   **a.** Using the basic set identities, establish the set identity

$$[C \cap (A \cup B)] \cup [(A \cup B) \cap C'] = A \cup B$$

(A, B, and C are any subsets of S.)

b. State the dual identity that you now know is true.

Table 3.1 summarizes the approaches to proving set identities.

Method	Comment
Draw a Venn diagram	Not a good plan because no one diagram fits all cases and it will not prove the general identity.
Establish set inclusion in each direction	Take an arbitrary member of one side and show it belongs to the other side, and conversely.
Use already proved identities	Be sure to match the pattern of the identity you want to use.

Table 3.1

Countable and Uncountable Sets

In a finite set S, we can always designate one element as the first member, s_1, another element as the second member, s_2, and so forth. If there are k elements in the set, then these can be listed in the order we have selected:

$s_1, s_2, \ldots, s_k$

This list represents the entire set.

If the set is infinite, we may still be able to select a first element s_1, a second element s_2, and so forth, so that the list

$s_1, s_2, s_3, \ldots$

represents all elements of the set. Every element of the set will eventually appear in this list. Such an infinite set is said to be **denumerable**. Both finite and denumerable sets are **countable** sets because we can count, or enumerate, all of their elements. Being countable does not mean that we can state the total number of elements in the set; rather, it means that we can say, "Here is a first one," "Here is a second one," and so on, through the set. There are, however, infinite sets that are **uncountable**. In an uncountable set, the set is so big that there is no way to count out the elements and get the whole set in the process. Before we prove that uncountable sets exist, let's look at some denumerable (countably infinite) sets.

EXAMPLE 19   The set $\mathbb{N}$ is denumerable.

To prove denumerability, we need only exhibit a counting scheme. For the set $\mathbb{N}$ of nonnegative integers, it is clear that

0, 1, 2, 3, ...

is an enumeration that will eventually include every member of the set.

PRACTICE 20   Prove that the set of even positive integers is denumerable.

EXAMPLE 20   The set $\mathbb{Q}^+$ of positive rational numbers is denumerable.

We assume that each positive rational number is written as a fraction of positive integers. We can write all such fractions having the numerator 1 in one row, all those having the numerator 2 in a second row, and so on:

$$
\begin{array}{cccccc}
1/1 & 1/2 & 1/3 & 1/4 & 1/5 & \cdots \\
2/1 & 2/2 & 2/3 & 2/4 & 2/5 & \cdots \\
3/1 & 3/2 & 3/3 & 3/4 & 3/5 & \cdots \\
4/1 & 4/2 & 4/3 & 4/4 & 4/5 & \cdots \\
\vdots & \vdots & \vdots & \vdots & \vdots & \ddots
\end{array}
$$

To show that the set of all fractions in this array is denumerable, we shall thread an arrow through the entire array, beginning with 1/1; following the arrow gives an enumeration of the set. Thus the fraction 1/3 is the fourth member in this enumeration:

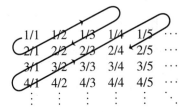

Therefore the set represented by the array is denumerable. Note that our path through the array must "spread out" from one corner. If we begin to follow just the first row, for example, or just the first column, we will never finish it to get on to other rows (or columns).

To obtain an enumeration of $\mathbb{Q}^+$, we use the enumeration of the set shown but eliminate any fractions not in lowest terms. This avoids the problem of listing both 1/2 and 2/4, for example, which represent the same positive rational. The enumeration of $\mathbb{Q}^+$ thus begins with

1/1, 2/1, 1/2, 1/3, 3/1, 4/1, ...

For example, we have eliminated 2/2, which reduces to 1/1.

PRACTICE 21   What is the 11th fraction in the above enumeration? What is the 11th positive rational?

Now let's show that there is an infinite set that is not denumerable. The proof technique that seems appropriate to prove that set A does *not* have property B is to assume that A does have property B and look for a contradiction. The proof in Example 21 is a very famous proof by contradiction known as **Cantor's diagonalization method**, after Georg Cantor, the nineteenth-century German mathematician known as the "father of set theory."

EXAMPLE 21
We will show that the set of all the real numbers between 0 and 1 is uncountable.

We will write such numbers in decimal form; thus any member of the set can be written as

$0.d_1d_2d_3 ...$

A number such as $0.24999999 ...$ can be written in alternative form as $0.2500000 ...$. In order to avoid writing the same element twice, we shall choose (arbitrarily) to always use the former representation and not the latter. Now let us assume that our set is countable. Therefore some enumeration of the set exists. We can depict an enumeration of the set as follows, where d_{ij} is the jth decimal digit in the ith number in the enumeration:

$0.d_{11}d_{12}d_{13} \quad \cdots$
$0.d_{21}d_{22}d_{23} \quad \cdots$
$0.d_{31}d_{32}d_{33} \quad \cdots$
$\qquad \vdots \quad \vdots \quad \vdots \quad \ddots$

We now construct a real number $p = 0.p_1p_2p_3 ...$ as follows: p_i is always chosen to be 5 if $d_{ii} \neq 5$ and 6 if $d_{ii} = 5$. Thus p is a real number between 0 and 1. For instance, if the enumeration begins with

$0.342134 ...$

$0.257001 ...$

$0.546122 ...$

$0.716525 ...$

then $d_{11} = 3$, $d_{22} = 5$, $d_{33} = 6$, and $d_{44} = 5$, so $p_1 = 5$, $p_2 = 6$, $p_3 = 5$, and $p_4 = 6$. Thus p begins with $0.5656 ...$.

If we compare p with the enumeration of the set, p differs from the first number at the first decimal digit, from the second number at the second decimal digit, from the third number at the third decimal digit, and so on. Therefore p does not agree with any of the representations in the enumeration. Furthermore, because p contains no 0s to the right of the decimal, it is not the alternative representation of any of the numbers in the enumeration. Therefore p is a real number between 0 and 1 different from any other number in the enumeration, yet the enumeration was supposed to include all members of the set. Here, then, is the contradiction, and the set of all real numbers between 0 and 1 is indeed uncountable. (Why do you suppose this proof is called a "diagonalization method"?)

Although it is interesting and perhaps surprising to learn that there are uncountable sets, we shall usually be concerned with countable sets. A computer, of course, can manage only finite sets. In the rest of this chapter, we, too, will limit our attention to finite sets and various ways to count their elements.

Section 3.1 Review

Techniques

- Describe sets by a list of elements and by a characterizing property.
- Prove that one set is a subset of another.
- Find the power set of a set.
- Check that the required properties for a binary or unary operation are satisfied.
- Form new sets by taking the union, intersection, complement, and cross product of sets.
- Prove set identities by showing set inclusion in each direction or using the basic set identities.
- Demonstrate the denumerability of certain sets.
- Use the Cantor diagonalization method to prove that certain sets are uncountable.

Main Ideas

Sets are unordered collections of objects that can be related (equal sets, subsets, etc.) or combined (union, intersection, etc.).

Certain standard sets have their own notation.

The power set of a set with n elements has 2^n elements.

Basic set identities exist (in dual pairs) and can be used to prove other set identities; once an identity is proved in this manner, its dual is also true.

Countable sets can be enumerated, and uncountable sets exist.

Exercises 3.1

★1. Let $S = \{2, 5, 17, 27\}$. Which of the following are true?

    a. $5 \in S$ **b.** $2 + 5 \in S$ **c.** $\emptyset \in S$ **d.** $S \in S$

2. Let $B = \{x \mid x \in \mathbb{Q} \text{ and } -1 < x < 2\}$. Which of the following are true?

    a. $0 \in B$ **b.** $-1 \in B$ **c.** $-0.84 \in B$ **d.** $\sqrt{2} \in B$

3. How many different sets are described below? What are they?

    $(2, 3, 4\}$ $\emptyset$

    $\{x \mid x \text{ is the first letter of cat, bat,} $ $\{x \mid x \text{ is the first letter of cat, bat,}$
       or apple$\}$ and apple$\}$

    $\{x \mid x \in \mathbb{N} \text{ and } 2 \leq x \leq 4\}$ $\{2, a, 3, b, 4, c\}$

    $\{a, b, e\}$ $\{3, 4, 2\}$

★4. Describe each of the following sets by listing its elements:

    a. $\{x \mid x \in \mathbb{N} \text{ and } x^2 < 25\}$
    b. $\{x \mid x \in \mathbb{N} \text{ and } x \text{ is even and } 2 < x < 11\}$
    c. $\{x \mid x \text{ is one of the first three U.S. presidents}\}$
    d. $\{x \mid x \in \mathbb{R} \text{ and } x^2 = -1\}$
    e. $\{x \mid x \text{ is one of the New England states}\}$
    f. $\{x \mid x \in \mathbb{Z} \text{ and } |x| < 4\}$ ($|x|$ denotes the absolute value function)

5. Describe each of the following sets by listing its elements:

    a. $\{x \mid x \in \mathbb{N} \text{ and } x^2 - 5x + 6 = 0\}$
    b. $\{x \mid x \in \mathbb{R} \text{ and } x^2 = 7\}$
    c. $\{x \mid x \in \mathbb{N} \text{ and } x^2 - 2x - 8 = 0\}$

6. Describe each of the following sets by giving a characterizing property:

    a. $\{1, 2, 3, 4, 5\}$
    b. $\{1, 3, 5, 7, 9, 11, ...\}$
    c. $\{\text{Melchior, Gaspar, Balthazar}\}$
    d. $\{0, 1, 10, 11, 100, 101, 110, 111, 1000, ...\}$

7. Describe each of the following sets:

    a. $\{x \mid x \in \mathbb{N} \text{ and } (\exists q)(q \in \{2, 3\} \text{ and } x = 2q)\}$
    b. $\{x \mid x \in \mathbb{N} \text{ and } (\exists y)(\exists z)(y \in \{0, 1\} \text{ and } z \in \{3, 4\} \text{ and } y < x < z)\}$
    c. $\{x \mid x \in \mathbb{N} \text{ and } (\forall y)(y \text{ even} \rightarrow x \neq y)\}$

★8. Given the description of a set A as $A = \{2, 4, 8, ... \}$, do you think $16 \in A$?

9. Let

$$A = \{x \mid x \in \mathbb{N} \text{ and } 1 < x < 50\}$$
$$B = \{x \mid x \in \mathbb{R} \text{ and } 1 < x < 50\}$$
$$C = \{x \mid x \in \mathbb{Z} \text{ and } |x| \geq 25\}$$

Which of the following statements are true?

    a. $A \subseteq B$ **b.** $17 \in A$ **c.** $A \subseteq C$
    d. $-40 \in C$ **e.** $\sqrt{3} \in B$ **f.** $\{0, 1, 2\} \subseteq A$
    g. $\varnothing \in B$ **h.** $\{x \mid x \in \mathbb{Z} \text{ and } x^2 > 625\} \subseteq C$

10. Let

$$R = \{1, 3, \pi, 4.1, 9, 10\} \qquad S = \{\{1\}, 3, 9, 10\}$$
$$T = \{1, 3, \pi\} \qquad\qquad U = \{\{1, 3, \pi\}, 1\}$$

Which of the following are true? For those that are not, why not?

    ★**a.** $S \subseteq R$ ★**b.** $1 \in R$ ★**c.** $1 \in S$
    ★**d.** $1 \subseteq U$ ★**e.** $\{1\} \subseteq T$ ★**f.** $\{1\} \subseteq S$
    g. $T \subset R$ **h.** $\{1\} \in S$ **i.** $\varnothing \subseteq S$
    j. $T \subseteq U$ **k.** $T \in U$ **l.** $T \notin R$
    m. $T \subseteq R$ **n.** $S \subseteq \{1, 3, 9, 10\}$

11. Let

$$A = \{a, \{a\}, \{\{a\}\}\} \qquad B = \{a\} \qquad C = \{\varnothing, \{a, \{a\}\}\}$$

Which of the following are true? For those that are not, why not?

a. $B \subseteq A$ b. $B \in A$ c. $C \subseteq A$
d. $\varnothing \subseteq C$ e. $\varnothing \in C$ f. $\{a, \{a\}\} \in A$
g. $\{a, \{a\}\} \subseteq A$ h. $B \subseteq C$ i. $\{\{a\}\} \subseteq A$

12. Let

$$A = \{x \mid x \in \mathbb{R} \text{ and } x^2 - 4x + 3 < 0\}$$

and

$$B = \{x \mid x \in \mathbb{R} \text{ and } 0 < x < 6\}$$

Prove that $A \subset B$.

★13. Let

$$A = \{(x, y) \mid (x, y) \text{ lies within 3 units of the point } (1, 4)\}$$

and

$$B = \{(x, y) \mid (x - 1)^2 + (y - 4)^2 \le 25\}$$

Prove that $A \subset B$.

14. Program QUAD finds and prints solutions to quadratic equations of the form $ax^2 + bx + c = 0$. Program EVEN lists all the even integers from $-2n$ to $2n$. Let Q denote the set of values output by QUAD and E denote the set of values output by EVEN.

a. Show that for $a = 1$, $b = -2$, $c = -24$, and $n = 50$, $Q \subseteq E$.
b. Show that for the same values of a, b, and c, but a value for n of 2, $Q \not\subseteq E$.

15. Let $A = \{x \mid \cos(x/2) = 0\}$ and $B = \{x \mid \sin x = 0\}$. Prove that $A \subseteq B$.

16. Which of the following are true for all sets A, B, and C?
★a. If $A \subseteq B$ and $B \subseteq A$, then $A = B$.   ★b. $\{\varnothing\} = \varnothing$
★c. $\{\varnothing\} = \{0\}$ ★d. $\varnothing \in \{\varnothing\}$
★e. $\varnothing \subseteq A$ ★f. $\varnothing \in A$
g. $\{\varnothing\} = \{\{\varnothing\}\}$
h. If $A \subset B$ and $B \subseteq C$, then $A \subset C$.
i. If $A \ne B$ and $B \ne C$, then $A \ne C$.
j. If $A \in B$ and $B \not\subseteq C$, then $A \notin C$.

17. Prove that if $A \subseteq B$ and $B \subseteq C$, then $A \subseteq C$.

18. Prove that if $A' \subseteq B'$, then $B \subseteq A$.

19. Find $\wp(S)$ for $S = \{a\}$.

★20. Find $\wp(S)$ for $S = \{1, 2, 3, 4\}$. How many elements do you expect this set to have?

21. Find $\wp(S)$ for $S = \{\emptyset\}$.

22. Find $\wp(S)$ for $S = \{\emptyset, \{\emptyset\}, \{\emptyset, \{\emptyset\}\}\}$.

23. Find $\wp(\wp(S))$ for $S = \{a, b\}$.

★24. What can be said about A if $\wp(A) = \{\emptyset, \{x\}, \{y\}, \{x, y\}\}$?

25. What can be said about A if $\wp(A) = \{\emptyset, \{a\}, \{\{a\}\}\}$?

26. Prove that if $A \subseteq B$, then $\wp(A) \subseteq \wp(B)$.

★27. Prove that if $\wp(A) = \wp(B)$, then $A = B$.

28. Solve for x and y.
   a. $(y, x + 2) = (5, 3)$ **b.** $(2x, y) = (16, 7)$ **c.** $(2x - y, x + y) = (-2, 5)$

29. **a.** Recall that ordered pairs must have the property that $(x, y) = (u, v)$ if and only if $x = u$ and $y = v$. Prove that $\{\{x\}, \{x, y\}\} = \{\{u\}, \{u, v\}\}$ if and only if $x = u$ and $y = v$. Therefore, although we know that $(x, y) \neq \{x, y\}$, we can define the ordered pair (x, y) as the set $\{\{x\}, \{x, y\}\}$.
   b. Show by an example that we cannot define the ordered triple (x, y, z) as the set $\{\{x\}, \{x, y\}, \{x, y, z\}\}$

30. Which of the following are binary or unary operations on the given sets? For those that are not, why not?
   ★**a.** $x \circ y = x + 1$; $S = \mathbb{N}$
   ★**b.** $x \circ y = x + y - 1$; $S = \mathbb{N}$

   ★**c.** $x \circ y = \begin{cases} x - 1 & \text{if } x \text{ is odd} \\ x & \text{if } x \text{ is even} \end{cases}$ $S = \mathbb{Z}$

   d. $x^{\#} = \ln x$; $S = \mathbb{R}$
   e. $x^{\#} = x^2$; $S = \mathbb{Z}$

   f.

$\circ$	1	2	3
1	1	2	3
2	2	3	4
3	3	4	5

   $S = \{1, 2, 3\}$

   g. $x \circ y = $ that fraction, x or y, with the smaller denominator; $S = $ set of all fractions
   h. $x \circ y = $ that person, x or y, whose name appears first in an alphabetical sort; $S = $ set of 10 people with different names

31. Which of the following are binary or unary operations on the given sets? For those that are not, why are not?

   a. $x \circ y = \begin{cases} 1/x & \text{if } x \text{ is positive} \\ 1/(-x) & \text{if } x \text{ is negative} \end{cases}$ $S = \mathbb{R}$

   b. $x \circ y = xy$ (concatenation); $S = $ set of all finite-length strings of symbols from the set $\{p, q, r\}$

   c. $x^\# = \lfloor x \rfloor$ where $\lfloor x \rfloor$ denotes the greatest integer less than or equal to x; $S = \mathbb{R}$

   d. $x \circ y = \min(x, y)$; $S = \mathbb{N}$

   e. $x \circ y = $ greatest common multiple of x and y; $S = \mathbb{N}$

   f. $x \circ y = x + y$; $S = $ the set of Fibonacci numbers

   g. $x^\# = $ the string that is the reverse of x; $S = $ set of all finite-length strings of symbols from the set $\{p, q, r\}$

   h. $x \circ y = x + y$; $S = \mathbb{R} - \mathbb{Q}$

32. How many different binary operations can be defined on a set with n elements? (*Hint:* Think about filling in a table.)

33. We have written binary operations in *infix* notation, where the operation symbol appears between the two operands, as in $A + B$. Evaluation of a complicated arithmetic expression is more efficient when the operations are written in *postfix* notation, where the operation symbol appears after the two operands, as in $AB +$. Many compilers change expressions in a computer program from infix to postfix form. One way to produce an equivalent postfix expression from an infix expression is to write the infix expression with a full set of parentheses, move each operator to replace its corresponding right parenthesis, and then eliminate all left parentheses. (Parentheses are not required in postfix notation.) Thus,

   $A * B + C$

   becomes, when fully parenthesized,

   $((A * B) + C)$

   and the postfix notation is $AB * C+$. Rewrite each of the following in postfix notation:

   a. $(A + B) * (C - D)$

   b. $A ** B - C * D$ (**denotes exponentiation)

   c. $A * C + B/(C + D * B)$

★34. Evaluate the following postfix expressions (see Exercise 33):

   a. $2\ 4 * 5 +$ **b.** $5\ 1 + 2 / 1 -$ **c.** $3\ 4 + 5\ 1 - *$

35. Let

   $A = \{p, q, r, s\}$
   $B = \{r, t, v\}$
   $C = \{p, s, t, u\}$

be subsets of $S = \{p, q, r, s, t, u, v, w\}$. Find

a. $B \cap C$ **b.** $A \cup C$ **c.** C'

d. $A \cap B \cap C$ **e.** $B - C$ **f.** $(A \cup B)'$

g. $A \times B$ **h.** $(A \cup B) \cap C'$

36. Let

$A = \{2, 4, 5, 6, 8)$
$B = \{1, 4, 5, 9\}$
$C = \{x \mid x \in \mathbb{Z} \text{ and } 2 \le x < 5\}$

be subsets of $S = \{0, 1, 2, 3, 4, 5, 6, 7, 8, 9\}$. Find

★**a.** $A \cup B$ ★**b.** $A \cap B$ ★**c.** $A \cap C$

d. $B \cup C$ **e.** $A - B$ **f.** A'

g. $A \cap A'$ ★**h.** $(A \cap B)'$ **i.** $C - B$

j. $(C \cap B) \cup A'$ **k.** $(B - A)' \cap (A - B)$ **l.** $(C' \cup B)'$

m. $B \times C$

37. Let

$A = \{a, \{a\}, \{\{a\}\}\}$
$B = \{\varnothing, \{a\}, \{a, \{a\}\}\}$
$C = \{a\}$

be subsets of $S = \{\varnothing, a, \{a\}, \{\{a\}\}, \{a, \{a\}\}\}$. Find

a. $A \cap C$ **b.** $B \cap C'$ **c.** $A \cup B$

d. $\varnothing \cap B$ **e.** $(B \cup C) \cap A$ **f.** $A' \cap B$

g. $\{\varnothing\} \cap B$

38. $A = \{x \mid x \text{ is the name of a former president of the United States}\}$
$B = \{\text{Adams, Hamilton, Jefferson, Grant}\}$
$C = \{x \mid x \text{ is the name of a state}\}$

Find

a. $A \cap B$ **b.** $A \cap C$ **c.** $B \cap C$

39. Let

$A = \{x \mid x \text{ is a word that appears before } dog \text{ in an English language dictionary}\}$
$B = \{x \mid x \text{ is a word that appears after } canary \text{ in an English language dictionary}\}$
$C = \{x \mid x \text{ is a word of more than four letters}\}$

Which of the following are true statements?

a. $B \subseteq C$

b. $A \cup B = \{x \mid x \text{ is a word in an English language dictionary}\}$

c. $cat \in B \cap C'$

d. $bamboo \in A - B$

★40. Consider the following subsets of $\mathbb{Z}$:

$$A = \{x \mid (\exists y)(y \in \mathbb{Z} \text{ and } y \geq 4 \text{ and } x = 3y)\}$$
$$B = \{x \mid (\exists y)(y \in \mathbb{Z} \text{ and } x = 2y)\}$$
$$C = \{x \mid x \in \mathbb{Z} \text{ and } |x| \leq 10\}$$

Using set operations, describe each of the following sets in terms of A, B, and C.

a. set of all odd integers
b. $\{-10, -8, -6, -4, -2, 0, 2, 4, 6, 8, 10\}$
c. $\{x \mid (\exists y)(y \in \mathbb{Z} \text{ and } y \geq 2 \text{ and } x = 6y)\}$
d. $\{-9, -7, -5, -3, -1, 1, 3, 5, 7, 9\}$
e. $\{x \mid (\exists y)(y \in \mathbb{Z} \text{ and } y \geq 5 \text{ and } x = 2y + 1)\} \cup \{x \mid (\exists y)(y \in \mathbb{Z} \text{ and } y \leq -5 \text{ and } x = 2y - 1)\}$

41. Let

$$A = \{x \mid x \in \mathbb{R} \text{ and } 1 < x \leq 3\}$$
$$B = \{x \mid x \in \mathbb{R} \text{ and } 2 \leq x \leq 5\}$$

Using set operations, describe each of the sets shown in terms of A and B.

a.
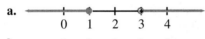

b.

c.

42. Consider the following subsets of the set of all students:

A = set of all computer science majors
B = set of all physics majors
C = set of all science majors
D = set of all female students

Using set operations, describe each of the following sets in terms of A, B, C, and D:

★**a.** set of all students not majoring in science
★**b.** set of all female physics majors
★**c.** set of all students majoring in both computer science and physics
d. set of all male computer science majors
e. set of all male students who are not physics majors
f. set of all science majors who are not computer science majors
g. set of all students who are female or science majors

43. An Internet search engine has the following set of URL references in its database:

A = automobiles for sale

with subsets

B = used cars
C = Fords
D = Buicks
E = pre-1995 models

You want to search for all references to used Fords or Buicks that are 1995 or later models. Write the set expression that represents your query.

44. Which of the following are true for all sets A, B, and C?
    ★**a.** $A \cup A = A$ **b.** $B \cap B = B$
    ★**c.** $(A \cap B)' = A' \cap B'$ **d.** $(A')' = A$
    ★**e.** $A - B = (B - A)'$ **f.** $(A - B) \cap (B - A) = \varnothing$
      g. If $A \cap B = \varnothing$, then $A \subset B$.   **h.** $B \times A = A \times B$
      i. $\varnothing \times A = \varnothing$ **j.** $\varnothing \cap \{\varnothing\} = \varnothing$
      k. $(A - B) \cup (B - C) = A - C$ **l.** $(A - C) \cap (A - B) = A - (B \cup C)$

45. For each of the following statements, find general conditions on sets A and B to make the statement true:
    ★**a.** $A \cup B = A$ **b.** $A \cap B = A$ **c.** $A \cup \varnothing = \varnothing$
      d. $B - A = \varnothing$ **e.** $A \cup B \subseteq A \cap B$

46. Prove that

    $$(A \cap B) \subseteq A$$

    where A and B are arbitrary sets.

47. Prove that

    $$A \subseteq (A \cup B)$$

    where A and B are arbitrary sets.

★48. Prove that $\wp(A) \cap \wp(B) = \wp(A \cap B)$ where A and B are arbitrary sets.

49. Prove that $\wp(A) \cup \wp(B) \subseteq \wp(A \cup B)$ where A and B are arbitrary sets.

50. Prove that if $A \cup B = A - B$, then $B = \varnothing$. (*Hint:* Do a proof by contradiction.)

★51. Prove that if $(A - B) \cup (B - A) = A \cup B$, then $A \cap B = \varnothing$. (*Hint:* Do a proof by contradiction.)

52. Prove that $(A \cap B) \cup C = A \cap (B \cup C)$ if and only if $C \subseteq A$.

53. A binary operation on sets called the *symmetric difference* is defined by
    $A \oplus B = (A - B) \cup (B - A)$.

    a. Draw a Venn diagram to illustrate $A \oplus B$.
    b. For $A = \{3, 5, 7, 9\}$ and $B = \{2, 3, 4, 5, 6\}$, what is $A \oplus B$?
    c. Prove that $A \oplus B = (A \cup B) - (A \cap B)$ for arbitrary sets A and B.
    d. For an arbitrary set A, what is $A \oplus A$? What is $\emptyset \oplus A$?
    e. Prove that $A \oplus B = B \oplus A$ for arbitrary sets A and B.
    f. For any sets A, B, and C, prove that $(A \oplus B) \oplus C = A \oplus (B \oplus C)$.

54. Which of the following are true for all sets A, B, and C?

    a. $A \cup (B \times C) = A \cup B) \times (A \cup C)$
    b. $A \times (B \cap C) = (A \times B) \cap (A \times C)$
    c. $A \times \emptyset = \emptyset$
    d. $\wp(A) \times \wp(A) = \wp(A^2)$
    e. $A \times (B \times C) = (A \times B) \times C$

55. Verify the basic set identities on page 171 by showing set inclusion in each direction. (We have already done 3a and 4a.)

56. A and B are subsets of a set S. Prove the following set identities by showing set inclusion in each direction:

    a. $(A \cup B)' = A' \cap B'$ ⎫
    b. $(A \cap B)' = A' \cup B'$ ⎭ De Morgan's Laws
    ★**c.** $A \cup (B \cap A) = A$ **d.** $(A \cap B')' \cup B = A' \cup B$
    e. $(A \cap B) \cup (A \cap B') = A$ **f.** $[A \cap (B \cup C)]' = A' \cup (B' \cap C')$

57. A, B, and C are subsets of a set S. Prove the following set identities using the basic set identities listed in this section. State the dual of each of these identities.

    ★**a.** $(A \cup B) \cap (A \cup B') = A$
    b. $([(A \cap C) \cap B] \cup [(A \cap C) \cap B']) \cup (A \cap C)' = S$
    c. $(A \cup C) \cap [(A \cap B) \cup (C' \cap B)] = A \cap B$

58. A is a subset of a set S. Prove the following set identities:

    a. $A \cup A = A$ **b.** $A \cap A = A$
    c. $A \cap \emptyset = \emptyset$ **d.** $A \cup S = S$
    ★**e.** $(A')' = A$

59. A, B, and C are subsets of a set S. Prove the following set identities by using previously proved identities, including those in Exercises 56–58.

    ★**a.** $A \cap (B \cup A') = B \cap A$ **b.** $(A \cup B) - C = (A - C) \cup (B - C)$
    c. $(A - B) - C = (A - C) - B$ **d.** $[(A' \cup B') \cap A']' = A$
    e. $(A - B) - C = (A - C) - (B - C)$
    f. $A - (A - B) = A \cap B$

60. The operation of set union can be defined as an n-ary operation for any integer $n \geq 2$.

   a. Give a definition similar to that for the union of two sets for

   $$A_1 \cup A_2 \cup \cdots \cup A_n$$

   b. Give a recursive definition for $A_1 \cup A_2 \cup \cdots \cup A_n$.

★61. Using the recursive definition of set union from Exercise 60(b), prove the generalized associative property of set union, which is that for any n with $n \geq 3$ and any p with $1 \leq p \leq n - 1$,

   $$(A_1 \cup A_2 \cup \cdots \cup A_p) \cup (A_{p+1} \cup A_{p+2} \cup \cdots \cup A_n) = A_1 \cup A_2 \cup \cdots \cup A_n$$

62. The operation of set intersection can be defined as an n-ary operation for any integer $n \geq 2$.

   a. Give a definition similar to that for the intersection of two sets for

   $$A_1 \cap A_2 \cap \cdots \cap A_n$$

   b. Give a recursive definition for $A_1 \cap A_2 \cap \cdots \cap A_n$.

63. Using the recursive definition of set intersection from Exercise 62(b), prove the generalized associative property of set intersection, which is that for any n with $n \geq 3$ and any p with $1 \leq p \leq n - 1$,

   $$(A_1 \cap A_2 \cap \cdots \cap A_p) \cap (A_{p+1} \cap A_{p+2} \cap \cdots \cap A_n) = A_1 \cap A_2 \cap \cdots \cap A_n$$

64. Prove that for subsets $A_1, A_2, \ldots, A_n$ and B of a set S, the following identities hold, where $n \geq 2$. (See Exercises 60 and 62.)

   ★**a.** $B \cup (A_1 \cap A_2 \cap \cdots \cap A_n) = (B \cup A_1) \cap (B \cup A_2) \cap \cdots \cap (B \cup A_n)$
   b. $B \cap (A_1 \cup A_2 \cup \cdots \cup A_n) = (B \cap A_1) \cup (B \cap A_2) \cup \cdots \cup (B \cap A_n)$

65. Prove that for subsets $A_1, A_2, \ldots, A_n$ of a set S, the following identities hold, where $n \geq 2$. (See Exercises 56, 60, and 62.)

   a. $(A_1 \cup A_2 \cup \cdots \cup A_n)' = A_1' \cap A_2' \cap \cdots \cap A_n'$
   b. $(A_1 \cap A_2 \cap \cdots \cap A_n)' = A_1' \cup A_2' \cup \cdots \cup A_n'$

66. The operations of set union and set intersection can be extended to apply to an infinite family of sets. We may describe the family as the collection of all sets A_i, where i takes on any of the values of a fixed set I. Here, I is called the *index set* for the family. The union of the family, $\bigcup_{i \in I} A_i$, is defined by

   $$\bigcup_{i \in I} A_i = \{x \mid x \text{ is a member of some } A_i\}$$

   The intersection of the family, $\bigcap_{i \in I} A_i$, is defined by

   $$\bigcap_{i \in I} A_i = \{x \mid x \text{ is a member of each } A_i\}$$

   a. Let $I = \{1, 2, 3, \ldots\}$, and for each $i \in I$, let A_i be the set of real numbers in the interval $(-1/i, 1/i)$. What is $\bigcup_{i \in I} A_i$? What is $\bigcap_{i \in I} A_i$?
   b. Let $I = \{1, 2, 3 \ldots\}$, and for each $i \in I$, let A_i be the set of real numbers in the interval $[-1/i, 1/i]$. What is $\bigcup_{i \in I} A_i$? What is $\bigcap_{i \in I} A_i$?

67. According to our use of the word "set," if A is a subset of the universal set S, then every element of S either does or does not belong to A. In other words, the probability of a member x of S being a member of A is either 1 (x is a member of A) or 0 (x is not a member of A). A is a *fuzzy set* if every $x \in S$ has a probability p, $0 \le p \le 1$, of being a member of A. The probability p associated with x is an estimate of the likelihood that x may belong to A when the actual composition of A is unknown. Set operations can be done on fuzzy sets as follows: If element x has probability p_1 of membership in A and probability p_2 of membership in B, then the probability of x being a member of $A \cup B$, $A \cap B$, and A' is, respectively, $\max(p_1, p_2)$, $\min(p_1, p_2)$, and $1 - p$.

Let S be a set of possible disease-causing agents, $S = \{\text{genetics, virus, nutrition, bacteria, environment}\}$. The fuzzy sets AIDS and ALZHEIMERS are defined as AIDS = {genetics, 0.2; virus, 0.8; nutrition, 0.1; bacteria, 0.4; environment, 0.3} and ALZHEIMERS = {genetics, 0.7; virus, 0.4; nutrition, 0.3; bacteria, 0.3; environment, 0.4}.

   a. Find the fuzzy set AIDS $\cup$ ALZHEIMERS.
   b. Find the fuzzy set AIDS $\cap$ ALZHEIMERS.
   c. Find the fuzzy set (AIDS)$'$.

68. The *principle of well-ordering* says that every nonempty set of positive integers has a smallest member. Prove that the first principle of mathematical induction, that is,

$$\left.\begin{array}{l} 1.\ P(1) \text{ is true} \\ 2.\ (\forall k)[P(k) \text{ true} \to P(k + 1) \text{ true}] \end{array}\right\} \to P(n) \text{ true for all positive integers } n$$

implies the principle of well-ordering. (*Hint:* Assume that the first principle of mathematical induction is valid, and use proof by contradiction to show that the principle of well-ordering is valid. Let T be a nonempty subset of the positive integers that has no smallest member. Let $P(n)$ be the property that every member of T is greater than n.)

★69. Prove that the principle of well-ordering (see Exercise 68) implies the second principle of mathematical induction. *Hint:* Assume that the principle of well-ordering is valid, and let P be a property for which

   $1'$. $P(1)$ true
   $2'$. $(\forall k)[P(r)$ true for all r, $1 \le r \le k \to P(k + 1)$ true]

Let T be the subset of the positive integers defined by

   $$T = \{t \mid P(t) \text{ is not true}\}$$

Show that T is the empty set.

★70. Prove that the set of odd positive integers is denumerable.

71. Prove that the set $\mathbb{Z}$ of all integers is denumerable.

★72. Prove that the set of all finite-length strings of the letter a is denumerable.

73. Prove that the set $\mathbb{Z} \times \mathbb{Z}$ is denumerable.

74. Use Cantor's diagonalization method to show that the set of all infinite sequences of positive integers is not countable.

75. Use Cantor's diagonalization method to show that the set of all infinite strings of the letters $\{a, b\}$ is not countable.

76. Explain why any subset of a countable set is countable.

★77. Explain why the union of any two denumerable sets is denumerable.

78. Sets can have sets as elements (see Exercise 10, for example). Let B be the set defined as follows:

$$B = \{S \mid S \text{ is a set and } S \notin S\}$$

Argue that both $B \in B$ and $B \notin B$ are true. This contradiction is called *Russell's paradox*, after the famous philosopher and mathematician Bertrand Russell, who stated it in 1901. (A carefully constructed axiomatization of set theory puts some restrictions on what can be called a set. All ordinary sets are still sets, but peculiar sets that get us into trouble, like B in this exercise, seem to be avoided.)

Section 3.2 Counting

Combinatorics is the branch of mathematics that deals with counting. Counting questions are important whenever we have finite resources (How much storage does a particular database consume? How many users can a given computer configuration support?) or whenever we are interested in efficiency (How many computations does a particular algorithm involve?).

Counting questions often boil down to finding how many members there are in some finite set. This seemingly trivial question can be difficult to answer. We have already answered some "how many" questions—How many rows are there in a truth table with n statement letters, and how many subsets are there in a set with n elements? (Actually, as we've noted, these can be thought of as the same question.)

Multiplication Principle

We solved the truth table question by drawing a tree of possibilities. This tree suggests a general principle that can be used to solve many counting problems. Before we state the general principle, we'll look at another tree example.

EXAMPLE 22   A child is allowed to choose one jellybean out of two jellybeans, one red and one black, and one gummy bear out of three gummy bears, yellow, green, and white. How many different sets of candy can the child have?

We can solve this problem by breaking the task of choosing candy into two sequential tasks of choosing the jellybean and then choosing the gummy bear. The tree of Figure 3.3 shows that there are $2 \times 3 = 6$ possible outcomes: {R, Y}, {R, G}, {R, W}, {B, Y}, {B, G}, and {B, W}.

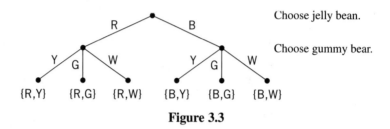

Choose jelly bean.

Choose gummy bear.

Figure 3.3

In this problem the sequence of events could be reversed; the child could choose the gummy bear first and the jellybean second, resulting in the tree of Figure 3.4, but the number of outcomes is the same ($3 \times 2 = 6$). Thinking of a sequence of successive events helps us solve the problem, but the sequencing is not a part of the problem since the set {R, Y} is the same as the set {Y, R}.

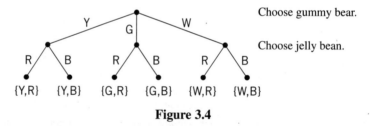

Choose gummy bear.

Choose jelly bean.

Figure 3.4

Example 22 illustrates that the total number of outcomes for a sequence of events can be obtained by multiplying the number of outcomes for the first event by the number of outcomes for the second. This idea is summarized in the *Multiplication Principle*.

Multiplication Principle
If there are n_1 possible outcomes for a first event and n_2 possible outcomes for a second event, there are $n_1 \cdot n_2$ possible outcomes for the sequence of the two events.

The Multiplication Principle can be extended by induction to apply to a sequence of any finite number of events. (See Exercise 64 at the end of this section.) The Multiplication Principle is useful whenever we want to count the total number of possible outcomes for a task that can be broken down into a sequence of successive subtasks.

EXAMPLE 23   The last part of your telephone number contains four digits. How many such four-digit numbers are there?

We can construct four-digit numbers by performing a sequence of subtasks: choose the first digit, then the second, the third, and finally the fourth. The first digit can be any one of the 10 digits from 0 to 9, so there are 10 possible outcomes for the first subtask. Likewise, there are 10 different possibilities each for the second digit, the third, and the fourth. Using the Multiplication Principle, we multiply the number of outcomes for each subtask in the sequence. Therefore there are $10 \cdot 10 \cdot 10 \cdot 10 = 10{,}000$ different numbers.

If an element cannot be used again—that is, if repetitions are not allowed—the number of possible outcomes for successive events will be affected.

EXAMPLE 24   Referring to Example 23, how many four-digit numbers are there if the same digit cannot be used twice?

Again we have the sequence of subtasks of selecting the four digits, but no repetitions are allowed. There are 10 choices for the first digit, but only 9 choices for the second because we can't use what we used for the first digit, and so on. There are $10 \cdot 9 \cdot 8 \cdot 7 = 5040$ different numbers.

EXAMPLE 25

    a. How many ways are there to choose three officers from a club of 25 people?
    b. How many ways are there to choose three officers from a club of 25 people if someone can hold more than one office?

In (a), there are three successive subtasks with no repetitions. The first subtask, choosing the first officer, has 25 possible outcomes. The second subtask has 24 outcomes, the third 23 outcomes. The total number of outcomes is $25 \cdot 24 \cdot 23 = 13{,}800$. In (b), the same three subtasks are done in succession, but repetitions are allowed. The total number of outcomes is $25 \cdot 25 \cdot 25 = 15{,}625$.

PRACTICE 22   If a man has four suits, eight shirts, and five ties, how many outfits can he put together?

EXAMPLE 26   For any finite set S, let $|S|$ denote the number of elements in S. If A and B are finite sets, then

$$|A \times B| = |A| \cdot |B|$$

$A \times B$ consists of all ordered pairs with first component from A and second component from B. Forming such ordered pairs can be thought of as the sequence of tasks of choosing the first component, for which there are $|A|$ outcomes, and then choosing the second component, for which there are $|B|$ outcomes. The result follows from the Multiplication Principle.

Addition Principle

Suppose we want to select a dessert from three pies and four cakes. In how many ways can this be done? There are two events, one with three outcomes (choosing a pie) and one with four outcomes (choosing a cake). However, we are not doing a sequence of two events here, since we are getting only one dessert, which must be chosen from the two disjoint sets of possibilities. The number of different outcomes is the total number of choices we have, $3 + 4 = 7$. This illustrates the *Addition Principle*.

> **REMINDER:**
> Use the Addition Principle only when the events are disjoint—have no common outcomes.

Addition Principle

If A and B are disjoint events with n_1 and n_2 possible outcomes, respectively, then the total number of possible outcomes for event "A or B" is $n_1 + n_2$.

The Addition Principle can be extended by induction to the case of any finite number of disjoint events. (See Exercise 65 at the end of this section.) The Addition Principle is useful whenever we want to count the total number of possible outcomes for a task that can be broken down into disjoint cases.

EXAMPLE 27

A customer wants to purchase a vehicle from a dealer. The dealer has 23 autos and 14 trucks in stock. How many selections does the customer have?

The customer wants to choose a car or truck. These are disjoint events; choosing an auto has 23 outcomes and choosing a truck has 14. By the Addition Principle, choosing a vehicle has $23 + 14 = 37$ outcomes. Notice the requirement that the outcomes for events A and B be disjoint sets. Thus, if a customer wanted to purchase a vehicle from a dealer who had 23 autos, 14 trucks, and 17 red vehicles in stock, we could not conclude that the customer had $23 + 14 + 17$ choices!   ●

EXAMPLE 28

Let A and B be disjoint finite sets. Then $|A \cup B| = |A| + |B|$.

Finding $|A \cup B|$ can be done by the disjoint cases of counting the number of elements in A, $|A|$ and the number of elements in B, $|B|$. By the Addition Principle, we sum these two numbers.   ●

EXAMPLE 29

If A and B are finite sets, then

$$|A - B| = |A| - |A \cap B|$$

and

$$|A - B| = |A| - |B| \text{ if } B \subseteq A$$

To prove the first equality, note that

$$(A - B) \cup (A \cap B) = (A \cap B') \cup (A \cap B)$$
$$= A \cap (B' \cup B)$$
$$= A \cap S$$
$$= A$$

so that $A = (A - B) \cup (A \cap B)$. Also, $A - B$ and $A \cap B$ are disjoint sets; therefore, by Example 28,

$$|A| = |(A - B) \cup (A \cap B)| = |A - B| + |A \cap B|$$

or

$$|A - B| = |A| - |A \cap B|$$

The second equation follows from the first, because if $B \subseteq A$, then $A \cap B = B$.   ●

Using the Principles Together

Frequently the Addition Principle is used in conjunction with the Multiplication Principle.

EXAMPLE 30   Referring to Example 22, suppose we want to find how many different ways the child can *choose* the candy, rather than the number of sets of candy the child can have. Then choosing a red jellybean followed by a yellow gummy bear is not the same as choosing a yellow gummy bear followed by a red jellybean. We can consider two disjoint cases—choosing jellybeans first or choosing gummy bears first. Each of these cases (by the Multiplication Principle) has six outcomes, so (by the Addition Principle) there are $6 + 6 = 12$ possible ways to choose the candy.   ●

EXAMPLE 31   How many four-digit numbers begin with a 4 or a 5?

We can consider the two disjoint cases—numbers that begin with 4 and numbers that begin with 5. Counting the numbers that begin with 4, there is 1 outcome for the subtask of choosing the first digit, then 10 possible outcomes for the subtasks of choosing each of the other three digits. Hence, by the Multiplication Principle there are $1 \cdot 10 \cdot 10 \cdot 10 = 1000$ ways to get a four-digit number beginning with 4. The same reasoning shows that there are 1000 ways to get a four-digit number beginning with 5. By the Addition Principle, there are $1000 + 1000 = 2000$ total possible outcomes.   ●

PRACTICE 23   If a woman has seven blouses, five skirts, and nine dresses, how many different outfits does she have?   ●

Often a counting problem can be solved in more than one way. Although the possibility of a second solution might seem confusing, it provides an excellent way to check our work; if two different ways of looking at the problem produce the same answer, it increases our confidence that we have analyzed the problem correctly.

EXAMPLE 32   Consider the problem of Example 31 again. We can avoid using the Addition Principle by thinking of the problem as four successive subtasks, where the first subtask, choosing the first digit, has two possible outcomes—choosing a 4 or choosing a 5. Then there are $2 \cdot 10 \cdot 10 \cdot 10 = 2000$ possible outcomes.   ●

EXAMPLE 33   How many three-digit integers (numbers between 100 and 999 inclusive) are even?

One solution notes that an even number ends in 0, 2, 4, 6, or 8. Taking these as separate cases, the number of three-digit integers ending in 0 can be found by choosing the three digits in turn. There are 9 choices, 1 through 9, for the first digit; 10 choices, 0 through 9, for the second digit; and 1 choice for the third digit, 0. By the Multiplication Principle, there are 90 numbers ending in 0. Similarly, there are 90 numbers ending in 2, 4, 6, and 8, so by the Addition Principle, there are 90 + 90 + 90 + 90 + 90 = 450 numbers.

Another solution takes advantage of the fact that there are only 5 choices for the third digit. By the Multiplication Principle, there are $9 \cdot 10 \cdot 5 = 450$ numbers.

For this problem, there is a third solution of the "serendipity" type we discussed in Section 2.1. There are $999 - 100 + 1 = 900$ three-digit integers. Half are even and half are odd, so 450 of them must be even.   ●

EXAMPLE 34   Suppose the last four digits of a telephone number must include at least one repeated digit. How many such numbers are there?

Although it is possible to do this problem by using the Addition Principle directly, it is difficult because there are so many disjoint cases to consider. For example, if the first two digits are alike but the third and fourth are different, there are $10 \cdot 1 \cdot 9 \cdot 8$ ways this can happen. If the first and third digit are alike but the second and fourth are different, there are $10 \cdot 9 \cdot 1 \cdot 8$ ways this can happen. If the first two digits are alike and the last two are also alike but different from the first two, there are $10 \cdot 1 \cdot 9 \cdot 1$ such numbers. Clearly, there are many other possibilities.

Instead, we solve the problem by noting that numbers with repetitions and numbers with no repetitions are disjoint sets whose union equals all four-digit numbers. By Example 28 we can find the number with repetitions by subtracting the number with no repetitions (5040, according to Example 24) from the total number (10,000, according to Example 23). Therefore, there are 4960 numbers with repetitions.   ●

Decision Trees

Trees such as those in Figures 3.3 and 3.4 illustrate the number of outcomes of an event based on a series of possible choices. Such trees are called **decision trees**. We shall see in Chapter 5 how decision trees are used in analyzing algorithms, but for now we shall use them to solve additional counting problems. The trees of Figures 3.3 and 3.4 led to the Multiplication Principle because the number of outcomes at any one level of the tree is the same throughout that level. In Figure 3.4, for example, level 2 of the tree shows two outcomes for each of the 3 branches formed at level 1. Less regular decision trees can still be used to solve counting problems where the Multiplication Principle does not apply.

EXAMPLE 35   Tony is pitching pennies. Each toss results in heads (H) or tails (T). How many ways can he toss the coin five times without having two heads in a row?

Figure 3.5 shows the decision tree for this problem. Each coin toss has 2 outcomes; the left branch is labeled H for heads, the right branch is labeled T for tails. Whenever an H appears on a branch, the next level can only contain a right (T) branch. There are 13 possible outcomes.

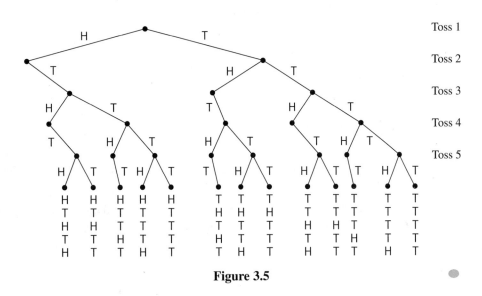

Figure 3.5

PRACTICE 24   Explain why the Multiplication Principle does not apply to Example 35.

PRACTICE 25   Draw the decision tree for the number of strings of *X*'s, *Y*'s, and *Z*'s with length 3 that do not have a *Z* following a *Y*.

Section 3.2 Review

Technique

- Use the Multiplication Principle, the Addition Principle, and decision trees for counting the number of objects in a finite set.

Main Ideas

The Multiplication Principle is used to count the number of possible outcomes for a sequence of events, each of which has a fixed number of outcomes.

The Addition Principle is used to count the number of possible outcomes for disjoint events.

The Multiplication and Addition principles are often used together.

Decision trees can be used to count the number of outcomes for a sequence of events where the number of outcomes for a given event is not constant but depends on the outcome of the preceding event.

Exercises 3.2

★1. A frozen yogurt shop allows you to choose one flavor (vanilla, strawberry, lemon, cherry, or peach), one topping (chocolate shavings, crushed toffee, or crushed peanut brittle), and one condiment (whipped cream or shredded coconut). How many different desserts are possible?

★2. In Exercise 1, how many dessert choices do you have if you are allergic to strawberries and chocolate?

3. A video game on a microcomputer is begun by making selections from each of three menus. The first menu (number of players) has four selections, the second menu (level of play) has eight, and the third menu (speed) has six. In how many configurations can the game be played?

4. A multiple choice exam has 20 questions, each with four possible answers, and 10 additional questions, each with five possible answers. How many different answer sheets are possible?

5. A user's password to access a computer system consists of three letters followed by two digits. How many different passwords are possible?

6. On the computer system of Exercise 5, how many passwords are possible if uppercase and lowercase letters can be distinguished?

★7. A telephone conference call is being placed from Central City to Booneville by way of Cloverdale. There are 45 trunk lines from Central City to Cloverdale and 13 from Cloverdale to Booneville. How many different ways can the call be placed?

8. A, B, C, and D are nodes on a computer network. There are two paths between A and C, two between B and D, three between A and B, and four between C and D. Along how many routes can a message from A to D be sent?

★9. How many Social Security numbers are possible?

10. An apartment building purchases a new lock system for its 175 units. A lock is opened by punching in a two-digit code. Has the apartment management made a wise purchase?

★11. A palindrome is a string of characters that reads the same forward and backward. How many five-letter English language palindromes are possible?

12. How many three-digit numbers less than 600 can be made using the digits 8, 6, 4, and 2?

13. A binary logical connective can be defined by giving its truth table. How many different binary logical connectives are there?

★14. In the original BASIC programming language, an identifier must be either a single letter or a letter followed by a single digit. How many identifiers are possible?

15. Three seats on the county council are to be filled, each with someone from a different party. There are four candidates running from the Concerned Environmentalist party, three from the Limited Development party, and two from the Friends of the Spotted Newt party. In how many ways can the seats be filled?

16. A president and vice-president must be chosen for the executive committee cf an organization. There are 17 volunteers from the Eastern Division and 24 volunteers from the Western Division. If both officers must come from the same division, in how many ways can the officers be selected?

★17. A dinner special allows you to select from five appetizers, three salads, four entrees, and three beverages. How many different dinners are there?

18. In Exercise 17, how many different dinners are there if you may have an appetizer or a salad but not both?

19. A new car can be ordered with a choice of 10 exterior colors; 7 interior colors; automatic, 3-speed, or 5-speed transmission; with or without air conditioning; with or without power steering; and with or without the option package that contains the power door lock and the rear window defroster. How many different cars can be ordered?

20. In Exercise 19, how many different cars can be ordered if the option package comes only on a car with an automatic transmission?

★21. In one state, automobile license plates must have two digits (no leading zeros) followed by one letter followed by a string of two to four digits (leading zeros are allowed). How many different plates are possible?

22. A customer at a fast-food restaurant can order a hamburger with or without mustard, ketchup, pickle, or onion; a fish sandwich with or without lettuce, tomato, or tartar sauce; and a choice of three kinds of soft drinks or two kinds of milk shakes. How many different orders are possible if a customer can order at most one hamburger, one fish sandwich, and one beverage but can order less?

★23. What is the value of *Count* after the following pseudocode has been executed?

```
Count = 0
for i = 1 to 5 do
    for Letter = 'A' to 'C' do
        Count = Count + 1
    end for
end for
```

24. What is the value of *Result* after the following pseudocode has been executed?

```
Result = 0
for Index = 20 down to 10 do
    for Inner 5 to 10 do
        Result = Result + 2
    end for
end for
```

Exercises 25–30 concern the set of three-digit integers (numbers between 100 and 999 inclusive).

25. How many are divisible by 5?

★**26.** How many are not divisible by 5?

27. How many are divisible by 4?

28. How many are divisible by 4 or 5?

29. How many are divisible by 4 and 5?

30. How many are divisible by neither 4 nor 5?

Exercises 31–40 concern the set of binary strings of length 8 (each character is either the digit 0 or the digit 1).

★**31.** How many such strings are there?

32. How many begin and end with 0?

★**33.** How many begin or end with 0?

34. How many have 1 as the second digit?

35. How many begin with 111?

36. How many contain exactly one 0?

37. How many begin with 10 or have a 0 as the third digit?

38. How many are palindromes? (See Exercise 11.)

★**39.** How many contain exactly seven 1s?

40. How many contain two or more 0s?

In Exercises 41–45, two dice are rolled, one black and one white.

41. How many different rolls are possible? (Note that a 4-black, 1-white result and a 1-black, 4-white result are two different outcomes.)

★**42.** How many rolls result in doubles (both dice showing the same value)?

43. How many rolls result in "snake eyes" (both dice showing 1)?

44. How many rolls result in a total of 7 or 11?

45. How many rolls occur in which neither die shows the value 4?

In Exercises 46–50, a customer is ordering a computer. The choices are 15", 17", 19", or 21" monitor; 266 MHz, 300 MHz, 333 MHz, 350 MHz, or 400 MHz processor; 6×, 8×, or 12× CD drive; 32 MB, 64 MB, or 128 MB of RAM; optional fax card; optional sound card.

46. How many different machine configurations are possible?

47. How many different machines can be ordered with a 350-MHz processor?

48. How many machines can be ordered with a 19" monitor but no sound card and no fax card?

49. How many machines can be ordered with no monitor?

50. How many machines can be ordered with a minimum 333-MHz processor and either a sound card or a fax card but not both?

In Exercises 51–60, a hand consists of 1 card drawn from a standard 52-card deck with flowers on the back and 1 card drawn from a standard 52-card deck with birds on the back.

★51. How many different hands are possible? (Note that a flower-ace-of-spades, bird-queen-of-hearts and a flower-queen-of-hearts, bird-ace-of-spades are two different outcomes.)

52. How many hands consist of a pair of aces?

53. How many hands contain all face cards?

★54. How many hands consist of two of a kind? (two aces, two jacks, etc.)

55. How many hands contain exactly one king?

56. How many hands have a face value of 5 (aces count as 1)?

57. How many hands have a face value of less than 5?

58. How many hands do not contain any face cards?

★59. How many hands contain at least one face card?

60. How many hands contain at least one king?

61. Draw a decision tree to find the number of binary strings of length 4 that do not have consecutive 0s.

★62. Voting on a certain issue is conducted by having everyone put a red, blue, or green slip of paper into a hat. Then the slips are pulled out one at a time. The first color to receive two votes wins. Draw a decision tree to find the number of ways in which the balloting can occur.

63. Draw a decision tree (use teams A and B) to find the number of ways the NBA playoffs can happen, where the winner is the first team to win 4 out of 7.

64. Use mathematical induction to extend the Multiplication Principle to a sequence of m events for any integer m, $m \geq 2$.

65. Use mathematical induction to extend the Addition Principle to m disjoint events for any integer m, $m \geq 2$.

Section 3.3   Principle of Inclusion and Exclusion; Pigeonhole Principle

In this section we'll discuss two more counting principles that can be used to solve combinatorics problems.

Principle of Inclusion and Exclusion

To develop the Principle of Inclusion and Exclusion, we first note that if A and B are any subsets of a universal set S, then $A - B$, $B - A$, and $A \cap B$ are mutually disjoint sets (see Figure 3.6). For example, if $x \in A - B$, then $x \notin B$; therefore $x \notin B - A$ and $x \notin A \cap B$. Also, something can be said about the union of these three sets.

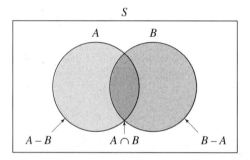

Figure 3.6

PRACTICE 26   What is another name for the set $(A - B) \cup (B - A) \cup (A \cap B)$?   ●

From Example 28 (extended to three disjoint finite sets),

$$|(A - B) \cup (B - A) \cup (A \cap B)| = |A - B| + |B - A| + |A \cap B| \qquad (1)$$

From Example 29,

$$|A - B| = |A| - |A \cap B|$$

and

$$|B - A| = |B| - |A \cap B|$$

Using these expressions in equation (1), along with the result of Practice 26, we get

$$|A \cup B| = |A| - |A \cap B| + |B| - |A \cap B| + |A \cap B|$$

or

$$|A \cup B| = |A| + |B| - |A \cap B| \qquad (2)$$

Equation (2) is the two-set version of the Principle of Inclusion and Exclusion. The name derives from the fact that when counting the number of elements in the union of A and B, we must "include" (count) the number of elements in A and the number of elements in B, but we must "exclude" (subtract) those elements in $A \cap B$ to avoid counting them twice.

PRACTICE 27   How does equation (2) relate to Example 28 of Section 3.2?   ●

EXAMPLE 36   A pollster queries 35 voters, all of whom support referendum 1, referendum 2, or both, and finds that 14 voters support referendum 1 and 26 support referendum 2. How many voters support both?

If we let A be the set of voters supporting referendum 1 and B be the set of voters supporting referendum 2, then we know that

$$|A \cup B| = 35 \qquad |A| = 14 \qquad |B| = 26$$

From equation (2),

$$|A \cap B| = |A| + |B| - |A \cup B| = 14 + 26 - 35 = 5$$

so 5 voters support both.   ●

Equation (2) can easily be extended to three sets, as follows:

$$\begin{aligned}
|A \cup B \cup C| = |A \cup (B \cup C)| &= |A| + |B \cup C| - |A \cap (B \cup C)| \\
&= |A| + |B| + |C| - |B \cap C| - |(A \cap B) \cup (A \cap C)| \\
&= |A| + |B| + |C| - |B \cap C| - (|A \cap B| + |A \cap C| - |A \cap B \cap C|) \\
&= |A| + |B| + |C| - |A \cap B| - |A \cap C| - |B \cap C| + |A \cap B \cap C|
\end{aligned}$$

Therefore the three-set version of the Principle of Inclusion and Exclusion is

$$\begin{aligned}
|A \cup B \cup C| = |A| + |B| + |C| - |A \cap B| \\
- |A \cap C| - |B \cap C| + |A \cap B \cap C|
\end{aligned} \qquad (3)$$

PRACTICE 28   Justify each of the equalities used in deriving equation (3).   ●

EXAMPLE 37   A group of students plan to order pizza. If 13 will eat sausage topping, 10 will eat pepperoni, 12 will eat extra cheese, 4 will eat both sausage and pepperoni, 5 will eat both pepperoni and extra cheese, 7 will eat both sausage and extra cheese, and 3 will eat all three toppings, how many students are in the group?

Let

$A = \{$students who will eat sausage$\}$
$B = \{$students who will eat pepperoni$\}$
$C = \{$students who will eat extra cheese$\}$

Then $|A| = 13$, $|B| = 10$, $|C| = 12$, $|A \cap B| = 4$, $|B \cap C| = 5$, $|A \cap C| = 7$, and $|A \cap B \cap C| = 3$. From Equation (3),

$$|A \cup B \cup C| = 13 + 10 + 12 - 4 - 5 - 7 + 3 = 22$$ ●

EXAMPLE 38   A produce stand sells only broccoli, carrots, and okra. One day the stand served 207 people. If 114 people purchased broccoli, 152 purchased carrots, 25 purchased okra, 64 purchased broccoli and carrots, 12 purchased carrots and okra, and 9 purchased all three, how many people purchased broccoli and okra?

Let

A = {people who purchased broccoli}
B = {people who purchased carrots}
C = {people who purchased okra}

Then $|A \cup B \cup C| = 207$, $|A| = 114$, $|B| = 152$, $|C| = 25$, $|A \cap B| = 64$, $|B \cap C| = 12$, and $|A \cap B \cap C| = 9$. From equation (3),

$$|A \cap C| = 114 + 152 + 25 - 64 - 12 + 9 - 207 = 17$$ ●

In equation (2), we add the number of elements in the single sets and subtract the number of elements in the intersection of two sets. In equation (3), we add the number of elements in the single sets, subtract the number of elements in the intersection of two sets, and add the number of elements in the intersection of three sets. This seems to suggest a pattern: If we have n sets, we should add the number of elements in the single sets, subtract the number of elements in the intersection of two sets, add the number of elements in the intersection of three sets, subtract the number of elements in the intersection of four sets, and so on. This leads us to the general form of the Principle of Inclusion and Exclusion.

Principle of Inclusion and Exclusion
Given the finite sets $A_1, \ldots, A_n$, $n \geq 2$, then

$$
\begin{aligned}
|A_1 \cup \cdots \cup A_n| = & \sum_{1 \leq i \leq n} |A_i| - \sum_{1 \leq i < j \leq n} |A_i \cap A_j| \\
& + \sum_{1 \leq i < j < k \leq n} |A_i \cap A_j \cap A_k| \\
& - \cdots + (-1)^{n+1} |A_1 \cap \cdots \cap A_n|
\end{aligned}
$$ (4)

In equation (4) the notation

$$\sum_{1 \leq i < j \leq n} |A_i \cap A_j|$$

for example, says to add together the number of elements in all the intersections of the form $A_i \cap A_j$ where i and j can take on any values between 1 and n as long as $i < j$. For $n = 3$, this gives $|A_1 \cap A_2|$ ($i = 1, j = 2$), $|A_1 \cap A_3|$ ($i = 1, j = 3$), and $|A_2 \cap A_3|$ ($i = 2, j = 3$). This agrees with equation (3), where $A_1 = A$, $A_2 = B$, and $A_3 = C$.

To prove the general form of the Principle of Inclusion and Exclusion, we use mathematical induction. Although the idea of the proof is straightforward, the notation is rather messy. The base case, $n = 2$, is just equation (2). We assume that equation (4) is true for $n = k$ and show that it is true for $n = k + 1$. We write

$$|A_1 \cup \cdots \cup A_{k+1}|$$
$$= |(A_1 \cup \cdots \cup A_k) \cup A_{k+1}|$$
$$= |A_1 \cup \cdots \cup A_k| + |A_{k+1}|$$
$$\quad - |(A_1 \cup \cdots \cup A_k) \cap A_{k+1}| \qquad \text{(by equation (2))}$$

$$\overset{①}{=} \sum_{1 \le i \le k} |A_i| - \sum_{1 \le i < j \le k} |A_i \cap A_j| + \sum_{1 \le i < j < m \le k} |A_i \cap A_j \cap A_m|$$

$$\overset{①}{} \quad - \cdots + (-1)^{k+1}|A_1 \cap \cdots \cap A_k| + |A_{k+1}|$$
$$\quad - |(A_1 \cap A_{k+1}) \cup \cdots \cup (A_k \cap A_{k+1})|$$

(by the inductive hypothesis and the distributive property)

$$= \sum_{1 \le i \le k+1} |A_i| - \overset{②}{\sum_{1 \le i < j \le k}} |A_i \cap A_j| + \overset{③}{\sum_{1 \le i < j < m \le k}} |A_i \cap A_j \cap A_m|$$

$$\overset{④}{} \quad - \cdots + (-1)^{k+1}|A_1 \cap \cdots \cap A_k|$$

$$\quad - \left(\overset{②}{\sum_{1 \le i \le k}} |A_i \cap A_{k+1}| - \overset{③}{\sum_{1 \le i < j \le k}} |A_i \cap A_j \cap A_{k+1}| + \cdots \right.$$

$$\quad + (-1)^k \overset{④}{\sum_{1 \le i < j < \cdots < m \le k}} \underbrace{|(A_i \cap A_{k+1}) \cap (A_j \cap A_{k+1}) \cap \cdots \cap (A_m \cap A_{k+1})|}_{k-1 \text{ terms}}$$

$$\quad \left. + (-1)^{k+1}|A_1 \cap \cdots \cap A_{k+1}| \right)$$

(by combining terms ① from above and using the inductive hypothesis on the k sets $A_1 \cap A_{k+1}, A_2 \cap A_{k+1}, \ldots, A_k \cap A_{k+1})$

$$= \sum_{1 \le i \le k+1} |A_i| - \sum_{1 \le i < j \le k+1} |A_i \cap A_j| + \sum_{1 \le i < j < m \le k+1} |A_i \cap A_j \cap A_m|$$

$$\quad - \cdots - (-1)^{k+1}|A_1 \cap \cdots \cap A_{k+1}|$$

(by combining like-numbered terms from above)

$$= \sum_{1 \le i \le k+1} |A_i| - \sum_{1 \le i < j \le k+1} |A_i \cap A_j| + \sum_{1 \le i < j < m \le k+1} |A_i \cap A_j \cap A_m|$$

$$\quad - \cdots + (-1)^{k+2}|A_1 \cap \cdots \cap A_{k+1}|$$

This completes the proof of equation (4).

Pigeonhole Principle

The Pigeonhole Principle has acquired its quaint name from the following idea: If more than k pigeons fly into k pigeonholes, then at least one hole will end up with more than one pigeon. Although this seems immediately obvious, we can belabor the point. Suppose each of the k pigeonholes contains at most one pigeon. Then there are at most k pigeons, not the more-than-k pigeons that supposedly flew in.

Now we'll state the Pigeonhole Principle in a less picturesque way.

> **Pigeonhole Principle**
> If more than k items are placed into k bins, then at least one bin contains more than one item.

By cleverly choosing items and bins, a number of interesting counting problems can be solved (see Example 7 of Chapter 2).

EXAMPLE 39   How many people must be in a room to guarantee that two people have last names that begin with the same initial?

There are 26 letters of the alphabet (bins). If there are 27 people, then there are 27 initials (items) to put into the 26 bins, so at least one bin will contain more than one last initial.   ●

PRACTICE 29   How many times must a single die be rolled in order to guarantee getting the same value twice?   ●

EXAMPLE 40   Prove that if 51 positive integers between 11 and 100 are chosen, then one of them must divide another.

Let the integers be $n_1, \ldots, n_{51}$. Each integer $n_i \geq 2$ can be written as a product of prime numbers (see Example 23 in Chapter 2), every prime number except 2 is odd, and the product of odd numbers is odd. Therefore for each i, $n_i = 2^{k_i} b_i$, where $k_i \geq 0$ and b_i is an odd number. Furthermore, $1 \leq b_i \leq 99$, and there are 50 odd integers between 1 and 99 inclusive. By the Pigeonhole Principle, $b_i = b_j$ for some i and j, so $n_i = 2^{k_i} b_i$ and $n_j = 2^{k_j} b_i$. If $k_i \leq k_j$, then n_i divides n_j; otherwise, n_j divides n_i.   ●

Section 3.3 Review

Techniques

- Use the Principle of Inclusion and Exclusion to find the number of elements in the union of sets.
- Use the Pigeonhole Principle to find the minimum number of elements to guarantee two with a duplicate property.

Main Idea

The Principle of Inclusion and Exclusion and the Pigeonhole Principle are additional counting mechanisms for sets.

Exercises 3.3

1. In a group of 42 tourists, everyone speaks English or French; there are 35 English speakers and 18 French speakers. How many speak both English and French?

★2. All the guests at a dinner party drink coffee or tea; 13 guests drink coffee, 10 drink tea, and 4 drink both coffee and tea. How many people are in this group?

3. Quality control in a factory pulls 40 parts with paint, packaging, or electronics defects from an assembly line. Of these, 28 had a paint defect, 17 had a packaging defect, 13 had an electronics defect, 6 had both paint and packaging defects, 7 had both packaging and electronics defects, and 10 had both paint and electronics defects. Did any part have all three types of defect?

4. In a group of 24 people who like rock, country, and classical music, 14 like rock, 17 like classical, 11 like both rock and country, 9 like rock and classical, 13 like country and classical, and 8 like rock, country, and classical. How many like country?

★5. Nineteen different mouthwash products make the following claims: 12 claim to freshen breath, 10 claim to prevent gingivitis, 11 claim to reduce plaque, 6 claim to both freshen breath and reduce plaque, 5 claim to both prevent gingivitis and freshen breath, and 5 claim to both prevent gingivitis and reduce plaque.

   a. How many products make all three claims?
   b. How many products claim to freshen breath but do not claim to prevent plaque?

6. From the 83 students who want to enroll in CS 320, 32 have completed CS 120, 27 have completed CS 180, and 35 have completed CS 215. Of these, 7 have completed both CS 120 and CS 180, 16 have completed CS 180 and CS 215, and 3 have completed CS 120 and CS 215. Two students have completed all three courses. The prerequisite for CS 320 is completion of one of CS 120, CS 180, or CS 215. How many students are not eligible to enroll?

7. Among a bank's 214 customers with checking or savings accounts, 189 have checking accounts, 73 have regular savings accounts, 114 have money market savings accounts, and 69 have both checking and regular savings accounts. No customer is allowed to have both regular savings and money market savings accounts.

   a. How many customers have both checking and money market savings accounts?
   b. How many customers have a checking account but no savings account?

★8. A survey of 150 college students reveals that 83 own automobiles, 97 own bikes, 28 own motorcycles, 53 own a car and a bike, 14 own a car and a motorcycle, 7 own a bike and a motorcycle, and 2 own all three.

   a. How many students own a bike and nothing else?
   b. How many students do not own any of the three?

9. At the beginning of this chapter you surveyed the subscribers to your newsletter in preparation for the release of your new software product.

   The results of your survey reveal that of the 87 subscribers, 68 have a Windows-based system available to them, 34 have a Unix system available, and 30 have access to a Mac. In addition, 19 have access to both Windows and Unix systems, 11 have access to both Unix systems and Macs, and 23 can use both Macs and Windows.

   Use the Principle of Inclusion and Exclusion to determine how many subscribers have access to all three types of systems.

10. You are developing a new bath soap, and you hire a public opinion survey group to do some market research for you. The group claims that in its survey of 450 consumers, the following were named as important factors in purchasing bath soap:

Odor	425
Lathering ease	397
Natural ingredients	340
Odor and lathering ease	284
Odor and natural ingredients	315
Lathering ease and natural ingredients	219
All three factors	147

    Should you have confidence in these results? Why or why not?

11. Write the expression for $|A \cup B \cup C \cup D|$ from equation (4).

12. Write an expression for the number of terms in the expansion of $|A_1 \cup \cdots \cup A_n|$ given by equation (4).

13. How many cards must be drawn from a standard 52-card deck to guarantee 2 cards of the same suit?

★14. If 12 cards are drawn from a standard deck, must at least 2 of them be of the same denomination?

15. A computerized dating service has a list of 50 men and 50 women. Names are selected at random; how many names must be chosen to guarantee one name of each gender?

16. A computerized housing service has a list of 50 men and 50 women. Names are selected at random; how many names must be chosen to guarantee two names of the same gender?

17. How many people must be in a group in order to guarantee that two people in the group have the same birthday (don't forget leap year)?

18. In a group of 25 people, must there be at least 3 who were born in the same month?

★19. Prove that if four numbers are chosen from the set $\{1, 2, 3, 4, 5, 6\}$, at least one pair must add up to 7. (*Hint:* Find all the pairs of numbers from the set that add to 7.)

20. How many numbers must be selected from the set $\{2, 4, 6, 8, 10, 12, 14, 16, 18, 20\}$ in order to guarantee that at least one pair adds up to 22? (See the hint for Exercise 19).

21. Let n be a positive number. Show that in any set of $n + 1$ numbers, there are at least two with the same remainder when divided by n.

Section 3.4   Permutations and Combinations

Permutations

Example 24 in Section 3.2 discussed the problem of counting all possibilities for the last four digits of a telephone number with no repeated digits. In this problem, the number 1259 is not the same as the number 2951 because the order of the four digits is important. An ordered arrangement of objects is called a **permutation**. Each of these numbers is a permutation of 4 distinct objects chosen from a set of 10 distinct objects (the digits). How many such permutations are there? The answer, found by using the Multiplication Principle, is $10 \cdot 9 \cdot 8 \cdot 7$—there are 10 choices for the first digit, then 9 for the next digit because repetitions are not allowed, 8 for the next digit, and 7 for the fourth digit. The number of permutations of r distinct objects chosen from n distinct objects is denoted by $P(n, r)$. Therefore the solution to the problem of the four-digit number without repeated digits can be expressed as $P(10, 4)$.

A formula for $P(n, r)$ can be written using the factorial function. For a positive integer n, **n factorial** is defined as $n(n - 1)(n - 2) \cdots 1$ and denoted by $n!$; also, $0!$ is defined to have the value 1. From the definition of $n!$, we see that

$$n! = n(n - 1)!$$

and that for $r < n$,

$$\frac{n!}{(n - r)!} = \frac{n(n - 1) \cdots (n - r + 1)(n - r)!}{(n - r)!}$$

$$= n(n - 1) \cdots (n - r + 1)$$

Using the factorial function,

$$P(10, 4) = 10 \cdot 9 \cdot 8 \cdot 7$$

$$= \frac{10 \cdot 9 \cdot 8 \cdot 7 \cdot 6 \cdot 5 \cdot 4 \cdot 3 \cdot 2 \cdot 1}{6 \cdot 5 \cdot 4 \cdot 3 \cdot 2 \cdot 1} = \frac{10!}{6!} = \frac{10!}{(10 - 4)!}$$

In general, $P(n, r)$ is given by the formula

$$P(n, r) = \frac{n!}{(n - r)!} \text{ for } 0 \leq r \leq n$$

EXAMPLE 41   The value of $P(7, 3)$ is

$$\frac{7!}{(7-3)!} = \frac{7!}{4!} = \frac{7 \cdot 6 \cdot 5 \cdot 4 \cdot 3 \cdot 2 \cdot 1}{4 \cdot 3 \cdot 2 \cdot 1} = 7 \cdot 6 \cdot 5 = 210$$

EXAMPLE 42   Three somewhat special cases that can arise when computing $P(n, r)$ are the two "boundary conditions" $P(n, 0)$ and $P(n, n)$, and also $P(n, 1)$. According to the formula,

$$P(n, 0) = \frac{n!}{(n-0)!} = \frac{n!}{n!} = 1$$

This can be interpreted as saying that there is only one ordered arrangement of zero objects—the empty set.

$$P(n, 1) = \frac{n!}{(n-1)!} = n$$

This formula reflects the fact that there are n ordered arrangements of one object. (Each arrangement consists of the one object, so this merely counts how many ways to get the one object.)

$$P(n, n) = \frac{n!}{(n-n)!} = \frac{n!}{0!} = n!$$

This formula states that there are $n!$ ordered arrangements of n distinct objects. (This merely reflects the Multiplication Principle—n choices for the first object, $n-1$ choices for the second object, and so on, with one choice for the nth object.)

EXAMPLE 43   The number of permutations of 3 objects, say a, b, and c, is given by $P(3, 3) = 3! = 3 \cdot 2 \cdot 1 = 6$. The 6 permutations of a, b, and c are

abc, acb, bac, bca, cab, cba

EXAMPLE 44   How many three-letter words (not necessarily meaningful) can be formed from the word "compiler" if no letters can be repeated? Here the arrangement of letters matters, and we want to know the number of permutations of three distinct objects taken from eight objects. The answer is $P(8, 3) = 8!/5! = 336$.

Note that we could have solved Example 44 just by using the Multiplication Principle—there are eight choices for the first letter, seven for the second, and six for the third, so the answer is $8 \cdot 7 \cdot 6 = 336$. $P(n, r)$ simply gives us a new way to think about the problem, as well as a compact notation.

EXAMPLE 45   Ten athletes compete in an Olympic event. Gold, silver, and bronze medals are awarded; in how many ways can the awards be made?

This is essentially the same problem as Example 44. Order matters; given three winners A, B, and C, the arrangement A—gold, B—silver, C—bronze is different than the arrangement C—gold, A—silver, B—bronze. So we want the number of ordered arrangements of 3 objects from a pool of 10, or $P(10, 3)$. Using the formula for $P(n, r)$, $P(10, 3) = 10!/7! = 10 \cdot 9 \cdot 8 = 720$.   ●

PRACTICE 30   In how many ways can a president and vice-president be selected from a group of 20 people?   ●

PRACTICE 31   In how many ways can six people be seated in a row of six chairs?   ●

Counting problems can have other counting problems as subtasks.

EXAMPLE 46   A library has 4 books on operating systems, 7 on programming, and 3 on data structures. Let's see how many ways these books can be arranged on a shelf, given that all books on the same subject must be together. We can think of this problem as a sequence of subtasks. First we consider the subtask of arranging the three subjects. There are 3! outcomes to this subtask, that is, 3! different orderings of subject matter. The next subtasks are arranging the books on operating systems (4! outcomes), then arranging the books on programming (7! outcomes), and finally arranging the books on data structures (3! outcomes). Thus, by the Multiplication Principle, the final number of arrangements of all the books is $(3!)(4!)(7!)(3!) = 4,354,560$.   ●

Combinations

Sometimes we want to select r objects from a set of n objects, but we don't care how they are arranged. Then we are counting the number of **combinations** of r distinct objects chosen from n distinct objects, denoted by $C(n, r)$. For each such combination, there are $r!$ ways to permute the r chosen objects. By the Multiplication Principle, the number of permutations of r distinct objects chosen from n objects is the product of the number of ways to choose the objects, $C(n, r)$, multiplied by the number of ways to arrange the objects chosen, $r!$. Thus,

$$C(n, r) \cdot r! = P(n, r)$$

or

$$C(n, r) = \frac{P(n, r)}{r!} = \frac{n!}{r!(n - r)!} \text{ for } 0 \leq r \leq n$$

Other notations for $C(n, r)$ are

$$_nC_r, \; C^n_r, \; \binom{n}{r}$$

EXAMPLE 47   The value of $C(7, 3)$ is

$$\frac{7!}{3!(7 - 3)!} = \frac{7!}{3!4!} = \frac{7 \cdot 6 \cdot 5 \cdot 4 \cdot 3 \cdot 2 \cdot 1}{3 \cdot 2 \cdot 1 \cdot 4 \cdot 3 \cdot 2 \cdot 1}$$

$$= \frac{7 \cdot 6 \cdot 5}{3 \cdot 2 \cdot 1} = 7 \cdot 5 = 35$$

From Example 41, the value of $P(7, 3)$ is 210, and $C(7, 3) \cdot (3!) = 35(6) = 210 = P(7, 3)$. ●

EXAMPLE 48   The special cases for $C(n, r)$ are $C(n, 0)$, $C(n, 1)$, and $C(n, n)$. The formula for $C(n, 0)$,

$$C(n, 0) = \frac{n!}{0!(n - 0)!} = 1$$

reflects the fact that there is only one way to choose zero objects from n objects: Choose the empty set.

$$C(n, 1) = \frac{n!}{1!(n - 1)!} = n$$

Here the formula indicates that there are n ways to select 1 object from n objects.

$$C(n, n) = \frac{n!}{n!(n - n)!} = 1$$

Here we see that there is only one way to select n objects from n objects, and that is to choose all of the objects. ●

In the formula for $C(n, r)$, suppose n is held fixed and r is increased. Then $r!$ increases, which tends to make $C(n, r)$ smaller, but $(n - r)!$ decreases, which tends to make $C(n, r)$ larger. For small values of r, the increase in $r!$ is not as great as the decrease in $(n - r)!$, and so $C(n, r)$ increases from 1 to n to larger values. At some point, however, the increase in $r!$ overcomes the decrease in $(n - r)!$, and the values of $C(n, r)$ decrease back down to 1 by the time $r = n$, as we calculated in Example 48. Figure 3.7a illustrates the rise and fall of the values of $C(n, r)$ for a fixed n. For $P(n, r)$, as n is held fixed and r is increased, $n - r$ and therefore $(n - r)!$ decreases, so $P(n, r)$ increases. Values of $P(n, r)$ for $0 \le r \le n$ thus increase from 1 to n to $n!$, as we calculated in Example 42. See Figure 3.7b; note the difference in the vertical scale of Figures 3.7a and 3.7b.

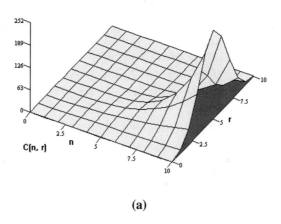

(a)

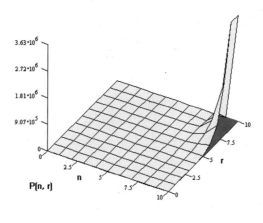

(b)

Figure 3.7

EXAMPLE 49   How many 5-card poker hands can be dealt from a 52-card deck? Here order does not matter because we simply want to know which cards end up in the hand. We want the number of ways to choose 5 objects from a pool of 52, which is a combinations problem. The answer is $C(52, 5) = 52!/(5!47!) = 2{,}598{,}960$.

Unlike earlier problems, the answer to Example 49 cannot easily be obtained by applying the Multiplication Principle. Thus, $C(n, r)$ gives us a way to solve new problems.

EXAMPLE 50   Ten athletes compete in an Olympic event; three will be declared winners. In how many ways can the winners be chosen?

Unlike Example 45, there is no order to the three winners, so we are simply choosing 3 objects out of 10. This is a combinations problem, not a permutations problem. The result is $C(10, 3) = 10!/(3!7!) = 120$. Notice that there are fewer ways to choose three winners (a combinations problem) than to award gold, silver, and bronze medals to three winners (a permutations problem—Example 45).   ●

PRACTICE 32   In how many ways can a committee of 3 be chosen from a group of 12 people?   ●

> **REMINDER:**
> In a counting problem, first ask yourself if order matters. If it does, it's a permutations problem. If not, it's a combinations problem.

Remember that the distinction between permutations and combinations lies in whether the objects are to be merely selected or both selected and ordered. If ordering is important, the problem involves permutations; if ordering is not important, the problem involves combinations. For example, Practice 30 is a permutations problem—two people are to be selected and ordered, the first as president, the second as vice-president—whereas Practice 32 is a combinations problem—three people are selected but not ordered.

In solving counting problems, $C(n, r)$ can be used in conjunction with the Multiplication Principle or the Addition Principle.

EXAMPLE 51   A committee of 8 students is to be selected from a class consisting of 19 freshmen and 34 sophomores.

    a. In how many ways can 3 freshmen and 5 sophomores be selected?
    b. In how many ways can a committee with exactly 1 freshman be selected?
    c. In how many ways can a committee with at most 1 freshman be selected?
    d. In how many ways can a committee with at least 1 freshman be selected?

Because the ordering of the individuals chosen is not important, these are combinations problems.

For part (a), we have a sequence of two subtasks, selecting freshmen and selecting sophomores. The Multiplication Principle should be used. (Thinking of a sequence of subtasks may seem to imply ordering, but it just sets up the levels of the decision tree, the basis for the Multiplication Principle. There is no ordering of the students.) Because there are $C(19, 3)$ ways to choose the freshmen and $C(34, 5)$ ways to choose the sophomores, the answer is

$$C(19, 3) \cdot C(34, 5) = \frac{19!}{3!16!} \cdot \frac{34!}{5!29!} = (969)(278{,}256)$$

For part (b), we again have a sequence of subtasks: selecting the single freshman and then selecting the rest of the committee from among the sophomores. There are $C(19, 1)$ ways to select the single freshman and $C(34, 7)$ ways to select the remaining 7 members from the sophomores. By the Multiplication Principle, the answer is

$$C(19, 1) \cdot C(34, 7) = \frac{19!}{1!(19-1)!} \cdot \frac{34!}{7!(34-7)!} = 19(5,379,616)$$

For part (c), we get at most 1 freshman by having exactly 1 freshman or by having 0 freshmen. Because these are disjoint events, we use the Addition Principle. The number of ways to select exactly 1 freshman is the answer to part (b). The number of ways to select 0 freshmen is the same as the number of ways to select the entire 8-member committee from among the 34 sophomores, $C(34, 8)$. Thus the answer is

$$C(19, 1) \cdot C(34, 7) + C(34, 8) = \text{some big number}$$

We can attack part (d) in several ways. One way is to use the Addition Principle, thinking of the disjoint possibilities as exactly 1 freshman, exactly 2 freshmen, and so on, up to exactly 8 freshmen. We could compute each of these numbers and then add them. However, it is easier to do the problem by counting all the ways the committee of 8 can be selected from the total pool of 53 people and then eliminating (subtracting) the number of committees with 0 freshmen (all sophomores). Thus the answer is

$$C(53, 8) - C(34, 8) \qquad \bullet$$

> **REMINDER:** "At least" counting problems are often best solved by subtraction.

The factorial function grows large quickly. A number like 100! cannot be computed on most calculators (or on most computers unless double-precision arithmetic is used), but expressions like

$$\frac{100!}{25!75!}$$

can nevertheless be computed by first canceling common factors.

Eliminating Duplicates

We mentioned earlier that counting problems can often be solved in different ways. Unfortunately, it is also easy to find so-called solutions that sound eminently reasonable but are incorrect. Usually they are wrong because they count something more than once (sometimes they overlook counting something entirely).

EXAMPLE 52

Consider again part (d) of Example 51, the number of ways to select a committee with at least 1 freshman. A bogus solution to this problem goes as follows: Think of a sequence of two subtasks, choosing a freshman and then choosing the rest of the committee. There are $C(19, 1)$ ways to choose 1 freshman. Once a freshman has been selected, that guarantees that at least 1 freshman will be on the committee, so we are free to choose the remaining 7 members of the committee from the remaining 52 people without any restrictions, giving us $C(52, 7)$ choices. By the Multiplication Principle, this gives $C(19, 1) \cdot C(52, 7)$. However, this is a bigger number than the correct answer.

The problem is this: Suppose Derek and Felicia are both freshmen. In one of the choices we have counted, Derek is the one guaranteed freshman, and we pick the rest of the committee in such a way that Felicia is on it along with six others. But we have also counted the option of making Felicia the guaranteed freshman and having Derek and the same six others be the rest of the committee. This is the same committee as before, and we have counted it twice.   ●

PRACTICE 33   A committee of two to be chosen from four math majors and three physics majors must include at least one math major. Compute the two values.

   a. $C(7, 2) - C(3, 2)$ (correct solution: all committees minus those with no math majors)
   b. $C(4, 1) \cdot C(6, 1)$ (bogus solution: choose one math major and then choose the rest of the committee)

   Note that $C(4, 1) \cdot C(6, 1) - C(4, 2)$ also gives the correct answer because $C(4, 2)$ is the number of committees with two math majors, and these are the committees counted twice in $C(4, 1) \cdot C(6, 1)$.   ●

EXAMPLE 53   **a.** How many distinct permutations can be made from the characters in the word FLORIDA?
   b. How many distinct permutations can be made from the characters in the word MISSISSIPPI?

   Part (a) is a simple problem of the number of ordered arrangements of seven distinct objects, which is 7!. However, the answer to part (b) is not 11! because the 11 characters in MISSISSIPPI are not all distinct. This means that 11! counts some of the same arrangements more than once (the same arrangement meaning that we cannot tell the difference between $MIS_1S_2ISSIPPI$ and $MIS_2S_1ISSIPPI$.)

   Consider any one arrangement of the characters. The four S's occupy certain positions in the string. Rearranging the S's within those positions would result in no distinguishable change, so our one arrangement has 4! look-alikes. In order to avoid overcounting, we must divide 11! by 4! to take care of all the ways of moving the S's around. Similarly, we must divide by 4! to take care of the four I's and by 2! to take care of the two P's. The number of distinct permutations is thus

   $$\frac{11!}{4!4!2!}$$ ●

   In general, suppose there are n objects of which a set of n_1 are indistinguishable from each other, another set of n_2 are indistinguishable from each other, and so on, down to n_k objects that are indistinguishable from each other. The number of distinct permutations of the n objects is

   $$\frac{n!}{(n_1!)(n_2!) \cdots (n_k!)}$$

PRACTICE 34   How many distinct permutations are there of the characters in the word MONGOOSES? ●

Permutations and Combinations with Repetitions

Our formulas for $P(n, r)$ and $C(n, r)$ assume that we arrange or select r objects out of the n available using each object only once. Therefore $r \le n$. Suppose, however, that the n objects are available for reuse as many times as desired. For example, we construct words using the 26 letters of the alphabet; the words may be as long as desired with letters used repeatedly. Or we may draw cards from a deck, replacing a card after each draw; we may draw as many cards as we like with cards used repeatedly. We can still talk about permutations or combinations of r objects out of n, but with repetitions allowed, r might be greater than n.

Counting the number of permutations of r objects out of n distinct objects with repetition is easy. We have n choices for the first object and, because we can repeat that object, n choices for the second object, n choices for the third, and so on. Hence, the number of permutations of r objects out of n distinct objects with repetition allowed is n^r.

To determine the number of combinations of r objects out of n distinct objects with repetition allowed, we use a rather clever idea.

EXAMPLE 54   A jeweler designing a pin has decided to use five stones chosen from diamonds, rubies, and emeralds. In how many ways can the stones be selected?

Because we are not interested in any ordered arrangement of the stones, this is a combinations problem rather than a permutations problem. We want the number of combinations of five objects out of three objects with repetition allowed. The pin might consist of one diamond, three rubies, and one emerald, for instance, or five diamonds. We can represent these possibilities by representing the stones chosen by five asterisks and placing markers between the asterisks to represent the distribution among the three types of gem. For example, we could represent the choice of one diamond, three rubies, and one emerald by

*|***|*

while the choice of five diamonds, no rubies, and no emeralds would be represented by

*****||

We are therefore looking at seven slots (the five gems and the two markers), and the different choices are represented by which of the seven slots are occupied by asterisks. We therefore count the number of ways to choose five items out of seven, which is $C(7, 5)$ or

$$\frac{7!}{5!2!}$$ ●

In general, if we use the same scheme to represent a combination of r objects out of n distinct objects with repetition allowed, there must be $n - 1$ markers to indicate the number of copies of each of the n objects. This gives $r + (n - 1)$ slots to fill, and we want to know the number of ways to select r of these. Therefore we want

$$C(r + n - 1, r) = \frac{(r + n - 1)!}{r!(r + n - 1 - r)!} = \frac{(r + n - 1)!}{r!(n - 1)!}$$

This agrees with the result in Example 54, where $r = 5, n = 3$.

PRACTICE 35   Six children choose one lollipop each from among a selection of red, yellow, and green lollipops. In how many ways can this be done? (We do not care which child gets which.)   ●

    We have discussed a number of counting techniques in this chapter. Table 3.2 summarizes the techniques you can apply in various circumstances, although there may be several legitimate ways to solve any one counting problem.

You want to count the number of ...	Technique to try
Subsets of an n-element set	Use formula 2^n.
Outcomes of successive events	Multiply the number of outcomes for each event.
Outcomes of disjoint events	Add the number of outcomes for each event.
Outcomes given specific choices at each step	Draw a decision tree and count the number of paths.
Elements in overlapping sections of related sets	Use Principle of Inclusion and Exclusion formula.
Ordered arrangements of r out of n distinct objects	Use $P(n, r)$ formula.
Ways to select r out of n distinct objects	Use $C(n, r)$ formula.
Ways to select r out of n distinct objects with repetition allowed	Use $C(r + n - 1, r)$ formula.

Table 3.2

Section 3.4 Review

Techniques

- Find the number of permutations of r distinct objects chosen from n distinct objects.
- Find the number of combinations of r distinct objects chosen from n distinct objects.
- Use permutations and combinations in conjunction with the Multiplication Principle and the Addition Principle.

- Find the number of distinct permutations of n objects that are not all distinct.
- Find the number of permutations of r objects out of n distinct objects when objects may be repeated.
- Find the number of combinations of r objects out of n distinct objects when objects may be repeated.

Main Ideas

There are formulas for counting various permutations and combinations of objects.

Care must be taken when analyzing a counting problem to avoid counting the same thing more than once or not counting some things at all.

Exercises 3.4

1. Compute the value of the following expressions.
    ★**a.** $P(7, 2)$ **b.** $P(8, 5)$ **c.** $P(6, 4)$ **d.** $P(n, n - 1)$

2. How many batting orders are possible for a nine-man baseball team?

3. The 14 teams in the local Little League are listed in the newspaper. How many listings are possible?

4. How many permutations of the characters in COMPUTER are there? How many of these end in a vowel?

★5. How many distinct permutations of the characters in ERROR are there? (Remember that the various R's cannot be distinguished from one another.)

6. In how many ways can six people be seated in a circle of six chairs? Only relative positions in the circle can be distinguished. (*Hint:* Think of taking any circle arrangement, cutting it open, and straightening it out to form a line.)

★7. In how many ways can first, second, and third prize in a pie-baking contest be given to 15 contestants?

8. **a.** Stock designations on an exchange are limited to three letters. How many different designations are there?
    b. How many different designations are there if letters cannot be repeated?

9. In how many different ways can you seat 11 men and 8 women in a row?

★10. In how many different ways can you seat 11 men and 8 women in a row if the men all sit together and the women all sit together?

11. In how many different ways can you seat 11 men and 8 women in a row if no 2 women are to sit together?

12. In how many different ways can you seat 11 men and 8 women around a circular table? (Only relative positions in the circle can be distinguished.)

13. In how many different ways can you seat 11 men and 8 women around a circular table if no 2 women are to sit together? (Only relative positions in the circle can be distinguished.)

14. Compute the value of the following expressions.

    ★**a.** $C(10, 7)$ **b.** $C(9, 2)$ **c.** $C(8, 6)$ **d.** $C(n, n - 1)$

15. Compute $C(n, n - 1)$. Explain why $C(n, n - 1) = C(n, 1)$.

★16. Quality control wants to test 25 microprocessor chips from the 300 manufactured each day. In how many ways can this be done?

17. A soccer team carries 18 players on the roster; 11 players make a team. In how many ways can the team be chosen?

★18. In how many ways can a jury of 5 men and 7 women be selected from a panel of 17 men and 23 women?

19. In how many ways can a librarian select 4 novels and 3 plays from a collection of 21 novels and 11 plays?

Exercises 20–23 deal with the following situation: Of a company's personnel, 7 work in design, 14 in manufacturing, 4 in testing, 5 in sales, 2 in accounting, and 3 in marketing. A committee of 6 people is to be formed to meet with upper management.

★20. In how many ways can the committee be formed if there is to be one member from each department?

21. In how many ways can the committee be formed if there must be exactly two members from manufacturing?

22. In how many ways can the committee be formed if the accounting department is not to be represented and marketing is to have exactly one representative?

★23. In how many ways can the committee be formed if manufacturing is to have at least two representatives?

Exercises 24–32 concern a 5-card hand from a standard 52-card deck.

24. How many hands contain four queens?

★25. How many hands contain three spades and two hearts?

26. How many hands contain all diamonds?

★27. How many hands contain a flush (five cards of the same suit)?

28. How many hands contain cards from all four suits?

29. How many hands consist of all face cards?

30. How many hands contain three of a kind (for example, three 5s)?

31. How many hands contain a full house (three of a kind and a pair)?

32. How many hands contain a straight (five consecutive cards, for example, ace, 2, 3, 4, 5)?

33. How many hands contain a straight flush (five consecutive cards, for example, ace, 2, 3, 4, 5 of the same suit)?

For Exercises 34–38, 14 copies of a code module are to be executed in parallel on identical processors organized into two communicating clusters, A and B. Cluster A contains 16 processors and cluster B contains 32 processors.

34. Find the number of ways to choose the processors.

35. Find the number of ways to choose the processors if all modules must execute on cluster B.

36. Find the number of ways to choose the processors if 8 modules are to be processed on cluster A and 6 on cluster B.

37. Find the number of ways to choose the processors if cluster A has 3 failed processors and cluster B has 2 failed processors.

★38. Find the number of ways to choose the processors if exactly two modules are to execute on cluster B.

For Exercises 39–42, a set of four coins is selected from a box containing five dimes and seven quarters.

★39. Find the number of sets of four coins.

40. Find the number of sets in which two are dimes and two are quarters.

41. Find the number of sets composed of all dimes or all quarters.

42. Find the number of sets with three or more quarters.

Exercises 43–46 concern a computer network with 60 switching nodes.

43. The network is designed to withstand the failure of any two nodes. In how many ways can such a failure occur?

★44. In how many ways can one or two nodes fail?

45. If one node has failed, in how many ways can seven nodes be selected without encountering the failed node?

46. If two nodes have failed, in how many ways can seven nodes be selected to include exactly one failed node?

In Exercises 47–50, a congressional committee of three is to be chosen from a set of five Democrats, three Republicans, and four independents.

★47. In how many ways can the committee be chosen?

48. In how many ways can the committee be chosen if it must include at least one independent?

★49. In how many ways can the committee be chosen if it cannot include both Democrats and Republicans?

50. In how many ways can the committee be chosen if it must have at least one Democrat and at least one Republican?

In Exercises 51–54, a hostess wishes to invite 6 dinner guests from a list of 14 friends.

★51. In how many ways can she choose her guests?

52. In how many ways can she choose her guests if six of them are boring and six of them are interesting, and she wants to have at least one of each?

53. In how many ways can she choose her guests if two of her friends dislike each other and neither will come if the other is present?

54. In how many ways can she choose her guests if two of her friends are very fond of each other and one won't come without the other?

★55. Twenty-five people, including Simon and Yuan, are candidates to serve on a committee of five. If the committee must include Simon or Yuan, in how many ways can the committee be selected?

56. A student must select 5 classes for the next semester from among 12, but 1 of the classes must be either American history or English literature. In how many ways can the student choose classes?

57. In a 5-card hand from a standard 52-card deck, how many ways are there to have exactly 4 aces and 1 club?

58. In a 5-card hand from a standard 52-card deck, how many ways are there to have exactly 3 jacks and 2 hearts?

★59. **a.** How many distinct permutations are there of the characters in the word HAWAIIAN?
   b. How many of these must begin with H?

60. **a.** How many distinct permutations are there of the characters in the word APALACHICOLA?
   b. How many of these must have both L's together?

61. A bookstore displays a shelf of five, three, and four copies, respectively, of the top three bestsellers. How many distinguishable arrangements of these books are there if books with the same title are not distinguishable?

62. The United Group for Divisive Action uses secret code words that are permutations of five characters. You learn that there are only 10 code words. What can you say about repeated characters in the code words?

★**63.** Five people in a dinner party each order an appetizer. If the choices are escargot, egg rolls, and nachos, in how many ways can the selections be made?

64. A florist has roses, carnations, lilies, and snapdragons in stock. How many different bouquets of one dozen flowers can be made?

65. Four friends each buy a pair of running shoes from among a store's selection of 14 kinds. In how many ways can the shoes be selected?

66. One bingo card is to be distributed to each of 12 players. In how many ways can this be done if there are 15 kinds of cards and repetitions are allowed?

★**67.** Six warehouses are each to receive one shipment of either paint, hammers, or shingles.
   a. In how many ways can this happen?
   b. In how many ways can this happen if no paint is shipped?
   c. In how many ways can this happen if there is at least one shipment of each item?

68. At a birthday party, a mother serves a cookie to each of 8 children. There are plenty of chocolate chip, peanut butter, and oatmeal cookies.
   a. In how many ways can each child get one cookie? (We don't care which child gets which kind.)
   b. In how many ways can each child get one cookie if at least one of each kind of cookie is given out?
   c. In how many ways can each child get one cookie if no one likes oatmeal cookies?
   d. In how many ways can each child get one cookie if two children insist on getting peanut butter?
   e. In how many ways can each child get one cookie if there are only two chocolate chip cookies?

★**69.** On Halloween, 10 apples are distributed to 7 children.
   a. In how many ways can this be done? (*Hint:* Although the problem says apples are distributed to children, think of assigning a child's name to each apple; a child's name can go to more than one apple.)
   b. In how many ways can this be done if each child is to receive at least one apple?

70. Eight identical antique pie safes are sold at a furniture auction to three bidders.
   a. In how many ways can this be done? (See the hint for Exercise 69.)
   b. In how many ways can this be done if bidder *A* gets only one pie safe?

★71. How many distinct nonnegative integer solutions are there to the equation

$$x_1 + x_2 + x_3 + x_4 = 10$$

where the solution

$$x_1 = 3, x_2 = 1, x_3 = 4, x_4 = 2$$

and the solution

$$x_1 = 4, x_2 = 2, x_3 = 3, x_4 = 1$$

are distinct? (*Hint*: Think of this problem as distributing 10 pennies to 4 children; then see the hint in Exercise 69.)

72. How many distinct nonnegative integer solutions are there to the equation

$$x_1 + x_2 + x_3 = 7$$

in which $x_i \geq 3$? (See the hint for Exercise 71.)

73. Prove that for $n \geq 2$, $P(n, 1) + P(n, 2) = n^2$. (The proof does not require induction, even though it sounds like a very likely candidate for induction.)

74. Prove that for any n and r with $0 \leq r \leq n$, $C(n, r) = C(n, n - r)$. Explain why this is intuitively true.

75. Prove that for any n and r with $0 \leq r \leq n$, $C(n, 2) = C(r, 2) + C(n - r, 2) + r(n - r)$.

76. A turtle begins at the upper left corner of an $n \times n$ grid and makes his way to the lower right corner. Along the way, he can only move right or down. The accompanying figure shows two possible paths in a 4×4 grid. How many possible paths can the turtle take?

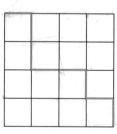 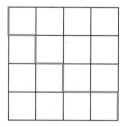

(*Hint*: Each path can be described by a sequence of R's (right moves) and D's (down moves). Find the number of ways to distribute the R's in such a sequence.)

Exercises 77 and 78 require computations of probabilities. Given a set of equally likely outcomes, the *probability* of a particular subset of these outcomes, called an *event*, is the number of ways the event can occur divided by the total number of outcomes. For example, the probability of rolling a 4 with one roll of a die is 1/6 because there are 6 possible outcomes, only 1 of which consists of rolling a 4.

77. In a 5-card hand drawn from a standard 52-card deck, what is the probability of getting the following?

   a. a flush (see Exercise 27)
   b. a full house (see Exercise 31)
   c. a straight (see Exercise 32)
   d. a straight flush (see Exercise 33)

78. To win the Connecticut Daily Numbers game, one must guess the correct sequence of three single-digit numbers. On both January 19 and January 20, 1998, the winning numbers were 8, 2, 8. What is the probability of this event?

Section 3.5 Binomial Theorem

The expression for squaring a binomial is a familiar one:

$$(a + b)^2 = a^2 + 2ab + b^2$$

This is a particular case of raising a binomial to a nonnegative integer power n. The formula for $(a + b)^n$ involves combinations of n objects. Before we prove this formula, we'll look at a historically interesting array of numbers that suggests a fact that we shall need in the proof.

Pascal's Triangle

Pascal's triangle is named for the seventeenth-century French mathematician Blaise Pascal (for whom the programming language Pascal is also named), although it was apparently known several centuries earlier. Row n of the triangle ($n \geq 0$) consists of all of the values $C(n, r)$ for $0 \leq r \leq n$. Thus the triangle looks like

	Row
$C(0, 0)$	0
$C(1, 0)\quad C(1, 1)$	1
$C(2, 0)\quad C(2, 1)\quad C(2, 2)$	2
$C(3, 0)\quad C(3, 1)\quad C(3, 2)\quad C(3, 3)$	3
$C(4, 0)\quad C(4, 1)\quad C(4, 2)\quad C(4, 3)\quad C(4, 4)$	4
$C(5, 0)\quad C(5, 1)\quad C(5, 2)\quad C(5, 3)\ \cdot\ C(5, 4)\quad C(5, 5)$	5
$\vdots$	$\vdots$
$C(n, 0)\quad C(n, 1)\qquad\qquad \dots \qquad\qquad C(n, n-1)\quad C(n, n)$	n

If we compute the numerical values of the expressions, we see that Pascal's triangle has the form

$$
\begin{array}{ccccccccccc}
 & & & & & 1 & & & & & \\
 & & & & 1 & & 1 & & & & \\
 & & & 1 & & 2 & & 1 & & & \\
 & & 1 & & 3 & & 3 & & 1 & & \\
 & 1 & & 4 & & 6 & & 4 & & 1 & \\
1 & & 5 & & 10 & & 10 & & 5 & & 1 \\
\end{array}
$$
$$\vdots$$

Observing this figure, it is clear that the outer edges are all 1s. But it also seems that any element not on the outer edge can be obtained by adding together the two elements directly above it in the preceding row (for example, the first 10 in the fifth row is below the first 4 and the 6 of the fourth row). If this is indeed always true, it means that

$$C(n, k) = C(n - 1, k - 1) + C(n - 1, k) \qquad \text{for} \qquad 1 \leq k \leq n - 1 \tag{1}$$

Equation (1) is known as **Pascal's formula**.

To prove Pascal's formula, we begin with the right side:

$$C(n - 1, k - 1) + C(n - 1, k) = \frac{(n - 1)!}{(k - 1)![n - 1 - (k - 1)]!} + \frac{(n - 1)!}{k!(n - 1 - k)!}$$

$$= \frac{(n - 1)!}{(k - 1)!(n - k)!} + \frac{(n - 1)!}{k!(n - 1 - k)!}$$

$$= \frac{k(n - 1)!}{k!(n - k)!} + \frac{(n - 1)!(n - k)}{k!(n - k)!}$$

(multiplying the first term by k/k and the second term by $(n - k)/(n - k)$)

$$= \frac{k(n - 1)! + (n - 1)!(n - k)}{k!(n - k)!}$$

(adding fractions)

$$= \frac{(n - 1)![k + (n - k)]}{k!(n - k)!}$$

(factoring the numerator)

$$= \frac{(n - 1)!(n)}{k!(n - k)!}$$

$$= \frac{n!}{k!(n - k)!}$$

$$= C(n, k)$$

Another, less algebraic way to prove Pascal's formula involves a counting argument; hence it is called a **combinatorial proof**. We want to compute $C(n, k)$, the number of ways to choose k objects from n objects. There are two disjoint categories of such choices—item 1 is one of the k objects or it is not. If item 1 is one of the

k objects, then the remaining $k - 1$ objects must come from the remaining $n - 1$ objects exclusive of item 1, and there are $C(n - 1, k - 1)$ ways for this to happen. If item 1 is not one of the k objects, then all k objects must come from the remaining $n - 1$ objects, and there are $C(n - 1, k)$ ways for this to happen. The total number of outcomes is the sum of the number of outcomes from these two disjoint cases.

Once we have Pascal's formula for our use, we can develop the formula for $(a + b)^n$, known as the *binomial theorem*.

Binomial Theorem and Its Proof

In the expansion of $(a + b)^2$,

$a^2 + 2ab + b^2$

the coefficients are 1, 2, and 1, which is row 2 in Pascal's triangle.

PRACTICE 36   Compute the expansion for $(a + b)^3$ and $(a + b)^4$ and compare the coefficients with rows 3 and 4 of Pascal's triangle.   ●

Looking at the coefficients in the expansion of $(a + b)^2$, $(a + b)^3$, and $(a + b)^4$ suggests a general result, which is that the coefficients in the expansion of $(a + b)^n$ look like row n in Pascal's triangle. This is indeed the binomial theorem.

Binomial Theorem
For every nonnegative integer n,

$$(a + b)^n = C(n, 0)a^n b^0 + C(n, 1)a^{n-1} b^1 + C(n, 2)a^{n-2} b^2 + \cdots$$
$$+ C(n, k)a^{n-k} b^k + \cdots + C(n, n - 1)a^1 b^{n-1} + C(n, n)a^0 b^n$$
$$= \sum_{k=0}^{n} C(n,k)a^{n-k}b^n$$

Because the binomial theorem is stated "for every nonnegative integer n," a proof by induction seems appropriate. For the basis step, $n = 0$, the theorem states

$(a + b)^0 = C(0, 0)a^0 b^0$

which is

$1 = 1$

Since this is certainly true, the basis step is satisfied.

As the inductive hypothesis, we assume

$(a + b)^k = C(k, 0)a^k b^0 + C(k, 1)a^{k-1}b^1 + \cdots + C(k, k - 1)a^1 b^{k-1} + C(k, k)a^0 b^k$

Now consider

$$(a + b)^{k+1} = (a + b)^k(a + b) = (a + b)^k a + (a + b)^k b$$
$$= [C(k, 0)a^k b^0 + C(k, 1)a^{k-1}b^1 + \cdots + C(k, k-1)a^1 b^{k-1}$$
$$+ C(k, k)a^0 b^k]a + [C(k, 0)a^k b^0 + C(k, 1)a^{k-1}b^1$$
$$+ \cdots + C(k, k-1)a^1 b^{k-1} + C(k, k)a^0 b^k]b$$

(by the inductive hypothesis)

$$= C(k, 0)a^{k+1}b^0 + C(k, 1)a^k b^1 + \cdots + C(k, k-1)a^2 b^{k-1}$$
$$+ C(k, k)a^1 b^k + C(k, 0)a^k b^1 + C(k, 1)a^{k-1}b^2$$
$$+ \cdots + C(k, k-1)a^1 b^k + C(k, k)a^0 b^{k+1}$$
$$= C(k, 0)a^{k+1}b^0 + [C(k, 0) + C(k, 1)]a^k b^1 + [C(k, 1) + C(k, 2)]a^{k-1}b^2$$
$$+ \cdots + [C(k, k-1) + C(k, k)]a^1 b^k + C(k, k)a^0 b^{k+1}$$

(collecting like terms)

$$= C(k, 0)a^{k+1}b^0 + C(k+1, 1)a^k b^1 + C(k+1, 2)a^{k-1}b^2$$
$$+ \cdots + C(k+1, k)a^1 b^k + C(k, k)a^0 b^{k+1}$$

(using Pascal's formula)

$$= C(k+1, 0)a^{k+1}b^0 + C(k+1, 1)a^k b^1 + C(k+1, 2)a^{k-1}b^2$$
$$+ \cdots + C(k+1, k)a^1 b^k + C(k+1, k+1)a^0 b^{k+1}$$

(because $C(k, 0) = 1 = C(k+1, 0)$ and $C(k, k) = 1 = C(k+1, k+1)$)

This completes the inductive proof of the binomial theorem.

The binomial theorem also has a combinatorial proof. Writing $(a + b)^n$ as $(a + b)(a + b) \cdots (a + b)$ (n factors), we know that the answer (using the distributive law of numbers) is the sum of all values obtained by multiplying each term in a factor by a term from every other factor. For example, using b as the term from k factors and a as the term from the remaining $n - k$ factors produces the expression $a^{n-k}b^k$. Using b from a different set of k factors and a from the $n - k$ remaining factors also produces $a^{n-k}b^k$. How many such terms are there? There are $C(n, k)$ different ways to select k factors from which to use b; hence there are $C(n, k)$ such terms. After adding these terms together, the coefficient of $a^{n-k}b^k$ is $C(n, k)$. As k ranges from 0 to n, the result of summing the terms is the binomial theorem.

Because of its use in the binomial theorem, the expression $C(n, r)$ is also known as a **binomial coefficient**.

Applying the Binomial Theorem

EXAMPLE 55   Using the binomial theorem, we can write the expansion of $(x - 3)^4$ as follows:

$$(x - 3)^4 = C(4, 0)x^4(-3)^0 + C(4, 1)x^3(-3)^1 + C(4, 2)x^2(-3)^2$$
$$+ C(4, 3)x^1(-3)^3 + C(4, 4)x^0(-3)^4$$
$$= x^4 + 4x^3(-3) + 6x^2(9) + 4x(-27) + 81$$
$$= x^4 - 12x^3 + 54x^2 - 108x + 81$$

PRACTICE 37   Expand $(x + 1)^5$ using the binomial theorem.

The binomial theorem tells us that term $k + 1$ in the expansion of $(a + b)^n$ is $C(n, k)a^{n-k}b^k$. This allows us to find individual terms in the expansion without computing the entire expression.

PRACTICE 38   What is the fifth term in the expansion of $(x + y)^7$?

By using various values for a and b in the binomial theorem, certain identities can be obtained.

EXAMPLE 56   Let $a = b = 1$ in the binomial theorem. Then

$$(1 + 1)^n = C(n, 0) + C(n, 1) + \cdots + C(n, k) + \cdots + C(n, n)$$

or

$$2^n = C(n, 0) + C(n, 1) + \cdots + C(n, k) + \cdots + C(n, n) \tag{2}$$

Actually, equation (2) can be proved on its own using a combinatorial proof. The number $C(n, k)$, the number of ways to select k items from a set of n items, can be thought of as the number of k-element subsets of an n-element set. The right side of equation (2) therefore represents the total number of all the subsets (of all sizes) of an n-element set. But we already know that the number of such subsets is 2^n.

Section 3.5 Review

Technique

- Use the binomial theorem to expand a binomial or to find a particular term in the expansion.

Main Ideas

The binomial theorem provides a formula for expanding a binomial without multiplying it out.

The coefficients of a binomial raised to a nonnegative integer power are combinations of n items as laid out in row n of Pascal's triangle.

Exercises 3.5

1. Expand the expression using the binomial theorem.
    ★**a.** $(a + b)^5$ **b.** $(x + y)^6$ ★**c.** $(a + 2)^5$
    d. $(a - 4)^4$ ★**e.** $(2x + 3y)^3$ **f.** $(3x - 1)^5$
    g. $(2p - 3q)^4$ **h.** $\left(3x + \dfrac{1}{2}\right)^5$

In Exercises 2–9, find the indicated term in the expansion.

2. The fourth term in $(a + b)^{10}$

3. The seventh term in $(x - y)^{12}$

★4. The sixth term in $(2x - 3)^9$

5. The fifth term in $(3a + 2b)^7$

★6. The last term in $(x - 3y)^8$

7. The last term in $(ab + 3x)^6$

★8. The third term in $(4x - 2y)^5$

9. The fourth term in $\left(3x - \dfrac{1}{2}\right)^8$

10. Use the binomial theorem (more than once) to expand $(a + b + c)^3$.

11. Expand $(1 + 0.1)^5$ in order to compute $(1.1)^5$.

★12. What is the coefficient of x^3y^4 in the expansion of $(2x - y + 5)^8$?

13. What is the coefficient of $x^5y^2z^2$ in the expansion of $(x + y + 2z)^9$?

14. Prove that

$$C(n + 2, r) = C(n, r) + 2C(n, r - 1) + C(n, r - 2) \text{ for } 2 \le r \le n$$

(*Hint:* Use Pascal's formula.)

15. Prove that

$$C(k, k) + C(k + 1, k) + \cdots + C(n, k) = C(n + 1, k + 1) \text{ for } 0 \le k \le n$$

(*Hint:* Use induction on n for a fixed, arbitrary k, as well as Pascal's formula.)

16. Use the binomial theorem to prove that

$$C(n, 0) - C(n, 1) + C(n, 2) - \cdots + (-1)^nC(n, n) = 0$$

★17. Use the binomial theorem to prove that

$$C(n, 0) + C(n, 1)2 + C(n, 2)2^2 + \cdots + C(n, n)2^n = 3^n$$

18. **a.** Use the binomial theorem to prove that

$$C(n, n) + C(n, n - 1)2 + C(n, n - 2)2^2 + \cdots + C(n, 1)2^{n-1} + C(n, 0)2^n = 3^n$$

b. Prove this result directly from Exercise 17.

19. **a.** Expand $(1 + x)^n$.

b. Differentiate both sides of the equation from part (a) with respect to x to obtain

$$n(1 + x)^{n-1} = C(n, 1) + 2C(n, 2)x + 3C(n, 3)x^2 + \cdots + nC(n, n)x^{n-1}$$

c. Prove that

$$C(n, 1) + 2C(n, 2) + 3C(n, 3) + \cdots + nC(n, n) = n2^{n-1}$$

d. Prove that

$$C(n, 1) - 2C(n, 2) + 3C(n, 3) - 4C(n, 4) + \cdots + (-1)^{n-1}nC(n, n) = 0$$

20. **a.** Prove that

$$\frac{2^{n+1} - 1}{n + 1} = C(n, 0) + \frac{1}{2} C(n, 1) + \frac{1}{3} C(n, 2) + \cdots + \frac{1}{n + 1} C(n, n)$$

  b. Prove that

$$\frac{1}{n + 1} = C(n, 0) - \frac{1}{2} C(n, 1) + \frac{1}{3} C(n, 2) + \cdots + (-1)^n \frac{1}{n + 1} C(n, n)$$

  (*Hint:* Integrate both sides of the equation from part (a) of Exercise 19.)

Chapter 3 Review

Terminology

Addition Principle (p. 190)
binary operation (p. 167)
binomial coefficient (p. 224)
binomial theorem (p. 223)
Cantor's diagonalization method (p. 175)
Cartesian product (cross product) of sets (p. 170)
closed set under an operation (p. 167)
combination (p. 207)
combinatorial proof (p. 222)
combinatorics (p. 187)

complement of a set (p. 170)
countable set (p. 173)
decision tree (p. 192)
denumerable set (p. 173)
disjoint sets (p. 170)
dual of a set identity (p. 172)
empty set (p. 163)
equal sets (p. 162)
intersection of sets (p. 169)
Multiplication Principle (p. 188)
n factorial (p. 205)
null set (p. 163)
ordered pair (p. 167)
Pascal's formula (p. 222)

Pascal's triangle (p. 221)
permutation (p. 205)
Pigeonhole Principle (p. 202)
power set (p. 166)
Principle of Inclusion and Exclusion (p. 200)
proper subset (p. 164)
set difference (p. 170)
subset (p. 163)
unary operation (p. 168)
uncountable set (p. 173)
union of sets (p. 169)
universal set (p. 169)
universe of discourse (p. 169)
well-defined operation (p. 167)

Self-Test

Answer the following true–false questions.

Section 3.1

1. The empty set is a proper subset of every set.

2. If A and B are disjoint sets, then $(A - B) \cup (B - A) = A \cup B$.

3. If a set has n elements, then its power set has 2^n elements.

4. If a binary operation $\circ$ on a set S is well-defined, then $x \circ y \in S$ for all x and y in S.

5. Cantor's diagonalization method is a way to prove that certain sets are denumerable.

Section 3.2

6. According to the Multiplication Principle, the number of outcomes for a sequence of tasks is the product of the number of outcomes for each separate task.

7. The Addition Principle finds the total number of branches of a decision tree.

8. The Addition Principle requires the tasks at hand to have disjoint sets of outcomes.

9. The Multiplication Principle says that the number of elements in $A \times B$ equals the number of elements in A times the number of elements in B.

10. Any problem that requires a decision tree for its solution cannot be solved by the Multiplication Principle.

Section 3.3

11. The Principle of Inclusion and Exclusion requires that A and B be disjoint sets in order to find the number of elements in $A \cup B$.

12. The Principle of Inclusion and Exclusion applied to two sets says that the number of elements in the union minus the number of elements in the intersection is the sum of the number of elements in each set.

13. The Principle of Inclusion and Exclusion applies to the union of any number of sets as long as at least one of them is finite.

14. The Pigeonhole Principle is a way to count the number of elements in the union of disjoint sets, or "pigeonholes."

15. The Pigeonhole Principle guarantees that if there are eight people in a room, at least two must have been born on the same day of the week.

Section 3.4

16. A permutation is an ordered arrangement of objects.

17. The number of combinations of r objects out of n, $r > 1$, is fewer than the number of permutations of r objects out of n.

18. To find the number of ways a subset of r objects can be selected from n objects, use the formula $P(n, r)$.

19. The number of permutations of the letters in a word with three sets of repeated letters is $n!/3$.

20. The formula $C(r + n - 1, r)$ computes the number of combinations of r objects out of n objects where objects may be used repeatedly.

Section 3.5

21. Pascal's triangle consists of rows that represent ways to arrange r out of n objects for various r.

22. Pascal's formula says that an "interior" number in Pascal's triangle is the sum of the two numbers directly above it in the triangle.

23. In the expansion of a binomial to the nth power, the kth term is found in row k of Pascal's triangle.

24. A combinatorial argument is one that is based upon counting techniques.

25. The coefficient of the seventh term in the expansion of $(a + b)^{12}$ is given by the expression $C(12, 6)$.

On the Computer

For Exercises 1–7, write a computer program that produces the desired output from the given input.

1. *Input*: Elements in a finite set S
   Output: Elements in $\wp(S)$
   Algorithm: Use recursion.

2. *Input*: Arithmetic expression in postfix notation (see Exercise 33 in Section 3.1)
   Output: Value of the expression

3. *Input*: Arithmetic expression in infix notation (see Exercise 33 in Section 3.1)
   Output: Postfix form of the expression
   Do this problem in two ways:
   a. Assume that the input is fully parenthesized.
   b. Do not assume that the input is fully parenthesized, but apply the proper order of precedence of operators within the program (order of precedence of operators is parenthesized expressions first, then exponentiation, then multiplication and division, then addition and subtraction).

4. *Input*: Values for n and r, $0 \leq r \leq n$
   Output: Value of $P(n, r)$

5. *Input*: Values for n and r, $0 \leq r \leq n$
   Output: Value of $C(n, r)$

6. *Input*: Value for n
   Output: All values of $C(n, r)$, $0 \leq r \leq n$

7. *Input*: Values for a, b, and n
   Output: Value of $(a + b)^n$
   a. Use the binomial theorem to compute your result.
   b. Compute $a + b$ and raise this value to the nth power; compare your answer with that of part (a).

8. *Input*: Values for a, b, n, and r, $1 \leq r \leq n + 1$
   Output: rth term in the expansion of $(a + b)^n$

9. Write a program that allows the user to enter a value for n, $1 \leq n \leq 10$, and then queries the user for the values needed on the right side of equation (4) of Section 3.3 (the Principle of Inclusion and Exclusion) and computes the value of $|A_1 \cup \cdots \cup A_n|$.

10. Write a program to generate a given number of rows of Pascal's triangle. Do this problem in two ways.
    a. Use the definition of Pascal's triangle (and perhaps use your answer to computer Exercise 5 above as a function).
    b. Use recursion and Pascal's formula.

Relations, Functions, and Matrices

4

Chapter Objectives

After studying this chapter, you will be able to:

- Identify ordered pairs related by a binary relation.
- Test a binary relation for the reflexive, symmetric, transitive, and antisymmetric properties.
- Find the reflexive, symmetric, and transitive closures of a binary relation.
- Recognize partial orderings and construct Hasse diagrams for them.
- Recognize an equivalence relation on a set and describe how it partitions the set into equivalence classes.
- Draw a PERT chart from a task table.
- Find the minimum time-to-completion and a critical path in a PERT chart.
- Extend a partial ordering on a finite set to a total ordering by doing a topological sort.
- Understand the entity-relationship model and the relational model for an enterprise.
- Perform restrict, project, and join operations in a relational database.
- Create relational database queries in the languages of relational algebra, SQL, and relational calculus.
- Determine whether a binary relation is a function.
- Test a function for the onto and one-to-one properties.
- Create composite functions.
- Decide whether a function has an inverse function and what the inverse function is.

- Manipulate cycle notation for permutation functions.
- Compute the number of functions, onto functions, and one-to-one functions from one finite set to another.
- Understand order of magnitude as a relative measure of function rate-of-growth.
- Perform matrix arithmetic on matrices of appropriate dimensions.
- Perform Boolean arithmetic operations on Boolean matrices of appropriate dimensions.

Your company has developed a program for use on a small parallel processing machine. According to the technical documentation, the program executes processes P1, P2, and P3 in parallel; these processes all need results from process P4, so they must wait for Process P4 to complete execution before they begin. Processes P7 and P10 execute in parallel but must wait until processes P1, P2, and P3 have finished. Process P4 requires results from P5 and P6 before it can begin execution. P5 and P6 execute in parallel. Processes P8 and P11 execute in parallel, but P8 must wait for process P7 to complete and P11 must wait for process P10 to complete. Process P9 must wait for results from P8 and P11. You have been assigned to convert the software for use on a single processor machine.

Question: In what order should the processes be executed?

Here various pairs of processes are related to one another by a "prerequisite" relation. This is a special case of a binary relation, a relationship between pairs of elements within a set. We will study the various properties of binary relations in Section 4.1. One type of binary relation is called a partial ordering; elements related by a partial ordering can be represented graphically. Another type of binary relation is an equivalence relation; elements related by an equivalence relation can be grouped into classes.

A topological sort extends a partial ordering to a total ordering. For a partial ordering of prerequisite tasks, a corresponding total ordering identifies the sequential order in which the tasks would have to be done, which is the solution to the parallel processing conversion problem. Topological sorting is presented in Section 4.2.

A generalization of a binary relation forms the basis for relational databases, considered in Section 4.3. Using operations of restrict, project, and join on the relations in a database, we can make various queries of the database.

A function is a special kind of binary relation. Functions as well as relations describe a number of real-world situations. Functions can also have special properties, as discussed in Section 4.4.

In Section 4.5, we consider matrices and develop an arithmetic for manipulating them. We shall later use matrices to represent relations and graphs.

Section 4.1   Relations

Binary Relations

If we learn that two people, Henrietta and Horace, are related, we understand that there is some family connection between them—that (Henrietta, Horace) stands out from other ordered pairs of people because there is a relationship (cousins, sister and brother, or whatever) that Henrietta and Horace satisfy. The mathematical analogue is to distinguish certain ordered pairs of objects from other ordered pairs because the components of the distinguished pairs satisfy some relationship that the components of the other pairs do not.

EXAMPLE 1

Remember (Section 3.1) that the Cartesian product of a set S with itself, $S \times S$ or S^2, is the set of all ordered pairs of elements of S. Let $S = \{1, 2, 3\}$; then

$$S \times S = \{(1, 1), (1, 2), (1, 3), (2, 1), (2, 2), (2, 3), (3, 1), (3, 2), (3, 3)\}$$

If we were interested in the relationship of equality, then $(1, 1)$, $(2, 2)$, $(3, 3)$ would be the distinguished elements of $S \times S$, that is, the only ordered pairs whose components are equal. If we were interested in the relationship of one number being less than another, we would choose $(1, 2)$, $(1, 3)$, and $(2, 3)$ as the distinguished ordered pairs of $S \times S$.   ●

In Example 1, we could pick out the distinguished ordered pairs (x, y) by saying that $x = y$ or that $x < y$. Similarly, the notation $x \, \rho \, y$ indicates that the ordered pair (x, y) satisfies a relation ρ. The relation ρ may be defined in words or simply by listing the ordered pairs that satisfy ρ.

EXAMPLE 2

Let $S = \{1, 2, 4\}$. On the set $S \times S = \{(1, 1), (1, 2), (1, 4), (2, 1), (2, 2), (2, 4), (4, 1), (4, 2), (4, 4)\}$, a relation ρ can be defined by $x \, \rho \, y$ if and only if $x = \frac{1}{2}y$, abbreviated $x \, \rho \, y \leftrightarrow x = \frac{1}{2}y$. Thus $(1, 2)$ and $(2, 4)$ satisfy ρ. Alternatively, the same ρ could be defined by saying that $\{(1, 2), (2, 4)\}$ is the set of ordered pairs satisfying ρ.   ●

As in Example 2, one way to define the binary relation ρ is to specify a subset of $S \times S$. Formally, this is the definition of a binary relation on a set.

> **Definition: Binary Relation on a Set S**
> Given a set S, a **binary relation on S** is a subset of $S \times S$ (a set of ordered pairs of elements of S).

Now that we know that a binary relation ρ is a subset, we see that

$$x \, \rho \, y \leftrightarrow (x, y) \in \rho$$

Generally, a binary relation is defined by describing the relation rather than by listing the ordered pairs. The description gives a characterizing property of elements of the relation; that is, it is a binary predicate satisfied by certain ordered pairs.

EXAMPLE 3   Let $S = \{1, 2\}$. Then $S \times S = \{(1, 1), (1, 2), (2, 1), (2, 2)\}$. Let ρ on S be given by the description $x \rho y \leftrightarrow x + y$ is odd. Then $(1, 2) \in \rho$, and $(2, 1) \in \rho$.   ●

EXAMPLE 4   Let $S = \{1, 2\}$. Then $S \times S = \{(1, 1), (1, 2), (2, 1), (2, 2)\}$. If ρ is defined on S by $\rho = \{(1, 1), (2, 1)\}$, then $1 \rho 1$ and $2 \rho 1$ hold, but not, for instance, $1 \rho 2$. Here ρ seems to have no obvious verbal description.   ●

In this section we will be concerned almost exclusively with binary relations on a single set, but more generally, relations can be defined on multiple sets.

Definition: Relations on Multiple Sets
Given two sets S and T, a **binary relation from S to T** is a subset of $S \times T$.
Given n sets $S_1, S_2, ..., S_n$, $n > 2$, an **n-ary relation on $S_1 \times S_2 \times \cdots \times S_n$** is a subset of $S_1 \times S_2 \times \cdots \times S_n$.

EXAMPLE 5   Let $S = \{1, 2, 3\}$ and $T = \{2, 4, 7\}$. Then the set

$$\{(1, 2), (2, 4), (2, 7)\}$$

consists of elements from $S \times T$. It is a binary relation from S to T.   ●

PRACTICE 1   For each of the following binary relations ρ on $\mathbb{N}$, decide which of the given ordered pairs belong to ρ.

    a. $x \rho y \leftrightarrow x = y + 1$; $(2, 2), (2, 3), (3, 3), (3, 2)$
    b. $x \rho y \leftrightarrow x$ divides y; $(2, 4), (2, 5), (2, 6)$
    c. $x \rho y \leftrightarrow x$ is odd; $(2, 3), (3, 4), (4, 5), (5, 6)$
    d. $x \rho y \leftrightarrow x > y^2$; $(1, 2), (2, 1), (5, 2), (6, 4), (4, 3)$ ●

If ρ is a binary relation on S, then ρ will consist of a set of ordered pairs of the form (s_1, s_2). A given first component s_1 or second component s_2 can be paired in various ways in the relation. The relation is **one-to-one** if each first component and each second component appears only once in the relation. The relation is **one-to-many** if some first component s_1 appears more than once; that is, one s_1 is paired with more than one second component. It is **many-to-one** if some second component s_2 is paired with more than one first component. Finally, it is **many-to-many** if at least one s_1 is paired with more than one second component and at least one s_2 is paired with more than one first component. Figure 4.1 illustrates these four possibilities. Note that not all values in S need be components in ordered pairs of ρ.

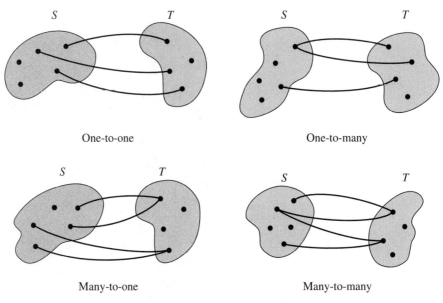

One-to-one   One-to-many

Many-to-one   Many-to-many

Figure 4.1

These ideas extend to relations from a set S to a set T. The relation of Example 5 is one-to-many because the first component 2 appears more than once; 2 from set S is associated with both 4 and 7 from set T.

PRACTICE 2   Identify each of these relations on S, where $S = \{2, 5, 7, 9\}$, as one-to-one, one-to-many, many-to-one, or many-to-many.

    a. $\{(5, 2), (7, 5), (9, 2)\}$
    b. $\{(2, 5), (5, 7), (7, 2)\}$
    c. $\{(7, 9), (2, 5), (9, 9), (2, 7)\}$ ●

Suppose B is the set of all binary relations on a given set S. If ρ and σ belong to B, then they are subsets of $S \times S$. As such, we can perform set operations of union, intersection, and complementation that result in new subsets of $S \times S$, that is, new binary relations, which we will denote by $\rho \cup \sigma$, $\rho \cap \sigma$, and ρ', respectively. Thus

$$x(\rho \cup \sigma)y \leftrightarrow x \rho y \text{ or } x \sigma y$$
$$x(\rho \cap \sigma)y \leftrightarrow x \rho y \text{ and } x \sigma y$$
$$x \rho' y \leftrightarrow \text{not } x \rho y$$

PRACTICE 3   Let ρ and σ be two binary relations on $\mathbb{N}$ defined by $x \, \rho \, y \leftrightarrow x = y$ and $x \, \sigma \, y \leftrightarrow x < y$. Give verbal descriptions for (a), (b), and (c); give a set description for (d).

         a. What is the relation $\rho \cup \sigma$?
         b. What is the relation ρ'?
         c. What is the relation σ'?
         d. What is the relation $\rho \cap \sigma$?

     The following facts about the operations $\cup$, $\cap$, and $'$ on relations are immediate consequences of the basic set identities found in Section 3.1. The set S^2 (which is, after all, a subset of S^2) is being viewed here as a binary relation on S.

1a. $\rho \cup \sigma = \sigma \cup \rho$ 1b. $\rho \cap \sigma = \sigma \cap \rho$
2a. $(\rho \cup \sigma) \cup \gamma = \rho \cup (\sigma \cup \gamma)$ 2b. $(\rho \cap \sigma) \cap \gamma = \rho \cap (\sigma \cap \gamma)$
3a. $\rho \cup (\sigma \cap \gamma) = (\rho \cup \sigma) \cap (\rho \cup \gamma)$ 3b. $\rho \cap (\sigma \cup \gamma) = (\rho \cap \sigma) \cup (\rho \cap \gamma)$
4a. $\rho \cup \varnothing = \rho$ 4b. $\rho \cap S^2 = \rho$
5a. $\rho \cup \rho' = S^2$ 5b. $\rho \cap \rho' = \varnothing$

Properties of Relations

A binary relation on a set S may have certain properties. For example, the relation ρ of equality on S, $(x, y) \in \rho \leftrightarrow x = y$, has three properties: (1) for any $x \in S$, $x = x$, or $(x, x) \in \rho$; (2) for any $x, y \in S$, if $x = y$, then $y = x$, or $(x, y) \in \rho \rightarrow (y, x) \in \rho$; and (3) for any $x, y, z \in S$, if $x = y$ and $y = z$, then $x = z$, or $[(x, y) \in \rho$ and $(y, z) \in \rho] \rightarrow (x, z) \in \rho$. These three properties make the equality relation reflexive, symmetric, and transitive.

REMINDER:
Reflexive: Every x is related to itself.
Symmetric: If x is related to y, then y is related to x.
Transitive: If x is related to y and y is related to z, then x is related to z.

Definition: Reflexive, Symmetric, and Transitive Relations
Let ρ be a binary relation on a set S. Then

     ρ is **reflexive** means $(\forall x)(x \in S \rightarrow (x, x) \in \rho)$
     ρ is **symmetric** means $(\forall x)(\forall y)(x \in S \wedge y \in S \wedge (x, y) \in \rho \rightarrow (y, x) \in \rho)$
     ρ is **transitive** means $(\forall x)(\forall y)(\forall z)(x \in S \wedge y \in S \wedge z \in S \wedge (x, y) \in \rho \wedge (y, z) \in \rho \rightarrow (x, z) \in \rho)$

EXAMPLE 6   Consider the relation $\leq$ on the set $\mathbb{N}$. This relation is reflexive because for any nonnegative integer x, $x \leq x$. It is also a transitive relation because for any nonnegative integers x, y, and z, if $x \leq y$ and $y \leq z$, then $x \leq z$. However, $\leq$ is not symmetric; $3 \leq 4$ does not imply $4 \leq 3$. In fact, for any $x, y \in \mathbb{N}$, if both $x \leq y$ and $y \leq x$, then $y = x$. This characteristic is described by saying that $\leq$ is antisymmetric.

Definition: Antisymmetric Relation

Let ρ be a binary relation on a set S. Then ρ is **antisymmetric** means

$$(\forall x)(\forall y)(x \in S \wedge y \in S \wedge (x, y) \in \rho \wedge (y, x) \in \rho \to x = y)$$

EXAMPLE 7

Let $S = \wp(\mathbb{N})$. Define a binary relation ρ on S by $A \rho B \leftrightarrow A \subseteq B$. Then ρ is reflexive because every set is a subset of itself. Also, ρ is transitive because if A is a subset of B and B is a subset of C, then A is a subset of C. Finally, ρ is antisymmetric because if A is a subset of B and B is a subset of A, then A and B are equal sets.   ●

All four relational properties involve the implication connective. The universal quantifiers mean that the implications must be true for arbitrary choices of variables. Recall that to prove an implication true, we assume that the antecedent is true and prove that the consequent must also be true. For the reflexive property, the antecedent just chooses an arbitrary element in S; the consequent says that this element must be related to itself. For a relation ρ on a set to be reflexive, then, every element in the set must be related to itself, which specifies certain ordered pairs that must belong to ρ.

However, in the symmetric, transitive, and antisymmetric properties, the antecedent does not say only that the elements are in S. To prove that a relation is symmetric, for example, we must show that if x and y are arbitrary elements in S and if in addition x is related to y, then it must be the case that y is related to x. This says that if certain ordered pairs are found in ρ, then certain other ordered pairs must also be in ρ in order for ρ to be symmetric. In other words, knowledge of the set S is critical to determining whether reflexivity holds, while to determine the other properties, it is sufficient just to look at the ordered pairs in ρ.

PRACTICE 4

Let $S = \{1, 2, 3\}$.

  a. If a relation ρ on S is reflexive, what ordered pairs must belong to ρ?
  b. If a relation ρ on S is symmetric, what ordered pairs must belong to ρ?
    (This is a trick question; see the answer in the back of the book.)
  c. If a relation ρ on S is symmetric and if $(a, b) \in \rho$, then what other ordered pair must belong to ρ?
  d. If a relation ρ on S is antisymmetric and if (a, b) and (b, a) belong to ρ, what must be true?
  e. Is the relation $\rho = \{(1, 2)\}$ on S transitive? (*Hint:* Remember the truth table for implication.)   ●

The properties of symmetry and antisymmetry for binary relations are not precisely opposites. Antisymmetric does not mean "not symmetric." A relation is not symmetric if some (x, y) belongs to the relation but (y, x) does not. More formally, not symmetric means

$$((\forall x)(\forall y)[x \in S \wedge y \in S \wedge (x, y) \in \rho \to (y, x) \in \rho])'$$
$$\leftrightarrow (\exists x)(\exists y)[x \in S \wedge y \in S \wedge (x, y) \in \rho \to (y, x) \in \rho]'$$
$$\leftrightarrow (\exists x)(\exists y)[(x \in S \wedge y \in S \wedge (x, y) \in \rho)' \vee (y, x) \in \rho]'$$
$$\leftrightarrow (\exists x)(\exists y)[(x \in S \wedge y \in S \wedge (x, y) \in \rho) \wedge (y, x) \notin \rho]$$

Relations can therefore be symmetric and not antisymmetric, antisymmetric and not symmetric, both, or neither.

The equality relation on a set S is both symmetric and antisymmetric. However, the equality relation on S (or a subset of this relation) is the only relation having both these properties. To illustrate, suppose ρ is a symmetric and antisymmetric relation on S, and let $(x, y) \in \rho$. By symmetry, it follows that $(y, x) \in \rho$. But by antisymmetry, $x = y$. Thus, only equal elements can be related. The relation $\rho = \{(1, 2), (2, 1), (1, 3)\}$ on the set $S = \{1, 2, 3\}$ is neither symmetric—$(1, 3)$ belongs but $(3, 1)$ does not—nor antisymmetric—$(1, 2)$ and $(2, 1)$ belong, but $1 \neq 2$.

PRACTICE 5   Test each binary relation on the given set S for reflexivity, symmetry, antisymmetry, and transitivity.

   a. $S = \mathbb{N}; x \rho y \leftrightarrow x + y$ is even
   b. $S = \mathbb{Z}^+$ (positive integers); $x \rho y \leftrightarrow x$ divides y
   c. $S = $ set of all lines in the plane; $x \rho y \leftrightarrow x$ is parallel to y or x coincides with y
   d. $S = \mathbb{N}; x \rho y \leftrightarrow x = y^2$
   e. $S = \{0, 1\}; x \rho y \leftrightarrow x = y^2$
   f. $S = \{x \mid x$ is a person living in Peoria$\}; x \rho y \leftrightarrow x$ is older than y
   g. $S = \{x \mid x$ is a student in your class$\}; x \rho y \leftrightarrow x$ sits in the same row as y
   h. $S = \{1, 2, 3\}; \rho = \{(1, 1), (2, 2), (3, 3), (1, 2), (2, 1)\}$

EXAMPLE 8   The discussion on recursion in Prolog (Section 1.5) noted that a recursive rule should be used when the predicate being described is one that is inherited from one object to the next. The predicate *in-food-chain* used there has this property because

$$in\text{-}food\text{-}chain(x, y) \land in\text{-}food\text{-}chain(y, z) \rightarrow in\text{-}food\text{-}chain(x, z)$$

Now we see that this is simply the transitive property.

Closures of Relations

If a relation ρ on a set S fails to have a certain property, we may be able to extend ρ to a relation ρ^* on S that does have that property. By "extend," we mean that the new relation ρ^* will contain all the ordered pairs in ρ plus the additional ordered pairs needed for the desired property to hold. Thus $\rho \subseteq \rho^*$. If ρ^* is the smallest such set, then ρ^* is called the closure of ρ with respect to that property.

> **Definition: Closure of a Relation**
> A binary relation ρ^* on a set S is the **closure of a relation** ρ on S with respect to property P if
>
> 1. ρ^* has property P.
> 2. $\rho \subseteq \rho^*$.
> 3. ρ^* is a subset of any other relation on S that includes ρ and has property P.

We can look for the **reflexive closure**, the **symmetric closure**, and the **transitive closure** of a relation on a set. Of course, if the relation already has a property, it is its own closure with respect to that property.

EXAMPLE 9   Let $S = \{1, 2, 3\}$ and $\rho = \{(1, 1), (1, 2), (1, 3), (3, 1), (2, 3)\}$. Then ρ is not reflexive, not symmetric, and not transitive. The closure of ρ with respect to reflexivity is

$$\{(1, 1), (1, 2), (1, 3), (3, 1), (2, 3), (2, 2), (3, 3)\}$$

This relation is reflexive and contains ρ. Furthermore, any reflexive relation on S would have to contain the new ordered pairs we've added—$(2, 2)$ and $(3, 3)$—so no smaller reflexive relation can exist; that is, any reflexive relation containing ρ must have the relation above as a subset.

The closure of ρ with respect to symmetry is

$$\{(1, 1), (1, 2), (1, 3), (3, 1), (2, 3), (2, 1), (3, 2)\}$$

Here it is also clear that we have added just those new pairs required—$(2, 1)$ and $(3, 2)$—for the relation to be symmetric.

For both reflexive closure and symmetric closure, we only had to inspect the ordered pairs already in ρ to find out what ordered pairs we needed to add (assuming we knew what the set S was). The reflexive or symmetric closure of the relation could be found in one step. Transitive closure may require a series of steps. Inspecting the ordered pairs in our example ρ, we see that we need to add $(3, 2)$ (because of $(3, 1)$ and $(1, 2)$); $(3, 3)$ (because of $(3, 1)$ and $(1, 3)$), and $(2, 1)$ (because of $(2, 3)$ and $(3, 1)$). This gives the relation

$$\{(1, 1), (1, 2), (1, 3), (3, 1), (2, 3), (3, 2), (3, 3), (2, 1)\}$$

However, this relation is still not transitive. Because of the new pair $(2, 1)$ and the old pair $(1, 2)$, we need to add $(2, 2)$. This gives the relation

$$\{(1, 1), (1, 2), (1, 3), (3, 1), (2, 3), (3, 2), (3, 3), (2, 1), (2, 2)\}$$

which is transitive and also the smallest transitive relation containing ρ. It is the transitive closure of ρ.

As in Example 9, one way to find the transitive closure of a relation is to inspect the ordered pairs in the original relation, add new pairs if necessary, inspect the resulting relation, add new pairs if necessary, and so on, until a transitive relation is achieved. This is a rather ad hoc procedure, and we will give a better algorithm in Chapter 6. There we shall also see that the transitive closure of a binary relation is related to "reachability in a directed graph," which has many applications.

PRACTICE 6   Does it make sense to look for the antisymmetric closure of a relation on a set? Why or why not?

PRACTICE 7   Find the reflexive, symmetric, and transitive closure of the relation

$$\{(a, a), (b, b), (c, c), (a, c), (a, d), (b, d), (c, a), (d, a)\}$$

on the set $S = \{a, b, c, d\}$.

For the rest of this section we shall concentrate on two types of binary relations that are characterized by which properties (reflexivity, symmetry, antisymmetry, and transitivity) they satisfy.

Partial Orderings

> **Definition: Partial Ordering**
> A binary relation on a set S that is reflexive, antisymmetric, and transitive is called a **partial ordering** on S.

From previous examples and Practice 5, we have the following instances of partial orderings:

On $\mathbb{N}$, $x \rho y \leftrightarrow x \leq y$.
On $\wp(\mathbb{N})$, $A \rho B \leftrightarrow A \subseteq B$.
On $\mathbb{Z}^+$, $x \rho y \leftrightarrow x$ divides y.
On $\{0, 1\}$, $x \rho y \leftrightarrow x = y^2$.

If ρ is a partial ordering on S, then the ordered pair (S, ρ) is called a **partially ordered set** (also known as a **poset**). We will denote an arbitrary, partially ordered set by $(S, \preccurlyeq)$; in any particular case, $\preccurlyeq$ has some definite meaning such as "less than or equal to," "is a subset of," "divides," and so on.

Let $(S, \preccurlyeq)$ be a partially ordered set, and let $A \subseteq S$. Then $\preccurlyeq$ is a set of ordered pairs of elements of S, some of which may be ordered pairs of elements of A. If we select from $\preccurlyeq$ the ordered pairs of elements of A, this new set is called the **restriction** of $\preccurlyeq$ to A and is a partial ordering on A. (Do you see why the three required properties still hold?) For instance, once we know that the relation "x divides y" is a partial ordering on $\mathbb{Z}^+$, we automatically know that "x divides y" is a partial ordering on $\{1, 2, 3, 6, 12, 18\}$.

We want to introduce some terminology about partially ordered sets. Let $(S, \preccurlyeq)$ be a partially ordered set. If $x \preccurlyeq y$, then either $x = y$ or $x \neq y$. If $x \preccurlyeq y$ but $x \neq y$, we write $x \prec y$ and say that x is a **predecessor** of y or y is a **successor** of x. A given y may have many predecessors, but if $x \prec y$ and there is no z with $x \prec z \prec y$, then x is an **immediate predecessor** of y.

PRACTICE 8   Consider the relation "x divides y" on $\{1, 2, 3, 6, 12, 18\}$.

   a. Write the ordered pairs (x, y) of this relation.
   b. Write all the predecessors of 6.
   c. Write all the immediate predecessors of 6.   ●

If S is finite, we can visually depict a partially ordered set $(S, \preccurlyeq)$ by using a **Hasse diagram**. Each of the elements of S is represented by a dot, called a **node**, or **vertex**, of the diagram. If x is an immediate predecessor of y, then the node for y is placed above the node for x and the two nodes are connected by a straight-line segment.

EXAMPLE 10   Consider $\wp(\{1, 2\})$ under the relation of set inclusion. This is a partially ordered set. (We already know that $(\wp(\mathbb{N}), \subseteq)$ is a partially ordered set.) The elements of $\wp(\{1, 2\})$ are $\varnothing$, $\{1\}$, $\{2\}$, and $\{1, 2\}$. The binary relation $\subseteq$ consists of the following ordered pairs:

$(\varnothing, \varnothing)$, $(\{1\}, \{1\})$, $(\{2\}, \{2\})$, $(\{1, 2\}, \{1, 2\})$, $(\varnothing, \{1\})$, $(\varnothing, \{2\})$,
$(\varnothing, \{1, 2\})$, $(\{1\}, \{1, 2\})$, $(\{2\}, \{1, 2\})$

The Hasse diagram of this partially ordered set appears in Figure 4.2. Note that although $\varnothing$ is not an immediate predecessor of $\{1, 2\}$, it is a predecessor of $\{1, 2\}$ (shown on the diagram by the chain of upward line segments connecting $\varnothing$ with $\{1, 2\}$). ●

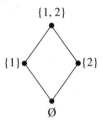

Figure 4.2

> **REMINDER:**
> Two nodes in a Hasse diagram should never be joined by a horizontal line.

PRACTICE 9   Draw the Hasse diagram for the relation "x divides y" on $\{1, 2, 3, 6, 12, 18\}$. ●

The Hasse diagram of a partially ordered set conveys all the information about the partial ordering. We can reconstruct the set of ordered pairs making up the partial ordering just by looking at the diagram. The lines in the diagram tell us immediate (predecessor, successor) pairs. We can fill in the rest by using the reflexive and transitive properties. Thus, given the Hasse diagram in Figure 4.3 of a partial ordering $\leq$ on a set $\{a, b, c, d, e, f\}$, we can conclude that $\leq$ is the set

$$\{(a, a), (b, b), (c, c), (d, d), (e, e), (f, f), (a, b)\ (a, c), (a, d), (a, e), (d, e)\}$$

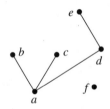

Figure 4.3

Two elements of S may be unrelated in a partial ordering of S. In Example 10, $\{1\}$ and $\{2\}$ are unrelated; so are 2 and 3, and 12 and 18 in Practice 9. In Figure 4.3, f is not related to any other element. A partial ordering in which every element of the set is related to every other element is called a **total ordering**, or **chain**. The Hasse diagram for a total ordering looks like Figure 4.4. The relation $\leq$ on $\mathbb{N}$ is a total ordering.

Figure 4.4

Again, let $(S, \preccurlyeq)$ be a partially ordered set. If there is a $y \in S$ with $y \preccurlyeq x$ for all $x \in S$, then y is a **least element** of the partially ordered set. A least element, if it exists, is unique. To show this, assume that y and z are both least elements. Then $y \preccurlyeq z$ because y is least and $z \preccurlyeq y$ because z is least; by antisymmetry, $y = z$. An element $y \in S$ is **minimal** if there is no $x \in S$ with $x < y$. In the Hasse diagram, a least element is below all others, while a minimal element has no elements below it. Similar definitions apply for greatest element and maximal elements.

PRACTICE 10   Define **greatest element** and **maximal element** in a partially ordered set $(S, \preccurlyeq)$. ●

EXAMPLE 11   In the partially ordered set of Practice 9, 1 is both least and minimal; 12 and 18 are both maximal, but there is no greatest element. ●

A least element is always minimal and a greatest element is always maximal, but the converses are not true (see Example 11). In a totally ordered set, however, a minimal element is the least element and a maximal element is the greatest element.

PRACTICE 11   Draw the Hasse diagram for a partially ordered set with four elements in which there are two minimal elements but no least element, two maximal elements but no greatest element, and each element is related to exactly two other elements. ●

REMINDER:
A partial ordering is antisymmetric; an equivalence relation is symmetric.

Partial orderings satisfy the properties of reflexivity, antisymmetry, and transitivity. Another type of binary relation, which we'll study next, satisfies a different set of properties.

Equivalence Relations

Definition: Equivalence Relation
A binary relation on a set S that is reflexive, symmetric, and transitive is called an **equivalence relation** on S.

We have already come across the following examples of equivalence relations:

On any set S, $x \rho y \leftrightarrow x = y$.
On $\mathbb{N}$, $x \rho y \leftrightarrow x + y$ is even.
On the set of all lines in the plane, $x \rho y \leftrightarrow x$ is parallel to y or coincides with y.
On $\{0, 1\}$, $x \rho y \leftrightarrow x = y^2$.
On $\{x \mid x$ is a student in your class$\}$, $x \rho y \leftrightarrow x$ sits in the same row as y.
On $\{1, 2, 3\}$, $\rho = \{(1, 1), (2, 2), (3, 3), (1, 2), (2, 1)\}$.

We can illustrate an important feature of an equivalence relation on a set by looking at $S = \{x \mid x$ is a student in your class$\}$, $x \, \rho \, y \leftrightarrow$ "x sits in the same row as y." Let's group together all those students in set S who are related to one another. We come up with Figure 4.5. We have partitioned the set S into subsets in such a way that everyone in the class belongs to one and only one subset.

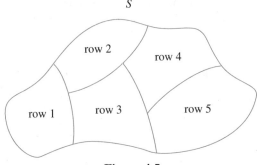

Figure 4.5

Definition: Partition of a Set
A **partition** of a set S is a collection of nonempty disjoint subsets of S whose union equals S.

Any equivalence relation, as we shall see, partitions the set on which it is defined. The subsets making up the partition, often called the **blocks** of the partition, are formed by grouping together related elements, as in the students in the classroom.

For ρ an equivalence relation on a set S and $x \in S$, we let $[x]$ denote the set of all members of S to which x is related, called the **equivalence class** of x. Thus,

$$[x] = \{y \mid y \in S \wedge x \, \rho \, y\}$$

EXAMPLE 12   In the case where $x \, \rho \, y \leftrightarrow$ "x sits in the same row as y," suppose that John, Chuck, José, Judy, and Ted all sit in row 3. Then [John] = {John, Chuck, José, Judy, Ted}. Also [John] = [Ted] = [Judy], and so on. These are not distinct classes, but the same class with multiple names. An equivalence class can take its name from any of the elements in it.   ●

Now we'll state the result about equivalence relations and partitions. For some practice with formal theorems and proofs, we'll give this result as a formal theorem, then analyze the structure of the proof and complete part of the proof.

Theorem on Equivalence Relations and Partitions
An equivalence relation ρ on a set S determines a partition of S, and a partition of a set S determines an equivalence relation on S.

Partial Proof: The theorem makes two separate statements:

a. An equivalence relation on S determines a partition of S.
b. A partition of S determines an equivalence relation on S.

To prove part a, we must show that the distinct equivalence classes of members of S under equivalence relation ρ satisfy the definition of a partition. To satisfy the definition of a partition, we must show that

i. The union of these distinct classes equals S.
ii. The distinct classes are disjoint.

To prove part a.i, we must show something about the union of the distinct equivalence classes formed by ρ. Equivalence classes are sets of elements of S, so their union is a set; let's denote this set by U. We must show that $U = S$, which is a set equality. To prove this set equality, we will prove set inclusion in each direction, i.e.,

1. $U \subseteq S$
2. $S \subseteq U$

For this part of the proof, we are finally down to two small statements that are easy to prove, as follows:

a.i.1: Let $x \in U$. Then x belongs to an equivalence class. Every equivalence class is a subset of S, so $x \in S$.

a.i.2: Let $x \in S$. Then $x \rho x$ (reflexivity of ρ); thus, $x \in [x]$, and every member of S belongs to some equivalence class, hence to the union of classes U.

This completes the proof of a.i. For a.ii, let $[x]$ and $[z]$ be two equivalence classes. We want to show that distinct classes are disjoint, or

$$[x] \neq [z] \rightarrow [x] \cap [z] = \varnothing \qquad \text{a.ii}$$

If we assume $[x] \neq [z]$, we must then show that $[x] \cap [z]$ does *not* contain anything, which might be hard to do. So instead, we'll prove the contrapositive of a.ii:

$$[x] \cap [z] \neq \varnothing \rightarrow [x] = [z] \qquad \text{contrapositive of a.ii}$$

Therefore, we assume that $[x] \cap [z] \neq \varnothing$ and that there is a $y \in S$ such that $y \in [x] \cap [z]$. What does this tell us?

$y \in [x] \cap [z]$	(assumption)
$y \in [x], y \in [z]$	(definition of $\cap$)
$x \rho y, z \rho y$	(definition of $[x]$ and $[z]$)
$x \rho y, y \rho z$	(symmetry of ρ)
$x \rho z$	(transitivity of ρ)

Now we can show that $[x] = [z]$ by proving set inclusion in each direction, i.e.,

3. $[z] \subseteq [x]$
4. $[x] \subseteq [z]$

To show (3), $[z] \subseteq [x]$, let $q \in [z]$ (we know $[z] \neq \varnothing$ because $y \in [z]$.) Then

$z \rho q$	(definition of $[z]$)
$x \rho z$	(from above)
$x \rho q$	(transitivity of ρ)
$q \in [x]$	(definition of $[x]$)
$[z] \subseteq [x]$	(definition of $\subseteq$)

Practice 12 asks for a proof of (4), $[x] \subseteq [z]$. Once this proof is supplied, it completes (3) and (4), which leads to the conclusion $[x] = [z]$. This completes the proof of the contrapositive of a.ii, and therefore proves a.ii, which in turn completes the proof of part a. Whew!

Practice 13 asks for a proof of part b.

End of Partial Proof

PRACTICE 12   For the foregoing argument, supply the proof that $[x] \subseteq [z]$.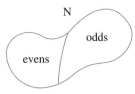

PRACTICE 13   Prove part b of the theorem. Given a partition of a set S, define a relation ρ by

$$x \, \rho \, y \leftrightarrow x \text{ is in the same subset of the partition as } y$$

and show that ρ is an equivalence relation on S; that is, show that ρ is reflexive, symmetric, and transitive.

EXAMPLE 13   The equivalence relation on $\mathbb{N}$ given by

$$x \, \rho \, y \leftrightarrow x + y \text{ is even}$$

partitions $\mathbb{N}$ into two equivalence classes. If x is an even number, then for any even number y, $x + y$ is even and $y \in [x]$. All even numbers form one class. If x is an odd number and y is any odd number, $x + y$ is even and $y \in [x]$. All odd numbers form the second class. The partition can be pictured as in Figure 4.6. Notice again that an equivalence class may have more than one name, or representative. In this example, $[2] = [8] = [1048]$, and so on; $[1] = [17] = [947]$, and so on.

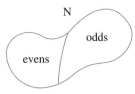

Figure 4.6

PRACTICE 14   For each of the following equivalence relations, describe the corresponding equivalence classes.

a. On the set of all lines in the plane, $x \, \rho \, y \leftrightarrow x$ is parallel to y or x coincides with y.

b. On the set $\mathbb{N}$, $x \, \rho \, y \leftrightarrow x = y$.

c. On $\{1, 2, 3\}$, $\rho = \{(1, 1), (2, 2), (3, 3), (1, 2), (2, 1)\}$.

Partitioning a set into equivalence classes is helpful because it is often convenient to go up one level of abstraction and treat the classes themselves as entities. We shall conclude this section with two examples where this is the case.

EXAMPLE 14   Let $S = \{a/b \mid a, b \in \mathbb{Z}, b \neq 0\}$. S is therefore the set of all fractions. Two fractions such as 1/2 and 2/4 are said to be equivalent. Formally, a/b is equivalent to c/d, denoted by $a/b \sim c/d$, if and only if $ad = bc$. We shall show that the binary relation $\sim$ on S is an equivalence relation. First, $a/b \sim a/b$ because $ab = ba$. Also, if $a/b \sim c/d$ then $ad = bc$, or $cb = da$ and $c/d \sim a/b$. Hence, $\sim$ is reflexive and symmetric. To show that ∼ is transitive, let $a/b \sim c/d$ and $c/d \sim e/f$. Then $ad = bc$ and $cf = de$. Multiplying the first equation by f and the second by b, we get $adf = bcf$ and $bcf = bde$. Therefore, $adf = bde$, or $af = be$ (why is it legitimate to divide by d here?). Thus, $a/b \sim e/f$, and ∼ is transitive. Some sample equivalence classes of S formed by this equivalence relation are

$$\left[\frac{1}{2}\right] = \left\{ \cdots, \frac{-3}{-6}, \frac{-2}{-4}, \frac{-1}{-2}, \frac{1}{2}, \frac{2}{4}, \frac{3}{6}, \cdots \right\}$$

$$\left[\frac{3}{10}\right] = \left\{ \cdots, \frac{-9}{-30}, \frac{-6}{-20}, \frac{-3}{-10}, \frac{3}{10}, \frac{6}{20}, \frac{9}{30}, \cdots \right\}$$

The set $\mathbb{Q}$ of rational numbers can be regarded as the set of all equivalence classes of S. A single rational number, such as [1/2], has many fractions representing it, although we customarily use the reduced fractional representation. When we add two rational numbers, such as [1/2] + [3/10], we look for representatives from the classes having the same denominator and add those representatives. Our answer is the class to which the resulting sum belongs, and we usually name the class by using a reduced fraction. Thus, to add [1/2] + [3/10], we represent [1/2] by 5/10 and [3/10] by 3/10. The sum of 5/10 and 3/10 is 8/10, and [8/10] is customarily named [4/5]. This procedure is so familiar that it is generally written as $1/2 + 3/10 = 4/5$; nonetheless, classes of fractions are being manipulated by means of representatives. ●

EXAMPLE 15

We shall define a binary relation of **congruence modulo 4** on the set $\mathbb{Z}$ of integers. An integer x is congruent modulo 4 to y, symbolized by $x \equiv_4 y$, or $x \equiv y \pmod 4$, if $x - y$ is an integral multiple of 4. Congruence modulo 4 is an equivalence relation on $\mathbb{Z}$. (Can you prove this?) To construct the equivalence classes, note that [0], for example, will contain all integers differing from 0 by a multiple of 4, such as 4, 8, −12, and so on. The distinct equivalence classes are

$[0] = \{ \ldots, -8, -4, 0, 4, 8, \ldots \}$

$[1] = \{ \ldots, -7, -3, 1, 5, 9, \ldots \}$

$[2] = \{ \ldots, -6, -2, 2, 6, 10, \ldots \}$

$[3] = \{ \ldots, -5, -1, 3, 7, 11, \ldots \}$ ●

There is nothing special about the choice of 4 in Example 15; we can give a definition for **congruence modulo n** for any positive integer n. This binary relation is always an equivalence relation. This equivalence relation and the resulting equivalence classes can be used for integer arithmetic on a computer. An integer is stored as a sequence of bits (0s and 1s) within a single memory location. Each computer allocates a fixed number of bits for a single memory location (this number varies depending on the architecture of the computer, i.e., how its memory space is laid out). The larger the integer, the more bits required to represent it. Therefore each machine has some limit on the size of the integers that it can store. Suppose that $n - 1$ is the maximum size and that x and y are integer values with $0 \le x \le n - 1$, $0 \le y \le n - 1$. If the sum $x + y$ exceeds the maximum size, it cannot be stored. As an alternative, the computer may perform **addition modulo n** and find the remainder r when $x + y$ is divided by n.

The equation

$$x + y = qn + r, 0 \leq r < n$$

symbolizes this division, where q is the quotient and r is the remainder. This equation may be written as

$$(x + y) - r = qn$$

which shows that $(x + y) - r$ is an integral multiple of n, or that $(x + y) \equiv r \pmod{n}$. The integer r may not be $x + y$, but it is in the equivalence class $[x + y]$, and since $0 \leq r < n$, it is also in the range of integers that can be stored. (The system may or may not issue an *integer overflow* message if $x + y$ is too large to store and addition modulo n must be used.) The situation is analogous to your car's odometer, which records mileage modulo 100,000; when mileage reaches 102,758, for example, it is displayed on the odometer as 2,758.

PRACTICE 15   What are the equivalence classes corresponding to the relation of congruence modulo 5 on $\mathbb{Z}$?   ●

PRACTICE 16   If 4 is the maximum integer that can be stored on a (micromicro) computer, what will be stored for the value $3 + 4$ if addition modulo 5 is used?   ●

Table 4.1 summarizes important features of partial orderings and equivalence relations.

Partial Orderings and Equivalence Relations					
Type of Binary Relation	Reflexive	Symmetric	Antisymmetric	Transitive	Important Feature
Partial ordering	Yes	No	Yes	Yes	Predecessors and successors
Equivalence relation	Yes	Yes	No	Yes	Determines a partition

Table 4.1

Section 4.1 Review

Techniques

● Test an ordered pair for membership in a binary relation.
● Test a binary relation for reflexivity, symmetry, antisymmetry, and transitivity.
● Find the reflexive, symmetric, and transitive closure of a relation.

- Draw the Hasse diagram for a partially ordered set.
- Find least, minimal, greatest, and maximal elements in a partially ordered set.
- Find the equivalence classes associated with an equivalence relation.

Main Ideas

A binary relation on a set S is formally a subset of $S \times S$; the distinctive relationship satisfied by the relation's members often has a verbal description as well.

Operations on binary relations on a set include union, intersection, and complementation.

Binary relations can have properties of reflexivity, symmetry, transitivity, and antisymmetry.

Finite partially ordered sets can be represented graphically.

An equivalence relation on a set defines equivalence classes, which may themselves be treated as entities. An equivalence relation on a set S determines a partition of S, and conversely.

Exercises 4.1

★1. For each of the following binary relations ρ on $\mathbb{N}$, decide which of the given ordered pairs belong to ρ.

   a. $x \rho y \leftrightarrow x + y < 7$; (1, 3), (2, 5), (3, 3), (4, 4)
   b. $x \rho y \leftrightarrow x = y + 2$; (0, 2), (4, 2), (6, 3), (5, 3)
   c. $x \rho y \leftrightarrow 2x + 3y = 10$; (5, 0), (2, 2), (3, 1), (1, 3)
   d. $x \rho y \leftrightarrow y$ is a perfect square; (1, 1), (4, 2), (3, 9), (25, 5)

2. Decide which of the given items satisfy the relation.

   a. ρ a binary relation on $\mathbb{Z}$, $x \rho y \leftrightarrow x = -y$; (1, −1), (2, 2), (−3, 3), (−4, −4)
   b. ρ a binary relation on $\mathbb{N}$, $x \rho y \leftrightarrow x$ is prime; (19, 7), (21, 4), (33, 13), (41, 16)
   c. ρ a binary relation on $\mathbb{Q}$, $x \rho y \leftrightarrow x \leq 1/y$; (1, 2), (−3, −5), (−4, 1/2), (1/2, 1/3)
   d. ρ a binary relation on $\mathbb{N} \times \mathbb{N}$, $(x, y) \rho (u, v) \leftrightarrow x + u = y + v$; ((1, 2), (3, 2)), ((4, 5), (0, 1))

3. For each of the following binary relations on $\mathbb{R}$, draw a figure to show the region of the plane it describes.

   ★**a.** $x \rho y \leftrightarrow y \leq 2$
   b. $x \rho y \leftrightarrow x = y - 1$
   c. $x \rho y \leftrightarrow x^2 + y^2 \leq 25$
   d. $x \rho y \leftrightarrow x \geq y$

4. For each of the accompanying figures, give the binary relation on $\mathbb{R}$ that describes the shaded area.

a.

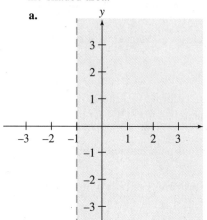

b.

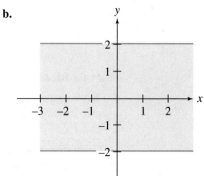

c.

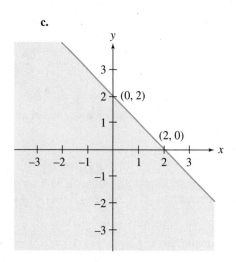

d.

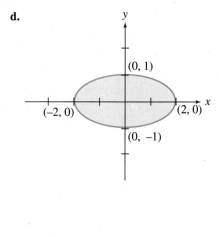

★5. Identify each relation on $\mathbb{N}$ as one-to-one, one-to-many, many-to-one, or many-to-many.

  a. $\rho = \{(1, 2), (1, 4), (1, 6), (2, 3), (4, 3)\}$
  b. $\rho = \{(9, 7), (6, 5), (3, 6), (8, 5)\}$
  c. $\rho = \{(12, 5), (8, 4), (6, 3), (7, 12)\}$
  d. $\rho = \{(2, 7), (8, 4), (2, 5), (7, 6), (10, 1)\}$

6. Identify each of the following relations on S as one-to-one, one-to-many, many-to-one, or many-to-many.

  a. $S = \mathbb{N}$
     $x \, \rho \, y \leftrightarrow x = y + 1$
  c. $S = \wp(\{1, 2, 3\})$
     $A \, \rho \, B \leftrightarrow |A| = |B|$

  b. S = set of all women in Vicksburg
     $x \, \rho \, y \leftrightarrow x$ is the daughter of y
  d. $S = \mathbb{R}$
     $x \, \rho \, y \leftrightarrow x = 5$

★7. Let ρ and σ be binary relations on $\mathbb{N}$ defined by $x\,\rho\,y \leftrightarrow$ "x divides y," $x\,\sigma\,y \leftrightarrow 5x \leq y$. Decide which of the given ordered pairs satisfy the following relations:

   a. $\rho \cup \sigma$; (2, 6), (3, 17), (2, 1), (0, 0)
   b. $\rho \cap \sigma$; (3, 6), (1, 2), (2, 12)
   c. ρ'; (1, 5), (2, 8), (3, 15)
   d. σ'; (1, 1), (2, 10), (4, 8)

8. Let $S = \{0, 1, 2, 4, 6\}$. Test the following binary relations on S for reflexivity, symmetry, antisymmetry, and transitivity.

   a. $\rho = \{(0, 0), (1, 1), (2, 2), (4, 4), (6, 6), (0, 1), (1, 2), (2, 4), (4, 6)\}$
   b. $\rho = \{(0, 1), (1, 0), (2, 4), (4, 2), (4, 6), (6, 4)\}$
   c. $\rho = \{(0, 1), (1, 2), (0, 2), (2, 0), (2, 1), (1, 0), (0, 0), (1, 1), (2, 2)\}$
   d. $\rho = \{(0, 0), (1, 1), (2, 2), (4, 4), (6, 6), (4, 6), (6, 4)\}$
   e. $\rho = \varnothing$

9. Let S be the set of people in the United States. Test the following binary relations on S for reflexivity, symmetry, antisymmetry, and transitivity.

   a. $x\,\rho\,y \leftrightarrow x$ is at least as tall as y.
   b. $x\,\rho\,y \leftrightarrow x$ is taller than y.
   c. $x\,\rho\,y \leftrightarrow x$ is the same height as y.
   d. $x\,\rho\,y \leftrightarrow x$ is a child of y.
   e. $x\,\rho\,y \leftrightarrow x$ is the husband of y.
   f. $x\,\rho\,y \leftrightarrow x$ is the spouse of y.
   g. $x\,\rho\,y \leftrightarrow x$ has the same parents as y.
   h. $x\,\rho\,y \leftrightarrow x$ is the brother of y.

10. Test the following binary relations on the given sets S for reflexivity, symmetry, antisymmetry, and transitivity.

   ★**a.** $S = \mathbb{Q}$
   $x\,\rho\,y \leftrightarrow |x| \leq |y|$
   ★**c.** $S = \mathbb{N}$
   $x\,\rho\,y \leftrightarrow x \cdot y$ is even
   e. $S =$ set of all squares in the plane
   $S_1\,\rho\,S_2 \leftrightarrow$ length of side of $S_1 =$ length of side of S_2
   f. $S =$ set of all finite-length strings of characters
   $x\,\rho\,y \leftrightarrow$ number of characters in $x =$ number of characters in y
   g. $S = \{0, 1, 2, 3, 4, 5\}$
   $x\,\rho\,y \leftrightarrow x + y = 5$
   h. $S = \wp(\{1, 2, 3, 4, 5, 6, 7, 8, 9\})$
   $A\,\rho\,B \leftrightarrow |A| = |B|$
   i. $S = \wp(\{1, 2, 3, 4, 5, 6, 7, 8, 9\})$
   $A\,\rho\,B \leftrightarrow |A| \neq |B|$
   j. $S = \mathbb{N} \times \mathbb{N}$
   $(x_1, y_1)\,\rho\,(x_2, y_2) \leftrightarrow x_1 \leq x_2$ and $y_1 \geq y_2$

   ★**b.** $S = \mathbb{Z}$
   $x\,\rho\,y \leftrightarrow x - y$ is an integral multiple of 3
   d. $S = \mathbb{N}$
   $x\,\rho\,y \leftrightarrow x$ is odd

11. Which of the binary relations of Exercise 10 are equivalence relations? For each equivalence relation, describe the associated equivalence classes.

12. For each case, think of a set S and a binary relation ρ on S (different from any in the examples or problems) satisfying the given conditions.
   a. ρ is reflexive and symmetric but not transitive.
   b. ρ is reflexive and transitive but not symmetric.
   c. ρ is not reflexive or symmetric but is transitive.
   d. ρ is reflexive but neither symmetric nor transitive.

13. Let ρ and σ be binary relations on a set S.
   a. If ρ and σ are reflexive, is $\rho \cup \sigma$ reflexive? Is $\rho \cap \sigma$ reflexive?
   b. If ρ and σ are symmetric, is $\rho \cup \sigma$ symmetric? Is $\rho \cap \sigma$ symmetric?
   c. If ρ and σ are antisymmetric, is $\rho \cup \sigma$ antisymmetric? Is $\rho \cap \sigma$ antisymmetric?
   d. If ρ and σ are transitive, is $\rho \cup \sigma$ transitive? Is $\rho \cap \sigma$ transitive?

14. Find the reflexive, symmetric, and transitive closure of each of the relations in Exercise 8.

★15. For each of the following relations, describe in words what the transitive closure relation would be.
   a. S = set of all buildings in a city
      $x \rho y \leftrightarrow x$ is one year older than y
   b. S = set of all males in Bulgaria
      $x \rho y \leftrightarrow x$ is the father of y
   c. S = set of cities in the United States
      $x \rho y \leftrightarrow$ you can drive from x to y in one day

16. Two additional properties of a binary relation ρ are defined as follows:

   ρ is *irreflexive* means $(\forall x)(x \in S \rightarrow (x, x) \notin \rho)$
   ρ is *asymmetric* means $(\forall x)(\forall y)(x \in S \wedge y \in S \wedge (x, y) \in \rho \rightarrow (y, x) \notin \rho)$

   ★a. Give an example of a binary relation ρ on set $S = \{1, 2, 3\}$ that is neither reflexive nor irreflexive.
   b. Give an example of a binary relation ρ on set $S = \{1, 2, 3\}$ that is neither symmetric nor asymmetric.
   c. Prove that if ρ is an asymmetric relation on a set S, then ρ is irreflexive.
   d. Prove that if ρ is an irreflexive and transitive relation on a set S, then ρ is asymmetric.
   e. Prove that if ρ is a nonempty, symmetric, and transitive relation on a set S, then ρ is not irreflexive.

17. Does it make sense to look for the closure of a relation with respect to the following properties? Why or why not?

    a. irreflexive property
    b. asymmetric property

18. Let S be an n-element set. How many different binary relations can be defined on S? (*Hint:* Recall the formal definition of a binary relation.)

19. Let ρ be a binary relation on a set S. For $A \subseteq S$, define

    $$\#A = \{x \mid x \in S \wedge (\forall y)(y \in A \to x \, \rho \, y)\}$$
    $$A\# = \{x \mid x \in S \wedge (\forall y)(y \in A \to y \, \rho \, x)\}$$

    a. Prove that if ρ is symmetric, then $\#A = A\#$.
    b. Prove that if $A \subseteq B$ then $\#B \subseteq \#A$ and $B\# \subseteq A\#$.
    c. Prove that $A \subseteq (\#A)\#$.
    d. Prove that $A \subseteq \#(A\#)$.

20. Draw the Hasse diagram for the following partial orderings:

    a. $S = \{a, b, c\}$
      $\rho = \{(a, a), (b, b), (c, c), (a, b), (b, c), (a, c)\}$
    ★**b.** $S = \{a, b, c\}$
      $\rho = \{(a, a), (b, b), (c, c), (d, d), (a, b), (a, c)\}$
    c. $S = \{\varnothing, \{a\}, \{a, b\}, \{c\}, \{a, c\}, \{b\}\}$
      $A \, \rho \, B \leftrightarrow A \subseteq B$

21. For Exercise 20, name any least elements, minimal elements, greatest elements, and maximal elements.

22. Let $(S, \preccurlyeq)$ be a partially ordered set, and let $A \subseteq S$. Prove that the restriction of $\preccurlyeq$ to A is a partial ordering on A.

23. **a.** Draw the Hasse diagram for the partial ordering "x divides y" on the set $\{2, 3, 5, 7, 21, 42, 105, 210\}$. Name any least elements, minimal elements, greatest elements, and maximal elements. Name a totally ordered subset with four elements.
    b. Draw the Hasse diagram for the partial ordering "x divides y" on the set $\{3, 6, 9, 18, 54, 72, 108, 162\}$. Name any least elements, minimal elements, greatest elements, and maximal elements. Name any unrelated elements.

★24. Draw the Hasse diagram for each of the two partially ordered sets.

    a. $S = \{1, 2, 3, 5, 6, 10, 15, 30\}$ **b.** $S = \wp(\{1, 2, 3\})$
      $x \, \rho \, y \leftrightarrow x$ divides y $A \, \rho \, B \leftrightarrow A \subseteq B$

    What do you notice about the structure of these two diagrams?

25. For each Hasse diagram of a partial ordering in the accompanying figure, list the ordered pairs that belong to the relation.

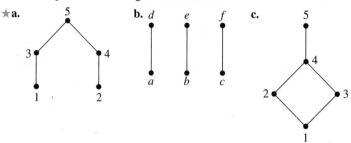

★a.   5   b. *d*   *e*   *f*   c.   5

3 • • 4   3 • 4

1   2   *a*   *b*   *c*   2 • • 3

1

26. Let (S, ρ) and (T, σ) be two partially ordered sets. A relation μ on $S \times T$ is defined by $(s_1, t_1) \, \mu \, (s_2, t_2) \leftrightarrow s_1 \, \rho \, s_2$ and $t_1 \, \sigma \, t_2$. Show that μ is a partial ordering on $S \times T$.

27. Let ρ be a binary relation on a set S. Then a binary relation called the *inverse of* ρ, denoted by ρ^{-1}, is defined by $x \, \rho^{-1} \, y \leftrightarrow y \, \rho \, x$.
   a. For $\rho = \{(1, 2), (2, 3), (5, 3), (4, 5)\}$ on the set $\mathbb{N}$, what is ρ^{-1}?
   b. Prove that if ρ is a reflexive relation on a set S, then ρ^{-1} is reflexive.
   c. Prove that if ρ is a symmetric relation on a set S, then ρ^{-1} is symmetric.
   d. Prove that if ρ is an antisymmetric relation on a set S, then ρ^{-1} is antisymmetric.
   e. Prove that if ρ is a transitive relation on a set S, then ρ^{-1} is transitive.
   f. Prove that if ρ is an irreflexive relation on a set S (see Exercise 16), then ρ^{-1} is irreflexive.
   g. Prove that if ρ is an asymmetric relation on a set S (see Exercise 16), then ρ^{-1} is asymmetric.

★28. Prove that if a binary relation ρ on a set S is reflexive and transitive, then the relation $\rho \cap \rho^{-1}$ is an equivalence relation (see Exercise 27 for the definition of ρ^{-1}).

29. a. Let (S, ρ) be a partially ordered set. Then ρ^{-1} can be defined as in Exercise 27. Show that (S, ρ^{-1}) is a partially ordered set, called the *dual* of (S, ρ).
   b. If (S, ρ) is a finite, partially ordered set with the Hasse diagram shown, draw the diagram of the dual of (S, ρ).

   c. Let (S, ρ) be a totally ordered set and let $X = \{(x, x) \mid x \in S\}$. Show that the set difference $\rho^{-1} - X$ equals the set ρ'.

30. A computer program is to be written that will generate a dictionary or the index for a book. We will assume a maximum length of n characters per word. Thus, we are given a set S of words of length at most n, and we want to produce a linear list of these words arranged in alphabetical order. There is a natural total ordering $\leqslant$ on alphabetic characters ($a < b$, $b < c$, etc.), and we shall assume our words contain only alphabetic characters. We want to define a total ordering $\leqslant$ on S called a *lexicographical ordering* that will arrange the members of S alphabetically. The idea is to compare two words X and Y character by character, passing over equal characters. If at any point the X-character alphabetically precedes the corresponding Y-character, then X precedes Y; if all characters in X are equal to the corresponding Y characters but we run out of characters in X before characters in Y, then X precedes Y. Otherwise, Y precedes X.

Formally, let $X = (x_1, x_2, \ldots, x_j)$ and $Y = (y_1, y_2, \ldots, y_k)$ be members of S with $j \leq k$. Let β (for blank) be a new symbol, and fill out X with $k - j$ blanks on the right. X can now be written $(x_1, x_2, \ldots, x_k)$. Let β precede any alphabetical character. Then $X \leqslant Y$ if

$$x_1 \neq y_1 \text{ and } x_1 \leqslant y_1$$

or

$$x_1 = y_1, x_2 = y_2, x_m = y_m \ (m \leq k)$$
$$x_{m+1} \neq y_{m+1} \text{ and } x_{m+1} \leqslant y_{m+1}$$

Otherwise, $Y \leqslant X$.

Note that because the ordering $\leqslant$ on alphabetical characters is a total ordering, if $Y \leqslant X$ by "otherwise," then there exists $m \leq k$ such that $x_1 = y_1$, $x_2 = y_2$, $\ldots$, $x_m = y_m$, $x_{m+1} \neq y_{m+1}$ and $y_{m+1} \leqslant x_{m+1}$.

a. Show that $\leqslant$ on S as defined above is a total ordering.

b. Apply the total ordering described to the words *boo*, *bug*, *be*, *bah*, and *bugg*. Note why each word precedes the next.

★31. Exercise 30 discusses a total ordering on a set of words of length at most n that will produce a linear list in alphabetical order. Suppose we want to generate a list of all the distinct words in a text (for example, a compiler must create a symbol table of variable names). As in Exercise 30, we shall assume that the words contain only alphabetical characters because there is a natural precedence relation already existing ($a < b$, $b < c$, etc.). If numeric or special characters are involved, they must be assigned a precedence relation with alphabetical characters (the collating sequence must be determined). If we list words alphabetically, it is a fairly quick procedure to decide whether a word currently being processed is new, but to fit the new word into place, all successive words must be moved one unit down the line. If the words are listed in the order in which they are processed, new words are simply tacked onto the end and no rearranging is necessary, but each word being processed has to be compared with each member of the list to determine if it is new. Thus, both logical linear lists have disadvantages.

We shall describe a listing process called a *binary tree search* that can usually determine quickly whether a word is new and, if it is, no juggling is required to fit it into place, thus combining the advantages of both methods above. Suppose we want to process the phrase "when in the course of human events." The first word in the text is used to label the first node of a graph. Once a node is labeled, it drops down a left and right arc, putting two unlabeled nodes below the one just labeled.

When the next word in the text is processed, it is compared with the first node. When the word being processed alphabetically precedes the label of a node, the left arc is taken; when the word follows the label alphabetically, the right arc is taken. The word becomes the label of the first unlabeled node it reaches. (If the word equals a node label, the next word in the text is processed.) This procedure continues for the entire text. Thus,

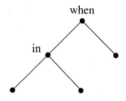

then

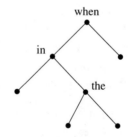

then

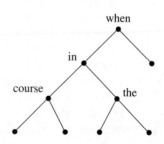

until finally

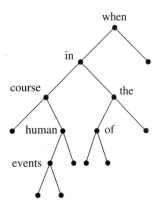

By traversing the nodes of this graph in the proper order (described by always processing the left nodes below a node first, then the node, then the right nodes below it), an alphabetical listing "course, events, human, in, of, the, when" is produced.

a. This type of graph is called a *tree*. (Unlabeled nodes and arcs to unlabeled nodes usually are not shown.) Turned upside down, it can be viewed as the Hasse diagram of a partial ordering $\leqslant$. What would be the least element? Would there be a greatest element? Which of the following would belong to $\leqslant$: (in, of), (the, of), (in, events), (course, of)?

Here the tree structure contains more information than the partial ordering, because we are interested in not only whether a word w_1 precedes a word w_2 but also whether w_2 is to the left or right of w_1.

b. Use the binary tree search to graph "Old King Cole was a merry old soul." Eliminate unlabeled nodes. Considering the graph (upside down) as the Hasse diagram of a partial ordering, name the maximal elements.

32. The alphabetical ordering defined in Exercise 30 can be applied to words of any finite length. If we define A^* to be the set of all finite-length "words" (strings of characters, not necessarily meaningful) from the English alphabet, then the alphabetical ordering on A^* has all words composed only of the letter a preceding all other words. Thus, all the words in the infinite list

$a, aa, aaa, aaaa, \ldots$

precede words such as "b" or "$aaaaaaab$." Therefore this list does not enumerate A^*, because we can never count up to any words with any characters other than a. However, the set A^* is denumerable. Prove this by ordering A^* by length (all words of length 1 precede all words of length 2, etc.) and then alphabetically ordering words of the same length.

★33. **a.** For the equivalence relation $\rho = \{(a, a), (b, b), (c, c), (a, c), (c, a)\}$, what is the set $[a]$? Does it have any other names?

b. For the equivalence relation $\rho = \{(1, 1), (2, 2), (1, 2), (2, 1), (1, 3), (3, 1), (3, 2), (2, 3), (3, 3), (4, 4), (5, 5), (4, 5), (5, 4)\}$, what is the set $[3]$? What is the set $[4]$?

c. For the equivalence relation of congruence modulo 2 on the set $\mathbb{Z}$, what is the set $[1]$?

d. For the equivalence relation of congruence modulo 5 on the set $\mathbb{Z}$, what is the set $[-3]$?

34. **a.** Given the partition $\{1, 2\}$ and $\{3, 4\}$ of the set $S = \{1, 2, 3, 4\}$, list the ordered pairs in the corresponding equivalence relation.

b. Given the partition $\{a, b, c\}$ and $\{d, e\}$ of the set $S = \{a, b, c, d, e\}$, list the ordered pairs in the corresponding equivalence relation.

35. Let S be the set of all books in the library. Let ρ be a binary relation on S defined by $x \rho y \leftrightarrow$ "the color of x's cover is the same as the color of y's cover." Show that ρ is an equivalence relation on S and describe the resulting equivalence classes.

★36. Let $S = \mathbb{N} \times \mathbb{N}$ and let ρ be a binary relation on S defined by $(x, y) \rho (z, w) \leftrightarrow y = w$. Show that ρ is an equivalence relation on S and describe the resulting equivalence classes.

37. Let $S = \mathbb{N} \times \mathbb{N}$ and let ρ be a binary relation on S defined by $(x, y) \rho (z, w) \leftrightarrow x + y = z + w$. Show that ρ is an equivalence relation on S and describe the resulting equivalence classes.

38. Let $S = \mathbb{N}$ and let ρ be a binary relation on S defined by $x \rho y \leftrightarrow x^2 - y^2$ is even. Show that ρ is an equivalence relation on S and describe the resulting equivalence classes.

★39. Let S be the set of all propositional wffs with n statement letters. Let ρ be a binary relation on S defined by $P \rho Q \leftrightarrow$ "$P \leftrightarrow Q$ is a tautology." Show that ρ is an equivalence relation on S and describe the resulting equivalence classes. (We have used the notation $P \Leftrightarrow Q$ for $P \rho Q$.)

40. Given two partitions π_1 and π_2 of a set S, π_1 is a *refinement* of π_2 if each block of π_1 is a subset of a block of π_2. Show that refinement is a partial ordering on the set of all partitions of S.

Exercises 41–50 all deal with partitions on a set.

41. Let P_n denote the total number of partitions of an n-element set, $n \geq 1$.

a. Find P_1.
b. Find P_3.
c. Find P_4.

★42. Let $S(n, k)$ denote the number of ways to partition a set of n elements into k blocks.

a. Find $S(3, 2)$.
b. Find $S(4, 2)$.

43. Prove that

$$P_n = \sum_{k=1}^{n} S(n, k)$$

44. Prove that for all $n \geq 1$, $S(n, k)$ satisfies the recurrence relation

$S(n, 1) = 1$

$S(n, n) = 1$

$S(n + 1, k + 1) = S(n, k) + (k + 1)S(n, k + 1)$ for $1 \leq k \leq n$

(*Hint:* Use a combinatorial proof instead of an inductive proof. Let x be a fixed but arbitrary member of a set with $n + 1$ elements, and put x aside. Partition the remaining set of n elements. A partition of the original set could be obtained either by adding $\{x\}$ as a separate block or by putting x in one of the existing blocks.)

★45. Use the formula of Exercise 44 to rework Exercise 42.

46. The numbers $S(n, k)$ are called *Stirling numbers*. The recurrence relation of Exercise 44 is similar to Pascal's formula, equation (1) of Section 3.5. Use this relation to compute the numeric values in the first five rows of *Stirling's triangle*, which begins

$$S(1, 1)$$
$$S(2, 1) \quad S(2, 2)$$
$$S(3, 1) \quad S(3, 2) \quad S(3, 3)$$
$$\vdots$$

★47. Find the number of ways to distribute 4 different-colored marbles among 3 identical containers so that no container is empty.

48. Find the number of ways in which 5 different jobs can be assigned to 3 identical processors so that each processor gets at least 1 job.

49. Let P_0 have the value 1. Prove that

$$P_n = \sum_{k=0}^{n-1} C(n - 1, k)P_k$$

(*Hint:* Use a combinatorial proof instead of an inductive proof. Let x be a fixed but arbitrary member of a set with n elements. In each term of the sum, $n - k$ represents the size of the partition block that contains x.)

50. **a.** Use the formula of Exercise 49 to compute P_1, P_2, P_3, and P_4, and compare your answers to those in Exercise 41.
    b. Use the formula of Exercise 43 and Stirling's triangle (Exercise 46) to compute P_1, P_2, P_3, and P_4.

51. Binary relations on a set S are ordered pairs of elements of S. More generally, an *n-ary relation on a set S* is a set of ordered *n*-tuples of elements of S. Decide which of the given items satisfy the relation.

   a. ρ a unary relation on $\mathbb{Z}$, $x \in \rho \leftrightarrow x$ is a perfect square
      25, 39, 49, 62

   b. ρ a ternary relation on $\mathbb{N}$, $(x, y, z) \in \rho \leftrightarrow x^2 + y^2 = z^2$
      (1, 1, 2), (3, 4, 5), (0, 5, 5), (8, 6, 10)

   c. ρ a 4-ary relation on $\mathbb{Z}$, $(x, y, z, w) \in \rho \leftrightarrow y = |x|$ and $w \geq x + z^2$
      $(-4, 4, 2, 0)$, $(5, 5, 1, 5)$, $(6, -6, 6, 45)$, $(-6, 6, 0, -2)$

Section 4.2   Topological Sorting

If ρ is a partial ordering on a set S, then some elements of S are predecessors of other elements. If S is a set of tasks that are to be done, then the idea of x as a predecessor of y can be interpreted literally to mean that task x must be done before task y. Thus partial orderings and Hasse diagrams are natural ways to represent problems in task scheduling.

EXAMPLE 16   Ernie and his brothers run a woodworking shop in the hills of New Hampshire that manufactures rocking chairs with padded cushion seats. The manufacturing process can be broken down into a number of tasks, some of which have certain other tasks as prerequisites. The following table shows the manufacturing tasks for a rocking chair, the prerequisite tasks, and the number of hours required to perform each task.

Task	Prerequisite Tasks	Hours to Perform
1. Selecting wood	None	3.0
2. Carving rockers	1	4.0
3. Carving seat	1	6.0
4. Carving back	1	7.0
5. Carving arms	1	3.0
6. Selecting fabric	None	1.0
7. Sewing cushion	6	2.0
8. Assembling back and seat	3, 4	2.0
9. Attaching arms	5, 8	2.0
10. Attaching rockers	2, 8	3.0
11. Varnishing	9, 10	5.0
12. Adding cushion	7, 11	0.5

We can define a partial ordering on the set of tasks by

$$x \leqslant y \leftrightarrow \text{task } x = \text{task } y \quad \text{or} \quad \text{task } x \text{ is a prerequisite to task } y$$

It is easy to see that this relation is reflexive, antisymmetric, and transitive. Also,

$$x < y \leftrightarrow \text{task } x \text{ is a prerequisite to task } y$$

In the Hasse diagram for this partial ordering, the nodes are tasks; we'll add to each node the information about the time to perform the task. Also, as is traditional, we'll orient the diagram so that if $x < y$, then x is to the left of y rather than below y. Thus the entire diagram runs from left to right rather than from bottom to top. Such a diagram for task scheduling is often called a **PERT (program evaluation and review technique)** chart, first developed for tracking the construction of Navy submarines but useful for managing any complex project with a number of subtasks. The PERT chart for manufacturing rocking chairs is shown in Figure 4.7, with task numbers substituted for task names and arrows pointing to a task from its prerequisite task(s). The numbers in parentheses indicate the time required to perform the task.

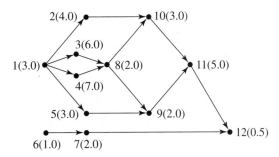

Figure 4.7

PRACTICE 17   Construct the PERT chart for building a house from the following task table.

Task	Prerequisite Tasks	Days to Perform
1. Clearing lot	None	4
2. Pouring pad	1	3
3. Doing framing	2	7
4. Shingling roof	3	6
5. Adding outside siding	3	4
6. Installing plumbing and wiring	4, 5	6
7. Hanging windows and doors	3	5
8. Installing wallboard	6	5
9. Painting interior	7, 8	5

A project represented by a PERT chart must begin with the tasks at the leftmost edge of the PERT chart and end with the tasks at the rightmost edge. An upper limit on the time required to complete the project can be obtained by adding the times for performing each task, but this does not take into account the fact that perhaps some tasks can be performed in parallel, such as tasks 2 through 5 in Example 16. To obtain the minimum time required to complete the project, we can move through the chart from left to right, computing for each node the minimum time to complete the work from the beginning through the work at that node. If a node x has multiple nodes as prerequisites, all the prerequisite tasks must be completed before we can begin work on x; thus we must add the time for task x to the maximum completion time of the prerequisite nodes.

EXAMPLE 17

Let's compute the time for completing each task in Example 16.

Task 1:   3.0
Task 2:   3.0 + 4.0 = 7.0
Task 3:   3.0 + 6.0 = 9.0
Task 4:   3.0 + 7.0 = 10.0
Task 5:   3.0 + 3.0 = 6.0
Task 6:   1.0
Task 7:   1.0 + 2.0 = 3.0
Task 8:   max(time to complete task 3, time to complete task 4)
            + time to perform task 8
            = max(9.0, 10.0) + 2.0 = 10.0 + 2.0 = 12.0
Task 9:   max(time to complete task 5, time to complete task 8)
            + time to perform task 9
            = max(6.0, 12.0) + 2.0 = 12.0 + 2.0 = 14.0
Task 10:   max(time to complete task 2, time to complete task 8)
            + time to perform task 10
            = max(7.0, 12.0) + 3.0 = 12.0 + 3.0 = 15.0
Task 11:   max(time to complete task 9, time to complete task 10)
            + time to perform task 11
            = max(14.0, 15.0) + 5.0 = 15.0 + 5.0 = 20.0
Task 12:   max(time to complete task 7, time to complete task 11)
            + time to perform task 12
            = max(3.0, 20.0) + 0.5 = 20.0 + 0.5 = 20.5

Therefore the minimum number of hours to manufacture a rocking chair is 20.5. From node 12, we can travel back in the chart, selecting at each point of multiple prerequisites the node that contributed the maximum value. This gives the sequence of nodes

12, 11, 10, 8, 4, 1

or, reversing this sequence,

1, 4, 8, 10, 11, 12

The sum of the times to perform each task in this sequence is 20.5. If any of these tasks takes longer to perform than its allotted time, the entire project will take longer than 20.5 hours. This sequence of nodes is a **critical path** through the

PERT chart—performing these tasks in the allotted time is critical to completing the entire project on time. ●

The critical path in a PERT chart represents the minimum time to completion of the entire project. If a task not on the critical path takes longer than its allotted time to perform, then the critical path may shift to include this node, because it then becomes the bottleneck slowing down completion of the total project. In a complex project, the critical path must continually be recomputed to determine where best to allocate resources to move the project forward.

PRACTICE 18   Compute the minimum time to completion and the nodes on the critical path for the house-building project of Practice 17. ●

Given a partial ordering ρ on a finite set, there is always a total ordering σ that is an extension of ρ, meaning that if $x \rho y$, then $x \sigma y$. The process of **topological sorting** finds such a total ordering from a partial ordering. This is indeed a sorting process in the sense that the objects end up being totally ordered, but since they must be partially ordered to begin with, it is a very specialized sorting process.

Recall that in a finite partially ordered set, an element is minimal if it has no predecessors. In a finite nonempty partially ordered set, at least one minimal element must exist. To see this, let x belong to the set. If x is not minimal, then there is a y in the set with $y \rho x$, $y \neq x$. If y is not minimal, then there is a z in the set with $z \rho y$, $z \neq y$, and so on. Because the set is finite, this process cannot go on indefinitely, so one such element must be minimal. A minimal element in a Hasse diagram has no elements below it; a minimal element in a PERT chart has no elements to its left.

The accompanying pseudocode algorithm for topological sorting operates on a partially ordered set (S, ρ). Minimal elements (picked at random if there is a choice of minimal elements at any stage) are repeatedly removed from the ordered set until the set is empty. Each removal of a minimal element leaves a finite partially ordered set, so that another minimal element may be found.

ALGORITHM   Topological Sort

TopSort(finite set S; partial ordering ρ on S)
//find a total ordering on S that is an extension of ρ
Local variable
integer i //enumerates tasks in total ordering
    $i = 1$
    while $S \neq \emptyset$
        pick a minimal element x_i from S;
        $S = S - \{x_i\}$
        $i = i + 1$
    end while
    //$x_1 < x_2 < x_3 < \cdots < x_n$ is now a total ordering that extends ρ
    write($x_1, x_2, x_3, \ldots, x_n$)
end function TopSort

The ordering $x_1 < x_2 < x_3 < \cdots < x_n$ produced by this algorithm is a total ordering. To see that it is an extension of ρ, suppose that $x_i \, \rho \, x_j$. Then x_i precedes x_j and x_i must be chosen as a minimal element and removed from the set before x_j can be chosen as a minimal element. Therefore $i < j$ and $x_i < x_j$.

EXAMPLE 18   One topological sort of the partial ordering of Example 16 is

6, 1, 7, 2, 3, 5, 4, 8, 10, 9, 11, 12

In Figure 4.7, either 6 or 1 is minimal and may be chosen as the first element. If 6 is chosen and removed from the set, then, as shown in Figure 4.8, either 1 or 7 is minimal. If 1 is then chosen and removed from the set (Figure 4.9), then 2, 3, 4, 5, and 7 are all minimal and any one can be chosen next. The process continues until all nodes have been chosen. If Ernie's brothers all move to the city and he is left to build rocking chairs alone, the topological sort gives an order in which he can perform tasks sequentially.

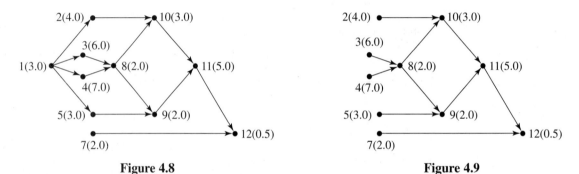

Figure 4.8   **Figure 4.9**

PRACTICE 19   Find another topological sort for the partial ordering of Example 16.

PRACTICE 20   Find a topological sort for the partial ordering of Practice 17.

The algorithm given here for topological sorting is still somewhat imprecise, as we have not given a mechanical method for finding a minimal element. Another algorithm will be described in Section 6.4.

Section 4.2 Review

Techniques

- Construct a PERT chart from a task table.
- Find the critical path in a PERT chart.
- Do a topological sort on a partially ordered set.

Main Ideas

PERT charts are diagrams of partially ordered sets representing tasks and prerequisites among tasks.

A topological sort extends a partial ordering on a finite set to a total ordering.

Exercises 4.2

1. The following tasks are required in order to assemble a bicycle. As the manufacturer, you must write a list of sequential instructions for the buyer to follow. Will the sequential order given below work? Give another sequence that could be used.

Task	Prerequisite Tasks
1. Tightening frame fittings	None
2. Attaching handle bars to frame	1
3. Attaching gear mechanism	1
4. Mounting tire on wheel assembly	None
5. Attaching wheel assembly to frame	1, 4
6. Installing brake mechanism	2, 3, 5
7. Adding pedals	6
8. Attaching seat	1
9. Adjusting seat height	7, 8

★2. Construct a PERT chart from the following task table.

Task	Prerequisite Tasks	Time to Perform
A	E	3
B	C, D	5
C	A	2
D	A	6
E	None	2
F	A, G	4
G	E	4
H	B, F	1

3. Construct a PERT chart from the following task table.

Task	Prerequisite Tasks	Time to Perform
1	2	4
2	3	2
3	8	5
4	3	2
5	4, 7	2
6	5, 1	1
7	3	3
8	None	5

★4. Compute the minimum time to completion and the nodes on the critical path for the problem in Exercise 2.

5. Compute the minimum time to completion and the nodes on the critical path for the problem in Exercise 3.

6. Do a topological sort on the partially ordered set shown in the accompanying figure.

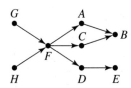

★7. Find a topological sort for the problem in Exercise 2.

8. Find a topological sort for the problem in Exercise 3.

9. Given the following task chart, find a total ordering in which the tasks can be performed sequentially.

Task	Prerequisite Tasks
1. Chop onions	9
2. Wash lettuce	11
3. Make dressing	11
4. Do stir fry	10
5. Toss salad	2, 3
6. Cut up chicken	None
7. Grate ginger	9
8. Chop bok choy	9
9. Marinate chicken	6
10. Heat wok	1, 7, 8, 11
11. Prepare rice	None

10. Recall the problem posed at the beginning of this chapter.

Your company has developed a program for use on a small parallel processing machine. According to the technical documentation, the program executes processes P1, P2, and P3 in parallel; these processes all need results from process P4, so they must wait for Process P4 to complete execution before they begin. Processes P7 and P10 execute in parallel but must wait until processes P1, P2, and P3 have finished. Process P4 requires results from P5 and P6 before it can begin execution. P5 and P6 execute in parallel. Processes P8 and P11 execute in parallel but P8 must wait for process P7 to complete, and process P11 must wait for P10 to complete. Process P9 must wait for results from P8 and P11. You have been assigned to convert the software for use on a single processor machine.

Use a topological sort to determine the order in which the processes should be executed sequentially.

Section 4.3   Relations and Databases

A **database** is a storehouse of associated information about some enterprise. The user of a database can certainly retrieve some specific fact stored in the database. But a well-designed database is more than simply a list of facts. The user can perform queries on the database to retrieve information not contained in any single fact. The whole becomes more than the sum of its parts.

To design a useful and efficient computerized database, it is necessary to model or represent the enterprise with which the database is concerned. A **conceptual model** attempts to capture the important features and workings of the enterprise. Considerable interaction with those who are familiar with the enterprise may be required to obtain all the information necessary to formulate the model.

Entity-Relationship Model

One high-level representation of an enterprise is the **entity-relationship model**. In this model, important objects, or **entities**, in the enterprise are identified, together with their relevant attributes or properties. Then the relationships between these various entities are noted. This information is represented graphically by an **entity-relation-ship diagram**, or **E-R diagram**. In an E-R diagram, rectangles denote entity sets, ellipses denote attributes, and diamonds denote relationships.

EXAMPLE 19   The Pet Lovers of America Club (PLAC) wants to set up a database. PLAC has bought mailing lists from commercial sources, and it is interested in those people who own pets and in some basic information about those pets, such as the name, type of pet (dog, cat, etc.), and the breed.

Figure 4.10 shows an E-R diagram for the PLAC enterprise. This diagram says that persons and pets are the entities. Persons have the attributes of *Name*, *Address*, *City*, and *State*. Pets have the attributes of *Pet-name*, *Pet-type*, and *Breed*. The diagram also shows that persons own pets. Thinking of the entities as sets, the Person set and the

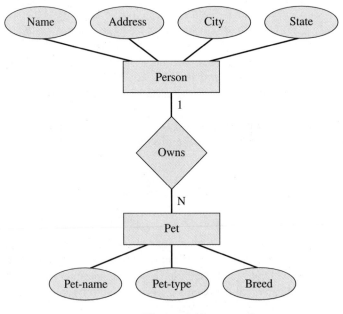

Figure 4.10

Pet set, the relationship "owns" is a binary relation from Person to Pet—the ownership relation is captured by (person, pet) ordered pairs. The "1" and "N" on the connecting lines indicate that this binary relation is one-to-many; that is, in this particular enterprise, one person can own many pets, but no pet has multiple owners. (Pets with multiple owners would result in a many-to-many relation.) Also, in this example, some persons may own no pets, and some pets may have no owners.

The fact that no pet has multiple owners is one of the "business rules" of the enterprise. Such business rules are important to identify when designing a database, because they can determine various features of the database, as we will see. ●

Relational Model

Another representation of an enterprise, called a **relational model**, can be developed from the E-R model. Both the entity sets and the relationships of the E-R model become relations (in the mathematical sense) in the relational model. The relations are described by tables. A **relational database** consists of collections of such tables.

An entity set table is named for the entity set. Each row in the table contains the values of the n attributes for a specific instance of that entity set. Thus the relational table may be thought of as a set of n-tuples (rows), and an individual row is called a **tuple**. True to the idea of a set, no duplicate tuples exist, and no ordering of the tuples is assumed. The ordering of the attributes is unimportant, except that consistency must be maintained; that is, each column in the table contains values for a specific attribute in all of the tuples.

More formally, a database relation is a subset of $D_1 \times D_2 \times \cdots \times D_n$, where D_i is the domain from which attribute A_i takes its values. This means that the database use of the word *relation* is consistent with our definition of an n-ary relation on multiple sets (p. 233).

EXAMPLE 20   The Person relation in the PLAC database might contain the following data:

Person			
Name	*Address*	*City*	*State*
Patrick, Tom	2425 Samset	Sarasota	FL
Smith, Mary	1121 Ridge Rd.	Rockville	IL
Collier, Jon	429 Via Rivio	Venice	IL
Jones, Kate	345 Forest St.	Cleveland	OH
Smith, Bob	1201 45th St.	Falls City	MA
White, Janet	110 Toledo Rd.	Brookville	GA
Garcia, Maria	24 E. 56th St.	New York City	NY

The four attributes for each tuple are *Name*, *Address*, *City*, and *State*. The Pet relation could be

	Pet	
Pet-name	*Pet-type*	*Breed*
Spot	Dog	Hound
Twinkles	Cat	Siamese
Lad	Dog	Collie
Lassie	Dog	Collie
Mohawk	Fish	Moorish idol
Tweetie	Bird	Canary
Tiger	Cat	Shorthair

Because there are no duplicate tuples in a relation, giving the value of all n attributes of a tuple clearly distinguishes the tuple from all others. However, there may be a minimal subset of the attributes that can be used to uniquely identify each tuple. This subset is called the **primary key** of the relation; if the subset consists of more than one attribute, then it is a **composite primary key**. In the table describing the relation, the primary key is underlined in the row of attribute names.

Another business rule of the PLAC enterprise is that all people have unique names; therefore *Name* is sufficient to identify each tuple and was chosen as the primary key in the Person relation. Note that for the Person relation as shown in this example *State* could not serve as a primary key because there are two tuples with *State* value "IL." However, just because *Name* has unique values in this instance does not in itself preclude the possibility of duplicate names. It is the business rule that determines that names will be unique. (There is no business rule that says that addresses or cities are unique, so neither of these attributes can serve as the primary key, even though there happen to be no duplicates in the Person relation shown.)

The assumption of unique names is a somewhat simplistic business rule. The primary key in a relation involving people is often an identifying number, such as a Social Security number, which is a convenient unique attribute. Because *Pet-name* is the primary key in the Pet relation of Example 20, we can surmise the even more surprising business rule that in the PLAC enterprise, all pets have unique names. A more realistic scenario would call for creating a unique attribute for each pet, sort of a pet Social Security number, to be used as the primary key. This key would have no counterpart in the real enterprise, so the database user would never need to see it; such a key is called a **blind key**.

Because each attribute domain D_i in a relational database is also assumed to contain a special null value, a given tuple could have a null value for one or more of its attributes. However, no attribute of the primary key should ever have a null (empty)

value. This **entity integrity** constraint merely confirms that each tuple must have a primary key value in order to distinguish that tuple and that all attribute values of the primary key are needed in order to identify a tuple uniquely.

An attribute in one relation (called the "child" relation) may have the same domain as the primary key attribute in another relation (called the "parent" relation). Such an attribute is called a **foreign key** (of the child relation) into the parent relation. A relation for a relationship between entities uses foreign keys to establish connections between those entities. There will be one foreign key in the relationship relation for each entity participating in the relationship,

EXAMPLE 21

The PLAC enterprise has identified the following instance of the Owns relationship. The *Name* attribute of Owns is a foreign key into the Person relation where *Name* is a primary key; *Pet-name* of Owns is a foreign key into the Pet relation, where *Pet-name* is a primary key. The first tuple establishes the Owns relationship between Bob Smith and Spot; that is, it indicates that Bob Smith owns Spot.

Owns	
Name	*Pet-name*
Smith, Bob	Spot
Smith, Mary	Twinkles
Jones, Kate	Lad
Jones, Kate	Lassie
Collier, Jon	Tweetie
White, Janet	Tiger

Persons who do not own pets are not represented in Owns, nor are pets with no owners. The primary key of Owns is *Pet-name*. Recall the business rule that no pet has multiple owners. If any pet could have multiple owners, the composite primary key *Name* / *Pet-name* would have to be used. ●

Sometimes a separate relationship table is not necessary. It is never necessary in a one-to-one relationship, and it can sometimes be removed in a one-to-many relationship, such as our example.

EXAMPLE 22   Because *Pet-name* in the Owns relation is a foreign key into the Pet relation, the two relations can be combined (using an operation called the *outer join over Pet-name*) to form the Pet-Owner relation.

	Pet-Owner		
Name	*Pet-name*	*Pet-type*	*Breed*
Smith, Bob	Spot	Dog	Hound
Smith, Mary	Twinkles	Cat	Siamese
Jones, Kate	Lad	Dog	Collie
Jones, Kate	Lassie	Dog	Collie
NULL	Mohawk	Fish	Moorish idol
Collier, Jon	Tweetie	Bird	Canary
White, Janet	Tiger	Cat	Shorthair

This Pet-Owner relation could replace both the Owns relation and the Pet relation with no loss of information. Pet-Owner contains a tuple with a null value for *Name*. This would not violate entity integrity because *Name* would not be a component of the primary key but instead would still be a foreign key into Person.   ●

Operations on Relations

Two unary operations that can be performed on relations are *restrict* and *project*. The **restrict** operation creates a new relation made up of those tuples of the original relation that satisfy a certain property. The **project** operation creates a new relation made up of certain attributes from the original relation, eliminating any duplicate tuples. The restrict and project operations can be thought of in terms of subsets. The restrict operation creates a subset of the rows that satisfy certain properties; the project operation creates a subset of the columns that represent certain attributes.

EXAMPLE 23   The operation

    Restrict Pet-Owner **where** *Pet-type* = "Dog" **giving** Dog-Owner

results in the relation Dog-Owner:

	Dog-Owner		
Name	*Pet-name*	*Pet-type*	*Breed*
Smith, Bob	Spot	Dog	Hound
Jones, Kate	Lad	Dog	Collie
Jones, Kate	Lassie	Dog	Collie

The operation

Project Pet-Owner **over** (*Name, Pet-type*) **giving** Preference

results in the relation Preference:

	Preference
Name	*Pet-type*
Smith, Bob	Dog
Smith, Mary	Cat
Jones, Kate	Dog
Collier, Jon	Bird
White, Janet	Cat

PRACTICE 21   Write the relation that results from the operation

Project Person **over** (*Name, State*) **giving** Locale

Because relations are sets of *n*-tuples, the binary operations of union, intersection, and set difference can be applied to two relations with the same basic structure. Thus in our example, two tables containing information about pet owners, both laid out with the same structure, could be intersected to produce a relation containing all the common 4-tuples.

Another binary operation, **join**, can be performed on two relations with a common attribute (column). This operation initially forms the Cartesian product of all *n*-tuples (rows) in the first relation with all *k*-tuples (rows) in the second relation. It views the result as a set of $(n + k)$-tuples and then restricts to the subset of those where the common attribute has the same value, writing the result as a set of $(n + k - 1)$-tuples (the common attribute is written only once). Join is therefore not really a separate operation but is defined as the result of doing a Cartesian product followed by a restrict.

EXAMPLE 24   The operation

Join Person **and** Pet-Owner **over** *Name* **giving** Listing

results in the Listing relation:

Listing						
Name	*Address*	*City*	*State*	*Pet-name*	*Pet-type*	*Breed*
Smith, Mary	1121 Ridge Rd.	Rockville	IL	Twinkles	Cat	Siamese
Collier, Jon	429 Via Rivio	Venice	IL	Tweetie	Bird	Canary
Jones, Kate	345 Forest St.	Cleveland	OH	Lad	Dog	Collie
Jones, Kate	345 Forest St.	Cleveland	OH	Lassie	Dog	Collie
Smith, Bob	1201 45th St.	Falls City	MA	Spot	Dog	Hound
White, Janet	110 Toledo Rd.	Brookville	GA	Tiger	Cat	Shorthair

The restrict, project, and join operations can be applied in various combinations to formulate queries that the user wishes to perform on the database. For example, suppose the query is

Give the names of all cats whose owners live in Illinois.   (1)

If the only existing relations are Person and Pet-Owner, the following sequence of operations will produce a relation that answers this query:

Restrict Pet-Owner **where** *Pet-type* = "Cat" **giving** Results1

Results1			
Name	*Pet-name*	*Pet-type*	*Breed*
Smith, Mary	Twinkles	Cat	Siamese
White, Janet	Tiger	Cat	Shorthair

Restrict Person **where** *State* = "IL" **giving** Results2

Results2			
Name	*Address*	*City*	*State*
Smith, Mary	1121 Ridge Rd.	Rockville	IL
Collier, Jon	429 Via Rivio	Venice	IL

Join Results2 **and** Results1 **over** *Name* **giving** Results3

Results3						
Name	*Address*	*City*	*State*	*Pet-name*	*Pet-type*	*Breed*
Smith, Mary	1121 Ridge Rd.	Rockville	IL	Twinkles	Cat	Siamese

Project Results3 **over** *Pet-name* **giving** Final_Results

Final_Results
Pet-name
Twinkles

EXAMPLE 25   **Relational algebra** is a theoretical relational database language in which the restrict, project, and join operations can be combined. The relational algebra equivalent of the sequence of operations we did to find the names of cats whose owners live in Illinois would be the statement

project(**join**(**restrict** Pet-Owner **where** *Pet-type* = "Cat") **and**
(**restrict** Person **where** *State* = "IL") **over** *Name*)
over *Pet-name* **giving** Final_Results   (2)

SQL is an international standard relational database language; the query above would appear as the following SQL statement, where the lines are numbered only for discussion purposes:

1. **SELECT** *Pet-name*
2. **FROM** Pet-Owner, Person
3. **WHERE** *Pet-Owner.Name = Person.Name*
4. **AND** *Pet-type = "Cat"*
5. **AND** *State = "IL"*;   (3)

SQL's **SELECT** statement can actually perform relational algebra restricts, projects, and joins, as shown here. Lines 4 and 5 represent the two restrict operations, line 3 represents the join, and line 1 represents the project operation. Line 2, of course, identifies the tables to be used. ●

Instead of using the relational algebra approach, in which the restrict, project, and join operations are used to process a query, we can use the relational calculus approach. In **relational calculus**, instead of specifying the operations to be done in order to process a query, we give a set-theoretic description of the desired result of the query. The description of the set may involve notation from predicate logic; remember that predicate logic is also called predicate calculus, hence the name relational calculus. Relational algebra and relational calculus are equivalent in their expressive power; that is, any query that can be formulated in one language can be formulated in the other.

EXAMPLE 26

The relational calculus expression for the query asking for the names of all cats whose owners live in Illinois is

Range of x is Pet-Owner
Range of y is Person
$\{x.Pet\text{-}name \mid x.Pet\text{-}type = \text{"Cat" and}$
    exists $y(y.Name = x.Name$ **and** $y.State = \text{"IL"})\}$ (4)

Here "Range of x is Pet-Owner" specifies the relation from which the tuple x may be chosen, and "Range of y is Person" specifies the relation from which the tuple y may be chosen. (The use of the term *range* is unfortunate. We are really talking about domain in the same sense we talked about the domain of an interpretation in predicate logic—the pool of potential values.) The notation "exists y" stands for the existential quantifier ($\exists y$). ●

Expressions (1) through (4) all represent the same query expressed in English language, relational algebra, SQL, and relational calculus, respectively.

PRACTICE 22

Using the relations Person and Pet-Owner, express the following query in relational algebra, SQL, and relational calculus form:

Give the names of all cities where dog-owners live. ●

Database Integrity

New information must be added to a database from time to time, obsolete information deleted, and changes or updates made to existing information. In other words, the database will be subjected to **add**, **delete**, and **modify** operations. An add operation can be carried out by creating a second relation table with the new information and performing a set union of the existing table and the new table. Delete can be accomplished by creating a second relation table with the tuples to be deleted and performing a set difference that subtracts the new table from the existing table. Modify can be achieved by a delete (of the old tuple) followed by an add (of the modified tuple).

These operations must be carried out so that the information in the database remains in a correct and consistent state that agrees with the business rules. Enforcing three "integrity rules" will help. **Data integrity** requires that the values for an attribute do indeed come from that attribute's domain. In our example, for instance, values for the *State* attribute of Person must be legitimate two-letter state abbreviations (or the null value). Entity integrity, as we discussed earlier, requires that no component of a primary key value be null. These integrity constraints clearly affect the tuples that can be added to a relation.

Referential integrity requires that any values for foreign keys into other relations either be null or have values that match values in the corresponding primary keys of those relations. The referential integrity constraint affects both add and delete operations (and therefore modify operations). For instance, we could not add a tuple to Pet-Owner with a non-null *Name* value that does not exist in the Person relation, because this would violate the Owns relation as a binary relation on Person × Pet. Also, if the Bob Smith tuple is deleted from the Person relation, then the Bob Smith tuple must be deleted from the Pet-Owner relation or the Name value "Bob Smith" changed to null (a business rule must specify which is to occur) so that Pet-Owner's foreign key *Name* does not violate referential integrity. This prevents the inconsistent state of a reference to Bob Smith in Pet-Owner when Bob Smith no longer exists as a "Person."

Section 4.3 Review

Techniques

- Carry out restrict, project, and join operations in a relational database.
- Formulate relational database queries using relational algebra, SQL, and relational calculus.

Main Ideas

A relational database uses mathematical relations, described by tables, to model objects and relationships in an enterprise.

The database operations of restrict, project, and join are operations on relations (sets of tuples).

Queries on relational databases can be formulated using the restrict, project, and join operations, SQL statements, or notations borrowed from set theory and predicate logic.

Exercises 4.3

Exercises 1–18 are all related to the same enterprise.

★1. A library maintains a database about its books. Information kept on authors includes the author's name and country of origin and the titles of that author's books. Information kept on books includes the title, ISBN, publisher, and subject. Authors and books are the entities in this enterprise, and "writes" is a relationship between these entities. Sketch an E-R diagram for the enterprise. In the absence of any business rules, what must be assumed about the binary relation "writes" regarding whether it is one-to-one, one-to-many, and so on?

2. Assume the business rule that authors are uniquely identified by their names. Does this change your answer to the last question of Exercise 1?

3. In a relational model of the library database, there is an author relation, a book relation, and a writes relation. Give the table heading for each of the relation tables, underlining the primary key. Explain your choice of primary keys. What implicit business rule (in addition to that of Exercise 2) underlies the choice of author attributes?

For Exercises 4–14, use the following relation tables and write the results of the operations.

	Author	
Name	*Country*	*Title*
Dorothy King	England	Springtime Gardening
Jon Nkoma	Kenya	Birds of Africa
Won Lau	China	Early Tang Paintings
Bert Kovalsco	U.S.	Baskets for Today
Jimmy Chan	China	Early Tang Paintings
Dorothy King	England	Autumn Annuals
Jane East	U.S.	Springtime Gardening

Book

Title	ISBN	Publisher	Subject
Springtime Gardening	816–35421–8	Harding	Nature
Early Tang Paintings	364–87547–8	Bellman	Art
Birds of Africa	115–67813–3	Loraine	Nature
Springtime Gardening	816–89335–8	Swift-Key	Nature
Baskets for Today	778–53705–7	Harding	Art
Autumn Annuals	414–88506–9	Harding	Nature

Writes

Name	Title	ISBN
Jimmy Chan	Early Tang Paintings	364–87547–8
Dorothy King	Autumn Annuals	414–88506–9
Jane East	Springtime Gardening	816–89335–8
Bert Kovalsco	Baskets for Today	778–53705–7
Won Lau	Early Tang Paintings	364–87547–8
Jon Nkoma	Birds of Africa	115–67813–3
Dorothy King	Springtime Gardening	816–35421–8

★4. **Restrict** Author **where** *Country* = "U.S." **giving** Results1.

5. **Restrict** Writes **where** *Name* = "Dorothy King" **giving** Results2.

6. **Restrict** Book **where** *Publisher* = "Bellman" or Publisher = "Swift-Key" **giving** Results3.

7. **Restrict** Book **where** *Publisher* = "Harding" and Subject = "Art" **giving** Results4.

★8. **Project** Author **over** (*Name, Title*) **giving** Results5.

9. **Project** Author **over** (*Name, Country*) **giving** Results6.

10. **Project** Book **over** (*Publisher, Subject*) **giving** Results7.

11. **Project** Book **over** (*Title, ISBN, Subject*) **giving** Results8.

★12. **Join** Book and Writes over *Title* and *ISBN* **giving** Results9.

13. **Join** Author and Writes over *Name* and *Title* **giving** Results10.

14. What would be wrong with doing a join of Author and Book over *Title*?

For Exercises 15–18, using the relation tables given, express each query in relational algebra, SQL, and relational calculus forms. Also give the result of each query.

★15. Give the titles of all books written by U.S. authors.

16. Give the names of all authors who publish with Harding.

17. Give the names of all authors who have written nature books.

18. Give the publishers of all art books whose authors live in the United States.

Section 4.4   Functions

In this section we discuss functions, which are really special cases of binary relations from a set S to a set T. This view of a function is a rather sophisticated one, however, and we will work up to it gradually.

Definition

Function is a common enough word even in nontechnical contexts. A newspaper may have an article on how starting salaries for this year's college graduates have increased over those for last year's graduates. The article might say something like "The salary increase varies depending on the degree program" or "The salary increase is a function of the degree program." It may illustrate this functional relationship with a graph like Figure 4.11. The graph shows that each degree program has some figure for the salary increase associated with it, that no degree program has more than one figure associated with it, and that both the physical sciences and the liberal arts have the same figure, 3%.

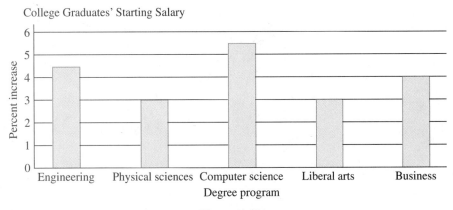

Figure 4.11

Of course, we also use mathematical functions in algebra and calculus. The equation $g(x) = x^3$ expresses a functional relationship between values for x and corresponding values that result when x is replaced in the equation by its values. Thus an x-value of 2 has the number $2^3 = 8$ associated with it. (This number is expressed as $g(2) = 8$.) Similarly, $g(1) = 1^3 = 1$, $g(-1) = (-1)^3 = -1$, and so on. For each *x*-value, the corresponding $g(x)$-value is unique. If we were to graph this function on a rectangular coordinate system, the points $(2, 8)$, $(1, 1)$, and $(-1, -1)$ would be points on the graph. If we allow *x* to take on any real-number value, the resulting graph is the continuous curve shown in Figure 4.12.

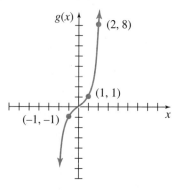

Figure 4.12

The function in the salary increase example could be described as follows. We set the stage by the diagram in Figure 4.13, which indicates that the function always starts with a given degree program and that a particular salary increase is associated with that degree program. The association itself is described by the set of ordered pairs {(engineering, 4.5%), (physical sciences, 3.0%), (computer science, 5.5%), (liberal arts, 3.0%), (business, 4.0%)}.

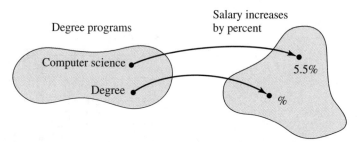

Figure 4.13

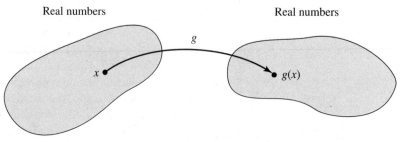

Figure 4.14

For the algebraic example $g(x) = x^3$, Figure 4.14 shows that the function always starts with a given real number and associates a second real number with it. The association itself is described by $\{(x, g(x)) \mid g(x) = x^3\}$, or simply $g(x) = x^3$. This set includes $(2, 8)$, $(1, 1)$, $(-1, -1)$, but because it is an infinite set, we cannot list all its members; we have to describe them.

From the above examples, we can conclude that there are three parts to a function: (1) a set of starting values, (2) a set from which associated values come, and (3) the association itself. The set of starting values is called the *domain* of the function, and the set from which associated values come is called the *codomain* of the function. Thus both the domain and codomain represent pools from which values may be chosen. (This is consistent with our use of the word *domain* when discussing predicate wffs in Section 1.2. There the domain of an interpretation is a pool of values that variables can assume and to which constant symbols may be assigned. Similarly, the domain D_i of an attribute A_i in a database relation, discussed in Section 4.3, is a pool of potential values for the attribute.)

The picture for an arbitrary function f is shown in Figure 4.15. Here f is a function from S to T, symbolized $f: S \rightarrow T$. S is the domain and T is the codomain. The association itself is a set of ordered pairs, each of the form (s, t) where $s \in S$, $t \in T$, and t is the value from T that the function associates with the value s from S; $t = f(s)$. Hence, the association is a subset of $S \times T$ (a binary relation from S to T). But the important property of this relation is that every member of S must have one and only one T-value associated with it, so every $s \in S$ will appear exactly once as the first component of an (s, t) pair. (This property does not prevent a given T-value from appearing more than once.)

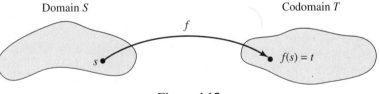

Domain S f Codomain T

s $f(s) = t$

Figure 4.15

We are now ready for the formal definition of a function.

Definitions: Terminology for Functions

Let S and T be sets. A **function** (**mapping**) f from S to T, $f: S \rightarrow T$, is a subset of $S \times T$ where each member of S appears exactly once as the first component of an ordered pair. S is the **domain** and T the **codomain** of the function. If (s, t) belongs to the function, then t is denoted by $f(s)$; t is the **image** of s under f, s is a **preimage** of t under f, and f is said to map s to t. For $A \subseteq S$, $f(A)$ denotes $\{f(a) \mid a \in A\}$.

A function from S to T is a subset of $S \times T$ with certain restrictions on the ordered pairs it contains. That is why we spoke of a function as a special kind of binary relation. By the definition of a function, a binary relation that is one-to-many (or many-to-many) cannot be a function. Also, each member of S must be used as a first component.

We have talked a lot about values from the sets S and T, but as our example of salary increases shows, these values are not necessarily numbers, nor is the association itself necessarily described by an equation.

PRACTICE 23   Which of the following are functions from the domain to the codomain indicated? For those that are not, why not?

    a. $f: S \to T$ where $S = T = \{1, 2, 3\}, f = \{(1, 1), (2, 3), (3, 1), (2, 1)\}$
    b. $g: \mathbb{Z} \to \mathbb{N}$ where g is defined by $g(x) = |x|$ (the absolute value of x)
    c. $h: \mathbb{N} \to \mathbb{N}$ where h is defined by $h(x) = x - 4$
    d. $f: S \to T$ where S is the set of all people in your hometown, T is the set of all Social Security numbers, and f associates with each person that person's Social Security number
    e. $g: S \to T$ where $S = \{1998, 1999, 2000, 2001\}$, $T = \{\$20,000, \$30,000, \$40,000, \$50,000, \$60,000\}$, and g is defined by the graph in Figure 4.16
    f. $h: S \to T$ where S is the set of all quadratic polynomials in x with integer coefficients, $T = \mathbb{Z}$, and h is defined by $h(ax^2 + bx + c) = b + c$
    g. $f: \mathbb{R} \to \mathbb{R}$ where f is defined by $f(x) = 4x - 1$
    h. $g: \mathbb{N} \to \mathbb{N}$ where g is defined by

$$g(x) = \begin{cases} x+3 & \text{if } x \geq 5 \\ x & \text{if } x \leq 5 \end{cases}$$

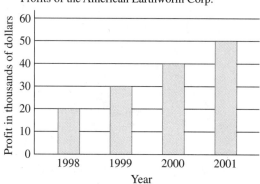

Profits of the American Earthworm Corp.

Figure 4.16

PRACTICE 24   **a.** For $f: \mathbb{Z} \to \mathbb{Z}$ given by $f(x) = x^2$, what is the image of -4?
    b. What are the preimages of 9 under the function in (a)?

EXAMPLE 27   When we studied recursive definitions in Section 2.4, we talked about sequences, where a sequence S was written as

    $S(1), S(2), S(3), ...$

Changing the notation to

    $f(1), f(2), f(3), ...$

we see that a sequence is nothing but a list of functional values for a function f whose domain is the positive integers. Indeed, the algorithms we gave for computing the values in such sequences were pseudocode that computes the function.

Also in Section 2.4, we talked about recursive operations such as a^n where a is a fixed nonzero real number and $n \geq 0$. This is also simply a function $f(n) = a^n$ whose domain is $\mathbb{N}$.

The definition of a function includes functions of more than one variable. We can have a function $f: S_1 \times S_2 \times \cdots \times S_n \to T$ that associates with each ordered n-tuple of elements $(s_1, s_2, \ldots, s_n)$, $s_i \in S_i$, a unique element of T.

EXAMPLE 28 $f: \mathbb{Z} \times \mathbb{N} \times \{1, 2\} \to \mathbb{Z}$ is given by $f(x, y, z) = x^y + z$. Then $f(-4, 3, 1) = (-4)^3 + 1 = -64 + 1 = -63$.

EXAMPLE 29   In Section 3.1 we defined a unary operation on a set S as associating a unique member of S, $x^\#$, with each member x of S. This means that a unary operation on S is a function with domain and codomain S. We also defined a binary operation $\circ$ on a set S as associating a unique member of S, $x \circ y$, with every (x, y) pair of elements of S. Therefore a binary operation on S is a function with domain $S \times S$ and codomain S.

Again, domain values and codomain values are not always numbers.

EXAMPLE 30   Let S be the set of all character strings of finite length. Then the association that pairs each string with the number of characters in the string is a function with domain S and codomain $\mathbb{N}$ (we allow the "empty string," which has zero characters).

EXAMPLE 31   Any propositional wff with n statement letters defines a function with domain $\{T, F\}^n$ and codomain $\{T, F\}$. The domain consists of all n-tuples of T–F values; with each n-tuple is associated a single value of T or F. The truth table for the wff gives the association. For example, if the wff is $A \vee B'$, then the truth table

A	B	B'	$A \vee B'$
T	T	F	T
T	F	T	T
F	T	F	F
F	F	T	T

says that the image of the 2-tuple (F, T) under this function is F. If we call this function w, then $w(F, T) = F$.

PRACTICE 25   Let the function defined by the wff $A \wedge (B \vee C')$ be denoted by f. What is $f(T, T, F)$? What is $f(F, T, F)$?

The next example defines two functions that are sometimes useful in analyzing algorithms.

EXAMPLE 32   The **floor function** $\lfloor x \rfloor$ associates with each real number x the greatest integer less than or equal to x. The **ceiling function** $\lceil x \rceil$ associates with each real number x the smallest integer greater than or equal to x. Thus $\lfloor 2.8 \rfloor = 2, \lceil 2.8 \rceil = 3, \lfloor -4.1 \rfloor = -5$, and $\lceil -4.1 \rceil = -4$. Both the floor function and the ceiling function are functions from $\mathbb{R}$ to $\mathbb{Z}$.   ●

PRACTICE 26
   a. Sketch a graph of the function $\lfloor x \rfloor$.
   b. Sketch a graph of the function $\lceil x \rceil$.   ●

We mentioned that the definition of a function $f: S \to T$ includes three parts—the domain sct S, the codomain set T, and the association itself. Is all this necessary? Why can't we simply write an equation, like $g(x) = x^3$, to define a function?

The quickest answer is that not all functional associations can be described by an equation (see Example 30, for instance). But there is more to it—let's limit our attention to situations where an equation can be used to describe the association, such as $g: \mathbb{R} \to \mathbb{R}$ where $g(x) = x^3$. Even in algebra and calculus, it is common to say "consider the function $g(x) = x^3$," implying that the equation *is* the function. Technically, the equation only describes a way to compute associated values. The function $h: \mathbb{R} \to \mathbb{R}$ given by $h(x) = x^3 - 3x + 3(x + 5) - 15$ is the same function as g since it contains the same ordered pairs. However, the equation is different in that it says to process any given x-value differently.

On the other hand, the function $f: \mathbb{Z} \to \mathbb{R}$ given by $f(x) = x^3$ is not the same function as g. The domain has been changed, which changes the set of ordered pairs. The graph of $f(x)$ would consist of discrete (separate) points (see Figure 4.17). Most of the functions in which we are interested have the latter feature. In a digital computer, information is processed in a series of distinct (discrete) steps. Even in situations where one quantity varies continuously with another, we approximate by taking data at discrete, small intervals, much as the graph of $g(x)$ (Figure 4.12) is approximated by the graph of $f(x)$ (Figure 4.17).

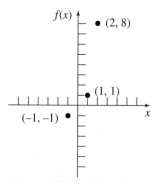

Figure 4.17

Finally, let's look at the function $k: \mathbb{R} \to \mathbb{C}$ given by $k(x) = x^3$. The equation and domain is the same as for $g(x)$; the codomain has been enlarged, but this does not affect the ordered pairs. Is this considered the same function as $g(x)$? It is not, but to see why, we'll have to wait until we discuss the *onto* property of functions. Then we shall see that g has the onto property while k does not, so we do not want to consider them the same function.

In summary, a complete definition of a function requires giving its domain, its codomain, and the association, where the association may be given by a verbal description, a graph, an equation, or a collection of ordered pairs.

Definition: Equal Functions

Two functions are **equal** if they have the same domain, the same codomain, and the same association of values of the codomain with values of the domain.

Suppose we are trying to show that two functions with the same domain and the same codomain are equal. Then we must show that the associations are the same. This can be done by showing that, given an arbitrary element of the domain, both functions produce the same associated value for that element; that is, they map it to the same place.

PRACTICE 27   Let $S = \{1, 2, 3\}$ and $T = \{1, 4, 9\}$. The function $f\colon S \to T$ is defined by $f = \{(1, 1), (2, 4), (3, 9)\}$. The function $g\colon S \to T$ is defined by the equation

$$g(n) = \frac{\displaystyle\sum_{k=1}^{n} (4k - 2)}{2}$$

Prove that $f = g$.   ●

Properties of Functions

Onto Functions

Let $f\colon S \to T$ be an arbitrary function with domain S and codomain T (see Figure 4.18). Part of the definition of a function is that every member of S has an image under f and that all the images are members of T; the set R of all such images is called the **range** of the function f. Thus, $R = \{f(s) \mid s \in S\}$, or $R = f(S)$. Clearly, $R \subseteq T$; the range R is shaded in Figure 4.19. If it should happen that $R = T$, that is, that the range coincides with the codomain, then the function is called an *onto* function.

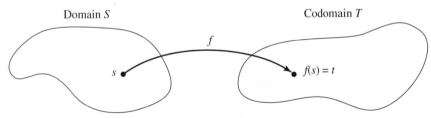

Figure 4.18

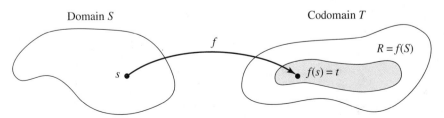

Figure 4.19

Definition: Onto (Surjective) Function

A function $f: S \to T$ is an **onto**, or **surjective**, function if the range of f equals the codomain of f.

In every function with range R and codomain T, $R \subseteq T$. To prove that a given function is onto, we must show that $T \subseteq R$; then it will be true that $R = T$. We must therefore show that an arbitrary member of the codomain is a member of the range, that is, that it is the image of some member of the domain. On the other hand, if we can produce one member of the codomain that is not the image of any member of the domain, then we have proved that the function is not onto.

EXAMPLE 33

The function $g: \mathbb{R} \to \mathbb{R}$ defined by $g(x) = x^3$ is an onto function. To prove that $g(x)$ is onto, let r be an arbitrary real number, and let $x = \sqrt[3]{r}$. Then x is a real number, so x belongs to the domain of g and $g(x) = \left(\sqrt[3]{r}\right)^3 = r$. Hence, any member of the codomain is the image under g of a member of the domain. The function $k: \mathbb{R} \to \mathbb{C}$ given by $k(x) = x^3$ is not surjective. There are many complex numbers (i, for example) that cannot be obtained by cubing a real number. Thus, g and k are not equal functions. ●

EXAMPLE 34

Let $f: \mathbb{Q} \to \mathbb{Q}$ be defined by $f(x) = 3x + 2$. To test whether f is onto, let $q \in \mathbb{Q}$. We want an $x \in \mathbb{Q}$ such that $f(x) = 3x + 2 = q$. When we solve this equation for x, we find that $x = (q - 2)/3$ is the only possible value and is indeed a member of $\mathbb{Q}$. Thus, q is the image of a member of $\mathbb{Q}$ under f, and f is onto. However, the function h: $\mathbb{Z} \to \mathbb{Q}$ defined by $h(x) = 3x + 2$ is not onto because there are many values $q \in \mathbb{Q}$, for example 0, for which the equation $3x + 2 = q$ has no integer solution. ●

PRACTICE 28   Which of the functions found in Practice 23 are onto functions? ●

PRACTICE 29   Suppose a function $f: \{T, F\}^n \to \{T, F\}$ is defined by a propositional wff P (see Example 31). Give the two conditions on P under each of which f will fail to be an onto function. ●

One-to-One Functions

The definition of a function guarantees a unique image for every member of the domain. A given member of the range may have more than one preimage, however. In our very first example of a function (salary increases), both physical sciences and liberal arts were preimages of 3%. This function was not one-to-one.

Definition: One-to-One (Injective) Function

A function $f: S \to T$ is **one-to-one**, or **injective**, if no member of T is the image under f of two distinct elements of S.

The one-to-one idea here is the same as for binary relations in general, as discussed in Section 4.1, except that every element of S must appear as a first component in an ordered pair. To prove that a function is one-to-one, we assume that there are elements s_1 and s_2 of S with $f(s_1) = f(s_2)$ and then show that $s_1 = s_2$. To prove that a function is not one-to-one, we produce a counterexample, an element in the range with two preimages in the domain.

EXAMPLE 35   The function $g: \mathbb{R} \to \mathbb{R}$ defined by $g(x) = x^3$ is one-to-one because if x and y are real numbers with $g(x) = g(y)$, then $x^3 = y^3$ and $x = y$. The function $f: \mathbb{R} \to \mathbb{R}$ given by $f(x) = x^2$ is not injective because, for example, $f(2) = f(-2) = 4$. However, the function $h: \mathbb{N} \to \mathbb{N}$ given by $h(x) = x^2$ is injective because if x and y are nonnegative integers with $h(x) = h(y)$, then $x^2 = y^2$; because x and y are both nonnegative, $x = y$. ●

PRACTICE 30   Which of the functions found in Practice 23 are one-to-one functions? ●

EXAMPLE 36   The floor function and the ceiling function of Example 32 are clearly not one-to-one. This is evident also in the graphs of these functions (Practice 26), which have a number of horizontal sections, indicating that many different domain values in R are mapped by the function to the same codomain value in $\mathbb{Z}$.

> **REMINDER:**
> To prove that a function is a bijection requires proving two things—onto and one-to-one.

Bijections

> **Definition: Bijective Function**
> A function $f: S \to T$ is **bijective** (a **bijection**) if it is both one-to-one and onto.

EXAMPLE 37   The function $g: \mathbb{R} \to \mathbb{R}$ given by $g(x) = x^3$ is a bijection. The function in part (g) of Practice 23 is a bijection. The function $f: \mathbb{R} \to \mathbb{R}$ given by $f(x) = x^2$ is not a bijection (not one-to-one), and neither is the function $k: \mathbb{R} \to \mathbb{C}$ given by $k(x) = x^3$ (not onto). ●

Composition of Functions

Suppose that f and g are functions with $f: S \to T$ and $g: T \to U$. Then for any $s \in S, f(s)$ is a member of T, which is also the domain of g. Thus, the function g can be applied to $f(s)$. The result is $g(f(s))$, a member of U; see Figure 4.20. Taking an arbitrary member s of S, applying the function f, and then applying the function g to $f(s)$ is the same as associating a unique member of U with s. In short, we have created a function $S \to U$, called the composition function of f and g and denoted by $g \circ f$ (Figure 4.21).

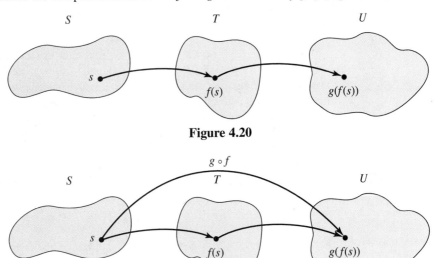

Figure 4.20

Figure 4.21

Definition: Composition Function

Let $f: S \to T$ and $g: T \to U$. Then the **composition function**, $g \circ f$, is a function from S to U defined by $(g \circ f)(s) = g(f(s))$.

Note that the function $g \circ f$ is applied right to left; function f is applied first and then function g.

The diagram in Figure 4.22 also illustrates the definition of the composition function. The corners indicate the domains and codomains of the three functions. The diagram says that, starting with an element of S, if we follow either path $g \circ f$ or path f followed by path g, we get to the same element in U. Diagrams illustrating that alternate paths produce the same effect are called **commutative diagrams**.

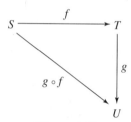

Figure 4.22

It is not always possible to take any two arbitrary functions and compose them; the domains and ranges have to be "compatible." For example, if $f: S \to T$ and $g: W \to Z$, where T and W are disjoint, then $(g \circ f)(s) = g(f(s))$ is undefined because $f(s)$ is not in the domain of g.

PRACTICE 31   Let $f: \mathbb{R} \to \mathbb{R}$ be defined by $f(x) = x^2$. Let $g: \mathbb{R} \to \mathbb{R}$ be defined by $g(x) = \lfloor x \rfloor$.

    a. What is the value of $(g \circ f)(2.3)$?
    b. What is the value of $(f \circ g)(2.3)$?

From Practice 31 we see that order is important in function composition, which should not be surprising. If you make a deposit in your checking account and then write a large check, the effect is not the same as if you write a large check and later make a deposit! Your bank is very sensitive to these differences.

Function composition preserves the properties of being onto and being one-to-one. Again, let $f: S \to T$ and $g: T \to U$, but also suppose that both f and g are onto functions. Then the composition function $g \circ f$ is also onto. Recall that $g \circ f: S \to U$, so we must pick an arbitrary $u \in U$ and show that it has a preimage under $g \circ f$ in S. Because g is surjective, there exists $t \in T$ such that $g(t) = u$. And because f is surjective, there exists $s \in S$ such that $f(s) = t$. Then $(g \circ f)(s) = g(f(s)) = g(t) = u$, and $g \circ f$ is an onto function.

PRACTICE 32   Let $f: S \to T$ and $g: T \to U$, and assume that both f and g are one-to-one functions. Prove that $g \circ f$ is a one-to-one function. (*Hint:* Assume that $(g \circ f)(s_1) = (g \circ f)(s_2)$.)

We have now proved the following theorem.

Theorem on Composing Two Bijections
The composition of two bijections is a bijection.

Inverse Functions

Bijective functions have another important property. Let $f: S \rightarrow T$ be a bijection. Because f is onto, every $t \in T$ has a preimage in S. Because f is one-to-one, that preimage is unique. We could associate with each element t of T a unique member of S, namely, that $s \in S$ such that $f(s) = t$. This association describes a function g, g: $T \rightarrow S$. The picture for f and g is given in Figure 4.23. The domains and codomains of g and f are such that we can form both $g \circ f: S \rightarrow S$ and $f \circ g: T \rightarrow T$. If $s \in S$, then $(g \circ f)(s) = g(f(s)) = g(t) = s$. Thus, $g \circ f$ maps each element of S to itself. The function that maps each element of a set S to itself, that is, that leaves each element of S unchanged, is called the **identity function** on S and denoted by i_S. Hence, $g \circ f = i_S$.

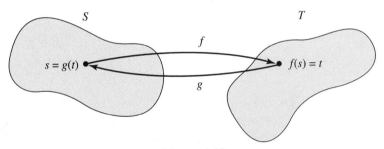

Figure 4.23

PRACTICE 33   Show that $f \circ g = i_T$.

We have now seen that if f is a bijection, $f: S \rightarrow T$, then there is a function g: $T \rightarrow S$ with $g \circ f = i_S$ and $f \circ g = i_T$. The converse is also true. To prove the converse, suppose $f: S \rightarrow T$ and there exists $g: T \rightarrow S$ with $g \circ f = i_S$ and $f \circ g = i_T$. We can prove that f is a bijection. To show that f is onto, let $t \in T$. Then $t = i_T(t) = (f \circ g)(t) = f(g(t))$. Because $g: T \rightarrow S$, $g(t) \in S$, and $g(t)$ is the preimage under f of t. To show that f is one-to-one, suppose $f(s_1) = f(s_2)$. Then $g(f(s_1)) = g(f(s_2))$ and $(g \circ f)(s_1) = (g \circ f)(s_2)$ implying $i_S(s_1) = i_S(s_2)$, or $s_1 = s_2$. Thus, f is a bijection.

Definition: Inverse Function
Let f be a function, $f: S \rightarrow T$. If there exists a function $g: T \rightarrow S$ such that $g \circ f = i_S$ and $f \circ g = i_T$, then g is called the **inverse function** of f, denoted by f^{-1}.

We have proved the following theorem.

Theorem on Bijections and Inverse Functions
Let $f: S \rightarrow T$. Then f is a bijection if and only if f^{-1} exists.

Actually, we have been a bit sneaky in talking about *the* inverse function of f. What we have shown is that if f is a bijection, this is equivalent to the existence of *an* inverse function. But it is easy to see that there is only one such inverse function. *When you want to prove that something is unique, the standard technique is to assume that there are two such things and then obtain a contradiction.* Thus, suppose f has two inverse functions, f_1^{-1} and f_2^{-1} (existence of either means that f is a bijection). Both f_1^{-1} and f_2^{-1} are functions from T to S; if they are not the same function, then they must act differently somewhere. Assume that there is a $t \in T$ such that $f_1^{-1}(t) \neq f_2^{-1}(t)$. Because f is one-to-one, it follows that $f(f_1^{-1}(t)) \neq f(f_2^{-1}(t))$, or $(f \circ f_1^{-1})(t) \neq (f \circ f_2^{-1})(t)$. But both $f \circ f_1^{-1}$ and $f \circ f_2^{-1}$ are i_T, so $t \neq t$, which is a contradiction. We are therefore justified in speaking of f^{-1} as *the* inverse function of f. If f is a bijection so that f^{-1} exists, then f is the inverse function for f^{-1}; therefore, f^{-1} is also a bijection.

PRACTICE 34 $f: \mathbb{R} \rightarrow \mathbb{R}$ given by $f(x) = 3x + 4$ is a bijection. Describe f^{-1}.   ●

We've introduced a lot of terminology about functions. Table 4.2 gives an informal summary of these terms.

Term	Meaning
function	Mapping from one set to another that associates with each member of the starting set exactly one member of the ending set
domain	Starting set for a function
codomain	Ending set for a function
image	Point that results from a mapping
preimage	Starting point for a mapping
range	Collection of all images of the domain
onto (surjective)	Range is the whole codomain; every codomain element has a preimage
one-to-one (injective)	No two elements in domain map to the same place
bijection	One-to-one and onto
identity function	Maps each element of a set to itself
inverse function	For a bijection, a new function that maps each codomain element back where it came from

Table 4.2

Permutation Functions

Bijections that map a set to itself are given a special name.

Definition: Permutations of a Set

For a given set A, $S_A = \{f \mid f: A \rightarrow A$ and f is a bijection$\}$. S_A is thus the set all bijections of set A into (and therefore onto) itself; such functions are called **permutations** of A.

If f and g both belong to S_A, then they each have domain = range = A. Therefore the composition functions $g \circ f$ is defined and maps $A \rightarrow A$. Furthermore, because f and g are both bijections, our theorem on composing bijections says that $g \circ f$ is a bijection, a (unique) member of S_A. Thus, function composition is a binary operation on the set S_A.

In Section 3.4 we described a permutation of objects in a set as being an ordered arrangement of those objects. Is this now a new use of the word "permutation?" Not exactly; permutation functions represent ordered arrangements of the objects in the domain. If $A = \{1, 2, 3, 4\}$, one permutation function of A, call it f, is given by $f = \{(1, 2), (2, 3), (3, 1), (4, 4)\}$. We can also describe function f in array form by listing the elements of the domain in a row and, directly beneath, the images of these elements under f. Thus,

$$f = \begin{pmatrix} 1 & 2 & 3 & 4 \\ 2 & 3 & 1 & 4 \end{pmatrix}$$

The bottom row is an ordered arrangement of the objects in the top row.

A shorter way to describe the permutation f shown above in array form is to use *cycle notation* and write $f = (1, 2, 3)$—understood to mean that f maps each element listed to the one on its right, the last element listed to the first, and an element of the domain not listed to itself. Here 1 maps to 2, 2 maps to 3, and 3 maps to 1. The element 4 maps to itself because it does not appear in the cycle. The cycle $(2, 3, 1)$ also represents f. It says that 2 maps to 3, 3 maps to 1, 1 maps to 2, and 4 maps to itself, the same information as before. Similarly, $(3, 1, 2)$ also represents f.

PRACTICE 35

a. Let $A = (1, 2, 3, 4, 5\}$, and let $f \in S_A$ be given in array form by

$$f = \begin{pmatrix} 1 & 2 & 3 & 4 & 5 \\ 4 & 2 & 3 & 5 & 1 \end{pmatrix}$$

Write f in cycle form.

b. Let $A = \{1, 2, 3, 4, 5\}$, and let $g \in S_A$ be given in cycle form by $g = (2, 4, 5, 3)$. Write g in array form. ●

If f and g are members of S_A for some set A, then $g \circ f \in S_A$, and the action of $g \circ f$ on any member of A is determined by applying function f and then function g. If f and g are cycles, $g \circ f$ is still computed the same way.

EXAMPLE 38   If $A = \{1, 2, 3, 4\}$ and $f, g \in S_A$ are given by $f = (1, 2, 3)$ and $g = (2, 3)$, then $g \circ f = (2, 3) \circ (1, 2, 3)$. But what does this composition function look like? Let's see what happens to element 1 of A. Working from right to left (first f, then g), $1 \to 2$ under f and then $2 \to 3$ under g, so $1 \to 3$ under $g \circ f$. Similarly, $2 \to 3$ under f and then $3 \to 2$ under g, so $2 \to 2$ under $g \circ f$. Testing what happens to 3 and 4, $3 \to 1$ and then $1 \to 1$, so $3 \to 1$, and $4 \to 4 \to 4$, so $4 \to 4$. We conclude that $g \circ f = (1, 3)$. ●

In Example 38, if we were to compute $f \circ g = (1, 2, 3) \circ (2, 3)$, we would get $(1, 2)$. (We already know that order is important in function composition.) If, however, f and g are members of S_A and f and g are **disjoint cycles**—the cycles have no elements in common—then $f \circ g = g \circ f$.

PRACTICE 36   Let $A = \{1, 2, 3, 4, 5\}$, and let f and g belong to S_A. Compute $g \circ f$ and $f \circ g$ for the following:

    a. $f = (5, 2, 3)$; $g = (3, 4, 1)$. Write the answers in cycle form.
    b. $f = (1, 2, 3, 4)$; $g = (3, 2, 4, 5)$. Write the answers in array form.
    c. $f = (1, 3)$; $g = (2, 5)$. Write the answers in array form. ●

Let $A = \{1, 2, 3, 4\}$ and consider the cycle $f \in S_A$ given by $f = (1, 2)$. If we compute $f \circ f = (1, 2) \circ (1, 2)$, we see that each element of A gets mapped to itself. The permutation that maps each element of A to itself is the identity function on A, i_A, also called the **identity permutation**.

If A is an infinite set, not every permutation of A can be written as a cycle. But even when A is a finite set, not every permutation of A can be written as a cycle; for example, the permutation $g \circ f$ of Practice 36(b) cannot be written as a cycle. However, every permutation on a finite set that is not the identity permutation can be written as a composition of one or more disjoint cycles. The permutation

$$\begin{pmatrix} 1 & 2 & 3 & 4 & 5 \\ 4 & 2 & 5 & 1 & 3 \end{pmatrix}$$

of Practice 36(b) is $(1, 4) \circ (3, 5)$ or $(3, 5) \circ (1, 4)$.

PRACTICE 37   Write

$$\begin{pmatrix} 1 & 2 & 3 & 4 & 5 & 6 \\ 2 & 4 & 5 & 1 & 3 & 6 \end{pmatrix}$$

as a composition of disjoint cycles. ●

Among the permutations of A, some will map certain elements of A to themselves, while others will so thoroughly mix elements around that no element in A is mapped to itself. A permutation on a set that maps no element to itself is called a **derangement**.

EXAMPLE 39   The permutation f on $A = \{1, 2, 3, 4, 5\}$ given in array form by

$$\begin{pmatrix} 1 & 2 & 3 & 4 & 5 \\ 2 & 5 & 4 & 1 & 3 \end{pmatrix}$$

is a derangement. Members of S_A that are *not* derangements, if written as a cycle or a product of cycles, will have at least one element of A that is not listed. Thus $g \in S_A$ defined as $g = (1, 4) \circ (3, 5)$ maps 2 to itself, so g is not a derangement. ●

How Many Functions?

Suppose that S and T are finite sets, say $|S| = m$ and $|T| = n$. What can we say about the number of functions with various properties that map S to T? First, let's just count the number of functions $f: S \rightarrow T$, assuming no special properties about the functions. The Multiplication Principle can be used here because we can think of defining a function by assigning an image to each of the m elements of S. This gives us a sequence of m tasks. Each task has n outcomes, because each element of S can map to any element in T. Therefore the number of functions is

$$\underbrace{n \times n \times n \times \cdots \times n}_{m \text{ factors}} = n^m$$

How many one-to-one functions are there from S to T? We must have $m \leq n$ or we can't have any one-to-one functions at all. (All the elements of S must be mapped to T, and if $m > n$, there are too many elements in S to allow for a one-to-one mapping. Actually, this is the Pigeonhole Principle at work.) We can again solve this problem by carrying out the sequence of tasks of assigning an image to each element in S, but this time we cannot use any image we have used before. By the Multiplication Principle, we have a product that begins with the factors

$$n(n - 1)(n - 2) \cdots$$

and must contain a total of m factors, so the result is

$$n(n - 1)(n - 2) \cdots [n - (m - 1)] = n(n - 1)(n - 2) \cdots (n - m + 1)$$
$$= \frac{n!}{(n - m)!} = P(n, m)$$

How many onto functions are there from S to T? This time we must have $m \geq n$ so that there are enough values in the domain to provide preimages for every value in the codomain. (By the definition of a function, an element in S cannot be a preimage of more than one element in T.) Our overall plan is to subtract the number of non-onto functions from the total number of functions, which we know. To count the number of non-onto functions, we'll use the Principle of Inclusion and Exclusion.

Enumerate the elements of set T as $t_1, \ldots, t_n$. For each i, $1 \leq i \leq n$, let A_i denote the set of functions from S to T that do not map anything to element t_i. (These sets are not disjoint, but every non-onto function belongs to at least one such set.) By the Principle of Inclusion and Exclusion, we can write

$$|A_1 \cup \cdots \cup A_n| = \sum_{1 \leq i \leq n} |A_i| - \sum_{1 \leq i < j \leq n} |A_i \cap A_j|$$
$$+ \sum_{1 \leq i < j < k \leq n} |A_i \cap A_j \cap A_k| - \cdots$$
$$+ (-1)^{n+1}|A_1 \cap \cdots \cap A_n| \qquad (1)$$

For any i, $|A_i|$ is the number of functions that do not map anything to t_i but have no other restrictions. By the Multiplication Principle, we can count the number of such functions by counting for each of the m domain elements its $n - 1$ possible images.

The result is that $|A_i| = (n - 1)^m$. Therefore the first summation in equation (1) adds together terms that are all of the same size. There is one such term for each distinct individual set A_i out of the n sets, so there are $C(n, 1)$ such terms.

For any i and j, $|A_i \cap A_j|$ is the number of functions that do not map anything to t_i or t_j, leaving $n - 2$ possible images for each of the m elements of S. Thus $|A_i \cap A_j| = (n - 2)^m$. The second summation adds one such term for each distinct group of two sets out of n, so there are $C(n, 2)$ of these.

A similar result holds for all the intersection terms. If there are k sets in the intersection, then there are $(n - k)^m$ functions in the intersection set and there are $C(n, k)$ distinct groups of k sets to form the intersection. Equation (1) can thus be written as

$$|A_1 \cup \cdots \cup A_n| = C(n, 1)(n - 1)^m - C(n, 2)(n - 2)^m + C(n, 3)(n - 3)^m$$
$$- \cdots + (-1)^{n+1}C(n, n)(n - n)^m \qquad (2)$$

Now the expression on the left of equation (2) represents all the functions that fail to map to at least one of the elements of T, that is, all the non-onto functions. If we subtract this from the total number of functions, which we know is n^m, we will have the number of onto functions. Thus the number of onto functions is

$$n^m - C(n, 1)(n - 1)^m + C(n, 2)(n - 2)^m - C(n, 3)(n - 3)^m$$
$$+ \cdots + (-1)^{n-1}C(n, n - 1)[n - (n - 1)]^m + (-1)^n C(n, n)(n - n)^m$$

where we've added the next-to-last term. The last term is zero, so the final answer is

$$n^m - C(n, 1)(n - 1)^m + C(n, 2)(n - 2)^m - C(n, 3)(n - 3)^m$$
$$+ \cdots + (-1)^{n-1}C(n, n - 1)(1)^m$$

We'll summarize these results.

Theorem on the Number of Functions with Finite Domains and Codomains
If $|S| = m$ and $|T| = n$, then

1. The number of functions $f: S \rightarrow T$ is n^m.
2. The number of one-to-one functions $f: S \rightarrow T$, assuming $m \leq n$, is

$$\frac{n!}{(n - m)!}$$

3. The number of onto functions $f: S \rightarrow T$, assuming $m \geq n$, is

$$n^m - C(n, 1)(n - 1)^m + C(n, 2)(n - 2)^m - C(n, 3)(n - 3)^m$$
$$+ \cdots + (-1)^{n-1}C(n, n - 1)(1)^m$$

EXAMPLE 40   Let $S = \{A, B, C\}$ and $T = \{a, b\}$. Find the number of functions from S onto T.

Here $m = 3$ and $n = 2$. By our theorem on the number of functions, there are

$$2^3 - C(2, 1)(1)^3 = 8 - 2 \cdot 1 = 6$$

such functions.

PRACTICE 38   One of the six onto functions in Example 40 can be pictured by the following diagram:

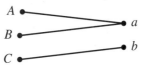

Draw diagrams for the remaining five onto functions.   ●

If A is a set with $|A| = n$, then the number of permutations of A is $n!$. This number can be obtained by any of three methods:

- a combinatorial argument (each of the n elements in the domain must map to one of the n elements in the range with no repetitions)
- thinking of such functions as permutations on a set with n elements and noting that $P(n, n) = n!$
- using result (2) in the previous theorem with $m = n$

We propose to count the number of derangements on A. Our plan is similar to the one we used in counting onto functions. We'll use the Principle of Inclusion and Exclusion to compute the number of permutations that are not derangements and then subtract this from the total number of permutation functions.

Enumerate the elements of set A as $a_1, \ldots, a_n$. For each i, $1 \leq i \leq n$, let A_i be the set of all permutations that leave a_i fixed. (These sets are not disjoint, but every permutation that is not a derangement belongs to at least one such set.) By the Principle of Inclusion and Exclusion, we can write

$$|A_1 \cup \cdots \cup A_n| = \sum_{1 \leq i \leq n} |A_i| - \sum_{1 \leq i < j \leq n} |A_i \cap A_j| + \sum_{1 \leq i < j < k \leq n} |A_i \cap A_j \cap A_k|$$
$$- \cdots + (-1)^{n+1}|A_1 \cap \cdots \cap A_n| \tag{3}$$

For any i, $|A_i|$ is the number of permutations that leave a_i fixed. By the Multiplication Principle we can count the number of such functions by counting for each of the n domain elements, beginning with a_i, its possible images. There is only one choice of where to map a_i because it must map to itself; the next element can map anywhere except to a_i, so there are $n - 1$ outcomes; the next element can map anywhere except the two images already used, so there are $n - 2$ outcomes, and so on. Continuing, there are

$$(1)(n - 1)(n - 2) \cdots (1) = (n - 1)!$$

elements in A_i for each i. Therefore the first summation in equation (3) adds together terms that are all of the same size. The number of such terms equals the number of ways to pick one set A_i out of the n such sets, or $C(n, 1)$.

In the second summation, the terms count the number of permutations on n elements that leave two of those elements fixed. There are

$$(1)(1)(n - 2) \cdots (1) = (n - 2)!$$

such functions in a given $A_i \cap A_j$, and $C(n, 2)$ ways to choose the two sets out of n. In general, if there are k sets in the intersection, then k elements must be held fixed, so there are $(n - k)!$ functions in the intersection set, and there are $C(n, k)$ ways to choose the k sets to form the intersection. Therefore equation (3) becomes

$$|A_1 \cup \cdots \cup A_n| = C(n, 1)(n - 1)! - C(n, 2)(n - 2)! + C(n, 3)(n - 3)!$$
$$- \cdots + (-1)^{n+1}C(n, n)(n - n)!$$

This expression represents all possible nonderangement permutations. We subtract this from the total number of permutation functions, which is $n!$:

$$n! - C(n, 1)(n - 1)! + C(n, 2)(n - 2)! - C(n, 3)(n - 3)! + \cdots + (-1)^n C(n, n)(n - n)!$$

Simplifying, we get

$$n! - \frac{n!}{1!(n - 1)!}(n - 1)! + \frac{n!}{2!(n - 2)!}(n - 2)! - \frac{n!}{3!(n - 3)!}(n - 3)!$$

$$+ \cdots + (-1)^n \frac{n!}{n!0!} 0!$$

$$= n! - \frac{n!}{1!} + \frac{n!}{2!} - \frac{n!}{3!} + \cdots + (-1)^n \frac{n!}{n!}$$

$$= n!\left[1 - \frac{1}{1!} + \frac{1}{2!} - \frac{1}{3!} + \cdots + (-1)^n \frac{1}{n!}\right] \qquad (4)$$

EXAMPLE 41   For $n = 3$, equation (4) says that the number of derangements is

$$3!\left(1 - \frac{1}{1!} + \frac{1}{2!} - \frac{1}{3!}\right) = \frac{3!}{2!} - \frac{3!}{3!} = 3 - 1 = 2$$

Written in array form, the two derangements are

$$\begin{pmatrix} 1 & 2 & 3 \\ 2 & 3 & 1 \end{pmatrix} \quad \text{and} \quad \begin{pmatrix} 1 & 2 & 3 \\ 3 & 1 & 2 \end{pmatrix}$$

Equivalent Sets

Definitions: Equivalent Sets and Cardinality
A set S is **equivalent** to a set T if there exists a bijection $f: S \to T$. Two sets that are equivalent have the same **cardinality**.

If S is equivalent to T, then all the members of S and T are paired off by f in a one-to-one correspondence. If S and T are finite sets, this pairing off can only happen when S and T are the same size. With infinite sets, the idea of size gets a bit fuzzy, because we can sometimes prove that a given set is equivalent to what seems to be a smaller set. Cardinality is the appropriate extension to infinite sets of the idea of size.

PRACTICE 39   Describe a bijection $f: \mathbb{Z} \to \mathbb{N}$, thus showing that $\mathbb{Z}$ is equivalent to $\mathbb{N}$ ($\mathbb{Z}$ and $\mathbb{N}$ have the same cardinality) even though $\mathbb{N} \subset \mathbb{Z}$.

If we have found a bijection between a set S and $\mathbb{N}$, we have established a one-to-one correspondence between the members of S and the nonnegative integers. We can then name the members of S according to this correspondence, writing s_0 for the value of S associated with 0, s_1 for the value of S associated with 1, and so on. Then the list

$s_0, s_1, s_2, ...$

includes all of the members of S. Since this list constitutes an enumeration of S, S is a denumerable set. Conversely, if S is denumerable, then a listing of the members of S exists and can be used to define a bijection between S and $\mathbb{N}$. Therefore a set is denumerable if and only if it is equivalent to $\mathbb{N}$.

For finite sets, we know that if S has n elements, then $\wp(S)$ has 2^n elements. Of course, $2^n > n$, and we cannot find a bijection between a set with n elements and a set with 2^n elements. Therefore S and $\wp(S)$ are not equivalent. This result is also true for infinite sets.

Cantor's Theorem
For any set S, S and $\wp(S)$ are not equivalent.

Proof: We will do a proof by contradiction and assume that S and $\wp(S)$ are equivalent. Let f be the bijection between S and $\wp(S)$. For any member s of S, $f(s)$ is a member of $\wp(S)$, so $f(s)$ is a set containing some members of S, possibly containing s itself. Now we define a set $X = \{x \in S \mid x \notin f(x)\}$. Because X is a subset of S, it is an element of $\wp(S)$ and therefore must be equal to $f(y)$ for some $y \in S$. Then y either is or is not a member of X. If $y \in X$, then by the definition of X, $y \notin f(y)$, but since $f(y) = X$, then $y \notin X$. On the other hand, if $y \notin X$, then since $X = f(y)$, $y \notin f(y)$, and by the definition of X, $y \in X$. In either case, there is a contradiction, and our original assumption is incorrect. Therefore S and $\wp(S)$ are not equivalent.
End of Proof

The proof of Cantor's theorem depends on the nature of set X, which was carefully constructed to provide the crucial contradiction. In this sense, the proof is similar to the diagonalization method (Example 21 in Chapter 3) used to prove the existence of an uncountable set. Indeed, the existence of an uncountable set can be shown directly from Cantor's theorem.

EXAMPLE 42

The set $\mathbb{N}$ is, of course, a denumerable set. By Cantor's theorem, the set $\wp(\mathbb{N})$ is not equivalent to $\mathbb{N}$ and is therefore not a denumerable set, although it is clearly infinite. ●

Order of Magnitude of Functions

Order of magnitude is a way of comparing the "rate of growth" of different functions. We know, for instance, that if we compute $f(x) = x$ and $g(x) = x^2$ for increasing values of x, the g-values will be larger than the f-values by an ever increasing amount. This difference in the rate of increase cannot be overcome by simply multiplying the f-values by some large constant; no matter how large a constant we choose, the g-values will eventually race ahead again. Our experience indicates that the f and

g functions seem to behave in fundamentally different ways with respect to their rates of growth. In order to characterize this difference formally, we define a binary relation on functions.

Let S be the set of all functions with domain and codomain the nonnegative real numbers. We can define a binary relation on S by

$$f \rho g \leftrightarrow \text{there exist positive constants } n_0, c_1, \text{ and } c_2 \text{ such that, for all } x \geq n_0,$$
$$c_1 g(x) \leq f(x) \leq c_2 g(x)$$

EXAMPLE 43

Let f and g be functions in S where $f(x) = 3x^2$ and $g(x) = 200x^2 + 140x + 7$. Let $n_0 = 2$, $c_1 = \dfrac{1}{100}$, and $c_2 = 1$. Then for $x \geq 2$,

$$\frac{1}{100}(200x^2 + 140x + 7) \leq 3x^2 \leq (1)(200x^2 + 140x + 7)$$

or

$$2x^2 + 1.4x + 0.07 \leq 3x^2 \leq 200x^2 + 140x + 7 \tag{5}$$

Therefore $f \rho g$. ●

PRACTICE 40

a. Verify inequality (5) for the following values of x: 2, 3, 4, 5. (Use a calculator.)
b. In Example 43, can n_0 have the value 1 if c_1 and c_2 remain the same?
c. Find a different set of three values n_0, c_1, and c_2 that will also work to show that $f \rho g$ in Example 43. ●

The relation ρ is an equivalence relation on S. For example, to prove that $f \rho f$, we can pick $n_0 = c_1 = c_2 = 1$ and have

$$(1)f(x) \leq f(x) \leq (1)f(x)$$

PRACTICE 41

a. Prove that ρ is symmetric.
b. Prove that ρ is transitive. ●

Given that ρ is an equivalence relation, it partitions S into equivalence classes. If f is in the same class as g, then f is said to have the same order of magnitude as g, denoted by $f = \Theta(g)$ and pronounced "f is order g." Because of symmetry, this also means that g is the same order of magnitude as f, or $g = \Theta(f)$. (The notation $f = \Theta(g)$ is a bit of a misuse of the equality symbol because $\Theta(g)$ is not some function identical to f. It is just a shorthand way of saying that $f \in [g]$ under the equivalence relation ρ defined above.)

Definition: Order of Magnitude
Let f and g be functions mapping nonnegative reals into nonnegative reals. Then f is the same **order of magnitude** as g, written $f = \Theta(g)$, if there exist positive constants n_0, c_1, and c_2 such that for $x \geq n_0$, $c_1 g(x) \leq f(x) \leq c_2 g(x)$.

We will usually try to find the simplest representative of a given equivalence class. Thus for the functions f and g of Example 43, we would say $f = \Theta(x^2)$ and $g = \Theta(x^2)$. A polynomial is always the order of magnitude of its highest-degree term; lower-order terms and all coefficients can be ignored. This is not surprising, since for large values of x, the highest-degree term will dominate the result.

PRACTICE 42   Prove (by finding appropriate constants that satisfy the definition of order of magnitude) that $f = \Theta(x^2)$ and $g = \Theta(x^2)$ for the functions f and g of Example 43.

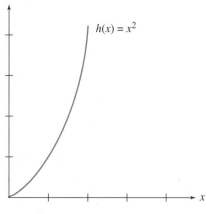

Figure 4.24

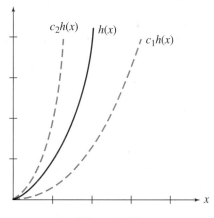

Figure 4.25

In order to understand more intuitively what these equivalence classes mean, we'll draw some graphs. Let $h(x) \in S$, where $h(x) = x^2$. Figure 4.24 shows the graph of $h(x)$. Now suppose we multiply the h-values by the two constants $c_1 = 1/2$ and $c_2 = 2$. The functions $c_1h(x)$ and $c_2h(x)$ are shown as dotted lines in Figure 4.25. These dotted lines form a kind of envelope around the $h(x)$ values, roughly tracing the shape of $h(x)$. Changing the value of the constants changes the width of the envelope but not the basic shape. If $h_1(x)$ is a function with $h_1 = \Theta(h)$, then there is some positive constant n_0 and some envelope around h such that for all domain values to the right of n_0, the h_1-values must fall within this envelope, as shown in Figure 4.26. Therefore the h_1-values can never stray too far from the h-values. The functions h_1 and h are roughly the same size—they are the same order of magnitude.

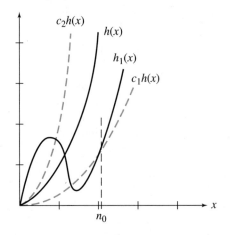

Figure 4.26

EXAMPLE 44   Let $f(x) = x$ and $h(x) = x^2$. Figure 4.27 illustrates that for the constants $c_1 = 1/2$ and $c_2 = 2$, f soon falls below the envelope. Reducing the c_1 constant (lowering the bottom edge of the envelope) only postpones the problem. Formally, we can do a proof by contradiction to show that f is not $\Theta(x^2)$. Suppose $f = \Theta(x^2)$. Then there exist constants n_0 and c_1 with $c_1x^2 \leq f(x)$ for $x \geq n_0$. But this would imply that $c_1x^2 \leq x$ or $c_1x \leq 1$ or $x \leq 1/c_1$ for all $x \geq n_0$. Because c_1 is fixed, we can always choose x large enough so that $x > 1/c_1$, which is a contradiction. Therefore $f(x) = x$ is not $\Theta(x^2)$.

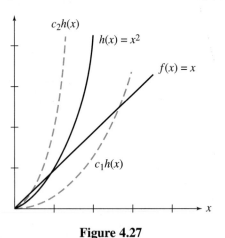

Figure 4.27

If we imagine functions representing various forms of transportation, then functions that are the same order of magnitude (belong to the same equivalence class) represent the same mode of transportation. One class represents travel on foot, another class represents travel by automobile, a third represents travel by air. Speeds within a given mode are about the same; ignoring coefficients and low-order terms amounts to ignoring the difference between walking and running, or between a Cavalier and a Corvette, or between a Cessna and a DC-10. Walking (at any speed) is distinctly different from driving, which is distinctly different from flying.

We can imagine a hierarchy of orders of magnitude. For example, the class $\Theta(x)$ is a lower order of magnitude than the class $\Theta(x^2)$, because functions that are $\Theta(x)$ eventually fall below functions that are $\Theta(x^2)$. Also, the class $\Theta(\log x)$ is a lower order of magnitude than $\Theta(x)$ (see Exercise 57 at the end of this section). In our transportation analogy, walking is slower than driving is slower than flying.

A sort of arithmetic can be developed using order of magnitude. For example, if $f_1(x) = x$ and $f_2(x) = x^2$, then the function $(f_1 + f_2)(x) = f_1(x) + f_2(x) = x + x^2 = \Theta(x^2)$. In general, if $f_1 = \Theta(g_1)$ and $f_2 = \Theta(g_2)$, then $f_1 + f_2 = \Theta(\max(g_1, g_2))$ (see Exercise 53). When expressed in abbreviated form, this leads to somewhat bizarre equations such as $\Theta(x) + \Theta(x^2) = \Theta(x^2)$ or $\Theta(x^2) + \Theta(x^2) = \Theta(x^2)$.

Order of magnitude is important in analysis of algorithms, which we discussed in Section 2.5. In analyzing an algorithm, we identify the important tasks the algorithm performs. Usually the number of times such tasks must be done in executing the algorithm will depend on the size of the input. For example, searching a list of n elements or sorting a list of n elements will require more work as n increases. Typically, we can express input size as a nonnegative integer, so the functions that express the amount of work will be functions with domain $\mathbb{N}$. We found in Section 2.5 that a sequential search of n elements requires n comparisons in the worst case, while a binary search requires $1 + \log n$ comparisons in the worst case (assuming n is a power of 2). Rather than compute the exact functions for the amount of work done, it is easier and often just as useful to settle for order-of-magnitude information. Sequential search is $\Theta(n)$ and binary search is $\Theta(\log n)$ in the worst case. Thus binary search is an order-of-magnitude improvement over sequential search.

To appreciate the effect of order of magnitude in evaluating algorithms, suppose we have two algorithms A and A' to do the same job but they differ in order of magnitude; say A is $\Theta(n)$ and A' is $\Theta(n^2)$. Even if each step in a computation takes only 0.0001 second, this difference will affect total computation time as n grows larger. The first two rows of Table 4.3 give total computation times for A and A' for various values of input length. Now suppose a third algorithm A'' exists whose order of magnitude is not even given by a polynomial function but by an exponential function, say 2^n. The total computation times for A'' are shown in the third row of Table 4.3.

Total Computation Time				
		Size of Input n		
Algorithm	Order	10	50	100
A	n	0.001 second	0.005 second	0.01 second
A'	n^2	0.01 second	0.25 second	1 second
A''	2^n	0.1024 second	3570 years	4×10^{16} centuries

Table 4.3

Note that the exponential case grows at a fantastic rate! Even if we assume that each computation step takes much less time than 0.0001 second, the relative growth rates between polynomial and exponential functions still follow this same pattern. Because of this immense growth rate, algorithms not of polynomial order are generally not useful for large values of n. In fact, problems for which no polynomial time algorithms exist are called **intractable**.

Sometimes algorithms that are not polynomial in the worst case may still be efficient for "average"—and useful—input cases. Nonetheless, in attempting to improve efficiency, we should ask whether a different algorithm of a lower order of magnitude exists before we worry about the details of finc-tuning a given algorithm.

If $f(n)$ represents the work done by an algorithm on an input of size n, it may be difficult to find a simple function g such that $f = \Theta(g)$. Remember that if we can find such a g, f and g are functions that eventually (for large enough n) have roughly the same shape. But we may still be able to find a function g that serves as an upper bound for f. In other words, while f may not have the same shape as g, f will never grow significantly faster than g. Formally, this is expressed by saying that $f = O(g)$ ("f is big O of g").

> **Definition: Big Oh**
> Let f and g be functions mapping nonnegative reals into nonnegative reals. Then f is **big oh** of g, written $f = O(g)$, if there exist positive constants n_0 and c such that for $x \geq n_0, f(x) \leq cg(x)$.

If $f = O(g)$, then g is a ceiling for f and gives us a worst-case picture of the growth of f. In Section 2.5, we learned that if $E(n)$ is the number of divisions required by the Euclidean algorithm to find $\gcd(a, b)$, where $b \leq a = n$, then $E(n) = O(\log n)$.

Section 4.4 Review

Techniques

- Test whether a given relation is a function.
- Test a function for being one-to-one or onto.
- Find the image of an element under function composition.
- Write permutations of a set in array or cycle form.
- Count the number of functions, one-to-one functions, and onto functions from one finite set to another.
- Determine whether two functions are the same order of magnitude.

Main Ideas

The concept of function, especially bijective function, is extremely important.

Composition of functions preserves bijectiveness.

The inverse function of a bijection is itself a bijection.

Permutations are bijections on a set.

Functions can be grouped into equivalence classes according to their order of magnitude, which is a measure of their growth rate.

Exercises 4.4

★**1.** The accompanying figure represents a function.

    a. What is the domain? What is the codomain? What is the range?

    b. What is the image of 5? of 8?

    c. What are the preimages of 9?

    d. Is this an onto function? Is it one-to-one?

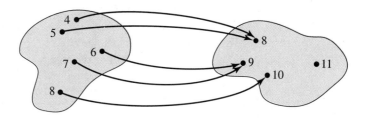

2. The accompanying figure illustrates various binary relations from $\mathbb{R}$ to $\mathbb{R}$. Which are functions? For those that are functions, which are onto? Which are one-to-one?

a.

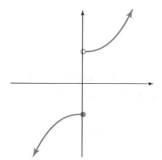

b.

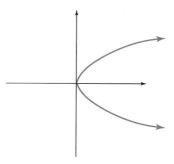

c.

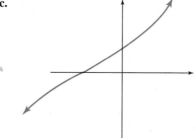

d.

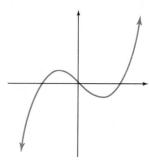

3. Using the equation $f(x) = 2x - 1$ to describe the functional association, write the function as a set of ordered pairs if the codomain is R and

   a. domain is $S = \{0, 1, 2\}$
   b. domain is $S = \{1, 2, 4, 5\}$
   c. domain is $S = \{\sqrt{7}, 1.5\}$

★4. If $f: \mathbb{Z} \to \mathbb{Z}$ is defined by $f(x) = 3x$, find $f(A)$ for

   a. $A = \{1, 3, 5\}$
   b. $A = \{x \mid x \in \mathbb{Z} \text{ and } (\exists y)(y \in \mathbb{Z} \text{ and } x = 2y)\}$

5. If $f: \mathbb{R} \to \mathbb{R}$ is defined by $f(x) = x^2$, describe

   a. $f(\mathbb{N})$
   b. $f(\mathbb{Z})$
   c. $f(\mathbb{R})$

6. Let $S = \{0, 2, 4, 6\}$ and $T = \{1, 3, 5, 7\}$. Determine whether each of the following sets of ordered pairs is a function with domain S and codomain T. If so, is it one-to-one? Is it onto?

   a. $\{(0, 2), (2, 4), (4, 6), (6, 0)\}$
   b. $\{(6, 3), (2, 1), (0, 3), (4, 5)\}$
   c. $\{(2, 3), (4, 7), (0, 1), (6, 5)\}$
   d. $\{(2, 1), (4, 5), (6, 3)\}$
   e. $\{(6, 1), (0, 3), (4, 1), (0, 7), (2, 5)\}$

7. For any bijections in Exercise 6, describe the inverse function.

8. Which of the following are functions from the domain to the codomain given? Which functions are one-to-one? Which functions are onto? Describe the inverse function for any bijective function.

   ★**a.** $f: \mathbb{Z} \to \mathbb{N}$ where f is defined by $f(x) = x^2 + 1$
   ★**b.** $g: \mathbb{N} \to \mathbb{Q}$ where g is defined by $g(x) = 1/x$
   ★**c.** $h: \mathbb{Z} \times \mathbb{N} \to \mathbb{Q}$ where h is defined by $h(z, n) = z/(n + 1)$
   ★**d.** $f: \{1, 2, 3\} \to \{p, q, r\}$ where $f = \{(1, q), (2, r), (3, p)\}$
   ★**e.** $g: \mathbb{N} \to \mathbb{N}$ where g is given by $g(x) = 2^x$
   ★**f.** $h: \mathbb{R}^2 \to \mathbb{R}^2$ where h is defined by $h(x, y) = (y + 1, x + 1)$
   g. $f: \mathbb{Z}^2 \to \mathbb{N}$ where f is defined by $f(x, y) = x^2 + 2y^2$
   h. $g: \mathbb{R} \to \mathbb{R}$ where g is defined by $g(x) = 1/\sqrt{(x + 1)}$
   i. $h: \mathbb{N}^3 \to \mathbb{N}$ where h is given by $h(x, y, z) = x + y - z$

9. Let $f: \mathbb{R} \to \mathbb{R}$ be defined by $f(x) = x^n$, where n is a fixed, positive integer. For what values of n is f bijective?

10. Let $S = \{x \mid x \in \mathbb{R} \text{ and } x \geq 1\}$, $T = \{x \mid x \in \mathbb{R} \text{ and } 0 < x \leq 1\}$. Find a function $f: S \to T$ that is a bijection.

11. Let $S = \{a, b, c, d\}$ and $T = \{x, y, z\}$.

    a. Give an example of a function from S to T that is neither onto nor one-to-one.

    b. Give an example of a function from S to T that is onto but not one-to-one.

    c. Can you find a function from S to T that is one-to-one?

12. What can be said about x if $\lfloor x \rfloor = \lceil x \rceil$?

★13. Prove that $\lfloor x \rfloor = -\lceil -x \rceil$.

14. Prove that $\lceil x \rceil + 1 = \lceil x + 1 \rceil$.

★15. Prove that if $2^k < n < 2^{k+1}$ then $k = \lfloor \log n \rfloor$ and $k + 1 = \lceil \log n \rceil$. (Here $\log n$ means $\log_2 n$.)

16. Prove that $\lfloor \log n \rfloor + 1 = \lceil \log(n + 1) \rceil$. (Here $\log n$ means $\log_2 n$.) (*Hint:* Let $2^k \leq n \leq 2^{k+1}$.)

17. Let S be a set and let A be a subset of S. The *characteristic function* of A is a function $c_A \colon S \to \{0, 1\}$ with $c_A(x) = 1$ exactly when $x \in A$.

    a. Let $S = \{1, 2, 3, 4, 5\}$ and $A = \{1, 3, 5\}$. Give the ordered pairs that belong to c_A.

    b. Prove that for any set S and any subsets A and B of S, $c_{A \cap B}(x) = c_A(x) \cdot c_B(x)$.

    c. Prove that $c_A(x) = 1 - c_{A'}(x)$.

    d. Is it true that for any set S and any subsets A and B of S, $c_{A \cup B}(x) = c_A(x) + c_B(x)$? Prove or give a counterexample.

18. *Ackermann's function*, mapping $\mathbb{N}^2$ to $\mathbb{N}$, is a function that grows very rapidly. It is given by

    $A(0, y) = 1$ for all $y \in \mathbb{N}$
    $A(1, 0) = 2$
    $A(x, 0) = x + 2$ for $x \geq 2$
    $A(x + 1, y + 1) = A(A(x, y + 1), y)$ for all $x \in \mathbb{N}, y \in \mathbb{N}$

    a. Find an equation $f(x)$ to describe $A(x, 1)$ for all $x \geq 1$.

    b. Find an equation $g(x)$ to describe $A(x, 2)$ for all $x \geq 1$.

    c. Compute the value of $A(4, 3)$.

19. Let $S = \{1, 2, 3, 4\}$, $T = \{1, 2, 3, 4, 5, 6\}$, and $U = \{6, 7, 8, 9, 10\}$. Also, let $f = \{(1, 2), (2, 4), (3, 3), (4, 6)\}$ be a function from S to T, and let $g = \{(1, 7), (2, 6), (3, 9), (4, 7), (5, 8), (6, 9)\}$ be a function from T to U. Write the ordered pairs in the function $g \circ f$.

★20. Let $f \colon \mathbb{N} \to \mathbb{N}$ be defined by $f(x) = x + 1$. Let $g \colon \mathbb{N} \to \mathbb{N}$ be defined by $g(x) = 3x$. Calculate the following:

    a. $(g \circ f)(5)$

    b. $(f \circ g)(5)$

    c. $(g \circ f)(x)$

    d. $(f \circ g)(x)$

    e. $(f \circ f)(x)$

    f. $(g \circ g)(x)$

21. **a.** Let $f: \mathbb{R} \to \mathbb{Z}$ be defined by $f(x) = \lfloor x \rfloor$. Let $g: \mathbb{Z} \to \mathbb{N}$ be defined by $g(x) = x^2$. What is $(g \circ f)(-4.7)$?

b. Let f map the set of books into the integers where f assigns to each book the page number of the last page. Let $g: \mathbb{Z} \to \mathbb{Z}$ be given by $g(x) = 2x$. What is $(g \circ f)(\text{this book})$?

c. Let f map strings of alphabetical characters and blank spaces into strings of alphabetical consonants where f takes any string and removes all vowels and all blanks. Let g map strings of alphabetical consonants into integers where g maps a string into the number of characters it contains. What is $(g \circ f)(\text{abraham lincoln})$?

22. The following functions map $\mathbb{R}$ to $\mathbb{R}$. Give an equation describing the composition functions $g \circ f$ and $f \circ g$ in each case.

$C \left(|E|, n-1 \right)$

a. $f(x) = 6x^3, g(x) = 2x$

b. $f(x) = (x - 1)/2, g(x) = 4x^2$

$\dfrac{n(n-1)}{2}$

c. $f(x) = \lceil x \rceil, g(x) = \lfloor x \rfloor$

23. Let $f: S \to T$ and $g: T \to U$ be functions.

a. Prove that if $g \circ f$ is one-to-one, so is f.

b. Prove that if $g \circ f$ is onto, so is g.

c. Find an example where $g \circ f$ is one-to-one but g is not one-to-one.

d. Find an example where $g \circ f$ is onto but f is not onto.

★24. For each of the following bijections $f: \mathbb{R} \to \mathbb{R}$, find f^{-1}.

a. $f(x) = 2x$ **b.** $f(x) = x^3$ **c.** $f(x) = (x + 4)/3$

25. **a.** Let f be a function, $f: S \to T$. If there exists a function $g: T \to S$ such that $g \circ f = i_S$, then g is called a *left inverse* of f. Show that f has a left inverse if and only if f is one-to-one.

b. Let f be a function, $f: S \to T$. If there exists a function $g: T \to S$ such that $f \circ g = i_T$, then g is called a *right inverse* of f. Show that f has a right inverse if and only if f is onto.

c. Let $f: \mathbb{N} \to \mathbb{N}$ be given by $f(x) = 3x$. Then f is one-to-one. Find two different left inverse functions for f.

d. Let $f: \mathbb{N}^+ \to \mathbb{N}^+$ be given by $f(x) = \left\lceil \dfrac{x}{2} \right\rceil$. Then f is onto. Find two different right inverse functions for f.

26. Let f and g be bijections, $f: S \to T$ and $g: T \to U$. Then f^{-1} and g^{-1} exist. Also, $g \circ f$ is a bijection from S to U. Show that $(g \circ f)^{-1} = f^{-1} \circ g^{-1}$.

★27. Let $A = \{1, 2, 3, 4, 5\}$. Write each of the following permutations on A in cycle form:

a. $f = \begin{pmatrix} 1 & 2 & 3 & 4 & 5 \\ 3 & 1 & 5 & 4 & 2 \end{pmatrix}$

b. $f = \{(1, 4), (2, 5), (3, 2), (4, 3), (5, 1)\}$

28. Let A $\{a, b, c, d\}$. Write each of the following permutations on A in array form:

a. $f = \{(a, c), (b, b), (c, d), (d, a)\}$ **b.** $f = (c, a, b, d)$

c. $f = (d, b, a)$ **d.** $f = (a, b) \circ (b, d) \circ (c, a)$

29. Let A be any set and let S_A be the set of all permutations of A. Let f, g, $h \in S_A$. Prove that the functions $h \circ (g \circ f)$ and $(h \circ g) \circ f$ are equal, thereby showing that we can write $h \circ g \circ f$ without parentheses to indicate grouping.

★30. Find the composition of the following cycles representing permutations on $A = \{1, 2, 3, 4, 5, 6, 7, 8\}$. Write your answer as a composition of one or more disjoint cycles.

  a. $(1, 3, 4) \circ (5, 1, 2)$
  b. $(2, 7, 8) \circ (1, 2, 4, 6, 8)$
  c. $(1, 3, 4) \circ (5, 6) \circ (2, 3, 5) \circ (6, 1)$ (By Exercise 29, we can omit parentheses indicating grouping.)
  d. $(2, 7, 1, 3) \circ (2, 8, 7, 5) \circ (4, 2, 1, 8)$

31. Find the composition of the following cycles representing permutations on $\mathbb{N}$. Write your answer as a composition of one or more disjoint cycles.

  a. $(3, 5, 2) \circ (6, 2, 4, 1) \circ (4, 8, 6, 2)$
  b. $(1, 5, 13, 2, 6) \circ (3, 6, 4, 13) \circ (13, 2, 6, 1)$
  c. $(1, 2) \circ (1, 3) \circ (1, 4) \circ (1, 5)$

32. Find a permutation on an infinite set that can't be written as a cycle.

33. The "pushdown store," or "stack," is a storage structure operating much like a set of plates stacked on a spring in a cafeteria. All storage locations are initially empty. An item of data is added to the top of the stack by a "push" instruction, which pushes any previously stored items further down in the stack. Only the top-most item on the stack is accessible at any moment, and it is fetched and removed from the stack by a "pop" instruction.

   Let's consider strings of integers that are an even number of characters in length; half the characters are positive integers, and the other half are zeros. We process these strings through a pushdown store as follows: As we read from left to right, the push instruction is applied to any nonzero integer, and a zero causes the pop instruction to be applied to the stack, thus printing the popped integer. Thus, processing the string 12030040 results in an output of 2314, and processing 12304000 results in an output of 3421. (A string such as 10020340 cannot be handled by this procedure, because we cannot pop two integers from a stack containing only one integer.) Both 2314 and 3421 can be thought of as permutations,

$$\begin{pmatrix} 1 & 2 & 3 & 4 \\ 2 & 3 & 1 & 4 \end{pmatrix} \quad \text{and} \quad \begin{pmatrix} 1 & 2 & 3 & 4 \\ 3 & 4 & 2 & 1 \end{pmatrix}$$

respectively, on the set $A = \{1, 2, 3, 4\}$.

  a. What permutation of $A = \{1, 2, 3, 4\}$ is generated by applying this procedure to the string 12003400?
  b. Name a permutation of $A = \{1, 2, 3, 4\}$ that cannot be generated from any string where the digits 1, 2, 3, and 4 appear in order, no matter where the zeros are placed.

★**34.** Let $S = \{2, 4, 6, 8\}$ and $T = \{1, 5, 7\}$.

    a. Find the number of functions from S to T.

    b. Find the number of surjective functions from S to T.

35. Let $S = \{P, Q, R\}$ and $T = \{k, l, m, n\}$.

    a. Find the number of functions from S to T.

    b. Find the number of injective functions from S to T.

36. **a.** For $|S| = 2, 3$, and 4, respectively, use the theorem on the number of functions to show that the number of one-to-one functions from S to S equals the number of onto functions from S to S.

    b. Argue that for $|S| = n$, $f: S \rightarrow S$ is one-to-one if and only if f is onto.

    c. Find an infinite set S and a function $f: S \rightarrow S$ such that f is one-to-one but not onto.

    d. Find an infinite set S and a function $f: S \rightarrow S$ such that f is onto but not one-to-one.

★**37.** Let $|S| = n$. Find:

    a. number of functions from S to S

    b. number of one-to-one functions from S to S

    c. number of functions from S onto S (see Exercise 36)

    d. number of permutations from S onto S

    e. number of derangements from S onto S

    f. Order the values obtained in parts (a) through (e) from smallest to largest and explain why this ordering is reasonable.

38. **a.** A system development project calls for 5 different tasks to be assigned to Maria, Jon, and Suzanne. In how many ways can this be done if each of the 3 workers must get at least 1 task?

    b. In how many ways can the projects be assigned if Maria must develop the test plan, which is 1 of the 5 tasks, but may do other tasks as well? (*Hint:* Consider the two cases where Maria does and does not do any of the other tasks.)

39. Prove that $S(m, n)$, the number of ways to partition a set of m elements into n blocks, is equal to $1/n!$ times the number of onto functions from a set with m elements to a set with n elements. (*Hint:* Compare Exercise 38 above with Exercise 48 of Section 4.1.)

40. Let $S = \{a, b, c, d\}$. How many elements are in S_A? How many of these are derangements? Write all the derangements in array form.

★**41.** In a programming class of 7 students, the instructor wants each student to modify the program from a previous assignment, but no student should work on his or her own program. In how many ways can the instructor assign programs to the students?

42. **a.** Find a calculus book and look up the Maclaurin series representation for the function e^x.

    b. Use the answer to part (a) to find a series representation for e^{-1}.

    c. Use a calculator to compute an approximate value for e^{-1}.

d. How can the answers to parts (b) and (c) help you approximate the number of derangements of n objects when n is large, say $n \geq 10$? (*Hint:* Look at equation (4) in this section.)

e. Apply this approach to Exercise 41 and compare the results.

f. Approximately how many derangements are there of 10 objects?

43. When a computer program is compiled, the compiler builds a *symbol table* to store information about the identifiers used in the program. A scheme is needed to quickly decide whether a given identifier has already been stored in the table and, if not, to store the new identifier. A *hash function* is often used to locate a position in the table at which to store information about an item. A hash function should be quick to compute and should ideally give a wide distribution of locations throughout the table.

For simplicity, assume that the items to be stored are integers and that the table can hold 17 items in positions 0–16. If n is the item to be stored, let the hash function $f(n)$ be $f(n) = n \bmod 17$, which is the integer remainder when n is divided by 17. Under this hashing function, the integer 23 hashes to 23 mod $17 = 6$, and 23 would be stored at location 6 in the table. However, 6, 40, and many other numbers also hash to 6. There must be some algorithm for *collision resolution* that tells how to proceed if a number to be stored hashes to an already occupied spot in the table. One technique is to use *linear probing*, which simply means that if a number hashes to an already occupied spot in the table, then it should be stored in the next open position, wrapping around to the top of the table if necessary.

a. Using the hash function and collision resolution scheme described, store the sequence of values 23, 14, 52, 40, 24, 18, 33, 58, 50. Give the location in the table at which each is stored.

b. After the table of part (a) has been filled, describe the process to search for 58 in the table. Describe the process to search (unsuccessfully) for 41 in the table.

c. Explain what problem can arise if an item stored in a hash table is later deleted.

44. Let $\mathcal{C}$ be a collection of sets, and define a binary relation ρ on $\mathcal{C}$ as follows: For $S, T \in \mathcal{C}$, $S \rho T \leftrightarrow$ "S is equivalent to T." Show that ρ is an equivalence relation on $\mathcal{C}$.

45. Group the following sets into equivalence classes according to the equivalence relation of Exercise 44:

$A = \{2, 4\}$
$B = \mathbb{N}$
$C = \{x \mid x \in \mathbb{N} \text{ and } (\exists y)(y \in \mathbb{N} \text{ and } x = 2 * y)\}$
$D = \{a, b, c, d\}$
$E = \wp(\{1, 2\})$
$F = \mathbb{Q}^+$

46. Let f be a function, $f: S \to T$.

a. Show that for all subsets A and B of S, $f(A \cap B) \subseteq f(A) \cap f(B)$.

b. Show that $f(A \cap B) = f(A) \cap f(B)$ for all subsets A and B of S if and only if f is one-to-one.

★47. By the definition of a function f from S to T, f is a subset of $S \times T$ where the image of every $s \in S$ under f is uniquely determined as the second component of the ordered pair (s, t) in f. Now consider any binary relation ρ from S to T. The relation ρ is a subset of $S \times T$ in which some elements of S may not appear at all as first components of an ordered pair and some may appear more than once. We can view ρ as a *nondeterministic function* from a subset of S to T. An $s \in S$ not appearing as the first component of an ordered pair represents an element outside the domain of ρ. For an $s \in S$ appearing once or more as a first component, ρ can select for the image of s any one of the corresponding second components.

Let $S = \{1, 2, 3\}$, $T = \{a, b, c\}$, and $U = \{m, n, o, p\}$. Let ρ be a binary relation on $S \times T$ and σ be a binary relation on $T \times U$ defined by

$$\rho = \{(1, a), (1, b), (2, b), (2, c), (3, c)\}$$
$$\sigma = \{(a, m), (a, o), (a, p), (b, n), (b, p), (c, o)\}$$

Thinking of ρ and σ as nondeterministic functions from S to T and T to U, respectively, we can form the composition $\sigma \circ \rho$, a nondeterministic function from S to U.

   a. What is the set of possible images of 1 under $\sigma \circ \rho$?
   b. What is the set of possible images of 2 under $\sigma \circ \rho$? of 3?

48. Let f be a function, $f: S \rightarrow T$.
   a. Define a binary relation ρ on S by $x \rho y \leftrightarrow f(x) = f(y)$. Prove that ρ is an equivalence relation.
   b. For $S = T = \mathbb{Z}$ and $f(x) = 3x^2$, what is [4] under the equivalence relation of part (a)?

★49. Prove, by finding constants that satisfy the definition of order of magnitude, that $f = \Theta(g)$ if $f(x) = x$ and $g(x) = 17x + 1$.

50. Prove, by finding constants that satisfy the definition of order of magnitude, that $f = \Theta(g)$ if $f(x) = 3x^3 - 7x$ and $g(x) = x^3/2$.

51. Prove that $\log(x^2 + 3)$ is $\Theta(\log x)$.

52. In this section, we noted that $h_1 = \Theta(h)$ implies that from some point on, h_1 is within an "envelope" of h. Can this envelope ever be entirely above or entirely below h? Explain.

53. Prove that if f_1 is a function that is $\Theta(g_1)$ and f_2 is a function that is $\Theta(g_2)$, then the function $f_1 + f_2$, defined by $(f_1 + f_2)(x) = f_1(x) + f_2(x)$, is $\Theta(\max(g_1, g_2))$, where $(\max(g_1, g_2))(x) = \max(g_1(x), g_2(x))$.

Exercises 54–58 require familiarity with ideas from calculus. As an alternative to the definition of order of magnitude, a limit test can be used to prove that $f = \Theta(g)$:

$$f = \Theta(g) \text{ if } \lim_{x \to \infty} \frac{f(x)}{g(x)} = p \qquad \text{where } p \text{ is a positive real number}$$

Thus $2x^2 + 7 = \Theta(x^2)$ because

$$\lim_{x \to \infty} \frac{2x^2 + 7}{x^2} = \lim_{x \to \infty} \frac{2 + \dfrac{7}{x^2}}{1} = \frac{2}{1} = 2$$

★54. Use the limit test to do Exercise 49 again.

55. Use the limit test to do Exercise 50 again.

If $\lim_{x \to \infty} f(x) = \infty$ and $\lim_{x \to \infty} g(x) = \infty$ and f and g are differentiable functions, then L'Hôpital's rule says that

$$\lim_{x \to \infty} \frac{f(x)}{g(x)} = \lim_{x \to \infty} \frac{f'(x)}{g'(x)}$$

As another limit test, if

$$\lim_{x \to \infty} \frac{f(x)}{g(x)} = 0$$

then the class $\Theta(f)$ is a lower order of magnitude than the class $\Theta(g)$.

★56. Use the second limit test to prove that the class $\Theta(x)$ is a lower order of magnitude than the class $\Theta(x^2)$.

57. Use the second limit test to prove that the class $\Theta(\log x)$ is a lower order of magnitude than the class $\Theta(x)$.

58. Use both limit tests to group the following functions into classes by order of magnitude and to order those classes.

$$17x \log x,\ 200 \log x,\ 2^x - x^2,\ \sqrt[4]{x},\ 10x^2 - 3x + 5,\ 420x,\ 41 \ln x^2$$

59. Use both limit tests to group the following functions into classes by order of magnitude and to order those classes. Here $\ln x$ is the natural log of x, $\log_e x$.

$$x,\ \sqrt{x},\ \log x,\ x^3,\ x \log x,\ 2x^3 + x,\ e^x,\ (\log x)^2,\ \ln x,\ x^3 + \log x$$

Section 4.5   Matrices

Terminology

Data about many kinds of problems can often be represented using a rectangular arrangement of values; such an arrangement is called a **matrix**. Thus

$$\mathbf{A} = \begin{bmatrix} 1 & 0 & 4 \\ 3 & -6 & 8 \end{bmatrix}$$

is a matrix with two rows and three columns. The **dimensions** of the matrix are the number of rows and columns; here $\mathbf{A}$ is a 2×3 matrix.

Elements of a matrix $\mathbf{A}$ are denoted by a_{ij}, where i is the row number of the element in the matrix and j is the column number. In the example matrix $\mathbf{A}$, $a_{23} = 8$ because 8 is the element in row 2, column 3, of $\mathbf{A}$.

EXAMPLE 45   Average temperatures in three different cities for each month can be neatly summarized in a 3×12 matrix. Here we interpret the 3 rows as the 3 cities and the 12 columns as the 12 months January–December. The average temperature in the third city in April, a_{34}, is 67.

$$\mathbf{A} = \begin{bmatrix} 23 & 26 & 38 & 47 & 58 & 71 & 78 & 77 & 69 & 55 & 39 & 33 \\ 14 & 21 & 33 & 38 & 44 & 57 & 61 & 59 & 49 & 38 & 25 & 21 \\ 35 & 46 & 54 & 67 & 78 & 86 & 91 & 94 & 89 & 75 & 62 & 51 \end{bmatrix} \quad \bullet$$

EXAMPLE 46   In Practice 2(c) of Section 4.1, the binary relation $\{(7, 9), (2, 5), (9, 9), (2, 7)\}$ was defined on the set $S = \{2, 5, 7, 9\}$. The matrix $\mathbf{R}$ below represents this binary relation by a 1 entry in position i,j if element i in set S is related to element j. (This assumes that an ordering has been imposed on the elements of set S; that is, 2 is element 1 in S, 5 is element 2, and so forth.)

$$\mathbf{R} = \begin{bmatrix} 0 & 1 & 1 & 0 \\ 0 & 0 & 0 & 0 \\ 0 & 0 & 0 & 1 \\ 0 & 0 & 0 & 1 \end{bmatrix} \quad \bullet$$

EXAMPLE 47   Solutions to many problems can be obtained by solving systems of linear equations. Suppose, for example, that you are placing an order for coffee beans for your sidewalk café. You want to order 70 pounds of beans, a mixture of Kona coffee beans and Colombian coffee beans. You are willing to spend $1180; Kona coffee costs $24 per pound, and Colombian coffee costs $14 per pound. How many pounds of each should you order?

The constraints in this problem are represented by the system of linear equations

$$x + y = 70$$
$$24x + 14y = 1180$$

The solution is $x = 20$, $y = 50$ (you can easily check that this is a solution). The matrix

$$\mathbf{A} = \begin{bmatrix} 1 & 1 \\ 24 & 14 \end{bmatrix}$$

is the **matrix of coefficients** for this system of linear equations. (See Exercise 15 at the end of this section to see how the matrix of coefficients is used to solve the system.) ●

PRACTICE 43   In the matrix

$$\mathbf{A} = \begin{bmatrix} 1 & 4 & -6 & 8 \\ 3 & 0 & 1 & -7 \end{bmatrix}$$

what is a_{23}? What is a_{24}? What is a_{13}? ●

In a matrix, the arrangement of the entries is significant. Therefore, for two matrices to be **equal** they must have the same dimensions and the same entries in each location.

EXAMPLE 48   Let

$$\mathbf{X} = \begin{bmatrix} x & 4 \\ 1 & y \\ z & 0 \end{bmatrix}$$

$$\mathbf{Y} = \begin{bmatrix} 3 & 4 \\ 1 & 6 \\ 2 & w \end{bmatrix}$$

If $\mathbf{X} = \mathbf{Y}$, then $x = 3$, $y = 6$, $z = 2$, and $w = 0$. ●

We will often be interested in square matrices, in which the number of rows equals the number of columns. If $\mathbf{A}$ is an $n \times n$ square matrix, then the elements $a_{11}, a_{22}, \ldots, a_{nn}$ form the **main diagonal** of the matrix. If the corresponding elements match when we think of folding the matrix along the main diagonal, then the matrix is symmetric about the main diagonal. In a **symmetric** matrix, $a_{ij} = a_{ji}$.

EXAMPLE 49   The square 3×3 matrix

$$\mathbf{A} = \begin{bmatrix} 1 & 5 & 7 \\ 5 & 0 & 2 \\ 7 & 2 & 6 \end{bmatrix}$$

is symmetric. The upper triangular part (the portion above the main diagonal) is a reflection of the lower triangular part. Note that $a_{21} = a_{12} = 5$. ●

A more general way to represent arrangements of data is the **array**. Arrays are n-dimensional arrangements of data, where n can be any positive integer. If $n = 1$, then the data are arranged in a single line, which is therefore a list or finite sequence of data items. This one-dimensional version of an array is called a **vector**. If $n = 2$, the array is a matrix. If $n = 3$, we can picture layers of two-dimensional matrices. For $n > 3$, we can formally deal with the array elements, but we can't really visualize the arrangement. The array data structure is available in many high-level programming languages; generally, the number of elements expected in each dimension of the array must be declared in the program. The array **X** of Example 48, for instance, would be declared as a 3×2 array—a two-dimensional array (matrix) with three elements in one dimension and two in the other (that is, three rows and two columns).

Matrix Operations

Although matrices are particular arrangements of individual elements, we can treat the matrices themselves as objects, just as we can treat sets of elements as objects. In each case we are abstracting up one level and looking at the *collection* as an entity, rather than looking at the individual elements making up the collection. We defined operations on sets (union, intersection, and so forth) that made sets useful for solving counting problems. We can define arithmetic operations on matrices whose entries are numerical. These operations make matrices interesting objects to study in their own right, but they also make matrices more useful for certain tasks such as solving systems of equations.

The first operation, called **scalar multiplication,** calls for multiplying each entry of a matrix by a fixed single number called a **scalar.** The result is a matrix with the same dimensions as the original matrix.

EXAMPLE 50   The result of multiplying matrix

$$A = \begin{bmatrix} 1 & 4 & 5 \\ 6 & -3 & 2 \end{bmatrix}$$

by the scalar $r = 3$ is

$$A = \begin{bmatrix} 3 & 12 & 15 \\ 18 & -9 & 6 \end{bmatrix}$$

●

Addition of two matrices **A** and **B** is only defined when **A** and **B** have the same dimensions; then it is simply a matter of adding the corresponding elements. Formally, if **A** and **B** are both $n \times m$ matrices, then $C = A + B$ is an $n \times m$ matrix with entries

$$c_{ij} = a_{ij} + b_{ij}$$

EXAMPLE 51   For

$$A = \begin{bmatrix} 1 & 3 & 6 \\ 2 & 0 & 4 \\ -4 & 5 & 1 \end{bmatrix} \qquad B = \begin{bmatrix} 0 & -2 & 8 \\ 1 & 5 & 2 \\ 2 & 3 & 3 \end{bmatrix}$$

the matrix $A + B$ is

$$A + B = \begin{bmatrix} 1 & 1 & 14 \\ 3 & 5 & 6 \\ -2 & 8 & 4 \end{bmatrix}$$

●

PRACTICE 44

For $r = 2$,

$$A = \begin{bmatrix} 1 & 7 \\ -3 & 4 \\ 5 & 6 \end{bmatrix} \qquad B = \begin{bmatrix} 4 & 0 \\ 9 & 2 \\ -1 & 4 \end{bmatrix}$$

find $rA + B$.   ●

Subtraction of matrices is defined by $A - B = A + (-1)B$.

In a **zero matrix**, all entries are 0. If we add an $n \times m$ zero matrix, denoted by **0**, to any $n \times m$ matrix A, the result is matrix A. We can symbolize this by the matrix equation

$$0 + A = A$$

This equation is true because of a similar equation that holds for all the individual numerical entries, $0 + a_{ij} = a_{ij}$. Other matrix equations are also true because of similar equations that hold for the individual entries.

EXAMPLE 52

If A and B are $n \times m$ matrices and r and s are scalars, the following matrix equations are true:

$$0 + A = A$$
$$A + B = B + A$$
$$(A + B) + C = A + (B + C)$$
$$r(A + B) = rA + rB$$
$$(r + s)A = rA + sA$$
$$r(sA) = (rs)A$$

To prove that $A + B = B + A$, for instance, it is sufficient to note that $a_{ij} + b_{ij} = b_{ij} + a_{ij}$ for each entry in matrices A and B.   ●

One might expect that in **multiplication** of matrices individual elements are simply multiplied, but the definition is more complicated than that. The definition of matrix multiplication is based on the use of matrices in mathematics to represent functions called linear transformations, which map points in the real-number plane to points in the real-number plane. Although we won't use matrices in this way, we will use the standard definition for matrix multiplication.

To compute A times B, $A \cdot B$, the number of columns in A must equal the number of rows in B. Thus we can compute $A \cdot B$ if A is an $n \times m$ matrix and B is an $m \times p$ matrix. The result is an $n \times p$ matrix. An entry in row i, column j of $A \cdot B$ is obtained by multiplying elements in row i of A by the corresponding elements in column j of B and adding the results. Formally, $A \cdot B = C$, where

$$c_{ij} = \sum_{k=1}^{m} a_{ik} b_{kj}$$

EXAMPLE 53   Let

$$A = \begin{bmatrix} 2 & 4 & 3 \\ 4 & -1 & 2 \end{bmatrix} \qquad B = \begin{bmatrix} 5 & 3 \\ 2 & 2 \\ 6 & 5 \end{bmatrix}$$

A is a 2×3 matrix and **B** is a 3×2 matrix, so the product $A \cdot B$ exists and is a 2×2 matrix **C**. To find element c_{11}, we multiply corresponding elements of row 1 of **A** and column 1 of **B** and add the results.

$$2(5) + 4(2) + 3(6) - 10 + 8 + 18 = 36$$

$$\begin{bmatrix} \boxed{2} & 4 & 3 \\ 4 & -1 & 2 \end{bmatrix} \begin{bmatrix} \boxed{5} & 3 \\ \boxed{2} & 2 \\ \boxed{6} & 5 \end{bmatrix} = \begin{bmatrix} 36 & \underline{} \\ \underline{} & \underline{} \end{bmatrix}$$

Element c_{12} is obtained by multiplying corresponding elements of row 1 of **A** and column 2 of **B** and adding the results.

$$\begin{bmatrix} \boxed{2} & 4 & 3 \\ 4 & -1 & 2 \end{bmatrix} \begin{bmatrix} 5 & \boxed{3} \\ 2 & \boxed{2} \\ 6 & \boxed{5} \end{bmatrix} = \begin{bmatrix} 36 & 29 \\ \underline{} & \underline{} \end{bmatrix}$$

The complete product is

$$\begin{bmatrix} 2 & 4 & 3 \\ 4 & -1 & 2 \end{bmatrix} \begin{bmatrix} 5 & 3 \\ 2 & 2 \\ 6 & 5 \end{bmatrix} = \begin{bmatrix} 36 & 29 \\ 30 & 30 \end{bmatrix}$$

PRACTICE 45   Compute $A \cdot B$ and $B \cdot A$ for

$$A = \begin{bmatrix} 1 & 4 \\ 6 & -2 \end{bmatrix} \qquad B = \begin{bmatrix} 3 & 6 \\ 3 & 4 \end{bmatrix}$$

From Practice 45 we see that even if **A** and **B** have dimensions so that both $A \cdot B$ and $B \cdot A$ are defined, $A \cdot B$ need not equal $B \cdot A$. There are, however, several matrix equations involving multiplication that are true.

EXAMPLE 54   Where **A**, **B**, and **C** are matrices of appropriate dimensions and r and s are scalars, the following matrix equations are true (the notation $A(B \cdot C)$ is shorthand for $A \cdot (B \cdot C)$):

$$A(B \cdot C) = (A \cdot B)C$$
$$A(B + C) = A \cdot B + A \cdot C$$
$$(A + B)C = A \cdot C + B \cdot C$$
$$rA \cdot sB = (rs)(A \cdot B)$$

Verifying these equations for matrices of particular dimensions is simple, if tedious.

The $n \times n$ matrix with 1s along the main diagonal and 0s elsewhere is called the **identity matrix**, denoted by **I**. If we multiply **I** times any $n \times n$ matrix **A**, we get **A** as the result. The equation

$$\mathbf{I} \cdot \mathbf{A} = \mathbf{A} \cdot \mathbf{I} = \mathbf{A}$$

holds.

PRACTICE 46   Let

$$\mathbf{I} = \begin{bmatrix} 1 & 0 \\ 0 & 1 \end{bmatrix} \qquad \mathbf{A} = \begin{bmatrix} a_{11} & a_{12} \\ a_{21} & a_{22} \end{bmatrix}$$

Verify that $\mathbf{I} \cdot \mathbf{A} = \mathbf{A} \cdot \mathbf{I} = \mathbf{A}$.

An $n \times n$ matrix **A** is **invertible** if there exists an $n \times n$ matrix **B** such that

$$\mathbf{A} \cdot \mathbf{B} = \mathbf{B} \cdot \mathbf{A} = \mathbf{I}$$

In this case **B** is called the **inverse** of **A**, denoted by $\mathbf{A}^{-1}$.

EXAMPLE 55   Let

$$\mathbf{A} = \begin{bmatrix} -1 & 2 & -3 \\ 2 & 1 & 0 \\ 4 & -2 & 5 \end{bmatrix} \qquad \mathbf{B} = \begin{bmatrix} -5 & 4 & -3 \\ 10 & -7 & 6 \\ 8 & -6 & 5 \end{bmatrix}$$

Then, following the rules of matrix multiplication, it can be shown (Practice 47) that $\mathbf{A} \cdot \mathbf{B} = \mathbf{B} \cdot \mathbf{A} = \mathbf{I}$, so $\mathbf{B} = \mathbf{A}^{-1}$.

PRACTICE 47   For the matrices **A** and **B** of Example 55
   a. Compute $\mathbf{A} \cdot \mathbf{B}$.
   b. Compute $\mathbf{B} \cdot \mathbf{A}$.

It is easy to write an algorithm for matrix multiplication by simply following the definition. A pseudocode version of the algorithm follows, where bracket notation $A[i, j]$ replaces the subscript notation a_{ij}.

ALGORITHM   **MatrixMultiplication**

//computes $n \times p$ matrix $\mathbf{A} \cdot \mathbf{B}$ for $n \times m$ matrix **A**, $m \times p$ matrix **B**
//stores result in **C**

for $i = 1$ **to** n **do**
   for $j = 1$ **to** p **do**
      $C[i, j] = 0$
      for $k = 1$ **to** m **do**
         $C[i, j] = C[i, j] + A[i, k] * B[k, j]$
      end for
   end for
end for
write out product matrix **C**

The computational steps done in this algorithm are multiplications and additions, one multiplication and one addition each time the statement $\mathbf{C}[i, j] = \mathbf{C}[i, j] + \mathbf{A}[i, k] * \mathbf{B}[k, j]$ is executed. This statement occurs within a triply nested loop and will be executed npm times. (Although this is quite obvious, it can also be justified by the Multiplication Principle as the number of possible outcomes of choosing indices i, j, and k.) If $\mathbf{A}$ and $\mathbf{B}$ are both $n \times n$ matrices, then there are $\Theta(n^3)$ multiplications and $\Theta(n^3)$ additions required. The total amount of work is therefore $\Theta(n^3) + \Theta(n^3) = \Theta(n^3)$.

Given the definition of matrix multiplication, it is hard to see how one could avoid $\Theta(n^3)$ steps in computing the product of two $n \times n$ matrices, but a sufficiently clever approach (which we won't go into) does yield an improvement.

Boolean Matrices

In Chapter 5 we shall be interested in matrices with only 0s and 1s as entries, called **Boolean matrices** (after George Boole, a nineteenth-century English mathematician; Boole also lent his name to *Boolean algebra*, which we shall consider later in this book). Matrix $\mathbf{R}$ of Example 46 is a Boolean matrix. We can define an operation of Boolean matrix multiplication $\mathbf{A} \times \mathbf{B}$ on Boolean matrices using Boolean multiplication and Boolean addition instead of regular multiplication and addition. These are defined as follows:

Boolean multiplication: $x \wedge y = \min(x, y)$

Boolean addition: $x \vee y = \max(x, y)$

PRACTICE 48   Fill in the following operation tables for Boolean multiplication and Boolean addition.

x	y	$x \wedge y$		x	y	$x \vee y$
1	1			1	1	
1	0			1	0	
0	1			0	1	
0	0			0	0	

Now take the tables from Practice 48 and substitute T for 1 and F for 0. They become the truth tables for conjunction and disjunction, respectively; for this reason, these operations are often called **Boolean and** (or **logical and**) and **Boolean or** (or **logical or**). This also explains the notation used for these operations. The operation of **Boolean matrix multiplication** $\mathbf{A} \times \mathbf{B}$ (on Boolean matrices of appropriate dimensions) is then defined by

$$c_{ij} = \bigvee_{k=1}^{m} (a_{ik} \wedge b_{kj})$$

We can also define two analogues of ordinary matrix addition (on Boolean matrices of the same dimensions): $\mathbf{A} \wedge \mathbf{B}$, where corresponding elements are combined using Boolean multiplication, and $\mathbf{A} \vee \mathbf{B}$, where corresponding elements are combined using Boolean addition.

EXAMPLE 56   Let $\mathbf{A}$ and $\mathbf{B}$ be Boolean matrices,

$$\mathbf{A} = \begin{bmatrix} 1 & 1 & 0 \\ 0 & 1 & 0 \\ 0 & 0 & 1 \end{bmatrix} \quad \mathbf{B} = \begin{bmatrix} 1 & 0 & 0 \\ 1 & 1 & 1 \\ 0 & 0 & 1 \end{bmatrix}$$

Then

$$\mathbf{A} \wedge \mathbf{B} = \begin{bmatrix} 1 & 0 & 0 \\ 0 & 1 & 0 \\ 0 & 0 & 1 \end{bmatrix} \quad \mathbf{A} \vee \mathbf{B} = \begin{bmatrix} 1 & 1 & 0 \\ 1 & 1 & 1 \\ 0 & 0 & 1 \end{bmatrix}$$

and the Boolean product $\mathbf{A} \times \mathbf{B}$ is

$$\mathbf{A} \times \mathbf{B} = \begin{bmatrix} 1 & 1 & 1 \\ 1 & 1 & 1 \\ 0 & 0 & 1 \end{bmatrix}$$

PRACTICE 49   in Example 56, does $\mathbf{A} \times \mathbf{B} = \mathbf{A} \cdot \mathbf{B}$?

PRACTICE 50   In Example 56, compute $\mathbf{B} \times \mathbf{A}$.

Section 4.5 Review

Techniques

- Add, subtract, multiply, and perform scalar multiplication on matrices.
- Perform Boolean and, or, and matrix multiplication on Boolean matrices.

Main Ideas

Matrices are rectangular arrangements of data that are used to represent information in tabular form.

Matrices have their own arithmetic, with operations of addition, subtraction, multiplication, and scalar multiplication.

Boolean matrices can be manipulated using Boolean operations of and, or, and Boolean multiplication.

Exercises 4.5

★1. For the matrix

$$A = \begin{bmatrix} 1 & 2 \\ 3 & 0 \\ -4 & 1 \end{bmatrix}$$

what is a_{12}? What is a_{31}?

2. Find x and y if

$$\begin{bmatrix} 1 & 3 \\ x & x + y \end{bmatrix} = \begin{bmatrix} 1 & 3 \\ 2 & 6 \end{bmatrix}$$

★3. Find x, y, z, and w if

$$\begin{bmatrix} x + y & 2x - 3y \\ z - w & z + 2w \end{bmatrix} = \begin{bmatrix} 4 & -7 \\ -6 & 6 \end{bmatrix}$$

4. If A is a symmetric matrix, find u, v, and w:

$$A = \begin{bmatrix} 2 & w & u \\ 7 & 0 & v \\ 1 & -3 & 4 \end{bmatrix}$$

5. If $r = 3$, $s = -2$,

$$A = \begin{bmatrix} 2 & 1 \\ -1 & 0 \\ 3 & 4 \end{bmatrix} \qquad B = \begin{bmatrix} 4 & 1 & 2 \\ 6 & -1 & 5 \\ 1 & 3 & 2 \end{bmatrix}$$

$$C = \begin{bmatrix} 2 & 4 \\ 6 & -1 \end{bmatrix} \qquad D = \begin{bmatrix} 4 & -6 \\ 1 & 3 \\ 2 & -1 \end{bmatrix}$$

compute the following (if possible).

★a. $A + D$
 b. $A - D$
 c. rB
 d. sC
★e. $A + rD$
 f. $B - rC$
 g. $r(A + D)$
 h. $r(sC)$

★**i.** **B · D**
 j. **D · C**
 k. **A · C**
 l. **C · A**
★**m.** $\mathbf{C}^2 = \mathbf{C} \cdot \mathbf{C}$
 n. **B · A + D**

6. For

$$\mathbf{A} = \begin{bmatrix} 3 & -1 \\ 2 & 5 \end{bmatrix}$$

$$\mathbf{B} = \begin{bmatrix} 4 & 1 \\ 2 & -1 \end{bmatrix}$$

$$\mathbf{C} = \begin{bmatrix} 6 & -5 \\ 2 & -2 \end{bmatrix}$$

compute the following.
   a. **A · B** and **B · A**
   b. **A(B · C)** and **(A · B)C**
 ★**c.** **A(B + C)** and **A · B + A · C**
   d. **(A + B)C** and **A · C + B · C**

7. If

$$\mathbf{A} = \begin{bmatrix} 2 & 3 \\ 4 & 1 \end{bmatrix} \qquad \mathbf{B} = \begin{bmatrix} x & 3 \\ y & 2 \end{bmatrix}$$

find x and y if **A · B = B · A**.

8. **a.** Prove that $\mathbf{I}^2 = \mathbf{I}$ for any identity matrix **I**.
   b. Prove that $\mathbf{I}^n = \mathbf{I}$ for any identity matrix **I** and any positive integer n.

9. Let **A** and **B** be $n \times n$ matrices.
   a. Prove that if **A** has one row consisting of all 0s, then so does **A · B**.
   b. Prove that if **B** has one column consisting of all 0s, then so does **A · B**.

10. An $n \times n$ matrix **A** is *diagonal* if all elements a_{ij} with $i \neq j$ are 0. For example, **A** below is a 3×3 diagonal matrix.

$$\mathbf{A} = \begin{bmatrix} 2 & 0 & 0 \\ 0 & 5 & 0 \\ 0 & 0 & -7 \end{bmatrix}$$

   a. Prove that if **A** and **B** are $n \times n$ diagonal matrices, then **A + B** is diagonal.
   b. Prove that that if **A** is an $n \times n$ diagonal matrix and r is a scalar, then $r\mathbf{A}$ is diagonal.
   c. Prove that if **A** and **B** are $n \times n$ diagonal matrices, then $\mathbf{A} \times \mathbf{B}$ is diagonal.

11.★**a.** Show that for

$$\mathbf{A} = \begin{bmatrix} 1 & 3 \\ 2 & 2 \end{bmatrix} \qquad \mathbf{B} = \begin{bmatrix} -1/2 & 3/4 \\ 1/2 & -1/4 \end{bmatrix}$$

$\mathbf{A} \cdot \mathbf{B} = \mathbf{B} \cdot \mathbf{A} = \mathbf{I}$, so $\mathbf{B} = \mathbf{A}^{-1}$.

★**b.** Show that

$$\mathbf{A} = \begin{bmatrix} 1 & 2 \\ 2 & 4 \end{bmatrix}$$

is not invertible.

c. Show that

$$\mathbf{A} = \begin{bmatrix} a_{11} & a_{12} \\ a_{21} & a_{22} \end{bmatrix}$$

is invertible with inverse

$$\mathbf{B} = \frac{1}{a_{11}\, a_{22} - a_{12}\, a_{21}} \begin{bmatrix} a_{22} & -a_{12} \\ -a_{21} & a_{11} \end{bmatrix}$$

if and only if $a_{11}\, a_{22} - a_{12}\, a_{21} \neq 0$.

12. Prove that if $\mathbf{A}$ is invertible and r is a non-zero scalar, then $r\mathbf{A}$ is invertible with $(r\mathbf{A})^{-1} = (1/r)\mathbf{A}^{-1}$.

13. Prove that if $\mathbf{A}$ is invertible and $\mathbf{A} \cdot \mathbf{B} = \mathbf{A} \cdot \mathbf{C}$, then $\mathbf{B} = \mathbf{C}$.

14. If $\mathbf{A}$ is an $n \times n$ invertible matrix, the following method can be used to find $\mathbf{A}^{-1}$:

1. Operate on A using any combination of the two operations
   i. Multiply all the elements in any one row of $\mathbf{A}$ by a non-zero scalar.
   ii. Add a scalar multiple of any row to any other row.
      until the resulting matrix is the $n \times n$ identity matrix $\mathbf{I}$.
2. At the same time, perform exactly the same sequence of operations on the $n \times n$ identity matrix $\mathbf{I}$.
3. The matrix that results from $\mathbf{I}$ after step 2 is $\mathbf{A}^{-1}$.

a. Use this method to find the inverse of matrix $\mathbf{A}$ in Exercise 11(a).
b. Use this method to find the inverse of matrix $\mathbf{A}$ in Example 55.

★15. Consider a system of n linear equations in n unknowns, such as the one from Example 47:

$$\begin{aligned} x + y &= 70 \\ 24x + 14y &= 1180 \end{aligned}$$

If

$$\mathbf{A} = \begin{bmatrix} 1 & 1 \\ 24 & 14 \end{bmatrix} \qquad \mathbf{X} = \begin{bmatrix} x \\ y \end{bmatrix} \qquad \mathbf{B} = \begin{bmatrix} 70 \\ 1180 \end{bmatrix}$$

then the system of equations can be represented in matrix form by

$$\mathbf{A} \cdot \mathbf{X} = \mathbf{B}$$

If $\mathbf{A}$, the matrix of coefficients, is invertible, then the solution to the system of equations is given by

$$\mathbf{X} = \mathbf{A}^{-1} \cdot \mathbf{B}$$

Make use of Exercise 14 to find $\mathbf{A}^{-1}$, and use this approach to solve the system of equations.

16. Solve the system of equations

$$x + 2y = -4$$
$$x + y = 5$$

using the method of Exercise 15.

17. The *transpose* of a matrix $\mathbf{A}$, $\mathbf{A}^T$, is obtained by interchanging its rows and columns. Thus, if we denote the element in row i, column j of $\mathbf{A}$ by $\mathbf{A}(i, j)$, then $\mathbf{A}^T(i, j) = \mathbf{A}(j, i)$.

a. Find $\mathbf{A}^T$ for

$$\mathbf{A} = \begin{bmatrix} 1 & 3 & 4 \\ 6 & -2 & 1 \end{bmatrix}$$

b. Prove that if $\mathbf{A}$ is a square matrix, then $\mathbf{A}$ is symmetric if and only if $\mathbf{A}^T = \mathbf{A}$.
c. Prove that $(\mathbf{A}^T)^T = \mathbf{A}$.
d. Prove that $(\mathbf{A} + \mathbf{B})^T = \mathbf{A}^T + \mathbf{B}^T$.
e. Prove that $(\mathbf{A} \cdot \mathbf{B})^T = \mathbf{B}^T \cdot \mathbf{A}^T$.

★18. Find two 2×2 matrices $\mathbf{A}$ and $\mathbf{B}$ such that $\mathbf{A} \cdot \mathbf{B} = \mathbf{0}$ but $\mathbf{A} \neq \mathbf{0}$ and $\mathbf{B} \neq \mathbf{0}$.

19. Find three 2×2 matrices $\mathbf{A}$, $\mathbf{B}$, and $\mathbf{C}$ such that $\mathbf{A} \cdot \mathbf{C} = \mathbf{B} \cdot \mathbf{C}$ but $\mathbf{A} \neq \mathbf{B}$.

20. If $\mathbf{A}$ and $\mathbf{B}$ are $n \times n$ matrices, is it always true that $(\mathbf{A} + \mathbf{B})^2 = \mathbf{A}^2 + 2(\mathbf{A} \cdot \mathbf{B}) + \mathbf{B}^2$? Will it ever be true?

21. The vector of real numbers $\mathbf{U} = [u_1 \; u_2]$ can be visualized on the real-number plane as an arrow from the origin to the point (u_1, u_2). The length of the arrow, also called the *magnitude* of the vector, is given by $\|\mathbf{U}\| = \sqrt{u_1^2 + u_2^2}$. The *dot product* of two such vectors, $\mathbf{U} \bullet \mathbf{V}$, is defined to be the real number $u_1 v_1 + u_2 v_2$. Show that if θ is the angle between $\mathbf{U}$ and $\mathbf{V}$, $0 \leq \theta \leq \pi$, then

$$\cos\theta = \frac{\mathbf{U} \bullet \mathbf{V}}{\|\mathbf{U}\| \cdot \|\mathbf{V}\|}$$

(*Hint:* Use the Law of Cosines.)

★22. For Boolean matrices

$$\mathbf{A} = \begin{bmatrix} 1 & 0 & 0 \\ 1 & 1 & 0 \\ 0 & 1 & 1 \end{bmatrix} \qquad \mathbf{B} = \begin{bmatrix} 1 & 0 & 1 \\ 0 & 1 & 1 \\ 1 & 1 & 1 \end{bmatrix}$$

find $\mathbf{A} \wedge \mathbf{B}$, $\mathbf{A} \vee \mathbf{B}$, $\mathbf{A} \times \mathbf{B}$, and $\mathbf{B} \times \mathbf{A}$.

23. For Boolean matrices $\mathbf{A}$ and $\mathbf{B}$, can it ever be the case that $\mathbf{A} \vee \mathbf{B} = \mathbf{A} \wedge \mathbf{B}$? If so, when?

24. For Boolean matrices $\mathbf{A}$ and $\mathbf{B}$, prove that $\mathbf{A} \vee \mathbf{B} = \mathbf{B} \vee \mathbf{A}$ and that $\mathbf{A} \wedge \mathbf{B} = \mathbf{B} \wedge \mathbf{A}$.

25. How many distinct symmetric $n \times n$ Boolean matrices are there?

★26. Prove that if a square matrix $\mathbf{A}$ of numbers is symmetric, then so is $\mathbf{A}^2$, where $\mathbf{A}^2 = \mathbf{A} \cdot \mathbf{A}$.

27. Prove that $\mathbf{A} \cdot \mathbf{A}^T$ is symmetric for any matrix $\mathbf{A}$ (see Exercise 17).

28. Let

$$\mathbf{A} = \begin{bmatrix} 1 & 1 \\ 1 & 0 \end{bmatrix}$$

For $n \geq 1$, let $F(n)$ equal the nth value in the Fibonacci sequence (see Example 30 in Chapter 2); let $F(0) = 0$. Prove that for any $n \geq 1$, $\mathbf{A}^n$ is given by

$$\begin{bmatrix} F(n+1) & F(n) \\ F(n) & F(n+1) \end{bmatrix}$$

Chapter 4 Review

Terminology

add to a database (p. 274)
addition modulo n
 (p. 245)
addition of matrices
 (p. 312)
antisymmetric relation
 (p. 236)
array (p. 311)
big oh (p. 300)
bijection (p. 285)
binary relation from S to
 T (p. 233)

binary relation on a set S
 (p. 232)
blind key (p. 268)
block (p. 242)
Boolean and (logical
 and) (p. 316)
Boolean matrix (p. 316)
Boolean matrix multipli-
 cation (p. 316)
Boolean or (logical or)
 (p. 316)
Cantor's theorem (p. 295)

cardinality of a set
 (p. 294)
ceiling function
 (p. 282)
chain (p. 240)
closure of a relation
 (p. 237)
codomain (p. 279)
commutative diagram
 (p. 286)
composition function
 (p. 286)

Self-Test   Answer the following true–false questions.

Section 4.1

1. In a one-to-many binary relation, at least one first component must appear in two different ordered pairs.

2. If an antisymmetric binary relation contains (x, y), then (y, x) will not belong to the relation.

3. A least element of a partially ordered set precedes all elements except itself.

4. An equivalence relation cannot also be a partial ordering.

5. A partial ordering on a set determines a partition of that set.

Section 4.2

6. If a task is not on the critical path in a PERT chart, then that task is optional.

7. A topological sort turns a partially ordered set into a totally ordered set.

8. If x precedes y after a topological sort on a finite partially ordered set, then x preceded y in the original partial ordering.

9. The times to complete parallel tasks are added together in determining a critical path in a PERT chart.

10. A given set of data results in a unique topological sort.

Section 4.3

11. A relation in a relational database is a set of n-tuples of attribute values.

12. A primary key in a relation is a minimum subset of attribute values that will uniquely identify each tuple.

13. The restrict operation can be achieved by doing a union followed by an intersection.

14. The join operation can be achieved by doing a Cartesian product followed by a restrict.

15. Deleting a tuple from a relation may result in additional deletions being done in order to satisfy data integrity.

Section 4.4

16. A binary relation on $S \times T$ that is not one-to-many or many-to-many is a function from S to T.

17. To prove that a function is onto, begin with an arbitrary element of the range and show that it has a preimage.

18. To show that a function is one-to-one, assume $f(s_1) = f(s_2)$ for some s_1, and s_2 in the domain and show that $s_1 = s_2$.

19. The composition of two permutation functions on a set is a permutation function on the set.

20. If f is $\Theta(g)$, then beyond some point the values for $f(x)$ must fall between $\frac{1}{2}g(x)$ and $2g(x)$.

Section 4.5

21. Two matrices that do not have the same dimensions cannot be added.

22. If **A** and **B** are square matrices, then $\mathbf{A} \cdot \mathbf{B} = \mathbf{B} \cdot \mathbf{A}$.

23. If **0** denotes the $n \times n$ zero matrix, then $\mathbf{A} \cdot \mathbf{0} = \mathbf{0} \cdot \mathbf{A}$ for every $n \times n$ matrix **A**.

24. The usual algorithm for matrix multiplication is $\Theta(n^3)$.

25. If **A** and **B** are square Boolean matrices, then $\mathbf{A} \times \mathbf{B} = \mathbf{B} \times \mathbf{A}$, where $\mathbf{A} \times \mathbf{B}$ denotes the Boolean product.

On the Computer

For Exercises 1–16, write a computer program that produces the desired output from the given input.

1. *Input*: The elements in a finite set S and a list of ordered pairs representing a binary relation on S
   Output: Statement indicating whether the relation is one-to-one, one-to-many, many-to-one, or many-to-many

2. *Input*: The elements in a finite set S and two lists of ordered pairs representing two binary relations on S
   Output: The ordered pairs in the union and in the intersection of the two relations, and the ordered pairs in the complements of each relation

3. *Input*: The elements in a finite set S and a list of ordered pairs representing a binary relation on S
   Output: Statement of which properties—reflexive, symmetric, transitive, and/or antisymmetric—the relation has

4. *Input*: The elements in a finite set S and a list of ordered pairs representing a binary relation on S
   Output: Reflexive, symmetric, and transitive closures of the relation

5. *Input*: The elements in a finite set S and a list of ordered pairs representing a partial ordering on S
   Output: A list of all minimal and maximal elements

6. *Input*: The elements in a finite set S and a list of ordered pairs representing a partial ordering on S
   Output: A list of any least or greatest elements
   Note that this task is more difficult than that in Exercise 5.

7. *Input*: The elements in a finite set S, a list of ordered pairs representing an equivalence relation on S, and an element x of S
   Output: The members of $[x]$

8. *Input*: Array representations of relation tables and appropriate input data for restrict, project, and join operations
   Output: Array representations of the resulting relation tables

9. *Input*: The elements in a finite set S and a list of ordered pairs representing a partial ordering on S
   Output: Sequence representing the total ordering that results from doing a topological sort
   Hint: Reuse some of your code from Exercise 5.

10. *Input*: The elements in a finite set S and in a finite set T, and a list of ordered pairs representing a binary relation on $S \times T$
    Output: An indication of whether the relation is a function from S to T and if so, whether it is onto or one-to-one or both

11. *Input*: The number of elements in two finite sets S and T
    Output: The number of functions from S to T, the number of one-to-one functions from S to T (or an indication that none exist), and the number of onto functions from S to T (or an indication that none exist)

12. *Input*: Two lists of ordered pairs representing functions f and g from S to S
    Output: List of ordered pairs representing the composition function $g \circ f$

13. *Input*: The elements in a finite set S and two lists that represent (in cycle form) permutations f and g on S
    Output: One or more lists that represent the composition function $g \circ f$ in cycle or product-of-cycle form

14. *Input*: The number of elements in a finite set S
    Output: The number of derangements on S

15. *Input*: n and the entries in two $n \times n$ matrices $\mathbf{A}$ and $\mathbf{B}$
    Output: Sum $\mathbf{A} + \mathbf{B}$ and products $\mathbf{A} \cdot \mathbf{B}$ and $\mathbf{B} \cdot \mathbf{A}$

16. *Input*: Dimensions of a matrix $\mathbf{A}$ and the entries in $\mathbf{A}$
    Output: $\mathbf{A}^T$ (see Exercise 17 in Section 4.4)

17. The *determinant* of an $n \times n$ matrix can be used in solving systems of linear equations, as well as for other purposes. The determinant of **A** can be defined in terms of minors and cofactors. The *minor* of element a_{ij} is the determinant of the $(n-1) \times (n-1)$ matrix obtained from **A** by crossing out the elements in row i and column j; denote this minor by M_{ij}. The *cofactor* of element a_{ij}, denoted by C_{ij}, is defined by

$$\mathbf{C}_{ij} = (-1)^{i+j} M_{ij}$$

The determinant of **A** is computed by multiplying all the elements in some fixed row of **A** by their respective cofactors and summing the results. For example, if the first row is used, then the determinant of **A** is given by

$$\sum_{k=1}^{n} (a_{1k})(C_{1k})$$

Write a program that, when given n and the entries in an $n \times n$ array **A** as input, computes the determinant of **A**. Use a recursive algorithm.

5 Graphs and Trees

Chapter Objectives

After studying this chapter, you will be able to:

- Understand and use the many terms associated with graphs, directed graphs, and trees.
- Appreciate the use of graphs, directed graphs, and trees as representation tools in a wide variety of contexts.
- Prove that two given graphs are isomorphic or give a reason why they are not.
- Use Euler's formula for a simple, connected, planar graph.
- Understand the role of the two specific graphs K_5 and $K_{3,3}$ in graph planarity.
- Prove elementary properties about graphs and trees.
- Use adjacency matrix and adjacency list representations for graphs and directed graphs.
- Do preorder, inorder, and postorder tree traversal.
- Use array and pointer representations for binary trees.
- Use decision trees to represent the steps a searching or sorting algorithm carries out.
- Build a binary search tree and conduct a binary tree search.
- Express the worst-case number of comparisons for searching or sorting on a list with n elements.
- Find Huffman codes for characters whose frequency of occurrence is given.

You work in the Information Systems Department at World Wide Widgets (WWW), the leading widget manufacturer. Widgets are extremely complex devices made up of an enormous number of very simple parts. Each part is one of the following types: Bolt (B), Component (C), Gear (G), Rod (R), or Screw (S). There are many different variations of each basic type. Part numbers are identified by a leading character of B, C, G, R, or S to identify the part type, followed by an 8-digit number. Thus

C00347289
B11872432
S45003781

are all legitimate part numbers. Using the Multiplication Principle, there are 5×10^8 different potential part numbers! WWW maintains a data file of the part numbers it uses, which, as it turns out, is most of the potential numbers. Most computers, including those at WWW, use the ASCII encoding scheme for converting characters into binary form, under which each character requires one byte (eight bits) of storage. Because each different part number consists of 9 characters, the WWW parts data file is approximately $9 \times 5 \times 10^8$ bytes, or 4.5 Gb.

Question: How can you compress this data file so it takes less storage space?

One answer to this question involves working with binary tree structures. A tree is a visual representation of data items and the connections between some of these items. It is a special case of a more general structure called a graph. Graphs or trees can be used to represent a surprising number of real-world situations—organization charts, road maps, transportation and communications networks, and so forth. Later we shall see other uses of graphs and trees to represent logic networks, finite-state machines, and formal-language derivations.

Graph theory is an extensive topic. Sections 5.1 and 5.2 present some of the considerable terminology connected with graphs and trees and some elementary results about these structures. To represent a graph or a tree in computer memory, data must be arranged in a way that preserves all the information contained in the visual representation. Several approaches to representing graphs and trees within a computer are discussed.

Decision trees are graphical representations of the activities of certain types of algorithms. In Section 5.3, decision trees are presented and used to find lower bounds on the worst-case behavior of searching and sorting algorithms. In Section 5.4, an algorithm is given for constructing binary trees that allow for data compression of large files.

Section 5.1   Graphs and Their Representations

Definitions of a Graph

One way to while away the hours on an airplane trip is to look at the literature in the seat pockets. This material almost always includes a map showing the routes of the airline you are flying, such as in Figure 5.1. All of this route information could be expressed in paragraph form; for example, there is a direct route between Chicago and Nashville but not between St. Louis and Nashville. However, the paragraph would be rather long and involved, and we would not be able to assimilate the information as quickly and clearly as we can from the map. There are many cases where "a picture is worth a thousand words."

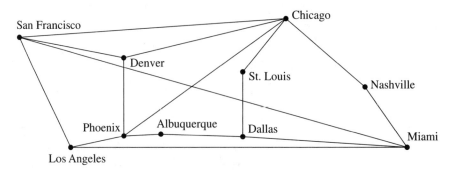

Figure 5.1

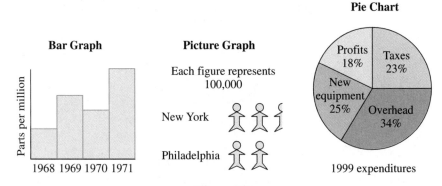

Figure 5.2

The term graph is often used informally for any visual representation of data, such as that in Figure 5.1; other forms include the bar graph, picture graph, and pie chart, which are shown in Figure 5.2. We have also talked about graphs of functions on rectangular coordinate systems. We will use two definitions of a graph; one relies on a visual representation like that of Figure 5.1, and the other is a more formal definition that actually says nothing about a visual representation.

Definition (Informal): Graph
A **graph** is a nonempty set of **nodes** (**vertices**) and a set of **arcs** (**edges**) such that each arc connects two nodes.

Our graphs will always have a finite number of nodes and arcs.

EXAMPLE 1   The set of nodes in the airline map of Figure 5.1 is {Chicago, Nashville, Miami, Dallas, St. Louis, Albuquerque, Phoenix, Denver, San Francisco, Los Angeles}. There are 16 arcs; Phoenix–Albuquerque is an arc (here we are naming an arc by the nodes it connects), Albuquerque–Dallas is an arc, and so on.

EXAMPLE 2   In the graph of Figure 5.3, there are five nodes and six arcs. Arc a_1 connects nodes 1 and 2, arc a_3 connects node 2 and 2, and so forth.

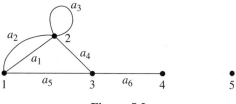

Figure 5.3

The informal definition of a graph works quite well if we have the visual representation of the graph before us to show which arcs connect which nodes. Without the picture, however, we need a concise way to convey this information. This leads to our second definition of a graph.

> **Definition (Formal): Graph**
> A **graph** is an ordered triple (N, A, g) where
>
> N = a nonempty set of **nodes** (**vertices**)
> A = a set of **arcs** (**edges**)
> g = a function associating with each arc a an *unordered* pair $x–y$ of nodes called the **endpoints** of a

EXAMPLE 3   For the graph of Figure 5.3, the function g associating arcs with endpoints performs the following mapping: $g(a_1) = 1–2$, $g(a_2) = 1–2$, $g(a_3) = 2–2$, $g(a_4) = 2–3$, $g(a_5) = 1–3$, and $g(a_6) = 3–4$.

PRACTICE 1   Sketch a graph having nodes $\{1, 2, 3, 4, 5\}$, arcs $\{a_1, a_2, a_3, a_4, a_5, a_6\}$, and function $g(a_1) = 1–2$, $g(a_2) = 1–3$, $g(a_3) = 3–4$, $g(a_4) = 3–4$, $g(a_5) = 4–5$, and $g(a_6) = 5–5$.

We might want the arcs of a graph to begin at one node and end at another, in which case we would use a *directed graph*.

> **Definition: Directed Graph**
> A **directed graph (digraph)** is an ordered triple (N, A, g) where
>
> N = a nonempty set of nodes
> A = a set of arcs
> g = a function associating with each arc a an *ordered* pair (x, y) of nodes where x is the **initial point** and y is the **terminal point** of a

In a directed graph, then, there is a direction associated with each arc.

EXAMPLE 4   Figure 5.4 shows a directed graph. There are 4 nodes and 5 arcs. The function g associating arcs with endpoints performs the mapping $g(a_1) = (1, 2)$, meaning that arc a_1 begins at node 1 and ends at node 2. Also, $g(a_3) = (1, 3)$, but $g(a_4) = (3, 1)$.

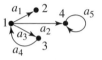

Figure 5.4   ●

Besides imposing direction on the arcs of a graph, we may want to modify the basic definition of a graph in other ways. We often want the nodes of a graph to carry identifying information, like the names of the cities in the map of airline routes. This would be a **labeled graph**. We may want to use a **weighted graph**, where each arc has some numerical value, or weight, associated with it. For example, we might want to indicate the distances of the various routes in the airline map.

In this book, the term "graph" will mean an undirected graph. To refer to a directed graph, we will always say "directed graph."

Applications of Graphs

Although the idea of a graph is very simple, an amazing number of situations have relationships between items that lend themselves to graphical representation. Not surprisingly, there are many graphs in this book. Graphical representations of partially ordered sets (Hasse diagrams) were introduced in Chapter 4. A PERT chart (e.g., Figure 4.7) is a directed graph. The E-R diagram (e.g., Figure 4.10) is a graph. The commutative diagram illustrating composition of functions (see Figure 4.22), is a directed graph. Chapter 7 will introduce logic networks and represent them as directed graphs. Directed graphs will also be used to describe finite-state machines in Chapter 8.

We saw that the airline route map was a graph. A representation of any network of transportation routes (a road map, for example), communications lines (as in a computer network), or product or service distribution routes such as natural gas pipelines or water mains is a graph. The chemical structure of a molecule is represented graphically.

PRACTICE 2   Draw the underlying graph in each of the following cases.

　　a. Figure 5.5 is a road map for part of Arizona.
　　b. Figure 5.6 is a representation of an ozone molecule with three oxygen atoms.

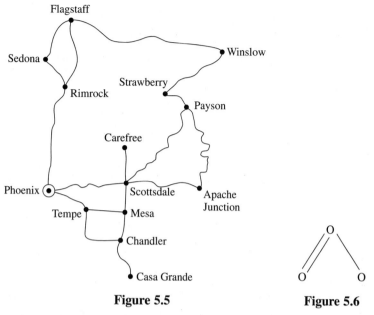

Figure 5.5

Figure 5.6

EXAMPLE 5   A high-level view of the information flow in a state automobile licensing office is prepared as the first step in developing a new computerized licensing system. Figure 5.7 shows the resulting directed graph, often called a **data flow diagram**.

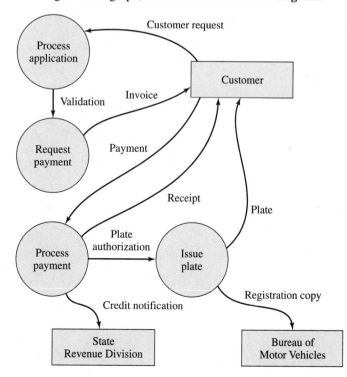

Figure 5.7

EXAMPLE 6   Figure 5.8 shows a graph representation of a local area network of computers in an office complex. In this "star topology," all machines communicate through a central server. The graph representation highlights one of the weaknesses of such a network design, namely its reliance on continued, dependable operation of the central server.

Figure 5.8

EXAMPLE 7   Neural networks, tools used in artificial intelligence for such tasks as pattern recognition, are represented by weighted directed graphs. Figure 5.9 shows a multilayer network consisting of input units, output units, and a "hidden layer" of units. Weights on the arcs of the graph are adjusted as the neural network "learns" how to recognize certain trial patterns.

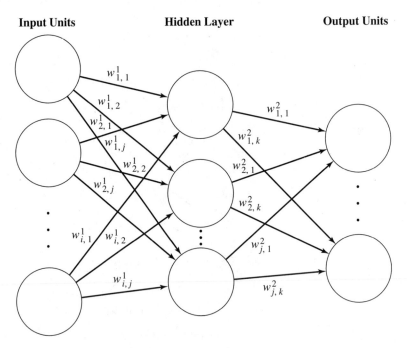

Figure 5.9

Graph Terminology

Before proceeding, we need some terminology about graphs. Surprisingly, although there is a large body of literature in graph theory, the terminology is not completely standard. Therefore other books may give slightly different variations of some of these terms.

Two nodes in a graph are **adjacent** if they are the endpoints associated with an arc. For example in the graph of Figure 5.3 (reproduced here), 1 and 3 are adjacent nodes but 1 and 4 are not. Node 2 is adjacent to itself. A **loop** in a graph is an arc with endpoints n–n for some node n; in Figure 5.3, arc a_3 is a loop with endpoints 2–2. A graph with no loops is **loop-free**. Two arcs with the same endpoints are **parallel arcs**; arcs a_1 and a_2 in Figure 5.3 are parallel. A **simple graph** is one with no loops or parallel arcs. An **isolated node** is adjacent to no other node; in Figure 5.3, 5 is an isolated node. The **degree** of a node is the number of arc ends at that node. In Figure 5.3, nodes 1 and 3 have degree 3, node 2 has degree 5, node 4 has degree 1, and node 5 has degree 0.

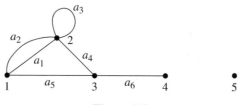

Figure 5.3

Because the function g that relates arcs to endpoints in the formal definition of a graph is indeed a function, each arc has a unique pair of endpoints. If g is a one-to-one function, then there is at most one arc associated with a pair of endpoints; such graphs have no parallel arcs. A **complete graph** is one in which any two distinct nodes are adjacent. In this case, g is almost an onto function—every pair x–y of distinct nodes is the image under g of an arc—but there does not have to be a loop at every node. Consequently, pairs of the form x–x need not have a preimage.

A **subgraph** of a graph consists of a set of nodes and a set of arcs that are subsets of the original node set and arc set, respectively, in which the endpoints of an arc must be the same nodes as in the original graph. In other words, it is a graph obtained by erasing part of the original graph and leaving the rest unchanged. Figure 5.10 shows two subgraphs of the graph in Figure 5.3. Note that the graph in Figure 5.10a is simple and also complete.

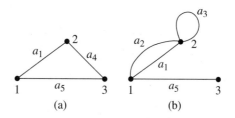

Figure 5.10

A **path** from node n_0 to node n_k is a sequence

$$n_0, a_0, n_1, a_1, \ldots, n_{k-1}, a_{k-1}, n_k$$

of nodes and arcs where for each i, the endpoints of arc a_i are n_i–n_{i+1}. In the graph of Figure 5.3, one path from node 2 to node 4 consists of the sequence 2, a_1, 1, a_2, 2, a_4, 3, a_6, 4. The **length** of a path is the number of arcs it contains; if an arc is used more than once, it is counted each time it is used. The length of the path just described from node 2 to node 4 is 4.

A graph is **connected** if there is a path from any node to any other node. The graphs in Figure 5.10a and Figure 5.10b are each connected, but the graph of Figure 5.3 is not connected. A **cycle** in a graph is a path from some node n_0 back to n_0 where no arc appears more than once in the path sequence, n_0 is the only node appearing more than once, and n_0 occurs only at the ends. (Nodes and arcs may be repeated in a path but not, except for node n_0, in a cycle.) In the graph of Figure 5.3,

$$1, a_1, 2, a_4, 3, a_5, 1$$

is a cycle. A graph with no cycles is **acyclic**.

PRACTICE 3   Refer to the graph created in Practice 1.

  a. Find two nodes that are not adjacent.
  b. Find a node adjacent to itself.
  c. Find a loop.
  d. Find two parallel arcs.
  e. Find the degree of node 3.
  f. Find a path of length 5.
  g. Find a cycle.
  h. Is this graph complete?
  i. Is this graph connected?

EXAMPLE 8   Figure 5.11 illustrates the simple, complete graphs with 1, 2, 3, and 4 vertices. The simple, complete graph with n vertices is denoted by K_n.

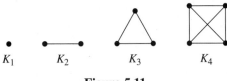

Figure 5.11

PRACTICE 4   Draw K_5.

Now consider the simple graph in Figure 5.12. It is not a complete graph because it is not true that every node is adjacent to every other node. However, the nodes can be divided into two disjoint sets, $\{1, 2\}$ and $\{3, 4, 5\}$, such that any two nodes chosen from the same set are not adjacent but any two nodes chosen one from each set are adjacent. Such a graph is a *bipartite complete graph*.

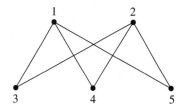

Figure 5.12

Definition: Bipartite Complete Graph
A graph is a **bipartite complete graph** if its nodes can be partitioned into two disjoint nonempty sets N_1 and N_2 such that two nodes x and y are adjacent if and only if $x \in N_1$ and $y \in N_2$. If $|N_1| = m$ and $|N_2| = n$, such a graph is denoted by $K_{m,n}$.

Figure 5.12 therefore illustrates $K_{2,3}$.

PRACTICE 5   Draw $K_{3,3}$.   ●

The concept of a path extends to a directed graph as we might expect: A **path** from node n_0 to node n_k, in a directed graph is a sequence

$$n_0, a_0, n_1, a_1, \ldots, n_{k-1}, a_{k-1}, n_k$$

where for each i, n_i is the initial point and n_{i+1} is the terminal point of a_i. If a path exists from node n_0 to node n_k, then n_k is **reachable** from n_0. The definition of a cycle also carries over to directed graphs.

EXAMPLE 9   In the directed graph of Figure 5.13, there are many paths from node 1 to node 3: 1, a_4, 3 and 1, a_1, 2, a_2, 2, a_2, 2, a_3, 3 are two possibilities. Node 3 is certainly reachable from 1. Node 1, however, is not reachable from any other node. The cycles in this graph are the loop a_2 and the path 3, a_5, 4, a_6, 3.

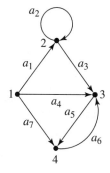

Figure 5.13   ●

We can prove some (fairly trivial) statements about graphs that follow directly from the definitions.

EXAMPLE 10   Prove that an acyclic graph is simple.

We'll use a proof by contraposition. If a graph is not simple, it has either parallel arcs or a loop. The parallel arcs and their endpoints, or the loop and its endpoints, then constitute a cycle, and the graph is not acyclic.   ●

Note that the converse to the statement in Example 10 is not true: Figure 5.10a is a simple graph, but it contains a cycle.

PRACTICE 6   **a.** Prove that every complete graph is connected.
b. Find a connected graph that is not complete.   ●

Isomorphic Graphs

Two graphs may appear quite different in their visual representation but still be the same graph according to our formal definition. The graphs in Figures 5.14 and 5.15 are the same—they have the same nodes, the same arcs, and the same arc-to-endpoint function. (In a representation of a graph, arcs can intersect at points that are not nodes of the graph.) The graph in Figure 5.16 is essentially the same graph as well. If we relabeled the nodes and arcs of the graph of Figure 5.14 by the following mappings, the graphs would be the same:

$$f_1: 1 \rightarrow a \qquad f_2: a_1 \rightarrow e_2$$
$$2 \rightarrow c \qquad\qquad a_2 \rightarrow e_1$$
$$3 \rightarrow b$$
$$4 \rightarrow d$$

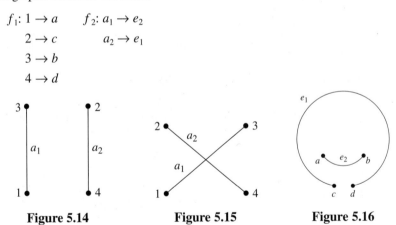

Figure 5.14   **Figure 5.15**   **Figure 5.16**

Structures that are the same except for relabeling are called *isomorphic* structures. To show that two structures are isomorphic, we must produce a relabeling (one-to-one, onto mappings between the elements of the two structures) and then show that the important properties of the structures are "preserved" (maintained) under this relabeling. In the case of graphs, the elements are nodes and arcs. The "important property" in a graph is which arcs connect which nodes.

The given mappings f_1 and f_2 are one-to-one, onto functions from the nodes and arcs, respectively, of the graph in Figure 5.14 to the nodes and arcs of the graph in Figure 5.16. Furthermore, if an arc a in the graph of Figure 5.14 has endpoints x–y, then the arc $f_2(a)$ in the graph of Figure 5.16 has endpoints $f_1(x)$–$f_1(y)$, and vice versa.

For example, arc a_1 in Figure 5.14 has endpoints 1–3, while its corresponding arc e_2 in Figure 5.16 has endpoints a–b, which are the nodes in Figure 5.16 that correspond to nodes 1 and 3 in Figure 5.14. We can formalize this idea.

Definition: Isomorphic Graphs

Two graphs $(N_1, A_1, g_1,)$ and (N_2, A_2, g_2) are **isomorphic** if there are bijections $f_1: N_1 \to N_2$ and $f_2: A_1 \to A_2$ such that for each arc $a \in A_1$, $g_1(a) = x$–y if and only if $g_2[f_2(a)] = f_1(x)$–$f_1(y)$.

EXAMPLE 11

The graphs shown in Figure 5.17 are isomorphic. The bijections that establish the isomorphism are partially given here:

$$f_1: 1 \to c \qquad f_2: a_1 \to e_1$$
$$2 \to e \qquad\quad a_2 \to e_4$$
$$3 \to d \qquad\quad a_3 \to e_2$$
$$4 \to b \qquad\quad \vdots$$
$$5 \to a$$

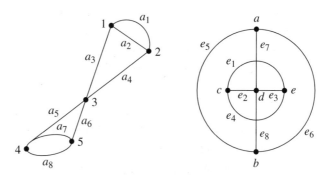

Figure 5.17

Using these bijections, $g_1(a_3) = 1$–3 and $g_2[f_2(a_3)] = g_2(e_2) = c$–$d = f_1(1)$–$f_1(3)$. This shows that the arc-to-endpoint relationship is preserved under the relabeling for the case of arc a_3. To prove that the graphs are isomorphic, we would have to complete the definition of the f_2 function and then demonstrate that the arc-to-endpoint relationship is preserved under these mappings by examining all possible cases. ●

PRACTICE 7

Complete the definition of the function f_2 in Example 11. ●

Graph isomorphism is easier to establish if we restrict our attention to simple graphs. If we can find an appropriate function f_1 mapping nodes to nodes, then a function f_2 mapping arcs to arcs is trivial because there is at most one arc between any pair of endpoints. Hence the following theorem is true.

Theorem on Simple Graph Isomorphism
Two simple graphs (N_1, A_1, g_1) and (N_2, A_2, g_2) are isomorphic if there is a bijection $f: N_1 \rightarrow N_2$ such that for any nodes n_i and n_j of N_1, n_i and n_j are adjacent if and only if $f(n_i)$ and $f(n_j)$ are adjacent. (The function f is called an **isomorphism** from graph 1 to graph 2.)

PRACTICE 8   Find an isomorphism from the graph of Figure 5.18a to that of Figure 5.18b.

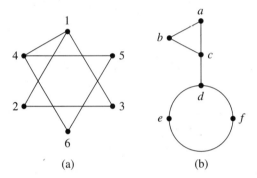

(a)   (b)

Figure 5.18

Proving that two graphs are isomorphic requires finding the bijection (or, for non-simple graphs, bijections) and then showing that the adjacency property (or arc-to-end-point relationship) is preserved. To prove that two graphs are not isomorphic, we must prove that the necessary bijection(s) do not exist. We could try all possible bijections (because there is a finite number of nodes and arcs, there is a finite number of bijections). However, this method would quickly get out of hand in graphs of any size at all. Instead, we can try to find some other reason that such bijections could not exist. Although this is not always an easy task, there are certain conditions under which it is clear that two graphs are not isomorphic (see Exercise 15). These include the following:

1. One graph has more nodes than the other.
2. One graph has more arcs than the other.
3. One graph has parallel arcs and the other does not.
4. One graph has a loop and the other does not.
5. One graph has a node of degree k and the other does not.
6. One graph is connected and the other is not.
7. One graph has a cycle and the other does not.

PRACTICE 9   Prove that the two graphs in Figure 5.19 are not isomorphic.

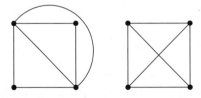

Figure 5.19

EXAMPLE 12 The two graphs of Figure 5.20 are not isomorphic. Note that each graph has six nodes and seven arcs. Neither has parallel arcs or loops. Both are connected. Both have three cycles, four nodes of degree 2, and two nodes of degree 3. Therefore none of the obvious nonisomorphism tests apply. However, the graph in Figure 5.20b has a node of degree 2 that is adjacent to two nodes of degree 3; this is not the case in Figure 5.20a, so the graphs are not isomorphic.

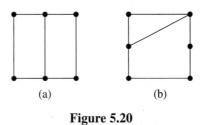

(a)　　　　(b)

Figure 5.20

Again, graphs that are isomorphic are considered to be "the same" regardless of cosmetic differences in how they are drawn or labeled, whereas nonisomorphic graphs have fundamental structural differences.

Planar Graphs

A **planar graph** is one that can be represented (on a sheet of paper, that is, in the plane) so that its arcs intersect only at nodes. The graph of Figure 5.14 is clearly planar. However, we know that it is isomorphic to the graph of Figure 5.15, so the graph of Figure 5.15 is also planar. The key word in the definition of a planar graph is that it *can* be drawn in a certain way.

PRACTICE 10 Prove that K_4 is a planar graph.

EXAMPLE 13 Consider K_5, the simple, complete graph with five vertices. We shall try to construct K_5 with no intersecting arcs by starting with some of the arcs and then adding as many new arcs as possible without crossing existing arcs. We'll first lay out five vertices and connect them as shown in Figure 5.21a. (Because all the vertices in K_n are symmetric, it doesn't matter how we label them.)

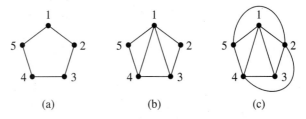

(a)　　　　(b)　　　　(c)

Figure 5.21

Next we connect 1 to 3 and 1 to 4, as shown in Figure 5.21b. Now 2 must be connected to both 4 and 5. This can be accomplished while still preserving the planarity of the graph by putting these new arcs on the outside, as in Figure 5.21c. The final connection is between nodes 3 and 5. But there is no way to draw an arc from node 3 to node 5 without crossing either the 2–4 arc or one or more of the interior arcs, such as 1–4.

We did have a choice of how to place arcs 1–3 and 1–4; we made them interior arcs. We could explore whether making these arcs exterior would change anything, but it turns out that it does not (see Practice 11). Thus it appears that K_5 is not a planar graph. However, we'd still like a proof of this with a firmer foundation—this sounds too much like an "I can't do it so it can't be done" argument. Such a proof will be given shortly.   ●

PRACTICE 11   Show that adding arcs 1–3 and 1–4 as exterior arcs when constructing K_5 still leads to a situation where arcs must intersect.   ●

PRACTICE 12   Present a construction-type argument that $K_{3,3}$ is not a planar graph.   ●

One fact about planar graphs was discovered by the eighteenth-century Swiss mathematician Leonhard Euler (pronounced "oiler"). A simple, connected, planar graph (when drawn in its planar representation, with no arcs crossing) divides the plane into a number of regions, including totally enclosed regions and one infinite exterior region. Euler observed a relationship between the number n of nodes, the number a of arcs, and the number r of regions in such a graph. This relationship is known as **Euler's formula**:

$$n - a + r = 2 \tag{1}$$

PRACTICE 13   Verify Euler's formula for the simple, connected, planar graph in Figure 5.18b.   ●

To prove Euler's formula, we shall do a proof by induction on a, the number of arcs. In the base case, $a = 0$ and the graph consists of a single node; the only region is the exterior region (Figure 5.22a). Here $n = 1$, $a = 0$, and $r = 1$, and equation (1) holds. Now assume that the formula holds for the planar representation of any simple, connected, planar graph with k arcs, and consider such a graph with $k + 1$ arcs. As usual, we must somehow relate the "$k + 1$ instance" to a "k instance" so that we can make use of the inductive hypothesis. Here we consider two cases for the graph with $k + 1$ arcs.

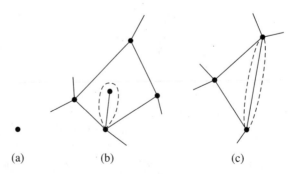

(a)   (b)   (c)

Figure 5.22

Case 1. The graph has a node of degree 1. Temporarily erase this node and its connecting arc (Figure 5.22b); this leaves a simple, connected, planar graph with k arcs, some number n of nodes, and some number r of regions for which (by the inductive hypothesis)

$$n - k + r = 2$$

In the original graph, there was one more arc and one more node but the same number of regions, so the appropriate formula is

$$(n + 1) - (k + 1) + r = 2$$

which, by the inductive hypothesis, is true.

Case 2. The graph has no nodes of degree 1. Then temporarily erase one arc that helps define an enclosed region (Figure 5.22c). (If no arcs help define an enclosed region, the graph is a chain and there is a node of degree 1.) This leaves a simple, connected, planar graph with k arcs, some number n of nodes, and some number r of regions for which (by the inductive hypothesis)

$$n - k + r = 2$$

In the original graph, there was one more arc and one more region, but the same number of nodes, so the appropriate formula is

$$n - (k + 1) + (r + 1) = 2$$

which, by the inductive hypothesis, is true.

PRACTICE 14   In the proof of Euler's formula, explain why in case 2 the arc to be erased must help define an enclosed region. Give two reasons.   ●

There are two consequences of Euler's formula if we place further restrictions on the graph. Suppose we require that the graph not only be simple, connected, and planar but also have at least three nodes. In a planar representation of such a graph, we can count the number of edges that are adjacent to (form the boundaries of) each region, including the exterior region. Arcs that are wholly interior to a region contribute two edges to that region; for example, if we trace the boundary of the interior region shown in Figure 5.22b, we travel six edges, including the arc out to the node of degree 1 and then back again. Arcs that separate two regions contribute one edge to each region. Therefore, if there are a arcs in the graph, the number of region edges is $2a$.

There are no regions with exactly one adjacent edge, because there are no loops in the graph. There are no regions with exactly two adjacent edges, because there are no parallel arcs and the graph consisting entirely of one arc joining two nodes (which would have two edges adjacent to the exterior region) is excluded. Therefore each region has at least three adjacent edges, so $3r$ is the minimum number of region edges. Thus

$$2a \geq 3r$$

or, from equation (1),

$$2a \geq 3(2 - n + a) = 6 - 3n + 3a$$

and finally

$$a \leq 3n - 6 \tag{2}$$

If a final restriction that there are no cycles of length 3 is placed on the graph, then each region has at least four adjacent edges, so $4r$ is the minimum number of region edges. This leads to the inequality

$$2a \geq 4r$$

which becomes

$$a \leq 2n - 4 \tag{3}$$

These results are summarized in the following theorem.

> **Theorem on the Number of Nodes and Arcs**
> For a simple, connected, planar graph with n nodes and a arcs:
>
> 1. If the planar representation divides the plane into r regions, then
>
> $$n - a + r = 2 \tag{1}$$
>
> 2. If $n \geq 3$, then
>
> $$a \leq 3n - 6 \tag{2}$$
>
> 3. If $n \geq 3$ and there are no cycles of length 3, then
>
> $$a \leq 2n - 4 \tag{3}$$

Note that inequality (3) places a tighter bound on the number of arcs than inequality (2), but an additional condition has been imposed on the graph.

We can use this theorem to prove that certain graphs are not planar.

EXAMPLE 14 K_5 is a simple, connected graph with 5 nodes (and 10 arcs). If it were a planar graph, inequality (2) of our theorem would hold, but $10 > 3(5) - 6$. Therefore, just as our construction argument showed, K_5 is not planar. $K_{3,3}$ is a simple, connected graph with 6 nodes (and 9 arcs). It has no cycles of length 3, since this would require two nodes in one of the two subsets to be adjacent. If it were a planar graph, inequality (3) would hold, but $9 > 2(6) - 4$. Therefore $K_{3,3}$ is not planar. ●

PRACTICE 15   Show that inequality (2) does hold for $K_{3,3}$, which shows that this inequality is a necessary but not sufficient condition for planarity in graphs with $n \geq 3$. ●

The nonplanar graphs K_5 and $K_{3,3}$ play a central role in all nonplanar graphs. To state what this role is, we need one more definition.

Definition: Homeomorphic Graphs

Two graphs are **homeomorphic** if both can be obtained from the same graph by a sequence of elementary subdivisions, in which a single arc x–y is replaced by two new arcs xv–vy connecting to a new node v.

EXAMPLE 15   The graphs in parts (b) and (c) of Figure 5.23 are homeomorphic because each can be obtained from the graph of Figure 5.23a by a sequence of elementary subdivisions. (However, neither can be obtained from the other by a sequence of elementary subdivisions.)

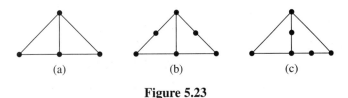

(a)   (b)   (c)

Figure 5.23

A graph that is planar cannot be turned into a nonplanar graph by elementary subdivisions, and a graph that is nonplanar cannot be turned into a planar graph by elementary subdivisions (see Exercise 26). As a result, homeomorphic graphs are either both planar or both nonplanar. The following theorem, due to the Polish mathematician Kuratowski, characterizes nonplanar graphs.

Kuratowski Theorem

A graph is nonplanar if and only if it contains a subgraph that is homeomorphic to K_5 or $K_{3,3}$.

We won't prove this theorem, although one direction is easy to see. If a graph has a subgraph homeomorphic to the nonplanar graphs K_5 or $K_{3,3}$, then the subgraph—and hence the entire graph—is nonplanar.

EXAMPLE 16   Figure 5.24a shows the "Petersen graph." We shall prove that this graph is not planar by finding a subgraph homeomorphic to $K_{3,3}$. By looking at the top of the graph, we can see that node a is adjacent to nodes $e, f,$ and b, none of which are adjacent to each other. Also, node e is adjacent to nodes d and j as well as a, and nodes $a, d,$ and j are not adjacent to each other. This information is incorporated in the graph of Figure 5.24b, which is also a subgraph of $K_{3,3}$. The arcs needed to complete $K_{3,3}$ are shown as dotted lines in Figure 5.23c. These arcs are not in the Petersen graph; for example, no *j–f* arc is present. However, there is a path in the Petersen graph from *j* to *f* using the intermediate node *h*, that is, *j–h* and *h–f*. Similarly, there are paths *j–g* and *g–b*, *d–i* and *i–f*, and *d–c* and *c–b*. Adding these paths to Figure 5.24b results in Figure 5.24d, which is a subgraph of the Petersen graph and is also obtainable from Figure 5.24c by a sequence of elementary subdivisions.

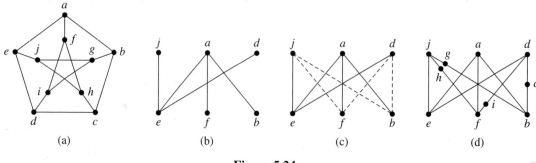

(a)   (b)   (c)   (d)

Figure 5.24

Computer Representation of Graphs

We have said that the major advantage of a graph is its visual representation of information. But for storage and manipulation within a computer, this information must be represented in other ways. The usual representations involve one of two data structures, either an adjacency matrix or an adjacency list.

Adjacency Matrix

Suppose a graph has *n* nodes, numbered $n_1, n_2, \ldots, n_n$. This numbering imposes an arbitrary ordering on the set of nodes; recall that a set is an unordered collection. However, this is done merely as a means to identify the nodes—no significance is attached to one node appearing before another in this ordering. Having ordered the nodes, we can form an $n \times n$ matrix where entry *i,j* is the number of arcs between nodes n_i and n_j. This matrix is called the **adjacency matrix A** of the graph with respect to this ordering. Thus,

$$a_{ij} = p \text{ where there are } p \text{ arcs between } n_i \text{ and } n_j$$

EXAMPLE 17   The adjacency matrix for the graph in Figure 5.25 with respect to the ordering 1, 2, 3, 4 is a 4×4 matrix.

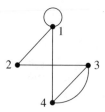

Figure 5.25

Entry 1,1 is a 1 due to the loop at node 1. All other elements on the main diagonal are 0. Entry 2,1 (second row, first column) is a 1 because there is one arc between node 2 and node 1, which also means that entry 1,2 is a 1. So far we have

$$A = \begin{bmatrix} 1 & 1 & - & - \\ 1 & 0 & - & - \\ - & - & 0 & - \\ - & - & - & 0 \end{bmatrix}$$

●

PRACTICE 16   Complete the adjacency matrix for Figure 5.25.   ●

The adjacency matrix in Practice 16 is symmetric, which will be true for the adjacency matrix of any undirected graph—if there are p arcs between n_i and n_j, there are certainly p arcs between n_j and n_i. The symmetry of the matrix means that only elements on or below the main diagonal need to be stored. Therefore, all the information contained in the graph in Figure 5.25 is contained in the "lower triangular" array shown, and the graph could be reconstructed from this array.

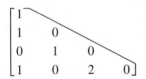

In a directed graph, the adjacency matrix **A** reflects the direction of the arcs. For a directed matrix,

$a_{ij} = p$ where there are p arcs from n_i to n_j

An adjacency matrix for a directed graph will not necessarily be symmetric, because an arc from n_i to n_j does not imply an arc from n_j to n_i.

EXAMPLE 18   Consider the directed graph of Figure 5.26.

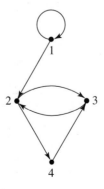

Figure 5.26

The adjacency matrix is

$$A = \begin{bmatrix} 1 & 1 & 0 & 0 \\ 0 & 0 & 1 & 1 \\ 0 & 1 & 0 & 0 \\ 0 & 0 & 1 & 0 \end{bmatrix}$$

In a simple weighted graph, the entries in the adjacency matrix can indicate the weight of an arc by the appropriate number rather than just indicating the presence of an arc by the number 1.

Adjacency List

Many graphs, far from being complete graphs, have relatively few arcs. Such graphs have **sparse** adjacency matrices; that is, the adjacency matrices contain many zeros. Yet if the graph has n nodes, it still requires n^2 data items to represent the adjacency matrix (or more than $n^2/2$ if a triangular matrix is used), even if many of these items are zero. Any algorithm or procedure in which every arc in the graph must be examined requires looking at all n^2 items in the matrix, since there is no way of knowing which entries are nonzero without examining them. To find all the nodes adjacent to a given node n_i requires scanning the entire ith row of the adjacency matrix, a total of n items.

A graph with relatively few arcs can be represented more efficiently by storing only the nonzero entries of the adjacency matrix. This representation consists of a list for each node of all the nodes adjacent to it. Pointers are used to get us from one item in the list to the next. Such an arrangement is called a **linked list**. There is an array of n pointers, one for each node, to get each list started. This **adjacency list** representation, although it requires extra storage for the pointers, may still be more efficient than an adjacency matrix. To find all the nodes adjacent to n_i requires traversing the linked list for n_i, which may have far fewer than the n elements we had to examine in the adjacency matrix. However, there are tradeoffs; if we want to determine whether one particular node n_j is adjacent to n_i, we may have to traverse all of n_i's linked list, whereas in the adjacency matrix we could access element i,j directly.

EXAMPLE 19   The adjacency list for the graph of Figure 5.25 contains a four-element array of pointers, one for each node. The pointer for each node points to an adjacent node, which points to another adjacent node, and so forth. The adjacency list structure is shown in Figure 5.27.

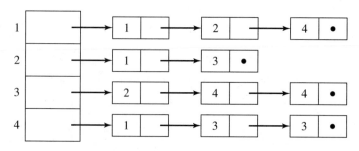

Figure 5.27

In the figure the dot indicates a **null pointer**, meaning that there is nothing more to be pointed to or that the end of the list has been reached. We have dealt with parallel arcs by listing a given node more than once on the adjacency list for n_i if there is more than one arc between n_i and that node.

PRACTICE 17   Draw the adjacency list representation for the graph shown in Figure 5.28.

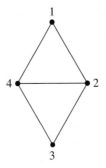

Figure 5.28

In an undirected graph, each arc is represented twice. If n_j is on the adjacency list of n_i, then n_i is also on the adjacency list of n_j. The adjacency list representation for a directed graph puts n_j on the list for n_i if there is an arc from n_i to n_j; n_i would not necessarily be on the adjacency list for n_j. For a labeled graph or a weighted graph, additional data items can be stored with the node name in the adjacency list.

EXAMPLE 20   Figure 5.29a shows a weighted directed graph. The adjacency list representation for this graph is shown in Figure 5.29b. For each record in the list, the first data item is the node, the second is the weight of the arc to that node, and the third is the pointer. Note that entry 4 in the array of startup pointers is null because there are no arcs that begin at node 4.

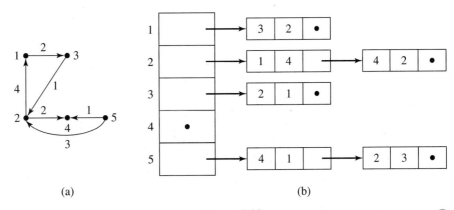

(a)   (b)

Figure 5.29

In a programming language that does not support pointers, we can still achieve the effect of an adjacency list by using a multicolumn array (or an array of records), where one column contains the nodes and another column contains the array index of the next node on the adjacency list—a "pseudopointer." The disadvantage of this approach is that the maximum amount of storage space that might be needed for an *n*-node graph must be set aside for the array; once we start to fill the array, new space cannot be dynamically created if we learn that there are still more adjacent nodes.

EXAMPLE 21   The array–pointer representation of the graph of Figure 5.29a is shown in Figure 5.30. A null pointer is indicated by an array index of 0. In this array row 2, representing node 2, has a pointer to index 7. At index 7 of the array, we find node 1 with weight 4, representing the arc of weight 4 from node 2 to node 1. The pointer to index 8 says that the adjacency list for node 2 has more entries. At index 8, we learn that there is an arc from 2 to 4 of weight 2, and that this completes the adjacency list for node 2.

	Node	Weight	Pointer
1			6
2			7
3			9
4			0
5			10
6	3	2	0
7	1	4	8
8	4	2	0
9	2	1	0
10	4	1	11
11	2	3	0

Figure 5.30

Section 5.1 Review

Techniques

- Use graph terminology.
- Prove or disprove that two graphs are isomorphic.
- Find a planar representation of a simple graph or prove that none exists.
- Construct adjacency matrices and adjacency lists for graphs and directed graphs.

Main Ideas

Diverse situations can be modeled by graphs.

Graphs can be represented in a computer by matrices or by linked lists.

Exercises 5.1

1. Give the function g that is part of the formal definition of the directed graph shown.

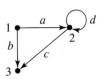

Exercise 1

2. Answer the following questions about the accompanying graph.
   a. Is the graph simple?
   b. Is the graph complete?
   c. Is the graph connected?
   d. Can you find two paths from 3 to 6?
   e. Can you find a cycle?
   f. Can you find an arc whose removal will make the graph acyclic?
   g. Can you find an arc whose removal will make the graph not connected?

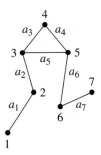

Exercise 2

★3. Sketch a picture of each of the following graphs:
   a. simple graph with three nodes, each of degree 2
   b. graph with four nodes, with cycles of length 1, 2, 3, and 4
   c. noncomplete graph with four nodes, each of degree 4

★4. Use the directed graph in the figure to answer the following questions.

  a. Which nodes are reachable from node 3?
  b. What is the length of the shortest path from node 3 to node 6?
  c. What is a path from node 1 to node 6 of length 8?

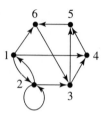

Exercise 4

5. **a.** Draw K_6.   **b.** Draw $K_{3,4}$. .

6. For each of the following characteristics, draw a graph or explain why such a graph does not exist:

  a. four nodes of degree 1, 2, 3, and 4, respectively
  b. simple, four nodes of degree 1, 2, 3, and 4, respectively
  c. four nodes of degree 2, 3, 3, and 4, respectively
  d. four nodes of degree 2, 3, 3, and 3, respectively

★7. Which of the graphs in the figure is not isomorphic to the others, and why?

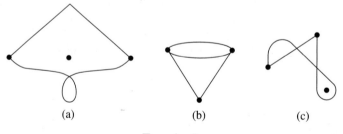

Exercise 7

8. Which of the graphs in the figure is not isomorphic to the others, and why?

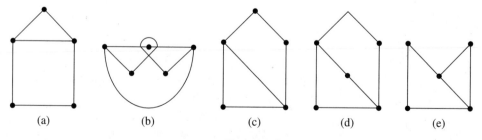

Exercise 8

For Exercises 9–14, decide if the two graphs are isomorphic. If so, give the function or functions that establish the isomorphism; if not, explain why.

★9.

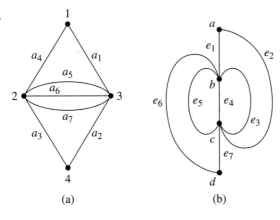

(a)   (b)

10.

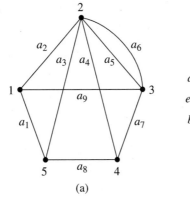

(a)

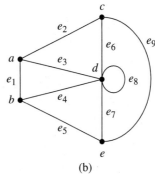

(b)

11.

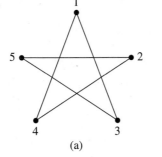

(a)

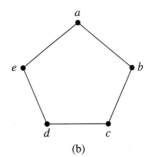

(b)

12.

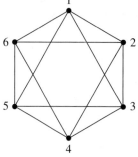

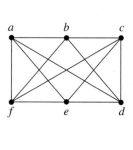

(a)   (b)

★13.

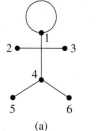   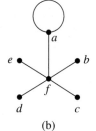

(a)   (b)

14.

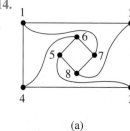   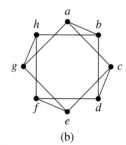

(a)   (b)

15. Prove that two graphs are not isomorphic if
    a. one has more nodes than the other.
    b. one has more arcs than the other.
    c. one has parallel arcs and the other does not.
    d. one has a loop and the other does not.
    e. one has a node of degree k and the other does not.
    f. one is connected and the other is not.
    g. one has a cycle and the other does not.

★16. Draw all the nonisomorphic, simple graphs with two nodes.

17. Draw all the nonisomorphic, simple graphs with three nodes.

18. Draw all the nonisomorphic, simple graphs with four nodes.

19. Find an expression for the number of arcs in K_n and prove that your expression is correct.

20. Verify Euler's formula for the simple, connected, planar graph in the accompanying figure.

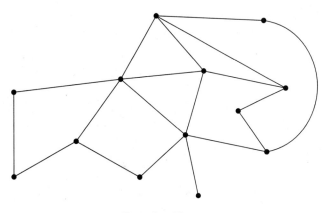

Exercise 20

★21. Prove that $K_{2,3}$ is a planar graph.

22. Prove that the graph of the accompanying figure is a planar graph.

Exercise 22

★23. If a simple, connected, planar graph has six nodes, all of degree 3, into how many regions does it divide the plane?

24. If all the nodes of a simple, connected, planar graph have degree 4 and the number of arcs is 12, into how many regions does it divide the plane?

25. Does Euler's formula (equation (1) of the theorem on the number of nodes and arcs) hold for nonsimple graphs? What about inequalities (2) and (3) of the theorem?

26. What is wrong with the following argument that claims to use elementary subdivisions to turn a nonplanar graph into a planar graph?

> In a nonplanar graph there must be two arcs a_i and a_j that intersect at a point v that is not a node. Do an elementary subdivision on a_i with an inserted node at v and an elementary subdivision on a_j with an inserted node at v. In the resulting graph, the point of intersection is a node. Repeat this process with any non-node intersections; the result is a planar graph.

For Exercises 27–30, determine if the graph is planar (by finding a planar representation) or nonplanar (by finding a subgraph homeomorphic to K_5 or $K_{3,3}$.

★27.

28.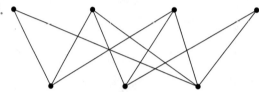

★29.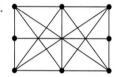

30.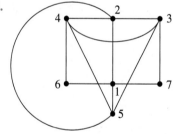

For Exercises 31–36, write the adjacency matrix for the graph in the specified figure.

★31.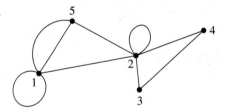

32.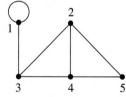

33.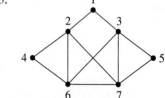

34.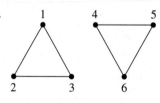

★35.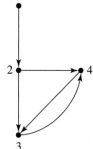

36.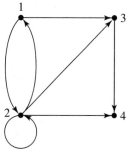

For Exercises 37–40, draw the graph represented by the adjacency matrix.

37.
$$\begin{bmatrix} 0 & 2 & 0 \\ 2 & 0 & 2 \\ 0 & 2 & 0 \end{bmatrix}$$

38.
$$\begin{bmatrix} 0 & 1 & 0 & 0 & 0 & 0 \\ 1 & 0 & 1 & 0 & 0 & 0 \\ 0 & 1 & 1 & 1 & 0 & 0 \\ 0 & 0 & 1 & 0 & 0 & 0 \\ 0 & 0 & 0 & 0 & 0 & 2 \\ 0 & 0 & 0 & 0 & 2 & 0 \end{bmatrix}$$

★39.
$$\begin{bmatrix} 0 & 1 & 1 & 1 & 0 \\ 1 & 0 & 0 & 0 & 1 \\ 1 & 0 & 0 & 0 & 1 \\ 1 & 0 & 0 & 0 & 1 \\ 0 & 1 & 1 & 1 & 0 \end{bmatrix}$$

40.
$$\begin{bmatrix} 0 & 1 & 0 & 0 & 1 \\ 1 & 0 & 1 & 0 & 0 \\ 0 & 1 & 0 & 1 & 0 \\ 0 & 0 & 1 & 0 & 1 \\ 1 & 0 & 0 & 1 & 0 \end{bmatrix}$$

★41. The adjacency matrix for an undirected graph is given in lower triangular form by

$$\begin{bmatrix} 2 & & & \\ 1 & 0 & & \\ 0 & 1 & 1 & \\ 0 & 1 & 2 & 0 \end{bmatrix}$$

Draw the graph.

42. The adjacency matrix for a directed graph is given by

$$\begin{bmatrix} 0 & 1 & 1 & 0 & 0 \\ 0 & 0 & 0 & 0 & 0 \\ 0 & 0 & 1 & 1 & 0 \\ 0 & 0 & 1 & 0 & 2 \\ 1 & 0 & 0 & 0 & 0 \end{bmatrix}$$

Draw the graph.

★43. Describe the graph whose adjacency matrix is I_n, the $n \times n$ identity matrix.

44. Describe the adjacency matrix for K_n, the simple, complete graph with n nodes.

45. Given the adjacency matrix **A** for a directed graph G, describe the graph represented by the adjacency matrix $\mathbf{A}^T$ (see Exercise 17 in Section 4.5).

For Exercises 46–51, draw the adjacency list representation for the indicated graph.

46. Exercise 31   **47.** Exercise 32   ★**48.** Exercise 33

49. Exercise 34   **50.** Exercise 35   **51.** Exercise 36

52. Refer to the accompanying graph.
   a. Draw the adjacency list representation.
   b. How many storage locations are required for the adjacency list? (A pointer takes one storage location.)
   c. How many storage locations would be required in an adjacency matrix for this graph?

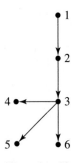

Exercise 52

★**53.** Draw the adjacency list representation for the weighted directed graph of the accompanying figure.

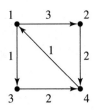

Exercise 53

★**54.** For the directed graph of Exercise 36, construct the array–pointer representation.

55. For the weighted directed graph of Exercise 53, construct the array–pointer representation.

56. Draw the undirected graph represented by the adjacency list in the accompanying figure.

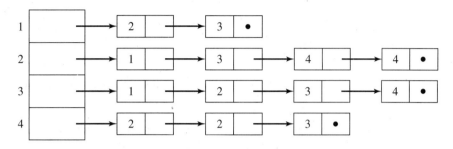

Exercise 56

57. Draw the directed graph represented by the adjacency list in the given figure.

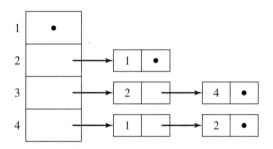

Exercise 57

Exercises 58–64 refer to the *complement* of a graph. If G is a simple graph, the complement of G, denoted G', is the simple graph with the same set of nodes as G, where nodes x–y are adjacent in G' if and only if they are not adjacent in G.

★**58.** Draw G' for the graph of Figure 5.18a.

59. Draw K_4'.

60. Show that if two simple graphs G_1 and G_2 are isomorphic, so are their complements G_1' and G_2'.

61. A simple graph is *self-complementary* if it is isomorphic to its complement. Prove that in a self-complementary graph with n nodes ($n > 1$), $n = 4k$ or $n = 4k + 1$ for some integer k. (*Hint:* Use the result of Exercise 19.)

★**62. a.** Prove that in any simple graph G with at least two nodes, if G is not connected, then G' is connected. (*Hint:* If G is not connected, then G consists of a collection of "disjoint" connected subgraphs.)
   b. Find a graph G where both G and G' are connected, thus showing that the converse of part (a) is false.

63. Given an adjacency matrix **A** for a simple graph G, describe the adjacency matrix for G'.

64. Prove that if $|N| \geq 11$ in a simple, connected graph G, then not both G and G' can be planar.

65. Prove that in any simple graph G with n nodes and a arcs, $2a \leq n^2 - n$.

★66. Prove that a simple, connected graph with n nodes has at least $n - 1$ arcs. (*Hint:* Show that this can be restated as "A simple, connected graph with m arcs has at most $m + 1$ nodes." Then use the second principle of induction on m.)

67. Prove that a simple graph with n nodes ($n \geq 2$) and more than $C(n - 1, 2)$ arcs is connected. (*Hint:* Use Exercises 62 and 66.)

Exercises 68–76 refer to the problem of *graph colorability*. The origin of graph coloring problems is a *map-coloring problem*: Suppose that a map of various countries, drawn on a sheet of paper, is to be colored so that no two countries with a common border have the same color. (We need not worry about countries that meet only at a point, and we shall assume that each country is "connected.") What is the minimum number of colors required to carry out this task for any map?

68. Show that a coloring of the accompanying map requires three colors and no more than three colors.

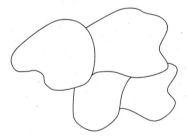

Exercise 68

69. Draw a map that requires four colors.

70. Associated with any map is a graph, called the *dual graph* for the map, formed as follows: Put one node in each region of the map and an arc between two nodes representing adjacent countries.

a. Draw the dual graph for the map of Exercise 68.

b. Draw the dual graph for the map in the accompanying figure.

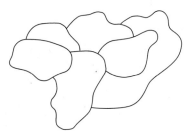

Exercise 70b

c. Draw a map for which the graph of the given figure would serve as the dual

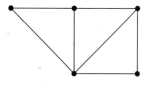

Exercise 70c

71. A *coloring* of a graph is an assignment of a color to each node of the graph in such a way that no two adjacent nodes have the same color. The *chromatic number* of a graph is the smallest number of colors needed to achieve a coloring. Find the chromatic number of the following graphs.

★**a.**   **b.**

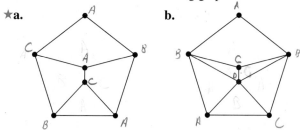

72. At least four colors are required to solve the general map-coloring problem (see Exercise 69). Because no one could produce a map requiring more than four colors, the conjecture was formulated that four colors are indeed sufficient. This conjecture became known as the *four-color problem*. It was first proposed to the mathematician Augustus De Morgan by one of his students in 1852, and it subsequently received much attention. It remained unproved, however, for over 100 years. In 1976 two mathematicians at the University of Illinois, Wolfgang Haken and Kenneth Appel, used a computer to work though a large number of cases in a proof by contradiction, thus verifying the four-color conjecture.

   The dual graph for a map (see Exercise 70), by the way it is constructed, will always be simple, connected, and planar. In addition, any simple, connected, planar graph can be viewed as the dual graph of a map. Restate the four-color conjecture in terms of the chromatic number (see Exercise 71) of a graph.

★73. Prove that in a simple, connected, planar graph with three or more nodes, there is at least one node with degree less than or equal to 5. (*Hint:* Use a proof by contradiction.)

74. (Challenging problem) The *five-color theorem* states that the chromatic number for any simple, connected, planar graph is at most 5. While the four-color theorem (Exercise 72) is very difficult to prove, the five-color theorem can be proved by induction on the number of nodes in the graph. Prove the five-color theorem, making use of the result in Exercise 73.

75. The six-color theorem can be proved as a map-coloring problem without using the dual graph. Instead of creating the dual graph, put nodes at the intersections of boundaries and straighten the boundaries of regions so that the problem of coloring the map shown in part (a) of the accompanying figure is represented by the problem of coloring the enclosed regions of the graph in part (b) of the figure. First assume that no country has a hole in it. Then the graph will be loop-free, planar, and connected. Also, every node will have degree at least 3.

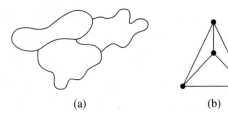

(a)   (b)

Exercise 75

★**a.** Show that the graph can be assumed to be simple by proving that if six colors are sufficient for coloring a simple graph, they are sufficient for a graph with parallel arcs as well. (*Hint:* Use temporary small countries at nodes.)
b. Prove that in a simple, connected, planar graph with R enclosed regions, $n - a + R = 1$.
c. Consider a simple, connected, planar graph and assume that every enclosed region has at least six edges adjacent to it. Show that $2a \leq 3n - 3$.
d. Now consider a simple, connected, planar graph where every node has degree at least 3. Show that such a graph has at least one enclosed region with no more than five adjacent edges.
e. Prove that six colors are sufficient to color any planar map where no country has a hole in it.
f. Prove that six colors are sufficient to color any planar map. (*Hint:* Cut some temporary slits in the map.)

76. Five political lobbyists are visiting seven members of Congress (labeled *A* through *G*) on the same day. The members of Congress the five lobbyists must see are

1. *A, B, D*
2. *B, C, F*
3. *A, B, D, G*
4. *E, G*
5. *D, E, F*

Each member of Congress will be available to meet with lobbyists for one hour. What is the minimum number of time slots that must be used to set up the one-hour meetings so that no lobbyist has a conflict? (*Hint:* Treat this as a graph-coloring problem.) What if lobbyist 3 discovers he or she does not need to see *B* and lobbyist 5 discovers he or she does not need to see *D*?

Section 5.2   Trees and Their Representations

Tree Terminology

A special type of graph called a *tree* turns out to be a very useful representation of data.

> **Definition: Tree**
> A **tree** is an acyclic, connected graph with one node designated as the **root** of the tree.

Figure 5.31 pictures two trees. Perversely, computer scientists like to draw trees with the root at the top. An acyclic connected graph with no designated root node is called a **nonrooted tree** or a **free tree**. (Again, terminology is nonstandard. Some books define trees as acyclic, connected graphs and then call them "rooted trees" when there is a designated root node.)

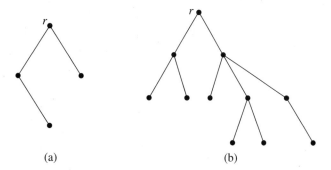

(a)   (b)

Figure 5.31

A tree can also be defined recursively. A single node is a tree (with that node as its root). If $T_1, T_2, \ldots, T_t$ are disjoint trees with roots $r_1, r_2, \ldots, r_t$, the graph formed by attaching a new node r by a single arc to each of $r_1, r_2, \ldots, r_t$ is a tree with root r. The nodes $r_1, r_2, \ldots, r_t$ are **children** of r, and r is a **parent** of $r_1, r_2, \ldots, r_t$.

Figure 5.32 shows the final step in the recursive construction of the tree in Figure 5.31b. It is often helpful to process a tree structure by working with it recursively, treating the subtrees as smaller tree objects.

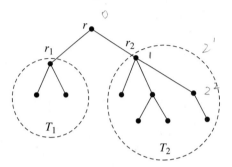

Figure 5.32

Because a tree is a connected graph, there is a path from the root to any other node in the tree; because the tree is acyclic, that path is unique. The **depth of a node** in a tree is the length of the path from the root to the node; the root itself has depth 0. The **depth (height) of the tree** is the maximum depth of any node in the tree; in other words, it is the length of the longest path from the root to any node. A node with no children is called a **leaf** of the tree; all nonleaves are **internal nodes**. A **forest** is an acyclic graph (not necessarily connected); thus a forest is a disjoint collection of trees. Figures 5.31a and 5.31b together form a forest.

Binary trees, where each node has at most two children, are of particular interest. In a binary tree, each child of a node is designated as either the **left child** or the **right child**. A **full binary tree** occurs when all internal nodes have two children and all leaves are at the same depth. Figure 5.33 shows a binary tree of height 4, and Figure 5.34 shows a full binary tree of height 3. A **complete binary tree** is an almost-full binary tree; the bottom level of the tree is filling from left to right but may not have its full complement of leaves. Figure 5.35 shows a complete binary tree of height 3. (Note that while a tree is a graph, a complete tree is not a complete graph!)

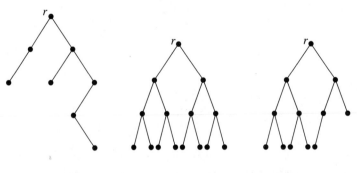

Figure 5.33   **Figure 5.34**   **Figure 5.35**

PRACTICE 18   Answer the following questions about the binary tree shown in Figure 5.36. (Assume that node 1 is the root.)

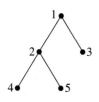

Figure 5.36

a. What is the height?   **b.** What is the left child of node 2?

c. What is the depth of node 5?

Applications of Trees

Decision trees were used to solve counting problems in Chapter 3 and will be used in Section 5.3 to help establish lower bounds on the work for certain algorithms. Exercise 31 of Section 4.1 describes the organization of data into a binary tree structure. By using these trees, a collection of records can be efficiently searched to locate a particular record or to determine that a record is not in the collection. Examples of such a search would be checking for a volume in a library, for a patient's medical record in a hospital, or for an individual's credit record at the bank. We shall also look at binary tree search in Section 5.3. The derivations of words in certain formal languages will be shown as trees in Chapter 8 (these are the parse trees generated by a compiler while analyzing a computer program).

A family tree is usually, indeed, a tree, although if there were intermarriages, it would be a graph but not a tree in the technical sense. (Information obtained from a family tree is not only interesting but also useful for research in medical genetics.) The organization chart indicating who reports to whom in a large company or other enterprise is usually a tree (see Figure 5.37).

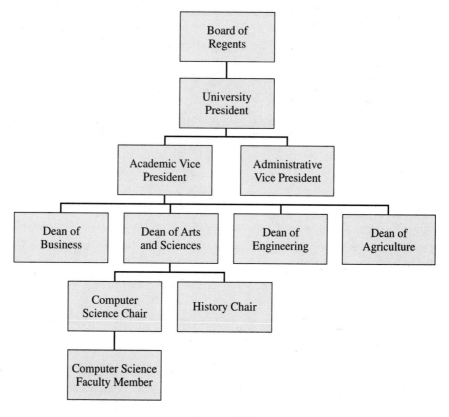

Figure 5.37

Files on your computer are organized in a hierarchical (treelike) structure, as are the topics in many on-line Help systems (see Figure 5.38, where there are a number of trees, each rooted at the left).

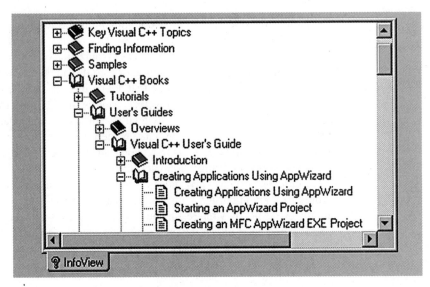

Figure 5.38

EXAMPLE 22   Algebraic expressions involving binary operations can be represented by labeled binary trees. The leaves are labeled as operands, and the internal nodes are labeled as binary operations. For any internal node, the binary operation of its label is performed on the expressions associated with its left and right subtrees. Thus the binary tree in Figure 5.39 represents the algebraic expression $(2 + x) - (y * 3)$.

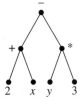

Figure 5.39

PRACTICE 19   What is the expression tree for $(2 + 3) * 5$?

Binary Tree Representation

Because a tree is also a graph, the representations discussed in Section 5.1 for graphs in general can also be used for trees. Binary trees, however, have special characteristics that we want to capture in the representation, namely, the identity of the left and right child. The equivalent of an adjacency matrix is a two-column array (or an array of records) where the data for each node is the left and right child of that node. The equivalent of the adjacency list representation is a collection of records with three fields containing, respectively, the current node, a pointer to the record for the left-child node, and a pointer to the record for the right-child node.

EXAMPLE 23   For the binary tree shown in Figure 5.40, the left child–right child array representation is given in Figure 5.41a. Zeros again indicate null pointers. The pointer representation is given in Figure 5.41b.

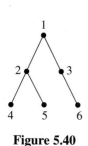

Figure 5.40

	Left child	Right child
1	2	3
2	4	5
3	0	6
4	0	0
5	0	0
6	0	0

(a)

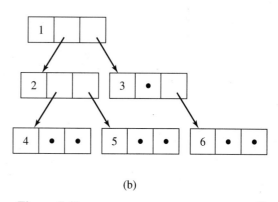

(b)

Figure 5.41

PRACTICE 20   **a.** Give the left child–right child array representation of the binary tree in Figure 5.42.
b. Give the pointer representation of the binary tree in Figure 5.42.

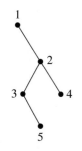

Figure 5.42

Tree Traversal Algorithms

If a tree structure is being used to store data, it is often helpful to have a systematic mechanism for writing out the data values stored at all the nodes. This can be accomplished by *traversing* the tree, that is, visiting each of the nodes in the tree structure. The three common **tree traversal** algorithms are preorder, inorder, and postorder traversal.

In these traversal methods, it is helpful to use the recursive view of a tree, where the root of a tree has branches down to roots of subtrees. We shall therefore assume that a tree T has a root r; any subtrees are labeled left to right as $T_1, T_2, \ldots, T_t$ (see

Figure 5.43). Because we are using a recursive definition of a tree, it will be easy to state the tree traversal algorithms in recursive form.

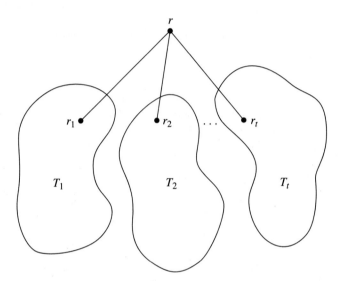

Figure 5.43

The terms *preorder, inorder,* and *postorder* refer to the order in which the root of a tree is visited compared to the subtree nodes. In **preorder traversal**, the root of the tree is visited first and then the subtrees are processed left to right, each in preorder.

ALGORITHM   Preorder

Preorder(tree T)
//Writes the nodes of a tree with root r in preorder

   write(r)
   for $i = 1$ to t **do**
      Preorder(T_i)
   end for

end Preorder

In **inorder traversal**, the left subtree is processed by an inorder traversal, then the root is visited, and then the remaining subtrees are processed from left to right, each in inorder. If the tree is a binary tree, the result is that the root is visited between processing of the two subtrees.

ALGORITHM   Inorder

Inorder(tree *T*)
//Writes the nodes of a tree with root *r* in inorder

    Inorder(T_1)
    write(*r*)
    for *i* = 2 to *t* **do**
        Inorder(T_i)
    end for

end Inorder

Finally, in **postorder traversal**, the root is visited last, after all subtrees have been processed from left to right in postorder.

ALGORITHM   Postorder

Postorder(tree *T*)
//Writes the nodes of a tree with root *r* in postorder

    for *i* = 1 to *t* **do**
        Postorder(T_i)
    end for
    write(*r*)

end Postorder

EXAMPLE 24

REMINDER:
For a binary tree:

Preorder traversal is root, left, right.

Inorder traversal is left, root, right.

Postorder traversal is left, right, root.

For the binary tree of Figure 5.44, the preorder traversal algorithm (root, left, right) says to write the root first, *a*, and then process the left subtree. At the left subtree, rooted at *b*, a preorder traversal writes the root, *b*, and moves again to the left subtree, which is the single node *d*. This single node is the root of a tree, so it is written out. Then *d*'s left subtree (empty) and *d*'s right subtree (empty) are traversed. Backing up to the tree rooted at *b*, its left subtree has been traversed, so now the right subtree is traversed, producing node *e*. The subtree rooted at *b* has now been completely traversed. Backing up to *a*, it is time to traverse *a*'s right subtree. A preorder traversal of the tree rooted at *c* causes *c* to be written, then traversal goes to *c*'s left subtree, which results in *f*, *h*, and *i* being written. Backing up to *c*, traversing *c*'s right subtree produces *g*. The subtree rooted at *c* has now been completely traversed, and the algorithm terminates. The preorder traversal produced

a, b, d, e, c, f, h, i, g

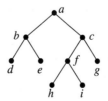

Figure 5.44

EXAMPLE 25   Using the tree of Figure 5.44 again, an inorder traversal (left, root, right), travels down to the farthest left subtree, rooted at d. An inorder traversal here traverses the left subtree (empty), writes out the root, d, then traverses the right subtree (empty). Backing up the tree to b, b's left subtree has been traversed, so it is time to write out the root, b. Proceeding then to b's right subtree, e is written. Backing up to a, a's left subtree has been traversed, so the root, a, is written out. Proceeding to a's right subtree, an inorder traversal says to go to the farthest left subtree first, which would cause the h to be written. After that, f and i are written, then the root c, then the right subtree of c, which is g. The nodes are therefore written as

$d, b, e, a, h, f, i, c, g$

A postorder traversal (left, right, root) would produce

$d, e, b, h\ i, f, g, c, a$ ●

EXAMPLE 26   Consider the tree shown in Figure 5.45, which is not a binary tree. A preorder traversal first writes the root a and then does a preorder traversal on the left subtree, rooted at b. The preorder traversal of this subtree writes out b and then proceeds to a preorder traversal of the left subtree of b, which is rooted at d. Node d is written and then a preorder traversal of the left subtree of d, which is rooted at i, is done. After writing out i, the traversal backs up to consider any other subtrees of d; there are none.

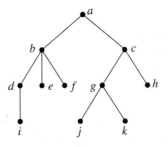

Figure 5.45

Backing up to b, there are other subtrees of b. Processing these left to right, nodes e and then f are written. All of the subtrees of b have now been traversed; backing up to node a to look for subtrees farther to the right reveals a subtree rooted at c. The algorithm writes out the root c, then moves to its leftmost subtree rooted at g, and writes out g. Processing the subtrees of g, j and k are written; then backing up to c, its remaining subtree is processed, producing h. Node c has no other subtrees; backing up to a, a has no other subtrees, and the algorithm terminates. The list of nodes in preorder traversal is

$a, b, d, i, e, f, c, g, j, k, h$

To do an inorder traversal of the tree in Figure 5.45, process left subtrees first. This leads down to node i, which has no subtrees. Therefore i is written out. Backing up to d, the left subtree of d has been traversed, so d is written out. Since node d has no further subtrees, the algorithm backs up to b. The left subtree of b has been processed, so b is written out and then its remaining subtrees are traversed, writing out e and f.

Backing up to *a*, *a* is written out, and then the right subtree of *a* is processed. This leads to nodes *j*, *g*, *k*, *c*, and *h*, in that order, and we are done. Thus the inorder list of nodes is

$$i, d, b, e, f, a, j, g, k, c, h$$

The following list of nodes results from a postorder traversal:

$$i, d, e, f, b, j, k, g, h, c, a$$

PRACTICE 21   Do a preorder, inorder, and postorder traversal of the tree in Figure 5.46.

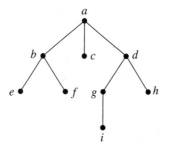

Figure 5.46

EXAMPLE 27   Example 22 showed how algebraic expressions can be represented as binary trees. If we do an inorder traversal of the expression tree, we retrieve the original algebraic expression. For the expression tree of Figure 5.47, for example, an inorder traversal gives the expression

$$(2 + x) * 4$$

where the parentheses are added as we complete the processing of a subtree. This form of an algebraic expression, where the operation symbol appears between the two operands, is called **infix notation**. Parentheses are necessary here to indicate the order of operations. Without parentheses, the expression becomes $2 + x * 4$, which is also an infix expression but, because of the order of precedence of multiplication over addition, is not what is intended.

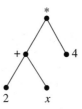

Figure 5.47

A preorder traversal of Figure 5.47 gives the expression

$$* + 2 \, x \, 4$$

Here the operation symbol precedes its operands. This form of an expression is called **prefix notation**, or **Polish notation**. The expression can be translated into infix form as follows:

$$* + 2\, x\, 4 \to * (2 + x)\, 4 \to (2 + x) * 4$$

A postorder traversal gives the expression

$$2\, x + 4\, *$$

where the operation symbol follows its operands. This form of an expression is called **postfix notation**, or **reverse Polish notation** (or just **RPN**). The expression can be translated into infix form as follows:

$$2\, x + 4\, * \to (2 + x)\, 4\, * \to (2 + x) * 4$$

Neither prefix nor postfix form requires parentheses to avoid ambiguity. These notations therefore provide more efficient, if less familiar, representations of algebraic expressions than infix notation. Such forms can be evaluated sequentially, with no need for "look-ahead" to locate parenthesized expressions. Compilers often change algebraic expressions in computer programs from infix to postfix notation for more efficient processing.

PRACTICE 22   Write the expression tree for

$$a + (b * c - d)$$

and write the expression in prefix and postfix notation.

Results about Trees

Trees are fertile ground (no pun intended) for proofs by induction, either on the number of nodes or arcs or on the height.

EXAMPLE 28   After drawing a few trees and doing some counting, it appears that the number of arcs in a tree is always 1 less than the number of nodes. More formally, it appears that

$n = 1, a = 0$ A tree with n nodes has $n - 1$ arcs.

Figure 5.48

We will prove this statement using induction on n, $n \geq 1$. For the base case, $n = 1$, the tree consists of a single node and no arcs (Figure 5.48), so the number of arcs is 1 less than the number of nodes.

Assume that any tree with k nodes has $k - 1$ arcs, and consider a tree with $k + 1$ nodes. We want to show that this tree has k arcs. Let x be a leaf of the tree (a leaf must exist since the tree is finite). Then x has a unique parent. Remove from the tree the node x and the single arc a connecting x and its parent (see Figure 5.49). The remaining graph is still a tree and has k nodes. Therefore, by the inductive hypothesis, it has $k - 1$ arcs, and the original graph, containing arc a, had $(k - 1) + 1 = k$ arcs. The proof is complete.

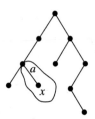

Figure 5.49

Notice that in the inductive proof of Example 28 we had to support the proof with many more words than we used in some of our early inductive proofs. In Example 15 in Chapter 2, for instance, the inductive proof that

$$1 + 2 + 2^2 + \cdots + 2^n = 2^{n+1} - 1$$

consisted mainly of manipulating the mathematical expressions in this equation, but now we have to do more verbal reasoning. Words are not only OK in a proof, they may form the major part of the proof.

The inductive proof of Example 28 differs in another way from proofs like that of Example 15 in Chapter 2. In those proofs, there was always a single term in the series (the last term) whose removal would lead to the "$P(k)$ case," the inductive hypothesis. In proofs involving trees with $k + 1$ nodes, which node should be removed to generate the $P(k)$ case? Usually the node to remove is not unique, but it is not completely arbitrary either. In the proof of Example 28, for instance, removing a nonleaf node from a tree with $k + 1$ nodes would result in a graph with k nodes, but not a tree with k nodes, so the inductive hypothesis would not apply.

PRACTICE 23   Prove that in any tree with n nodes, the total number of arc ends is $2n - 2$. Use induction on the number of nodes.   ●

EXAMPLE 29   Sometimes a clever observation can take the place of an inductive proof. The problem of Practice 23 can be solved by noting that each node of the tree except the root has a parent, and the arc connecting a node to its parent has 2 arc ends. There are $n - 1$ such nodes and therefore $2(n - 1)$ arc ends.   ●

Section 5.2 Review

Techniques

- Construct expression trees.
- Construct array and pointer representations for binary trees.
- Conduct preorder, inorder, and postorder traversals of a tree.

Main Ideas

Binary trees can be represented by arrays and by linked structures.

Recursive procedures exist to systematically visit every node of a binary tree.

Exercises 5.2

★1. Sketch a picture of each of the following trees:
   a. tree with five nodes and depth 1
   b. full binary tree of depth 2
   c. tree of depth 3 where each node at depth i has $i + 1$ children

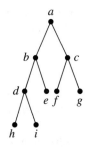

Exercise 2

2. Answer the following questions about the accompanying graph with node *a* as the root.

   a. Is this a binary tree?
   b. Is it a full binary tree?
   c. Is it a complete binary tree?
   d. What is the parent of *e*?
   e. What is the right child of *e*?
   f. What is the depth of *g*?
   g. What is the height of the tree?

In Exercises 3–6, draw the expression tree.

   3. $[(x - 2) * 3] + (5 + 4)$

★4. $[(2 * x - 3 * y) + 4 * z] + 1$

   5. $1 - (2 - [3 - (4 - 5)])$

   6. $[(6 \div 2) * 4] + [(1 + x) * (5 + 3)]$

★7. Write the left child–right child array representation for the binary tree in the figure.

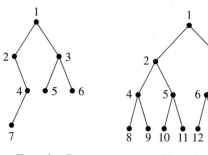

Exercise 7   **Exercise 8**

8. Write the left child–right child array representation for the binary tree in the figure.

★9. Draw the binary tree represented by the left child–right child representation of the figure. (1 is the root.)

	Left child	Right child
1	2	3
2	4	0
3	5	0
4	6	7
5	0	0
6	0	0
7	0	0

Exercise 9

10. Draw the binary tree represented by the left child–right child representation of the figure. (1 is the root.)

	Left child	Right child
1	2	0
2	3	4
3	0	0
4	5	6
5	0	0
6	0	0

Exercise 10

11. Write the left child–right child array representation for the binary search tree that is created by processing the following list of words: "All Gaul is divided into three parts" (see Exercise 31 of Section 4.1). Also store the name of each node.

★12. The figure represents a binary tree in which the left child and parent of each node are given. Draw the binary tree. (1 is the root.)

	Left child	Parent
1	2	0
2	4	1
3	0	1
4	0	2
5	0	2
6	0	3

Exercise 12

13. The figure represents a tree (not necessarily binary), where for each node, the left-most child and the closest right sibling of that node are given. Draw the tree. (1 is the root.)

	Left child	Right sibling
1	2	0
2	5	3
3	0	4
4	8	0
5	0	6
6	0	7
7	0	0
8	0	0

Exercise 13

14. **a.** For the tree in the figure, write the leftmost child–right sibling array representation described in Exercise 13.

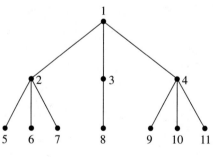

Exercise 14a

b. Now draw the binary tree that results from treating the answer to part (a) as a left child–right child binary tree representation. An arbitrary tree can thus be thought of as having a binary tree representation.

15. The binary tree in the figure is the representation of a general tree (as in part (b) of Exercise 14). Draw the tree.

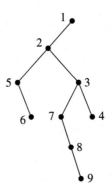

Exercise 15

For Exercises 16–21, write the list of nodes resulting from a preorder traversal, an inorder traversal, and a postorder traversal of the tree.

★16.

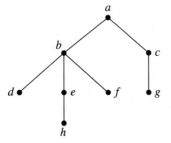

17.

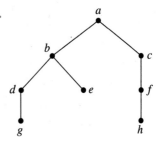

18.

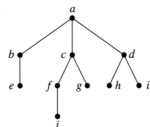

19.

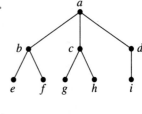

★20.

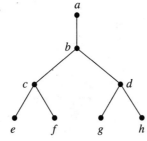

21.

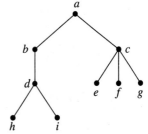

★22. Write in prefix and postfix notation: $3/4 + (2 - y)$

23. Write in prefix and postfix notation: $(x * y + 3/z)\ 4$

24. Write in infix and postfix notation: $-*+2\ 3\ 6\ x\ 7$

25. Write in infix and postfix notation: $-+-x\ y\ z\ w$

★26. Write in prefix and infix notation: $4\ 7\ x\ -*z\ +$

27. Write in prefix and infix notation: $x\ 2\ w + y\ z\ *-/$

28. Draw a tree whose preorder traversal is

    a, b, c, d, e

    and whose inorder traversal is

    b, a, d, c, e

29. Draw a tree whose inorder traversal is

    $f, a, g, b, h, d, i, c, j, e$

    and whose postorder traversal is

    $f, g, a, h, i, d, j, e, c, b$

★30. Find an example of a tree whose inorder and postorder traversals yield the same list of nodes.

31. Find two different trees that have the same list of nodes under a preorder traversal.

★32. Prove that a simple graph is a nonrooted tree if and only if there is a unique path between any two nodes.

33. Let G be a simple graph. Prove that G is a nonrooted tree if and only if G is connected and if the removal of any single arc from G makes G unconnected.

34. Let G be a simple graph. Prove that G is a nonrooted tree if and only if G is connected and the addition of one arc to G results in a graph with exactly one cycle.

35. Prove that a tree with n nodes, $n \geq 2$, has at least two nodes of degree 1.

★36. Prove that a binary tree has at most 2^d nodes at depth d.

37. **a.** Draw a full binary tree of height 2. How many nodes does it have?
    b. Draw a full binary tree of height 3. How many nodes does it have?
    c. Conjecture how many nodes there are in a full binary tree of height h.
    d. Prove your conjecture. (*Hint:* Use Exercise 36.)

38. **a.** Prove that a full binary tree with x internal nodes has $2x + 1$ total nodes.
    b. Prove that a full binary tree with x internal nodes has $x + 1$ leaves.
    c. Prove that a full binary tree with n nodes has $(n - 1)/2$ internal nodes and $(n + 1)/2$ leaves.

39. Prove that the number of leaves in any binary tree is 1 more than the number of nodes with two children.

★40. Find an expression for the height of a complete binary tree with n nodes. (*Hint: Use Exercise 37.*)

41. Prove that in the pointer representation of a binary tree with n nodes there are $n + 1$ null pointers. (*Hint: Use Exercise 39*).

42. Let E be the external path length of a tree, that is, the sum of the path lengths to all of the leaves. Let I be the internal path length, that is, the sum of the path lengths to all of the internal nodes. Let i be the number of internal nodes. Prove that in a binary tree where all internal nodes have two children, $E = I + 2i$.

43. Find the chromatic number of a tree (see Section 5.1, Exercise 71).

In Exercises 44 and 45, two trees are *isomorphic* if there is a bijection $f: N_1 \rightarrow N_2$, where f maps the root of one tree to the root of the other and where $f(y)$ is a child of $f(x)$ in the second tree when y is a child of x in the first tree. Thus in the figure shown, the two trees are isomorphic graphs but not isomorphic trees (in (a) the root has two children and in (b) it does not). These are the only two nonisomorphic trees with three nodes.

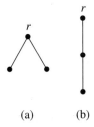

(a)   (b)

Exercises 44 and 45

★44. Show that there are four nonisomorphic trees with four nodes.

45. Show that there are nine nonisomorphic trees with five nodes.

46. In Figure 5.36, assume that node 4 is the root, and redraw the tree with the root at the top.

47. In Figure 5.36, assume that node 2 is the root, and redraw the tree with the root at the top.

Section 5.3 Decision Trees

We used decision trees in Chapter 3 to solve counting problems. Figure 5.50 shows the tree used in Example 35 of Chapter 3 to represent the various possibilities for five coin tosses under the constraint that two heads in a row do not occur. Each internal node of the tree represents an action (a coin toss), and the arcs to the children of internal nodes represent the outcomes of that action (heads or tails). The leaves of the tree represent the final outcomes, that is, the different ways that five tosses could occur.

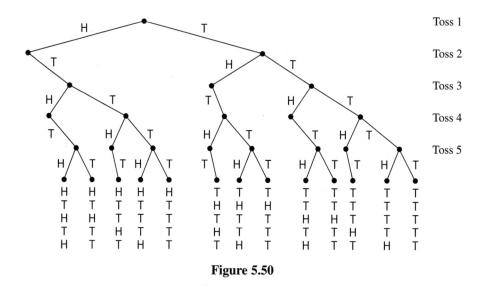

Figure 5.50

Although we have used decision trees, we haven't given a formal definition of what a decision tree is.

Definition: Decision Tree
A **decision tree** is a tree in which the internal nodes represent actions, the arcs represent outcomes of an action, and the leaves represent final outcomes.

Sometimes useful information can be obtained by using a decision tree to represent the activities of a real algorithm; actions that the algorithm performs take place at internal nodes, the children of an internal node represent the next action taken, based on the outcome of the previous action, and the leaves represent some sort of circumstance that can be inferred upon algorithm termination. Note that, unlike the trees we talked about in Section 5.2, a decision tree is not a data structure; that is, the nodes of the tree have no data values associated with them. Nor are the algorithms we are representing necessarily acting upon a tree structure. In fact, we will use decision trees in this section to learn more about algorithms for searching and sorting, and these algorithms act on lists of data items.

Searching

A search algorithm either finds a target element x within a list of elements or determines that x is not in the list. Such an algorithm generally works by making successive comparisons of x to the list items. We have already seen two such algorithms, sequential search and binary search. We can model the activities of these algorithms by using decision trees. The nodes represent the actions of comparing x to the list items, where the comparison of x to the ith element in the list is denoted by $x:L(i)$.

Sequential search only distinguishes between two possible outcomes of a comparison of x to $L(i)$. If $x = L(i)$, the algorithm terminates because x has been found in the list. If $x \neq L(i)$, the next comparison performed is $x:L(i + 1)$, regardless of whether x was less than or greater than $L(i)$. The leaves of this decision tree correspond to the final outcomes, where either x is one of the list items or x is not in the list.

EXAMPLE 30   Figure 5.51 shows the decision tree for the sequential search algorithm acting on a sorted list of five elements.

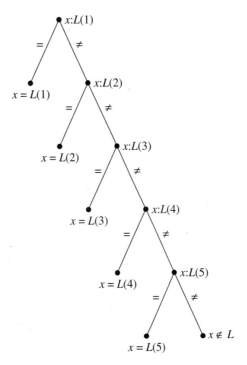

Figure 5.51

From the decision tree for a given search algorithm, we can see that the number of comparisons required to reach any particular outcome (leaf of the tree) is the number of internal nodes from the root to that leaf. This number equals the length of the path from the root to that leaf. The worst case, that is, the maximum number of comparisons, is the maximum length of any such path, which is the depth of the tree. Because every decision tree for sequential search looks like Figure 5.51, it is clear that the depth of such a tree, for an n-element list, is n. This agrees with what we already know, namely, that the worst case for sequential search on a list of n elements is n.

The decision tree for the binary search algorithm is more interesting. Binary search acts on a sorted list and distinguishes between three possible outcomes of the comparison:

$x = L(i)$: algorithm terminates, x has been found

$x < L(i)$: algorithm proceeds to the left half of the list

$x > L(i)$: algorithm proceeds to the right half of the list

We shall follow the usual custom and not write the leaf that corresponds to the "middle branch," $x = L(i)$ (see Exercise 21 for a discussion of the consequences of this convention). If $x < L(i)$, the next comparison the algorithm performs is found at the left child of this node; if $x > L(i)$, the algorithm's next comparison is found at the right child. If no child exists, the algorithm terminates because x is not in the list. The tree we've described is a binary tree whose leaves represent all the possible outcomes where x is not in the list. There are many more failure leaves in binary search than in sequential search, because binary search indicates *how* x fails to be in the list (e.g., $x < L(1)$ or $L(1) < x < L(2)$).

EXAMPLE 31   Figure 5.52 shows the decision tree for the binary search algorithm acting on a list of eight elements.

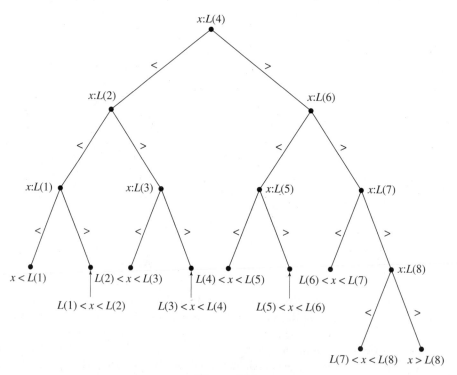

Figure 5.52

The worst case, that is, the maximum number of comparisons, will again be the depth of the tree, which is 4 in Figure 5.52. In Chapter 2 we solved a recurrence relation to get the worst-case behavior for binary search where n is a power of 2 and found this to be $1 + \log n$ (remember that we are using base 2 logarithms). Note that $1 + \log 8 = 4$, so the decision tree agrees with our previous result. The restriction of n to a power of 2 made the arithmetic of solving the recurrence relation simpler. If n is not a power of 2, then the depth of the tree is given by the expression $1 + \lfloor \log n \rfloor$.

PRACTICE 24   **a.** Draw the decision tree for the binary search algorithm on a set of five elements.
b. Find the depth of the tree and compare to $1 + \lfloor \log 5 \rfloor$.

Lower Bounds on Searching

We have used decision trees to represent the actions of two particular search algorithms. Such a tree could be used to represent the actions of any algorithm that solves the search problem by comparing the target element to the list elements. The internal nodes of such a tree would represent the comparisons done, and the depth of the tree would be the worst-case number of comparisons over all possible cases. What can be said about such a tree when we don't know the particulars of the algorithm involved? We can say that x must be compared to every element in the list at least once (perhaps more than once if the algorithm is quite stupid). For if there is some list element that escapes being compared to x, the algorithm cannot say whether that element equals x and thus cannot decide with certainty whether x belongs to the list. Comparisons are internal nodes in the decision tree. Therefore, if m is the number of internal nodes in the decision tree T_1 for any search algorithm acting on an n-element list, then $m \geq n$.

Before proceeding further with decision trees, we need some additional facts about binary trees in general. The number of nodes at each level in a *full* binary tree follows a geometric progression: 1 node at level 0, 2^1 nodes at level 1, 2^2 nodes at level 2, and so on. In a full binary tree of depth d, the total number of nodes is therefore

$$1 + 2 + 2^2 + 2^3 + \cdots + 2^d = 2^{d+1} - 1$$

(see Example 15 of Chapter 2). A full binary tree has the maximum number of nodes for a given depth of any binary tree. This gives us fact 1:

1. Any binary tree of depth d has at most $2^{d+1} - 1$ nodes.

Fact 2, which we'll prove momentarily, is

2. Any binary tree with m nodes has depth $\geq \lfloor \log m \rfloor$.

To prove fact 2, we'll use a proof by contradiction. Suppose a binary tree has m nodes and depth $d < \lfloor \log m \rfloor$. Then $d \leq \lfloor \log m \rfloor - 1$. From fact 1,

$$m \leq 2^{d+1} - 1 \leq 2^{(\lfloor \log m \rfloor - 1)+1} - 1$$
$$= 2^{\lfloor \log m \rfloor} - 1 \leq 2^{\log m} - 1 = m - 1$$

or

$$m \leq m - 1$$

a contradiction. Therefore $d \geq \lfloor \log m \rfloor$.

Now back to decision trees representing search algorithms on n-element lists. Temporarily strip the leaves from tree T_1 (with m internal nodes) to create a new tree T_2 with m nodes, $m \geq n$. By fact 2, T_2 has depth $d \geq \lfloor \log m \rfloor \geq \lfloor \log n \rfloor$. Therefore tree T_1 has depth $\geq \lfloor \log n \rfloor + 1$. Since the depth of the decision tree gives the worst-case number of comparisons, we can state the following theorem.

Theorem on the Lower Bound for Searching

Any algorithm that solves the search problem for an n-element list by comparing the target element x to the list items must do at least $\lfloor \log n \rfloor + 1$ comparisons in the worst case.

This gives us a lower bound on the number of comparisons required in the worst case for any algorithm that uses comparisons to solve the search problem. Since binary search does no more work than this required minimum amount, binary search is an **optimal algorithm** in its worst-case behavior.

Binary Tree Search

The binary search algorithm requires that data already be sorted. Arbitrary data can be organized into a structure called a *binary search tree*, which can then be searched using a different algorithm called *binary tree search*. To build a binary search tree, the first item of data is made the root of the tree. Successive items are inserted by comparing them to existing nodes, beginning with the root. If the item is less than a node, the next node tested is the left child; otherwise it is the right child. When no child node exists, the new item becomes the child.

EXAMPLE 32   The data items

5, 8, 2, 12, 10, 14, 9

are to be organized into a binary search tree. Figure 5.53 shows the successive states of constructing the tree.

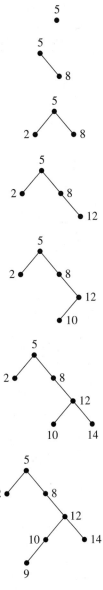

Figure 5.53

A **binary search tree**, by the way it is constructed, has the property that the value at each node is greater than all values in its left subtree (the subtree rooted at its left child) and less than all values in its right subtree. A **binary tree search** compares item x with a succession of nodes beginning with the root. If x equals the node item, the algorithm terminates; if x is less than the item, the left child is checked next; if x is greater than the item, the right child is checked next. If no child exists, the algorithm terminates because x is not in the list. Thus the binary search tree, except for the leaves, becomes the decision tree for the binary tree search algorithm. (Here is a case where the algorithm itself is described in terms of a tree.) The worst-case number of comparisons equals the depth of the tree plus 1 (for the missing leaves). However, a binary search tree for a given set of data is not unique; the tree (and hence the depth of the tree) depends on the order in which the data items are inserted into the tree.

EXAMPLE 33   The data in Example 32 entered in the order

9, 12, 10, 5, 8, 2, 14

produces the binary search tree of Figure 5.54.

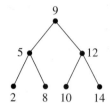

Figure 5.54

The actions performed in a binary tree search certainly resemble those in the "regular" binary search algorithm; in both cases the procedure is to make a comparison and, if unsuccessful, to look left or right (in the tree if it is a binary tree search or in the list if it is a binary search). It is possible to order the data for a binary tree search such that the search tree built from these data matches the decision tree (minus the leaves) for a binary search of the same data in sorted order. This is illustrated in Example 33 (note that the tree was not built from data items in sorted order). Here the binary search tree has the minimum depth and requires the least amount of work in the worst case.

The depth of a binary search tree for a given set of data items can vary. The depth of the tree in Figure 5.53 is 4, while that of Figure 5.54 is 2. Thus the worst-case number of comparisons to search for an item can also vary. The tree-building process can be modified to keep the tree more "balanced," that is, short and wide rather than tall and skinny; such a modification reduces the depth of the tree and therefore the search time. Of course, we know from the theorem on the lower bound for searching that a certain minimum amount of work is required no matter how clever we are in building the tree.

PRACTICE 25   **a.** Construct the binary search tree for the data of Example 32 entered in the order

12, 9, 14, 5, 10, 8, 2

b. What is the depth of the tree?

Sorting

Decision trees can also model algorithms that sort a list of items by a sequence of comparisons between two items from the list. The internal nodes of such a decision tree are labeled $L(i){:}L(j)$ to indicate a comparison of list item i to list item j. To simplify our discussion, let's assume that the list does not contain duplicate items. Then the outcome of such a comparison is either $L(i) < L(j)$ or $L(i) > L(j)$. If $L(i) < L(j)$, the algorithm proceeds to the comparison indicated at the left child of this node; if $L(i) > L(j)$, the algorithm proceeds to the right child. If no child exists, the algorithm terminates because the sorted order has been determined. The tree is a binary tree, and the leaves represent the final outcomes, that is, the various sorted orders.

EXAMPLE 34   Figure 5.55 shows the decision tree for a sorting algorithm acting on a list of three elements. This algorithm is not particularly astute because it ignores the transitive property of $<$ and therefore performs some unnecessary comparisons. The leaves of the tree indicate the various final outcomes, including two cases (marked with an X) that result from contradictory information. For example, one X results from the following inconsistent sequence of outcomes: $L(1) < L(2)$, $L(2) < L(3)$, $L(1) > L(3)$.

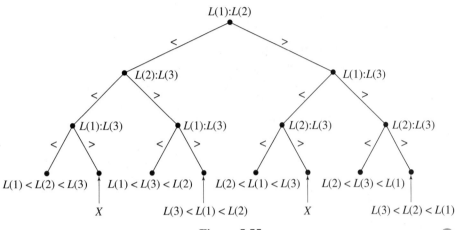

Figure 5.55

PRACTICE 26

Draw the decision tree that would result if the algorithm of Example 34 were modified to eliminate unnecessary comparisons.

A decision tree argument can also be used to establish a lower bound on the worst-case number of comparisons required to sort a list of n elements. As we did for the search problem, let us see what we can say about a decision tree for sorting based on comparisons, regardless of the algorithm it represents. The leaves of such a tree represent the final outcomes, that is, the various ordered arrangements of the n items. There are $n!$ such arrangements, so if p is the number of leaves in the decision tree, then $p \geq n!$. The worst case will equal the depth of the tree. But it is also true that if the tree has depth d, then $p \leq 2^d$ (Exercise 36 of Section 5.2). Taking the base 2 logarithm of both sides of this inequality, we get $\log p \leq d$ or, because d is an integer, $d = \lceil \log p \rceil$. Finally, we obtain

$$d = \lceil \log p \rceil \geq \lceil \log n! \rceil$$

This proves the following theorem.

Theorem on the Lower Bound for Sorting
Any algorithm that sorts an n-element list by comparing pairs of items from the list must do at least $\lceil \log n! \rceil$ comparisons in the worst case.

It can be shown (Exercise 23 at the end of this section) that $\log n! = \Theta(n \log n)$. Therefore we have proved that sorting n elements by comparing pairs of list items is bounded below by $\Theta(n \log n)$, whereas searching by comparing the target element to the list items is bounded below by $\Theta(\log n)$. As expected, it takes more work to sort than to search.

Section 5.3 Review

Techniques

- Draw decision trees for sequential search and binary search on n-element lists.
- Create a binary search tree.

Main Ideas

Decision trees represent the sequences of possible actions for certain algorithms.

Analysis of a generic decision tree for algorithms that solve a certain problem may lead to lower bounds on the minimum amount of work needed to solve the problem in the worst case.

The task of searching an n-element list for a target value x, if done by comparing x to elements in the list, requires at least $\lfloor \log n \rfloor + 1$ comparisons in the worst case.

The task of sorting an n-element list, if done by comparing pairs of list elements, requires at least $\lceil \log n! \rceil$ comparisons in the worst case.

Exercises 5.3

★1. Draw the decision tree for sequential search on a list of three elements.

2. Draw the decision tree for sequential search on a list of six elements.

3. Draw the decision tree for binary search on a list of seven elements. What is the depth of the tree?

4. Draw the decision tree for binary search on a list of four elements. What is the depth of the tree?

★5. Consider a search algorithm that compares an item with the last element in a list, then the first element, then the next-to-last element, then the second element, and so on. Draw the decision tree for searching a six-element sorted list. What is the depth of the tree? Does it appear that this is an optimal algorithm in the worst case?

6. Consider a search algorithm that compares an item with an element one-third of the way through the list; based on that comparison, it then searches either the first one-third or the second two-thirds of the list. Draw the decision tree for searching a nine-element sorted list. What is the depth of the tree? Does it appear that this is an optimal algorithm in the worst case?

★7. **a.** Given the data

9, 5, 6, 2, 4, 7

construct the binary search tree. What is the depth of the tree?

b. Find the average number of comparisons done to search for an item that is known to be in the list using binary tree search on the tree of part (a). (*Hint:* Find the number of comparisons for each of the items.)

8. **a.** Given the data

   g, d, r, s, b, q, c, m

   construct the binary search tree. What is the depth of the tree?
   b. Find the average number of comparisons done to search for an item that is known to be in the list using binary tree search on the tree of part (a). (*Hint:* Find the number of comparisons for each of the items.)

9. **a.** For a set of six data items, what is the minimum worst-case number of comparisons a search algorithm must perform?
   b. Given the set of data items {*a, d, g, i, k, s*}, find an order in which to enter the data so that the corresponding binary search tree has the minimum depth.

10. **a.** For a set of nine data items, what is the minimum worst-case number of comparisons a search algorithm must perform?
    b. Given the set of data items {4, 7, 8, 10, 12, 15, 18, 19, 21}, find an order in which to enter the data so that the corresponding binary search tree has the minimum depth.

★11. An inorder tree traversal of a binary search tree produces a listing of the tree nodes in alphabetical or numerical order. Construct a binary search tree for "To be or not to be, that is the question," and then do an inorder traversal.

12. Construct a binary search tree for "In the high and far off times the Elephant, O Best Beloved, had no trunk," and then do an inorder traversal. (See Exercise 11.)

★13. Use the theorem on the lower bound for sorting to find lower bounds on the number of comparisons required in the worst case to sort lists of the following sizes:

    a. 4   **b.** 8   **c.** 16

14. Contrast the number of comparisons required for selection sort and merge sort in the worst case with the lower bounds found in Exercise 13 (see Exercise 15 in Section 2.5). What are your conclusions?

Exercises 15–20 concern the problem of identifying a counterfeit coin (one that is two heavy or too light) from a set of *n* coins. A balance scale is used to weigh a group of any number of coins from the set against a like number of coins from the set. The outcome of such a comparison is that group A weighs less than, the same as, or more than group B. A decision tree representing the sequence of comparisons done will thus be a *ternary tree*, where an internal node can have three children.

15. One of five coins is counterfeit and is lighter than the other four. The problem is to identify the counterfeit coin.
    a. What is the number of final outcomes (the number of leaves in the decision tree)?
    b. Find a lower bound on the number of comparisons required to solve this problem in the worst case.
    c. Devise an algorithm that meets this lower bound (draw its decision tree).

★**16.** One of five coins is counterfeit and is either too heavy or too light. The problem is to identify the counterfeit coin and determine whether it is heavy or light.

   a. What is the number of final outcomes (the number of leaves in the decision tree)?

   b. Find a lower bound on the number of comparisons required to solve this problem in the worst case.

   c. Devise an algorithm that meets this lower bound (draw its decision tree).

17. One of four coins is counterfeit and is either too heavy or too light. The problem is to identify the counterfeit coin but not to determine whether it is heavy or light.

   a. What is the number of final outcomes (the number of leaves in the decision tree)?

   b. Find a lower bound on the number of comparisons required to solve this problem in the worst case.

   c. Devise an algorithm that meets this lower bound (draw its decision tree).

18. One of four coins is counterfeit and is either too heavy or too light. The problem is to identify the counterfeit coin and determine whether it is heavy or light.

   a. What is the number of final outcomes (the number of leaves in the decision tree)?

   b. Find a lower bound on the number of comparisons required to solve this problem in the worst case.

   c. Prove that no algorithm exists that can meet this lower bound. (*Hint:* The first comparison can be made with either two coins or four coins. Consider each case.)

19. Devise an algorithm to solve the problem of Exercise 18 using three comparisons in the worst case.

20. One of eight coins is counterfeit and is either too heavy or too light. The problem is to identify the counterfeit coin and determine whether it is heavy or light.

   a. What is the number of final outcomes (the number of leaves in the decision tree)?

   b. Find a lower bound on the number of comparisons required to solve this problem in the worst case.

   c. Devise an algorithm that meets this lower bound (draw its decision tree).

★**21.** In the decision tree for the binary search algorithm (and the binary tree search algorithm), we have counted each internal node as one comparison. For example, the top of Figure 5.52 looks like

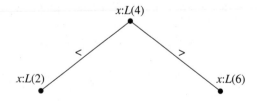

To get to either of the child nodes of the root, we have assumed that one comparison has been done. However, the outcome of the comparison at each internal node is really a three-way branch:

x = node element

$x <$ node element

$x >$ node element

Think about how this three-way branch would be implemented in most programming languages, and write a more accurate expression than $1 + \lfloor \log n \rfloor$ for the number of comparisons in the worst case.

22. Our existing binary search algorithm (Chapter 2, Example 40) contains the pseudocode instruction

find the index k of the middle item in the list $L(i) - L(j)$

after which the target x is compared to the list item at index k, the "middle item." Suppose that this instruction is replaced by

if $i = j$ **then**
    $k = j$
else
    $k = i + 1$
end if

a. Draw the decision tree that results from using the modified algorithm on a sorted list with $n = 8$.
b. Give the exact number of comparisons required (see Exercise 21) in the worst case for $n = 8$.
c. Give a worst-case order-of-magnitude expression for the number of comparisons required as a function of n, and justify your expression. Comment on the use of this algorithm as opposed to the original binary search algorithm, which is $\Theta(\log n)$.

23. To prove that $\log n! = \Theta(n \log n)$, we can use the definition of order of magnitude (see Section 4 of Chapter 4) and show that there exist positive constants n_0, c_1, and c_2 such that for $n \geq n_0$, $c_1(n \log n) \leq \log n! \leq c_2(n \log n)$.

a. Show that for $n \geq 1$, $\log n! \leq n \log n$. (*Hint:* Use the definition of $n!$ and properties of logarithms.)
b. Show that for $n \geq 4$, $\log n! \geq (1/4)(n \log n)$. (*Hint:* Use the definition of $n!$ and properties of logarithms, but stop at $\log \lceil n/2 \rceil$.)

Section 5.4   Huffman Codes

Problem and Trial Solution

Character data consist of letters of the alphabet (both uppercase and lowercase), punctuation symbols, and other keyboard symbols such as @ and %. Computers store character data in binary form, as a sequence of 0s and 1s. The usual approach is to fix some length n so that 2^n is as large as the number of distinct characters and to encode each distinct character as a particular sequence of n bits. Each character must be encoded into its fixed binary sequence, and then the binary sequence must be decoded when the character is to be displayed. The most common encoding scheme is ASCII (American Standard Code for Information Interchange), which uses $n = 8$, so that each character requires 8 bits to store. But whatever value is chosen for n, each character requires the same amount of storage space.

Suppose a collection of character data to be stored in a file in binary form is large enough that the amount of storage required is a consideration. Suppose also that the file is archival in nature, and its contents will not often be changed. Then it may be worthwhile to invest some extra effort in the encoding process if the amount of storage space required for the file could be reduced.

Rather than using a fixed number of bits per character, an encoding scheme could use a variable number of bits and store frequently occurring characters as sequences with fewer bits. In order to store all the distinct characters, some sequences will still have to be long, but if the longer sequences are used for characters that occur less frequently, the overall storage required should be reduced. This approach requires knowledge of the particular file contents, which is why it is best suited for a file whose contents will not be frequently changed. We shall study such a **data compression** or **data compaction** scheme here, because it is best described as a series of actions taken on binary trees.

EXAMPLE 35

As a trivial example, suppose that a collection of data contains 50,000 instances of the six characters a, c, g, k, p, and ?, which occur with the following percent frequencies:

Character	a	c	g	k	p	?
Frequency	48	9	12	4	17	10

Because six distinct characters must be stored, the fixed-length scheme would require at a minimum three bits for each character ($2^3 = 8 \geq 6$). The total storage required would then be $50{,}000 * 3 = 150{,}000$ bits. Suppose instead that the following encoding scheme is used:

Character	a	c	g	k	p	?
Encoding scheme	0	1101	101	1100	111	100

Then the storage requirement (number of bits) is

$$50{,}000(0.48 * 1 + 0.09 * 4 + 0.12 * 3 + 0.04 * 4 + 0.17 * 3 + 0.10 * 3) = 108{,}500$$

which is roughly two-thirds of the previous requirement.   ●

In the fixed-length storage scheme with n bits for each character, the long string of bits within the encoded file can be broken up into the code for successive characters by simply looking at n bits at a time. This makes it easy to decode the file. In the variable-length code, there must be a way to tell when the sequence for one character ends and the sequence for another character begins.

PRACTICE 27   Using the variable-length code of Example 35, decode each of the following strings:

a. 11111111010100
b. 1101010101100
c. 100110001101100   ●

In Practice 27 the strings can be broken into the representation of characters in only one way. As each new digit is considered, the possibilities are narrowed as to which character is being represented until the character is uniquely identified by the end of that character's representation. There is never any need to guess at what the character might be and then backtrack if our guess proves wrong. This ability to decode unique-ly without false starts and backtracking comes about because the code is an example of a **prefix code**. In a prefix code, the code for any character is never the prefix of the code for any other character. (A prefix code is therefore an "antiprefix" code!)

EXAMPLE 36   Consider the code

Character	a	b	c
Encoding scheme	01	101	011

which is not a prefix code. Given the string 01101, it could represent either ab (01–101) or ca (011–01). Furthermore, in processing the string 011011 digit by digit as a computer would do, the decoding could begin with ab (01–101) and only encounter a mismatch at the last digit. Then the process would have to go all the way back to the first digit in order to recognize cc (011–011).   ●

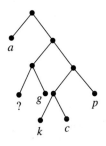

Figure 5.56

In our approach to prefix codes, we shall build binary trees with the characters as leaves. Once the tree is built, a binary code can be assigned to each character by simply tracing the path from the root to that leaf, using 0 for a left branch and 1 for a right branch. Because no leaf precedes any other leaf on some path from the root, the code will be a prefix code. The binary tree for the code of Example 35 is shown in Figure 5.56.

Suppose a code tree T exists, with leaves representing characters. For any leaf i, its depth $d(i)$ in T equals the number of bits in the code for the corresponding character. Let $f(i)$ denote the percentage frequency of that character in the data to be stored, and let S be the total number of characters to be stored. Then, just as in Example 35, the total bits required is given by the expression

$$S * \left[\sum_{\text{all leaves } i} (d(i)f(i)) \right]$$

We seek to build an optimal tree T, one for which the expression

$$E(T) = \sum_{\text{all leaves } i} (d(i)f(i)) \tag{1}$$

is a minimum and hence the file size is a minimum.

This process could be done by trial and error, because there is only a finite number of characters and thus only a finite number of ways to construct a tree and assign characters to its leaves. However, the finite number quickly becomes very large! Instead we shall use the algorithm known as **Huffman encoding**.

Huffman Encoding Algorithm

Suppose, then, that we have m characters in a file and we know the percentage frequency of each character. The algorithm to build the tree works by maintaining a list L of nodes that are roots of binary trees. Initially L will contain m roots, each labeled with the frequency of one of the characters; the roots will be ordered according to increasing frequency, and each will have no children.

A pseudocode description of the algorithm follows.

ALGORITHM   HuffmanTree

HuffmanTree (node list L; integer m)
//Each of the m nodes in L has an associated frequency f, and L is
//ordered by increasing frequency; algorithm builds the Huffman tree

    for $i = 1$ to $m - 1$ **do**
        create new node z
        let x, y be the first two nodes in L //minimum frequency nodes
        $f(z) = f(x) + f(y)$
        insert z in order into L
        left child of z = node x
        right child of z = node y //x and y are no longer in L
    end for

end HuffmanTree

When this algorithm terminates, L consists of just one node, which is the root of the final binary tree. Codes can then be assigned to each leaf of the tree by tracing the path from the root to the leaf and accumulating 0s for left branches and 1s for right branches. By the way the tree is constructed, every internal node will have exactly two children.

EXAMPLE 37   We'll use algorithm *HuffmanTree* to build the tree of Figure 5.56, which is based on the data of Example 35. L initially contains the six nodes, ordered by frequency:

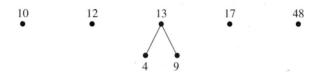

Following the algorithm, we enter the **for** loop for the first time. The x and y nodes are those with frequencies 4 and 9, respectively. A new node z with frequency $4 + 9 = 13$ is created and inserted in order into L, with the x node as its left child and the y node as its right child The new L looks like the following:

This process is repeated four more times. The resulting L at each stage follows:

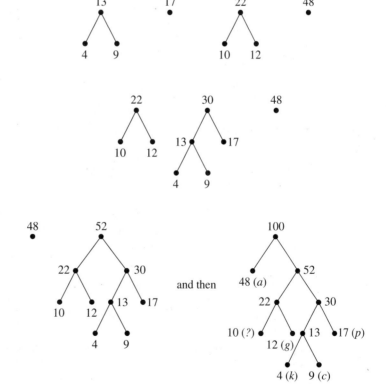

At this point the tree is complete and the codes can be assigned. The code for c, for example, is 1101 (right branch, right branch, left branch, right branch).

PRACTICE 28   Construct the Huffman tree for the following characters and frequencies:

Character	w	q	h	e
Frequency	10	12	20	58

●

PRACTICE 29   Find the Huffman codes for the characters of Practice 28.   ●

Table 5.1 shows the steps required to use Huffman encoding/decoding for data compression.

Encoding Step 1	On the original CLEARTEXT file, perform a frequency analysis; that is, create a file FREQUENCY that contains data of the form a—18 b—7 and so forth.
Encoding Step 2	Using CLEARTEXT and FREQUENCY, create a file CODETABLE that contains the Huffman code for each character, i.e., a—001 b—1110 and so forth.
Encoding Step 3	Using CLEARTEXT and CODETABLE, create a file called CODED that contains the compressed data.
Decoding	Using CODED and CODETABLE, decode the data to recover CLEARTEXT.

Table 5.1

The CODED file is the data-compressed version of CLEARTEXT, and presumably requires less storage space. However, the CODETABLE file must also be stored in order to be able to decode the file.

Justification

Although the algorithm to construct the Huffman tree T is easy enough to describe, we must justify that it gives us the minimum possible value for $E(T)$.

First, if we have an optimal tree T for m characters, the nodes with the lowest frequencies can always be assumed to be the left and right children of some node. To prove this, label the two nodes with the lowest frequencies x and y. If x and y are not siblings in the tree, then find two siblings p and q at the lowest level of the tree, and consider the case where x and y are not at that level (see Figure 5.57a). Because $f(x)$ is one of the two smallest values, we know that $f(x) \leq f(p)$. If $f(x) < f(p)$, then interchanging x and p in the tree would result in a new tree T' with $E(T') < E(T)$

(Figure 5.57b: the larger frequency is now at a lesser depth—see Exercise 13a), but this would contradict the fact that T was optimal. Therefore $f(x) = f(p)$, and x and p can be interchanged in the tree with no effect on $E(T)$. Similarly, y and q can be interchanged, resulting in Figure 5.57c, in which x and y are siblings. If x or y are at the same level as p and q to begin with, they can certainly be interchanged with p or q without affecting $E(T)$ (Figure 5.57d).

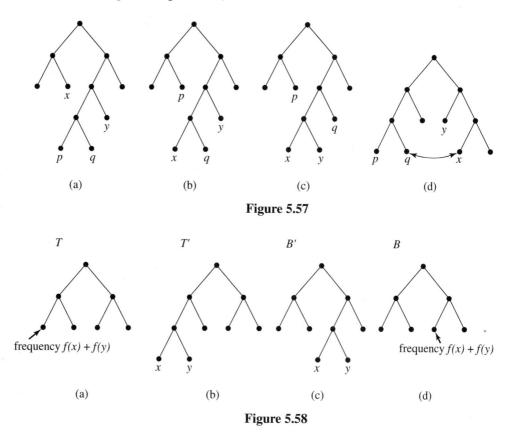

(a)   (b)   (c)   (d)

Figure 5.57

(a)   (b)   (c)   (d)

Figure 5.58

Now again let $f(x)$ and $f(y)$ be the minimum frequencies, and suppose we have a tree T that is optimal for the other frequencies together with the sum $f(x) + f(y)$ (Figure 5.58a). This sum will be the frequency of a leaf node; create a tree T' that has this node as an interior node with children x and y having frequencies $f(x)$ and $f(y)$ (Figure 5.58b). T' will be optimal for frequencies $f(x), f(y)$, and the rest. The proof of this fact begins with some optimal tree B' for frequencies $f(x), f(y)$, and the rest. We know such an optimal tree exists (since it could be found by trial and error), and from the preceding paragraph, we can assume that x and y are siblings in B' (Figure 5.58c). Now create a tree B by stripping nodes x and y from B' and giving frequency $f(x) + f(y)$ to their parent node, now a leaf (Figure 5.58d). Because T is optimal for the other frequencies together with $f(x) + f(y)$, we have

$$E(T) \le E(B) \tag{1}$$

But the difference between $E(B)$ and $E(B')$ is one arc each for x and y; that is, $E(B') = E(B) + f(x) + f(y)$ (see Exercise 13b). Similarly, we have $E(T') = E(T) + f(x) + f(y)$. Thus, if we add $f(x) + f(y)$ to both sides of (1), we get

$$E(T') \le E(B') \tag{2}$$

Because B' was optimal, it cannot be the case that $E(T') < E(B')$, so $E(T') = E(B')$, and T' is optimal.

Finally, a tree with a single node whose frequency is the sum of all the frequencies is trivially optimal for that sum. We can repeatedly split up this sum and drop down children in such a way that we end up with the Huffman tree. By the previous paragraph, each such tree, including the final Huffman tree, is optimal.

EXAMPLE 38   If we applied the process that preserves optimality to the tree of Figure 5.56, we would begin with a single node with frequency 100 and "grow" that tree downward, as shown in Figure 5.59.

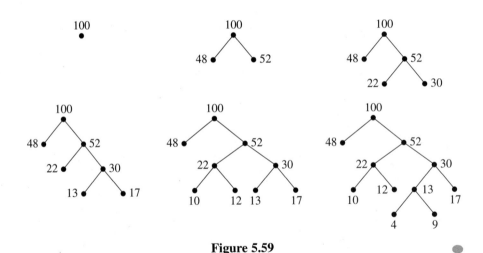

Figure 5.59

Section 5.4 Review

Technique

- Find Huffman codes, given a set of characters and their frequencies.

Main Idea

Given the frequency of characters in a collection of data, a binary encoding scheme can be found that minimizes the number of bits required to store the data but still allows for easy decoding.

Exercises 5.4

★1. Given the codes

Character	a	e	i	o	u
Encoding scheme	00	01	10	110	111

decode the sequences

a. 11011011101
b. 1000110111
c. 010101

2. Given the codes

Character	b	h	q	w	%
Encoding scheme	1000	1001	0	11	101

decode the sequences

a. 1000,1001,101,101
b. 11110
c. 01001111000

3. Given the codes

Character	a	p	w	(	)
Encoding scheme	001	1010	110	1111	1110

decode the sequences

a. 111110101101110001
b. 1010001110
c. 1111111100111101110

4. Given the nonprefix codes

Character	1	3	5	7	9
Encoding scheme	1	111	101	10	10101

give all possible decodings of the sequence 111110101.

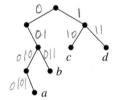

Exercise 5

5. Write the Huffman codes for a, b, c, and d in the binary tree shown.

6. Write the Huffman codes for r, s, t, u in the binary tree shown.

Exercise 6

★7. a. Construct the Huffman tree for the following characters and frequencies:

Character	c	d	g	m	r	z
Frequency	28	25	6	20	3	18

b. Find the Huffman codes for these characters.

8. **a.** Construct the Huffman tree for the following characters and frequencies:

Character	b	n	p	s	w
Frequency	6	32	21	14	27

   b. Find the Huffman codes for these characters.

9. **a.** Construct the Huffman tree for the following characters and frequencies:

Character	a	z	t	e	c
Frequency	27	12	15	31	15

   b. Find the Huffman codes for these characters.

★10. Someone does a global substitution on the text file of Exercise 9, replacing all instances of "*z*" with "*sh*"; find the new Huffman codes.

11. Consider the following paragraph:

   However, in my thoughts I could not sufficiently wonder at the intrepidity of these diminutive mortals who durst venture to mount and walk upon my body, while one of my hands was at liberty, without trembling at the very sight of so prodigious a creature as I must appear to them.[1]

   If this paragraph were to be compressed using a Huffman code, what single character, aside from punctuation or uppercase characters, would be apt to have one of the longest codes? Which would have one of the shortest?

12. Recall the problem posed at the beginning of this chapter:

   You work in the Information Systems Department at World Wide Widgets, the leading widget manufacturer. Part numbers are identified by a leading character of B, C, G, R, or S to identify the part type, followed by an 8-digit number. Thus

   C00347289
   B11872432
   S45003781

   are all legitimate part numbers. WWW maintains a data file of the part numbers it uses, which, as it turns out, is most of the potential numbers.
   *   How can you compress this data file so it takes less storage space than the approximately 4.5 Gb required using the ASCII encoding scheme of eight bits per character?*

[1]*From* Gulliver's Travels *by Jonathan Swift, London, 1726.*

a. Running a frequency count on the WWW data file reveals the following information:

Character	B	C	G	R	S	0	1	2	3	4	5	6	7	8	9
Frequency	2	5	1	2	1	18	13	7	12	9	6	11	7	2	4

Construct a Huffman code for these characters.

b. Compute the space requirements of the compressed file as a percent of the uncompressed file.

★**13.** In the justification that the Huffman algorithm produces an optimal tree, the following two assertions were made. Prove that each is true.

a. $E(T') < E(T)$
b. $E(B') = E(B) + f(x) + f(y)$

Chapter 5 Review

Self-Test

Answer the following true–false questions.

Section 5.1

1. A connected graph has an arc between any two nodes.

2. If graph G_1 is isomorphic to graph G_2, then a node of degree 5 in G_1 will be mapped to a node of degree 5 in G_2.

3. No matter how a planar graph is drawn, its arcs will intersect only at nodes.

4. If part of the adjacency list representation of a graph contains

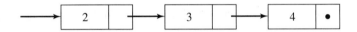

   it means that node 2 is adjacent to node 3 and that node 3 is adjacent to node 4.

5. The adjacency matrix of a directed graph is not symmetric.

Section 5.2

6. The depth of any node in a tree is less than or equal to the height of the tree.

7. Because a tree is a graph, a complete tree is also a complete graph.

8. In the left child–right child array representation of a binary tree, any row of the array that corresponds to a leaf will have all zero entries.

9. Postorder traversal of an expression tree results in an algebraic expression in reverse Polish notation.

10. In preorder tree traversal, the root is always the first node visited.

Section 5.3

11. The root of a decision tree for the binary search algorithm acting on a sorted list of 11 items would represent the comparison of the target element with the sixth list item.

12. Searching for any target element x in a list of n elements requires at least $1 + \lfloor \log n \rfloor$ comparisons.

13. A binary tree search is done with a target element of 14 on a binary search tree whose root has the value 10; the right subtree will be searched next.

14. A binary search tree is unique for any given set of data.

15. A decision tree for sorting n elements must have a depth of at least $n!$.

Section 5.4

16. The ASCII encoding scheme requires 8 bits to store each character.

17. In a prefix code, each code word is the prefix of another code word.

18. In a Huffman code, characters that occur most frequently have the most 0s in their binary string representation.

19. The maximum number of bits for any encoded character using a Huffman code will be the depth of the Huffman tree.

20. In order to be able to decode an encoded file, a frequency count from the original file must be stored along with the encoded file.

On the Computer

For Exercises 1–4, write a computer program that produces the desired output from the given input.

1. *Input*: Adjacency list for a graph
   Output: Adjacency matrix for the graph

2. *Input*: Adjacency matrix for a graph
   Output: Adjacency list for the graph

3. *Input*: Adjacency list for a graph and the name of a node *n* in the graph
   Output: Adjacency list for the graph with node *n* and its associated arcs removed

4. *Input*: List of *n* characters and their (integer) frequencies
   Output: Huffman code for the characters
   (*Hint:* Maintain a sorted linked list of records that represent the roots of binary trees. Initially there will be *n* such records, each with no children; at the end, there will be one such record, the root of the Huffman tree.)

5. Write a program that allows the user to enter a list of integers and constructs a binary search tree with those integers as nodes. The user can then enter one integer at a time, and the program will do a binary tree search and indicate whether the given integer is in the list.

6. Write a program that allows the user to enter a list of integers and then constructs a binary search tree with those integers as nodes. The user can then enter the type of traversal desired (inorder, preorder, or postorder), and the program will write out the nodes in the appropriate order.

7. Write a program that carries out the first three steps in Table 5.1. That is, beginning with a text file, the program should produce a frequency count file, then a code table file, then an encoded version of the original file. Write a second program that uses the encoded file and the code table, and recreates the original file.

Graph Algorithms

Chapter Objectives

After studying this chapter, you will be able to:

- Convert between adjacency matrix, adjacency relation, and directed graph representations.
- Use the reachability matrix of a directed graph to determine whether one node is reachable from another.
- Compute the reachability matrix of a directed graph either directly or by using Warshall's algorithm.
- Test a graph for the existence of an Euler path (solve the highway inspector problem).
- Understand the Hamiltonian circuit problem (and the traveling salesman problem) and how they are fundamentally different from the Euler path problem.
- Use Dijkstra's algorithm to find the shortest path between two nodes in a simple, weighted, connected graph.
- Use Prim's algorithm to find the minimal spanning tree in a simple, weighted, connected graph.
- Carry out depth-first search and breadth-first search in a simple, connected graph.
- Understand how depth-first search can be used to test for reachability in a directed graph, perform a topological sort on a partially ordered set represented by a directed graph, and find the connected components of an unconnected graph.
- Find the articulation points in a simple, connected graph.

You are the network administrator for a wide-area backbone network that serves your company's many offices across the country. Messages travel through the network by being routed from point to point until they reach their destination. Each node in the network therefore acts as a switching station to forward messages to other nodes according to a routing table maintained at each node. Some connections in the network carry heavy traffic, while others are less used. Traffic may vary

with the time of day; in addition, new nodes occasionally come on line and existing nodes may go off line. Therefore you must periodically provide each node with
updated information so that it can forward messages along the most efficient (that
is, the least heavily traveled) route.

Question: How can you compute the routing table for each node?

If the network described is viewed as a graph, your task as network administrator
is to find the "shortest" path from one node to another in the graph. Because
graphs have so many applications, there is a great deal of interest in finding efficient algorithms to answer certain questions about graphs, directed graphs, or
trees, and to perform certain tasks on them, such as finding shortest paths. All
graph algorithms use one of the convenient representations (adjacency matrix or
adjacency list) presented in Chapter 5.

This chapter covers many of the "classical" graph algorithms. Section 6.1 first
relates directed graphs to binary relations and reachability in a graph to the transitive closure of a binary relation. Then two different algorithms pertaining to reachability are given.

In Section 6.2 we shall look at algorithms that answer two historically interesting questions about graphs. These questions are known as the *highway inspector problem* and the *traveling salesman problem*. The highway inspector problem
asks whether there is a path through a given graph that uses each arc exactly once,
thereby providing an efficient way for a highway inspector to check all roads without going over the same road twice. The traveling salesman problem asks whether
there is a cycle in a given graph that visits each node of the graph and, if so, which
such cycle requires the minimum distance to travel. Recall that a cycle is a path
that ends where it started and does not use any other node more than once; thus
such a cycle would provide an efficient way for a salesperson to visit all cities in
the sales territory only once and end up at home.

Section 6.3 provides algorithmic solutions to the two problems of finding the
minimum path between two nodes in a simple, connected graph and of minimizing the number of arcs used to connect all nodes in a simple, connected graph.
Section 6.4 discusses algorithms for traversing simple graphs—"visiting" all the
nodes in some systematic way. Finally, Section 6.5 presents an algorithm to identify relatively isolated sections of a graph.

Section 6.1 Directed Graphs and Binary Relations; Warshall's Algorithm

In this section we confine our attention to (unweighted) directed graphs with no
parallel arcs. (In a directed graph, two arcs from node a to node b would be parallel,
but one arc from a to b and another from b to a are not parallel arcs.) Consider the
adjacency matrix of the graph (assuming some arbitrary ordering of the n nodes,
which we always assume when discussing the adjacency matrix of a graph). This will
be an $n \times n$ matrix, not necessarily symmetric. Furthermore, because there are no
parallel arcs in the graph, the adjacency matrix will be a Boolean matrix, that is, a
matrix whose only elements are 0s and 1s. Conversely, given an $n \times n$ Boolean matrix,

we can reconstruct the directed graph that the matrix represents, and it will have no parallel arcs. Thus there is a one-to-one correspondence, which we can picture as

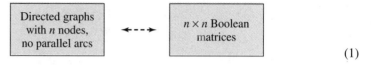

$$\boxed{\begin{array}{c}\text{Directed graphs}\\\text{with } n \text{ nodes,}\\\text{no parallel arcs}\end{array}} \quad \longleftarrow\text{-}\text{-}\longrightarrow \quad \boxed{\begin{array}{c}n \times n \text{ Boolean}\\\text{matrices}\end{array}} \tag{1}$$

Now we will see how binary relations tie in to this correspondence.

Directed Graphs and Binary Relations

Suppose G is a directed graph with n nodes and no parallel arcs. Let N be the set of nodes. If (n_i, n_j) is an ordered pair of nodes, then there either is or is not an arc in G from n_i to n_j. We can use this property to define a binary relation on the set N:

$$n_i \, \rho \, n_j \leftrightarrow \text{there is an arc in } G \text{ from } n_i \text{ to } n_j$$

This relation is the **adjacency relation** of the graph.

EXAMPLE 1   For the directed graph of Figure 6.1, the adjacency relation is $\{(1, 2), (1, 3), (3, 3), (4, 1), (4, 2), (4, 3)\}$.

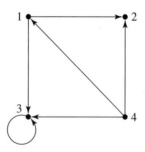

Figure 6.1

Conversely, if ρ is a binary relation on a set N, we can define a directed graph G with N as the set of nodes, and an arc from n_i to n_j if and only if $n_i \, \rho \, n_j$. G will have no parallel arcs.

EXAMPLE 2   For the set $N = \{1, 2, 3, 4\}$ and the binary relation $\{(1, 4), (2, 3), (2, 4), (4, 1)\}$ on N, we obtain the associated directed graph shown in Figure 6.2.

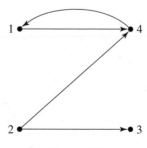

Figure 6.2

We now have another one-to-one correspondence:

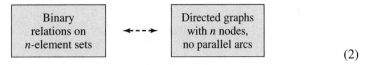

$$(2)$$

Of course, a one-to-one correspondence means the existence of a bijection. If function composition is carried out on the bijections in (1) and (2), the result is a bijection that gives us a one-to-one correspondence between binary relations and matrices. Thus we have three equivalent sets:

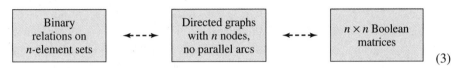

$$(3)$$

An item from any of the three sets has corresponding representations in the other two sets.

PRACTICE 1   Give the collection of ordered pairs in the adjacency relation for the following Boolean matrix; also draw the directed graph:

$$\begin{bmatrix} 0 & 0 & 0 & 0 \\ 1 & 1 & 0 & 0 \\ 1 & 0 & 0 & 1 \\ 0 & 0 & 0 & 0 \end{bmatrix}$$

Recall the reflexive, symmetric, antisymmetric, and transitive properties of a binary relation on a set that we studied in Chapter 4. If a binary relation on a set N has a certain property, this will be reflected in the corresponding graph and the corresponding Boolean matrix. Conversely, certain characteristics of a directed graph or of a Boolean matrix imply certain properties of the corresponding adjacency relation.

EXAMPLE 3   If ρ is a reflexive relation on a set N, then for each $n_i \in N$, $n_i \rho n_i$. In the corresponding directed graph there will be a loop at each node, and in the corresponding Boolean matrix there will be 1s on the main diagonal.

PRACTICE 2   Explain why the corresponding binary relation is not antisymmetric for the directed graph in Figure 5.13.

In Chapter 4 we represented partial orderings on a set by using a Hasse diagram. How does this representation differ from the directed graph representation? The Hasse diagram is a simplification of the directed graph representation. Suppose that G is the directed graph representation of a partial ordering. Because a partial ordering is reflexive, G will have a loop at each node. We can eliminate these loops in the Hasse diagram without losing any information because we know that each node has a loop; that is, each node is related to itself. Because a partial ordering is transitive, if $a \rho b$ and $b \rho c$, then $a \rho c$. In the directed graph there would be an arc from a to b, an arc from b to c, and an arc from a to c. In the Hasse diagram we can eliminate the arc from a to c without losing any information if we keep the transitive property in mind.

Finally, the Hasse diagram is not a directed graph at all, but we did impose the convention that if a is an immediate predecessor of b, then node a will appear below node b in the Hasse diagram. Thus we could achieve a directed graph from the Hasse diagram by making all the arc directions point upward. The antisymmetry property prevents any potential conflict where node a should be below node b and node b should be below node a.

In Chapter 4 we also noted set operations that could be performed on two binary relations ρ and σ on a set N, $\rho \cup \sigma$ and $\rho \cap \sigma$. The relation $\rho \cup \sigma$ is the union of the ordered pairs in ρ or σ, while $\rho \cap \sigma$ is the intersection of the ordered pairs in ρ and σ. Let $\mathbf{R}$ and $\mathbf{S}$ be the Boolean matrices for ρ and σ, respectively. The Boolean matrix for $\rho \cup \sigma$ will have a 1 in position i,j if and only if there is a 1 in position i,j of $\mathbf{R}$ or a 1 in position i,j of $\mathbf{S}$. Each entry in the Boolean matrix for $\rho \cup \sigma$ is thus the maximum of the two corresponding entries in $\mathbf{R}$ and $\mathbf{S}$, so the Boolean matrix for $\rho \cup \sigma$ is $\mathbf{R} \vee \mathbf{S}$ (see the discussion of Boolean matrix operations in Section 4.5). Similarly, the Boolean matrix for $\rho \cap \sigma$ will have a 1 in position i,j if and only if there is a 1 in position i,j of both $\mathbf{R}$ and $\mathbf{S}$. Therefore the Boolean matrix for $\rho \cap \sigma$ is $\mathbf{R} \wedge \mathbf{S}$.

Reachability

The "reachability" property has an interesting interpretation in each of the three equivalent forms in (3)—directed graph, adjacency relation, and adjacency matrix. We already have a definition for this term for directed graphs from Section 5.1, which we'll restate now.

> **Definition: Reachable Node**
> In a directed graph, node n_j is **reachable** from node n_i if there is a path from n_i to n_j.

EXAMPLE 4   In the directed graph of Figure 6.2, node 3 is not reachable from node 4 or node 1. Node 1 is reachable from node 2 by the path 2–4–1.   ●

In a system modeled by a directed graph (a data flow diagram, for example) with a "start node," any node that is unreachable from the start node can never affect the system and thus can be eliminated. If the directed graph represents something like airline routes or communication paths in a computer network, it would be undesirable to have some node be unreachable from some other node. Thus the ability to test reachability has very practical applications.

The adjacency matrix $\mathbf{A}$ of a directed graph G with n nodes and no parallel arcs will have a 1 in position i,j if there is an arc from n_i to n_j. This would be a path of length 1 from n_i to n_j. The adjacency matrix by itself therefore tells us about a limited form of reachability, via length-1 paths. However, let us perform the Boolean matrix multiplication $\mathbf{A} \times \mathbf{A}$. We'll denote this product by $\mathbf{A}^{(2)}$ to distinguish it from $\mathbf{A}^2$, the result of $\mathbf{A} \cdot \mathbf{A}$ using ordinary matrix multiplication. Recalling from Section 4.5 the definition of Boolean matrix multiplication, the i,j entry of $\mathbf{A}^{(2)}$ is given by

$$\mathbf{A}^{(2)}[i, j] = \bigvee_{k=1}^{n} (a_{ik} \wedge a_{kj}) \tag{4}$$

If a term such as $a_{i2} \wedge a_{2j}$ in this sum is 0, then either $a_{i2} = 0$ or $a_{2j} = 0$ (or both), and there is either no path of length 1 from n_i to n_2 or no path of length 1 from n_2 to n_j (or both). Thus there are no paths of length 2 from n_i to n_j passing through n_2. If $a_{i2} \wedge a_{2j}$ is not 0, then both $a_{i2} = 1$ and $a_{2j} = 1$. Then there is a path of length 1 from n_i to n_2 and a path of length 1 from n_2 to n_j, so there is a path of length 2 from n_i to n_j passing through n_2. A path of length 2 from n_i to n_j will exist if and only if there is a path of length 2 passing through at least one of the nodes from 1 to n, that is, if and only if at least one of the terms in the sum (4) is 1 and therefore $\mathbf{A}^{(2)}[i, j] = 1$. Therefore the entries in $\mathbf{A}^{(2)}$ tell us about reachability via length-2 paths.

PRACTICE 3   Find $\mathbf{A}$ for the graph of Figure 6.2 and compute $\mathbf{A}^{(2)}$. What does the 2,1 entry indicate?

The matrix $\mathbf{A}^{(2)}$ indicates the presence or absence of length-2 paths. We might surmise that this result holds for arbitrary powers and path lengths.

Theorem on Boolean Adjacency Matrices and Reachability
If $\mathbf{A}$ is the Boolean adjacency matrix for a directed graph G with n nodes and no parallel arcs, then $\mathbf{A}^{(m)}[i, j] = 1$ if and only if there is a path of length m from node n_i to node n_j.

Proof: A proof by induction on m is called for. We have already shown the result true for $m = 1$ (and $m = 2$). Suppose that $\mathbf{A}^{(p)}[i, j] = 1$ if and only if there is a path of length p from n_i to n_j. We know

$$\mathbf{A}^{(p+1)}[i, j] = \bigvee_{k=1}^{n} (\mathbf{A}^{(p)}[i, k] \wedge a_{kj})$$

This will equal 1 if and only if at least one term, say $\mathbf{A}^{(p)}[i, q] \wedge a_{qj} = 1$, or $A^{(p)}[i, q] = 1$ and $a_{qj} = 1$. This will be true if and only if there is a path of length p from n_i to n_q (by the inductive hypothesis) and there is a path of length 1 from n_q to n_j, which means there is a path of length $p + 1$ from n_i to n_j.

PRACTICE 4   From the graph of Figure 6.2, what would you expect for the value of entry 2,1 in $\mathbf{A}^{(4)}$? Compute $\mathbf{A}^{(4)}$ and check this value. ●

If node n_j is reachable from node n_i, it is by a path of some length. Such a path will be shown by a 1 as the i,j entry in $\mathbf{A}$ or $\mathbf{A}^{(2)}$ or $\mathbf{A}^{(3)}$ and so on, but we cannot compute an infinite number of matrix products. Fortunately, there is a limit to how far in this list we have to look. If there are n nodes in the graph, then any path with n or more arcs, and therefore $n + 1$ or more nodes, must have a repeated node. This is a consequence of the Pigeonhole Principle—there are n "bins" (distinct nodes) into which we are putting more than n objects (the nodes in a path with n or more arcs). The section of a path lying between the repeated nodes is a cycle. If $n_i \neq n_j$, the cycle can be eliminated to make a shorter path; then if a path exists from n_i to n_j, there will be such a path of length at most $n - 1$. If $n_i = n_j$, then the cycle could be the entire path from n_i to n_i with maximum length n; although we could eliminate this cycle (noting that any node may be considered reachable from itself), we shall retain it to show that a nontrivial path does exist from n_i to n_i.

Consequently, whether $n_i = n_j$ or $n_i \neq n_j$, we need never look for a path from n_i to n_j of length greater than n. Therefore to determine reachability, we need only consult element i,j in $\mathbf{A}, \mathbf{A}^{(2)}, \dots, \mathbf{A}^{(n)}$. Alternatively, we can define a **reachability matrix R** by

$$\mathbf{R} = \mathbf{A} \vee \mathbf{A}^{(2)} \vee \cdots \vee \mathbf{A}^{(n)}$$

Then n_j is reachable from n_i if and only if entry i,j in $\mathbf{R}$ is positive.

We now see how reachability in a graph can be expressed in terms of the adjacency matrix. How is reachability represented in terms of the adjacency relation that corresponds to the graph?

If ρ is the adjacency relation for a graph G, we let $\rho^{\mathbf{R}}$ denote the binary relation of reachability; that is, $(n_i, n_j) \in \rho^{\mathbf{R}}$ exactly when there is a path in G from n_i to n_j. Then we can show that $\rho^{\mathbf{R}}$ is the transitive closure of ρ. Recall from the definition of closure of a relation that the transitive closure of ρ is a relation that is transitive, contains ρ, and is a subset of any transitive relation containing ρ.

To see that $\rho^{\mathbf{R}}$ is transitive, let (n_i, n_j) and (n_j, n_k) belong to $\rho^{\mathbf{R}}$. Then there is a path in G from n_i to n_j and a path in G from n_j to n_k. Therefore there is a path in G from n_i to n_k, and (n_i, n_k) belongs to $\rho^{\mathbf{R}}$. To see that $\rho^{\mathbf{R}}$ contains ρ, let (n_i, n_j) belong to ρ. Then there is an arc from n_i to n_j in G, which means there is a path of length 1 from n_i to n_j, and (n_i, n_j) belongs to $\rho^{\mathbf{R}}$. Finally, suppose σ is any transitive relation on the nodes of G that includes ρ, and let (n_i, n_j) belong to $\rho^{\mathbf{R}}$. This means there is a path from n_i to n_j using, say, nodes $n_i, n_x, n_y, \dots, n_w, n_j$. Then there is an arc from each node in this path to the next, and the ordered pairs $(n_i, n_x), (n_x, n_y), \dots, (n_w, n_j)$ all belong to ρ, and therefore all belong to σ. Because σ is transitive, (n_i, n_j) belongs to σ, and $\rho^{\mathbf{R}}$ is a subset of σ. Therefore $\rho^{\mathbf{R}}$ is the transitive closure of ρ.

To summarize, corresponding to the three equivalent representations of adjacency relation ρ, directed graph G, and adjacency matrix $\mathbf{A}$, we have that

(n_i, n_j) belongs to the transitive closure of ρ	$\leftrightarrow$ n_j reachable from n_i in G $\leftrightarrow$	$R[i, j] = 1$ where $\mathbf{R} = \mathbf{A} \vee \mathbf{A}^{(2)} \vee \cdots \vee \mathbf{A}^{(n)}$

EXAMPLE 5   Let G be the directed graph in Figure 6.3; G has 5 nodes.

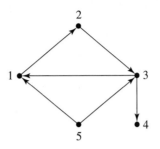

Figure 6.3

The adjacency matrix $\mathbf{A}$ for G is

$$\mathbf{A} = \begin{bmatrix} 0 & 1 & 0 & 0 & 0 \\ 0 & 0 & 1 & 0 & 0 \\ 1 & 0 & 0 & 1 & 0 \\ 0 & 0 & 0 & 0 & 0 \\ 1 & 0 & 1 & 0 & 0 \end{bmatrix}$$

The adjacency relation ρ is $\rho = \{(1, 2), (2, 3), (3, 1), (3, 4), (5, 1), (5, 3)\}$.

The successive powers of $\mathbf{A}$ are

$$\mathbf{A}^{(2)} = \begin{bmatrix} 0 & 0 & 1 & 0 & 0 \\ 1 & 0 & 0 & 1 & 0 \\ 0 & 1 & 0 & 0 & 0 \\ 0 & 0 & 0 & 0 & 0 \\ 1 & 1 & 0 & 1 & 0 \end{bmatrix} \quad \mathbf{A}^{(3)} = \begin{bmatrix} 1 & 0 & 0 & 1 & 0 \\ 0 & 1 & 0 & 0 & 0 \\ 0 & 0 & 1 & 0 & 0 \\ 0 & 0 & 0 & 0 & 0 \\ 0 & 1 & 1 & 0 & 0 \end{bmatrix}$$

$$\mathbf{A}^{(4)} = \begin{bmatrix} 0 & 1 & 0 & 0 & 0 \\ 0 & 0 & 1 & 0 & 0 \\ 1 & 0 & 0 & 1 & 0 \\ 0 & 0 & 0 & 0 & 0 \\ 1 & 0 & 1 & 1 & 0 \end{bmatrix} \quad \mathbf{A}^{(5)} = \begin{bmatrix} 0 & 0 & 1 & 0 & 0 \\ 1 & 0 & 0 & 1 & 0 \\ 0 & 1 & 0 & 0 & 0 \\ 0 & 0 & 0 & 0 & 0 \\ 1 & 1 & 0 & 1 & 0 \end{bmatrix}$$

These matrices indicate, for example, that there is a path of length 2 from 2 to 1 because $\mathbf{A}^{(2)}[2, 1] = 1$ (the path is 2–3–1), and there is a path of length 4 from 5 to 3 because $\mathbf{A}^{(4)}[5, 3] = 1$ (the path is 5–3–1–2–3), but there is no path of length 3 from 1 to 3 because $\mathbf{A}^{(3)}[1, 3] = 0$.

The reachability matrix $\mathbf{R}$ is the Boolean sum of $\mathbf{A}$, $\mathbf{A}^{(2)}$, $\mathbf{A}^{(3)}$, $\mathbf{A}^{(4)}$, and $\mathbf{A}^{(5)}$:

$$\mathbf{R} = \begin{bmatrix} 1 & 1 & 1 & 1 & 0 \\ 1 & 1 & 1 & 1 & 0 \\ 1 & 1 & 1 & 1 & 0 \\ 0 & 0 & 0 & 0 & 0 \\ 1 & 1 & 1 & 1 & 0 \end{bmatrix}$$

> **REMINDER:**
> Don't try to compute matrix products without writing the two matrices side by side; you'll surely make a mistake.

The 1 values in R indicate that there are paths in G from nodes 1, 2, 3, and 5 to every node except 5, but no path from node 4 to anywhere, which can be confirmed by looking at Figure 6.3.

We have proved that the 1 entries in $\mathbf{R}$ mark the ordered pairs of nodes that belong to the transitive closure of ρ. The transitive closure will therefore be the following set of ordered pairs:

$$\{(1, 1), (1, 2), (1, 3), (1, 4), (2, 1), (2, 2), (2, 3), (2, 4),$$
$$(3, 1), (3, 2), (3, 3), (3, 4), (5, 1), (5, 2), (5, 3), (5, 4)\}$$

Beginning with ρ and following the ad hoc procedure described in Chapter 4 for finding the transitive closure of a relation, we see that to obtain transitivity, we must first add the pairs (1, 3), (2, 1), (2, 4), (3, 2), (5, 2), and (5, 4). Reviewing the new set, we see that we must also add (1, 1), (1, 4), (2, 2), and (3, 3). The resulting collection of ordered pairs is transitive (and agrees with what we obtained earlier). ●

PRACTICE 5
Compute $\mathbf{R}$ for the directed graph of Figure 6.2. What information does column 2 convey? ●

In Chapter 4, we promised a better algorithm to find the transitive closure of a relation. Here it is: Write the binary relation in adjacency matrix form and compute

$$\mathbf{R} = \mathbf{A} \vee \mathbf{A}^{(2)} \vee \cdots \vee \mathbf{A}^{(n)}$$

How much work is required to carry out this algorithm? The expression for $\mathbf{R}$ indicates that Boolean matrix operations are to be done, but matrix operations in turn require Boolean **and** and Boolean **or** operations on matrix elements. We shall therefore use Boolean **and** and Boolean **or** as the measure of work. In Section 4.5, we noted that ordinary matrix multiplication of two $n \times n$ matrices requires $\Theta(n^3)$ multiplications and additions; by a similar argument, Boolean matrix multiplication of two $n \times n$ Boolean matrices requires $\Theta(n^3)$ Boolean **and/or** operations. The algorithm to compute $\mathbf{R}$ requires $n - 1$ Boolean matrix multiplications (to find the products $\mathbf{A}^{(2)}$, $\mathbf{A}^{(3)}, \dots , \mathbf{A}^{(n)}$). To compute $n - 1$ such products requires $(n - 1)\Theta(n^3) = \Theta(n^4)$ Boolean operations. To compute $\mathbf{C} \vee \mathbf{D}$ where $\mathbf{C}$ and $\mathbf{D}$ are two $n \times n$ Boolean matrices requires n^2 Boolean **or** operations. To compute $\mathbf{R}$, $n - 1$ such matrix operations are required, so $(n - 1)n^2 = \Theta(n^3)$ Boolean **or** operations are performed. The total amount of work is $\Theta(n^4) + \Theta(n^3) = \Theta(n^4)$.

Next we'll discuss a more efficient algorithm for computing the transitive closure of a relation (or the reachability matrix of a graph).

Warshall's Algorithm

For a graph G with n nodes, Warshall's algorithm computes a sequence of $n + 1$ matrices $\mathbf{M}_0, \mathbf{M}_1, \mathbf{M}_2, \dots , \mathbf{M}_n$. For each k, $0 \le k \le n$, $\mathbf{M}_k[i, j] = 1$ if and only if there is a path in G from n_i to n_j whose interior nodes (i.e., nodes that are not the endpoints of the path) come only from the set of nodes $\{n_1, n_2, \dots , n_k\}$.

Let us examine the "end conditions." When $k = 0$, the set $\{n_1, n_2, \ldots, n_0\}$ is the empty set, so $\mathbf{M}_0[i, j] = 1$ if and only if there is a path in G from n_i to n_j whose interior nodes come from the empty set; that is, there are no interior nodes. The path from n_i to n_j must then consist only of the endpoints and one connecting arc, so n_i and n_j are adjacent nodes. Thus $\mathbf{M}_0 = \mathbf{A}$. The other end condition occurs when $k = n$. Then the set $\{n_1, n_2, \ldots, n_n\}$ consists of all the nodes in G so there is really no restriction at all on the interior nodes in the path, and $\mathbf{M}_n[i, j] = 1$ if and only if there is a path from n_i to n_j, which means that $\mathbf{M}_n = \mathbf{R}$.

Therefore Warshall's algorithm begins with $\mathbf{A} = \mathbf{M}_0$ and successively computes $\mathbf{M}_1, \mathbf{M}_2, \ldots, \mathbf{M}_n = \mathbf{R}$. This computation can be defined inductively. The base case is to let $\mathbf{M}_0 = \mathbf{A}$. Now assume that $\mathbf{M}_k$ has been computed, and consider how to compute $\mathbf{M}_{k+1}$ or, more specifically, $\mathbf{M}_{k+1}[i, j]$. We have $\mathbf{M}_{k+1}[i, j] = 1$ if and only if there is a path from n_i to n_j whose interior nodes come only from the set $\{n_1, n_2, \ldots, n_{k+1}\}$. This can happen in two ways:

1. All of the interior nodes come from $\{n_1, n_2, \ldots, n_k\}$, in which case $\mathbf{M}_k[i, j] = 1$. We should therefore carry forward any 1 entries in $\mathbf{M}_k$ into $\mathbf{M}_{k+1}$.

2. Node n_{k+1} is an interior node. We can assume that n_{k+1} is an interior node only once, because cycles can be eliminated from a path. Then there must be a path from n_i to n_{k+1} whose interior nodes come from $\{n_1, n_2, \ldots, n_k\}$ and a path from n_{k+1} to n_j whose interior nodes come from $\{n_1, n_2, \ldots, n_k\}$. This means that $\mathbf{M}_k[i, k + 1] = 1$ and $\mathbf{M}_k[k + 1, j] = 1$, which is to say that $\mathbf{M}_k[i, k + 1] \wedge \mathbf{M}_k[k + 1, j] = 1$; this condition can be tested because our assumption is that $\mathbf{M}_k$ has already been computed.

In the following pseudocode version of Warshall's algorithm, the initial value of matrix $\mathbf{M}$ is $\mathbf{A}$. Each pass through the outer loop computes the next matrix in the sequence $\mathbf{M}_1, \mathbf{M}_2, \ldots, \mathbf{M}_n = \mathbf{R}$.

ALGORITHM   **Warshall**

Warshall($n \times n$ Boolean matrix $\mathbf{M}$)
//Initially, $\mathbf{M}$ = adjacency matrix of a directed graph G with no parallel arcs

> **for** $k = 0$ to $n - 1$ **do**
>> **for** $i = 1$ to n **do**
>>> **for** $j = 1$ to n **do**
>>>> $\mathbf{M}[i, j] = \mathbf{M}[i, j] \vee (\mathbf{M}[i, k + 1] \wedge \mathbf{M}[k + 1, j])$
>>> **end for**
>> **end for**
> **end for**
> //at termination, $\mathbf{M}$ = reachability matrix of G
> **end** Warshall

This gives a nice neat description of Warshall's algorithm, which can be implemented as computer code rather easily. These steps are confusing to do by hand, however, requiring some bookkeeping to keep track of all the indices. We can write the algorithm more informally, making it easier to do manually. Suppose again that matrix $\mathbf{M}_k$ in the sequence exists and we are trying to write row i of the next matrix in the sequence. This means that for the various values of j, we must evaluate the expression

$$\mathbf{M}[i, j] \vee (\mathbf{M}[i, k + 1] \wedge \mathbf{M}[k + 1, j]) \tag{5}$$

If entry $\mathbf{M}[i, k + 1]$ is 0, then $\mathbf{M}[i, k + 1] \wedge \mathbf{M}[k + 1, j] = 0$ for all j. Expression (5) then reduces to

$$\mathbf{M}[i, j] \vee 0 = \mathbf{M}[i, j]$$

In other words, row i of the matrix remains unchanged. If, on the other hand, entry $\mathbf{M}[i, k + 1]$ is 1, then $\mathbf{M}[i, k + 1] \wedge \mathbf{M}[k + 1, j] = \mathbf{M}[k + 1, j]$ for all j. Expression (5) then becomes

$$\mathbf{M}[i, j] \vee \mathbf{M}[k + 1, j]$$

In other words, row i of the matrix becomes the Boolean **or** of the current row i and the current row $k + 1$.

Table 6.1 describes the (informal) steps to compute entries in $\mathbf{M}_{k+1}$ from matrix $\mathbf{M}_k$.

---

1. Consider column $k + 1$ in $\mathbf{M}_k$.

---

2. For each row with a 0 entry in this column, copy that row to $\mathbf{M}_{k+1}$.

---

3. For each row with a 1 entry in this column, **or** that row with row $k + 1$ and write the resulting row in $\mathbf{M}_{k+1}$.

Table 6.1

EXAMPLE 6   For the graph of Example 5, the initial matrix $\mathbf{M}_0$ is the adjacency matrix.

$$\mathbf{M}_0 = \begin{bmatrix} 0 & 1 & 0 & 0 & 0 \\ 0 & 0 & 1 & 0 & 0 \\ 1 & 0 & 0 & 1 & 0 \\ 0 & 0 & 0 & 0 & 0 \\ 1 & 0 & 1 & 0 & 0 \end{bmatrix}$$

We know $\mathbf{M}_0$ and we want to compute $\mathbf{M}_1$. Using step 1 of Table 6.1, we consider column 1 of $\mathbf{M}_0$. Using step 2 of Table 6.1, rows 1, 2, and 4 of $\mathbf{M}_0$ contain 0s in column 1, so these rows get copied directly to $\mathbf{M}_1$:

$$\mathbf{M}_1 = \begin{bmatrix} 0 & 1 & 0 & 0 & 0 \\ 0 & 0 & 1 & 0 & 0 \\ & & & & \\ 0 & 0 & 0 & 0 & 0 \\ & & & & \end{bmatrix}$$

Now we finish up by using step 3 of Table 6.1 Row 3 of column 1 of M_0 contains a 1, so row 3 of M_0 is **or**-ed with row 1 of M_0 and the result becomes the new row 3:

$$M_1 = \begin{bmatrix} 0 & 1 & 0 & 0 & 0 \\ 0 & 0 & 1 & 0 & 0 \\ 1 & 1 & 0 & 1 & 0 \\ 0 & 0 & 0 & 0 & 0 \end{bmatrix}$$

Row 5 of column 1 of M_0 contains a 1, so row 5 is **or**-ed with row 1 and the result becomes the new row 5:

$$M_1 = \begin{bmatrix} 0 & 1 & 0 & 0 & 0 \\ 0 & 0 & 1 & 0 & 0 \\ 1 & 1 & 0 & 1 & 0 \\ 0 & 0 & 0 & 0 & 0 \\ 1 & 1 & 1 & 0 & 0 \end{bmatrix}$$

To compute the entries in M_2, consider column 2. Rows 2 and 4 (the 0 positions in column 2) will be copied unchanged. Row 1 will be **or**-ed with row 2 to give the new row 1, row 3 will be **or**-ed with row 2 to give the new row 3, and row 5 will be **or**-ed with row 2 to give the new row 5:

$$M_2 = \begin{bmatrix} 0 & 1 & 1 & 0 & 0 \\ 0 & 0 & 1 & 0 & 0 \\ 1 & 1 & 1 & 1 & 0 \\ 0 & 0 & 0 & 0 & 0 \\ 1 & 1 & 1 & 0 & 0 \end{bmatrix}$$

M_3 is computed in a similar fashion:

$$M_3 = \begin{bmatrix} 1 & 1 & 1 & 1 & 0 \\ 1 & 1 & 1 & 1 & 0 \\ 1 & 1 & 1 & 1 & 0 \\ 0 & 0 & 0 & 0 & 0 \\ 1 & 1 & 1 & 1 & 0 \end{bmatrix}$$

M_4 and M_5 will be the same as M_3; row 4 is all 0s, so any row that gets **or**-ed with it will be unchanged, and column 5 is all 0s so all rows are copied directly. In terms of the graph, no new 1 entries are produced because there are no paths from 4 to any node or from any node to 5. Thus $M_3 = M_4 = M_5 = R$ as computed in Example 5. Note, however, that the matrices computed by Warshall's algorithm, except for A and R, do not agree with the matrices that are powers of A used in our previous algorithm for R. ●

Each pass through the outer loop of Warshall's algorithm modifies in place the matrix that existed at the end of the previous pass. Warshall's algorithm requires no additional storage for other matrices, even though we wrote down new matrices in our example. There is one more point we need to check. Because we are modifying the (only) matrix as we go along, during any one pass through the outer loop, some of the entries will belong to $\mathbf{M}_{k+1}$ while others will still belong to $\mathbf{M}_k$, Specifically, on pass $k + 1$, we may consider $\mathbf{M}[i, k + 1] \wedge \mathbf{M}[k + 1, j]$ in expression (5), where these values have already been computed on this pass and therefore represent $\mathbf{M}_{k+1}[i, k + 1]$ and $\mathbf{M}_{k+1}[k + 1, j]$ rather than the values $\mathbf{M}_k[i, k + 1]$ and $\mathbf{M}_k[k + 1, j]$ we used in our justification for this algorithm. Can there be a case where the values $\mathbf{M}_{k+1}[i, k + 1]$ and $\mathbf{M}_{k+1}[k + 1, j]$ are 1, so that a 1 value goes into $\mathbf{M}_{k+1}[i, j]$, whereas the values $\mathbf{M}_k[i, k + 1]$ and $\mathbf{M}_k[k + 1, j]$ are 0? No—if $\mathbf{M}_{k+1}[i, k + 1] = 1$, there is a path from n_i to n_{k+1} with interior nodes drawn from the set $\{n_1, n_2, \ldots, n_{k+1}\}$. However, because n_{k+1} is an endpoint and cycles can be eliminated, there must also be a path with interior nodes drawn from the set $\{n_1, n_2, \ldots, n_k\}$ so that $\mathbf{M}_k[i, k + 1] = 1$. A similar argument holds for $\mathbf{M}_{k+1}[k + 1, j]$.

PRACTICE 6   Use Warshall's algorithm (formally or informally) to compute $\mathbf{R}$ for the graph of Figure 6.2. Compare your answer with that for Practice 5.   ●

How much work does Warshall's algorithm require as measured by the number of Boolean **and/or** operations? Consider the formal algorithm. The single assignment statement in the algorithm lies within a triply nested loop; it will be executed n^3 times. Each execution of the assignment statement requires one **and** and one **or**; therefore, the total amount of work is $2n^3 = \Theta(n^3)$. Recall that our previous algorithm for computing $\mathbf{R}$ was an $\Theta(n^4)$ algorithm.

Section 6.1 Review

Techniques

- Find any two of the adjacency relation, directed graph, or adjacency matrix representations, given the third.
- Compute the reachability matrix $\mathbf{R}$ for a graph G (or, equivalently, find the transitive closure of the adjacency relation on G) by using the formula $\mathbf{R} = \mathbf{A} \vee \mathbf{A}^{(2)} \vee \cdots \vee \mathbf{A}^{(n)}$ and by using Warshall's algorithm.

Main Ideas

There is a one-to-one correspondence between a directed graph G with no parallel arcs, the adjacency relation on G, and the adjacency matrix for G (with respect to some arbitrary ordering of the nodes).

The reachability matrix of a graph G also represents the transitive closure of the adjacency relation on G.

The reachability matrix for a graph can be computed with $\Theta(n^4)$ Boolean **and/or** operations by summing powers of the adjacency matrix $\mathbf{A}$ or with $\Theta(n^3)$ Boolean **and/or** operations by using Warshall's algorithm.

Exercises 6.1

★1. Find the adjacency matrix and adjacency relation for the graph in the figure.

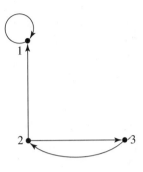

Exercise 1

2. Find the adjacency matrix and adjacency relation for the graph in the figure.

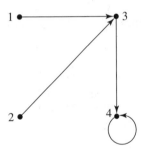

Exercise 2

★3. Find the corresponding directed graph and adjacency relation for the following adjacency matrix.

$$\mathbf{A} = \begin{bmatrix} 0 & 0 & 0 & 1 & 0 \\ 0 & 0 & 0 & 0 & 1 \\ 0 & 0 & 0 & 0 & 0 \\ 0 & 1 & 0 & 0 & 0 \\ 0 & 0 & 1 & 1 & 0 \end{bmatrix}$$

4. Find the corresponding directed graph and adjacency relation for the following adjacency matrix.

$$\mathbf{A} = \begin{bmatrix} 0 & 0 & 0 & 0 & 0 \\ 1 & 0 & 0 & 0 & 1 \\ 0 & 1 & 0 & 1 & 0 \\ 0 & 1 & 0 & 0 & 0 \\ 0 & 0 & 0 & 1 & 0 \end{bmatrix}$$

5. Given the adjacency relation $\rho = \{(1, 4), (1, 5), (1, 6), (6, 2), (6, 3), (6, 5)\}$ on the set $N = \{1, 2, 3, 4, 5, 6\}$, find the corresponding directed graph and adjacency matrix.

6. Given the adjacency relation $\rho = \{(2, 1), (3, 2), (3, 3), (3, 4), (4, 5), (6, 3), (6, 6)\}$ on the set $N = \{1, 2, 3, 4, 5, 6\}$, find the corresponding directed graph and adjacency matrix.

7. Describe a property of a directed graph whose adjacency matrix is symmetric.

★8. Describe the directed graph whose adjacency matrix has all 1s in row 1 and column 1, and 0s elsewhere.

9. Describe the directed graph whose adjacency matrix has 1s in positions $(i, i + 1)$ for $1 \le i \le n - 1$, a 1 in position $(n, 1)$, and 0s elsewhere.

10. Describe a property of the adjacency matrix of a graph whose adjacency relation is antisymmetric.

11. Adjacency relations ρ and σ have the following associated adjacency matrices $\mathbf{R}$ and $\mathbf{S}$. Find the adjacency matrices associated with the relations $\rho \cup \sigma$ and $\rho \cap \sigma$.

$$\mathbf{R} = \begin{bmatrix} 0 & 1 & 1 & 0 \\ 0 & 0 & 0 & 1 \\ 1 & 1 & 0 & 0 \\ 1 & 0 & 0 & 1 \end{bmatrix} \qquad \mathbf{S} = \begin{bmatrix} 0 & 1 & 0 & 0 \\ 0 & 0 & 1 & 0 \\ 1 & 0 & 0 & 1 \\ 1 & 0 & 0 & 0 \end{bmatrix}$$

★12. The two directed graphs in the figure have adjacency relations ρ and σ. Draw the graphs associated with the relations $\rho \cup \sigma$ and $\rho \cap \sigma$.

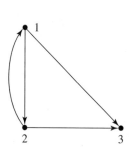
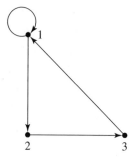

Exercise 12

13. Let $\mathbf{A}$ be the matrix

$$\mathbf{A} = \begin{bmatrix} 1 & 1 & 0 & 1 \\ 0 & 0 & 1 & 0 \\ 1 & 0 & 0 & 1 \\ 1 & 0 & 1 & 0 \end{bmatrix}$$

Find the products $\mathbf{A}^2$ and $\mathbf{A}^{(2)}$.

14. The definition of a *connected graph* can be extended to directed graphs. Describe the reachability matrix **R** for a connected, directed graph.

★15. For the graph of the figure, write the reachability matrix **R** by simply inspecting the graph.

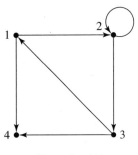

Exercise 15

16. For the graph of the figure, write the reachability matrix **R** by simply inspecting the graph.

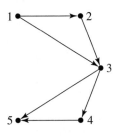

Exercise 16

For Exercises 17–22, compute the reachability matrix **R** by using the formula $\mathbf{R} = \mathbf{A} \vee \mathbf{A}^{(2)} \vee \cdots \vee \mathbf{A}^{(n)}$.

17. Exercise 1   18. Exercise 2   ★19. Exercise 3

20. Exercise 4   21. Exercise 5   22. Exercise 6

For Exercises 23–28, compute the reachability matrix **R** by using Warshall's algorithm.

★23. Exercise 1   24. Exercise 2   25. Exercise 3

26. Exercise 4   ★27. Exercise 5   28. Exercise 6

Exercises 29–32 use regular matrix multiplication to obtain information about a graph.

★29. Let G be a directed graph, possibly with parallel arcs, and let $\mathbf{A}$ be its adjacency matrix. Then $\mathbf{A}$ may not be a Boolean matrix. Prove that the i,j entry of matrix $\mathbf{A}^2$ is the number of paths of length 2 from node i to node j.

30. Let $\mathbf{A}$ be the adjacency matrix of a directed graph G, possibly with parallel arcs. Prove that the i,j entry of matrix $\mathbf{A}^n$ gives the number of paths of length n from node i to node j.

31. For the graph G of the figure, count the number of paths of length 2 from node 1 to node 3. Check by computing $\mathbf{A}^2$.

Exercise 31

32. For the graph G of the figure, count the number of paths of length 4 from node 1 to node 5. Check by computing $\mathbf{A}^4$.

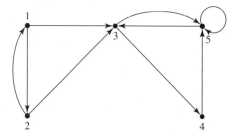

Exercise 32

33. Let ρ be a binary relation defined on the set $\{0, \pm1, \pm2, \pm4, \pm16\}$ by $x \rho y \leftrightarrow y = x^2$. Draw the associated directed graph.

Section 6.2   Euler Path and Hamiltonian Circuit

Euler Path Problem

The Euler path problem (the highway inspector problem) originated many years ago. Swiss mathematician Leonhard Euler (pronounced "oiler") (1707–1783) was intrigued by a puzzle popular among the townsfolk of Königsberg (an East Prussian city later called Kaliningrad, which is in Russia). The river flowing through the city branched around an island. Various bridges crossed the river as shown in Figure 6.4.

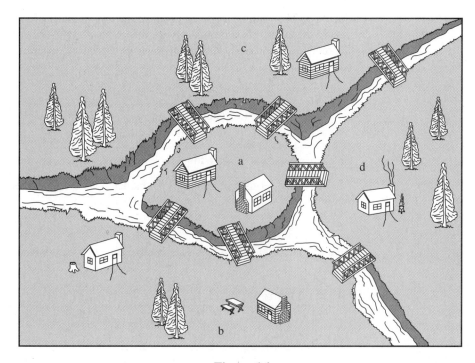

Figure 6.4

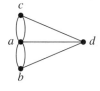

Figure 6.5

The puzzle was to decide whether a person could walk through the city crossing each bridge only once. It is possible to answer the question by trial and error, listing (or walking) all possible routes, so some dedicated Königsberger could have solved this particular puzzle. Euler's idea was to represent the situation as a graph (see Figure 6.5), where the bridges are arcs and the land masses (labeled *a* through *d*) are nodes. He then solved—by a better mechanism than trial and error—the general question of when an Euler path exists in any graph.

> **Definition: Euler Path**
> An **Euler path** in a graph G is a path that uses each arc of G exactly once.

PRACTICE 7

(a)

(b)

Figure 6.6

Do Euler paths exist for either graph in Figure 6.6? (Use trial and error to answer. This is the old children's game of whether you can trace the whole graph without lifting your pencil and without retracing any arcs.) ●

For this discussion we shall assume that all graphs are connected, since an Euler path generally cannot exist otherwise. Whether an Euler path exists in a given graph hinges on the degrees of its nodes. A node is **even** if its degree is even and **odd** if its degree is odd. It turns out that every graph has an even number of odd nodes. To see this, choose any graph and let N be the number of odd nodes in it, $N(1)$ the number of nodes of degree 1, $N(2)$ the number of nodes of degree 2, and so on. Then the sum S of the degrees of all the nodes of the graph is

$$S = 1 \cdot N(1) + 2 \cdot N(2) + 3 \cdot N(3) + \cdots + k \cdot N(k) \tag{1}$$

for some k. This sum is, in fact, a count of the total number of arc ends in the graph. Because the number of arc ends is twice the number of arcs, S is an even number. We shall reorganize equation (1) to group together terms for odd nodes and terms for even nodes:

$$S = \underbrace{2 \cdot N(2) + 4 \cdot N(4) + \cdots + 2m \cdot N(2m)}_{\text{even nodes}}$$

$$\underbrace{+ 1 \cdot N(1) + 3 \cdot N(3) + \cdots + (2n + 1) \cdot N(2n + 1)}_{\text{odd nodes}}$$

The sum of the terms representing even nodes is an even number. If we subtract it from both sides of the equation, we get a new equation

$$S' = 1 \cdot N(1) + 3 \cdot N(3) + \cdots + (2n + 1) \cdot N(2n + 1) \tag{2}$$

where S' (the difference of two even numbers) is an even number. Now if we rewrite equation (2) as

$$S' = \underbrace{1 + 1 + \cdots + 1}_{N(1) \text{ terms}} + \underbrace{3 + 3 + \cdots + 3}_{N(3) \text{ terms}} + \cdots$$

$$\underbrace{+ (2n + 1) + (2n + 1) + \cdots + (2n + 1)}_{N(2n + 1) \text{ terms}}$$

we see that there are N terms altogether in the sum (the number of odd nodes) and that each term is an odd number. For the sum of N odd numbers to be even, N must be even. (Can you prove this?) We have thus proved the following theorem.

Theorem on Odd Nodes in a Graph
The number of odd nodes in any graph is even.

Now suppose a graph has an odd node n of degree $2k + 1$ and that an Euler path exists in the graph but does not start at n. Then for each arc we use to enter n, there is another unused arc for leaving n until we have used k pairs of arcs. The next time we enter n, there is no new arc on which to leave. Thus, if our path does not begin at n, it must end at n. The path either begins at n or it does not, and in the latter case it ends at n, so the path either begins or ends at this arbitrary odd node. Therefore, if there are more than two odd nodes in the graph, there can be no path. Thus, there are two possible cases where an Euler path may exist—on a graph with no odd nodes or on one with two odd nodes.

Consider the graph with no odd nodes. Pick any node m and begin an Euler path. Whenever you enter a different node, you will always have another arc on which to exit until you get back to m. If you have used up every arc of the graph, you are done. If not, there is some node m' of your path with unused arcs. Then construct an Euler path beginning and ending at m', much as you did the previous section of path, using all new arcs. Attach this cycle as a side trip on the original path. If you have now used up every arc of the graph, you are done. If not, continue this process until every arc has been covered.

If there are exactly two odd nodes, an Euler path can be started beginning at one odd node and ending at the other. If the path has not covered all of the arcs, extra cycles can be patched in as in the previous case.

We now have a complete solution to the Euler path problem.

Theorem on Euler Paths

An Euler path exists in a connected graph if and only if there are either no odd nodes or two odd nodes. For the case of no odd nodes, the path can begin at any node and will end there; for the case of two odd nodes, the path must begin at one odd node and end at the other.

PRACTICE 8   Using the preceding theorem, work Practice 7 again.   ●

PRACTICE 9   Is the Königsberg walk possible?   ●

The theorem on Euler paths is actually an algorithm to determine if an Euler path exists on an arbitrary connected graph. To make it look more like an algorithm, we'll rewrite it in pseudocode. The essence of the algorithm is to count the number of nodes adjacent to each node and to determine whether this is an odd or an even number. If there are too many odd numbers, an Euler path does not exist.

In the accompanying algorithm (algorithm *EulerPath*), the input is a connected graph represented by an $n \times n$ adjacency matrix $\mathbf{A}$. The variable *total* keeps track of the number of odd nodes found in the graph. The degree of any particular node, *degree*, is found by adding the numbers in that node's row of the adjacency matrix. The function *odd* results in a value "true" if and only if the argument is an odd integer.

ALGORITHM   **EulerPath**

EulerPath ($n \times n$ matrix $\mathbf{A}$)
//Determines whether an Euler path exists in a connected graph with
//adjacency matrix $\mathbf{A}$

Local variables:
integer *total*   //number of odd nodes so far found
integer *degree*   //the degree of a node
integer i, j //array indices

  total = 0
  $i = 1$
  while *total* $<= 2$ and $i <= n$ **do**
    degree = 0

(continued)

ALGORITHM   EulerPath *(continued)*

> **for** $j = 1$ to n **do**
>> $degree = degree + A[i, j]$ //find degree of node i (*)
>
> **end for**
> **if** odd(*degree*) **then**
>> $total = total + 1$ //another odd degree node found
>
> **end if**
> $i = i + 1$
> **end while**
> **if** $total > 2$ **then**
>> write ("No Euler path exists")
>
> **else**
>> write ("Euler path exists")
>
> **end if**
> **end** EulerPath

EXAMPLE 7   The adjacency matrix for the graph of Figure 6.6a follows.

$$\begin{bmatrix} 0 & 2 & 1 & 0 & 0 \\ 2 & 0 & 1 & 0 & 0 \\ 1 & 1 & 0 & 1 & 1 \\ 0 & 0 & 1 & 0 & 2 \\ 0 & 0 & 1 & 2 & 0 \end{bmatrix}$$

When the algorithm first enters the **while** loop, *total* is 0 and i is 1. Then *degree* is initialized to 0. Within the **for** loop, the values of row 1 of the adjacency matrix are added in turn to *degree*, resulting in a value for *degree* of 3. The *odd* function applied to *degree* returns the value "true," so the value of *total* is increased from 0 to 1; one node of odd degree has been found. Then i is incremented to 2. Neither the bounds on *total* nor the bounds on the array size have been exceeded, so the **while** loop executes again, this time for row 2 of the array. Once again, *degree* is found to be odd, so the value of *total* is changed to 2. When the **while** loop is executed for row 3 of the array, the value of *degree* is even (4), so *total* does not change, and the **while** loop is executed again with $i = 4$. Row 4 again produces an odd value for *degree*, so *total* is raised to 3. This terminates the **while** loop. The bad news is written that there is no Euler path because the number of odd nodes exceeds 2.   ●

PRACTICE 10   Write the adjacency matrix for the Königsberg walk problem and trace the execution of algorithm *EulerPath*.   ●

Let us analyze algorithm *EulerPath*. The important operation done by the algorithm is an examination of the elements of the adjacency matrix, which occurs at line (*). In the worst case, the **while** loop in the algorithm is executed n times, once for each row. Within the **while** loop, the **for** loop, containing line (*), is executed n times, once for each column. *EulerPath* is therefore an $\Theta(n^2)$ algorithm in the worst case.

At the cost of some extra decision logic, we could modify the algorithm because we never have to examine the last row of the matrix. We know from the theorem on Euler paths that the total number of odd nodes is even. If the number of odd nodes after processing the next-to-last row is odd, then the last row must represent an odd node; if that number is even, then the last row must represent an even node. This modification results in $(n - 1)n$ elements to examine in the worst case, which is still $\Theta(n^2)$.

If we represented the graph G by an adjacency list rather than an adjacency matrix, then the corresponding version of the algorithm would have to count the length of the adjacency list for each node and keep track of how many are of odd length. There would be n adjacency lists to examine, just as there were n rows of the adjacency matrix to examine, but the length of each adjacency list might be shorter than n, the length of a row of the matrix. It is possible to reduce the order of magnitude below n^2 if the number of arcs in the graph is small, but the worst case is still $\Theta(n^2)$.

Hamiltonian Circuit Problem

Another famous mathematician, William Rowan Hamilton (1805–1865), posed a problem in graph theory that sounds very much like Euler's. He asked how to tell whether a graph has a **Hamiltonian circuit**, a cycle using every node of the graph.

PRACTICE 11   Do Hamiltonian circuits exist for the graphs of Figure 6.6? (Use trial and error to answer.)   ●

Like the Euler path problem, the Hamiltonian circuit problem can be solved for a given graph by trial and error. The algorithm is as follows: Start from one node of the graph and try some path by choosing various arcs. If the path results in a repeated node, it is not a cycle, so throw it away and try a different path. If the path can be completed as a cycle, then see whether it visited every node; if not, throw it away and try a different path. Continue in this fashion until all possible paths have been tried or a Hamiltonian circuit has been found. This will involve some careful record keeping so that no path is tried more than once. The trial-and-error approach is theoretically possible—but it is practically impossible! In all but the smallest of graphs, there will simply be too many paths to try.

Euler found a simple, efficient algorithm to determine, for an arbitrary graph, if an Euler path exists. Although the Hamiltonian circuit problem sounds very similar to the Euler path problem, there is a basic difference. No efficient algorithm has ever been found to determine if a Hamiltonian circuit exists. In fact, there is some evidence (see Section 8.3) to suggest that no such algorithm will ever be found.

In certain types of graphs we can easily determine whether a Hamiltonian circuit exists. For example, a complete graph with $n > 2$ has a Hamiltonian circuit because for any node on the path, there is always an arc to travel to any unused node and finally an arc to return to the starting point. In general, however, we cannot make this determination readily.

Suppose we are dealing with a weighted graph. If a Hamiltonian circuit exists for the graph, can we find one with minimum weight? This is the traveling salesman problem. Once again it can be solved using trial and error by tracing all possible paths and keeping track of the weights of those paths that are Hamiltonian circuits, but, again, this is not an efficient algorithm. (Incidentally, the traveling salesman problem for visiting all 48 capitals of the contiguous United States has been solved—a total of 10,628 miles is required!)

Section 6.2 Review

Technique

- Using algorithm *EulerPath*, determine whether an Euler path exists in a graph.

Main Ideas

There is a simple criterion for determining whether Euler paths exist in a graph, but no such criterion for whether Hamiltonian circuits exist.

An algorithm that is $\Theta(n^2)$ in the worst case can determine the existence of an Euler path in a connected graph with n nodes.

Exercises 6.2

1. Rework Example 3 of Chapter 2 using the theorem on Euler paths.

For Exercises 2–9, determine whether the graph in the specified figure has an Euler path by using the theorem on Euler paths.

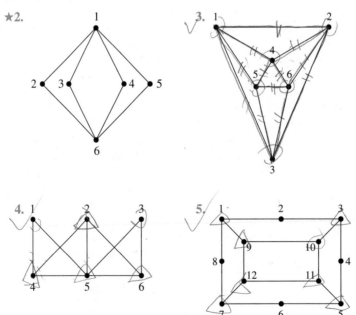

6.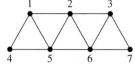

★7.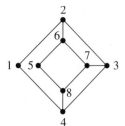

8.

9.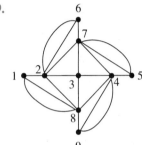

★10. Draw the adjacency matrix for the graph of Exercise 2. In applying algorithm *EulerPath*, what is the value of *total* after the second pass through the **while** loop?

11. Draw the adjacency matrix for the graph of Exercise 4. In applying algorithm *EulerPath*, what is the value of *total* after the fourth pass through the **while** loop?

★12. Draw the adjacency matrix for the graph of Exercise 6. In applying algorithm *EulerPath*, what is the value of *i* after the **while** loop is exited?

13. Draw the adjacency matrix for the graph of Exercise 7. In applying algorithm *EulerPath*, what is the value of *i* after the **while** loop is exited?

For Exercises 14–21, decide by trial and error whether Hamiltonian circuits exist for the graphs of the given exercise.

★14. Exercise 2   **15.** Exercise 3   **16.** Exercise 4   **17.** Exercise 5

18. Exercise 6   **★19.** Exercise 7   **20.** Exercise 8   **21.** Exercise 9

22. Find an example of an unconnected graph that has an Euler path. (*Hint:* Because this seems intuitively contradictory, you should look for a trivial case.)

★23. Prove that any graph with a Hamiltonian circuit is connected.

24. Consider a simple, complete graph with n nodes. Testing for a Hamiltonian circuit by trial and error could be done by selecting a fixed starting node and then generating all possible paths from that node of length n.

    a. How many paths of length n are there if repetition of arcs and nodes is allowed?

    b. How many paths of length n are there if repetition of arcs and nodes is allowed but an arc may not be used twice in succession?

    c. How many paths of length n are there if nodes and arcs cannot be repeated except for the starting node? (These are the Hamiltonian circuits.)

25. Is it possible to walk in and out of each room in the house shown in the accompanying figure so that each door of the house is used exactly once? Why or why not?

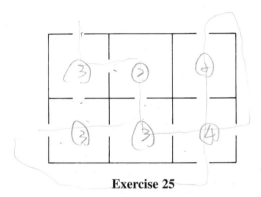

Exercise 25

★26. Recall that K_n denotes the simple, complete graph of order n.

    a. For what values of n does an Euler path exist in K_n?

    b. For what values of n does a Hamiltonian circuit exist in K_n?

27. Recall that $K_{m,n}$ denotes a bipartite, complete graph with $m + n$ nodes.

    a. For what values of m and n does an Euler path exist in $K_{m,n}$?

    b. For what values of m and n does a Hamiltonian circuit exist in $K_{m,n}$?

28. Consider a connected graph with $2n$ odd vertices, $n \geq 2$. By the theorem on Euler paths, an Euler path does not exist for this graph.

    a. What is the minimum number of disjoint Euler paths, each traveling some of the arcs of the graph, necessary to travel each arc exactly once?

    b. Show that the minimum number is sufficient.

★29. Prove that a Hamiltonian circuit always exists in a connected graph where every node has degree 2.

Section 6.3   Shortest Path and Minimal Spanning Tree

Shortest-Path Problem

Assume that we have a simple, weighted, connected graph, where the weights are positive. Then a path exists between any two nodes x and y. Indeed, there may be many such paths. The question is, How do we find a path with minimum weight? Because weight often represents distance, this problem has come to be known as the "shortest-path" problem. It is an important problem to solve for a computer or communications network, where information at one node must be routed to another node in the most efficient way possible, or for a transportation network, where products in one city must be shipped to another.

The traveling salesman problem is a minimum-weight path problem with such severe restrictions on the nature of the path that such a path may not exist at all. In the shortest-path problem, we put no restrictions (other than minimum weight) on the nature of the path, and because the graph is connected, we know that such a path exists. For this reason we may hope for an efficient algorithm to solve the problem, even though no such algorithm is known for the traveling salesman problem. Indeed such an algorithm does exist.

The shortest-path algorithm known as Dijkstra's algorithm works as follows. We want to find the minimum-distance path from a given node x to a given node y. We build a set (we'll call it IN) that initially contains only x but grows as the algorithm proceeds. At any given time IN contains every node whose shortest path from x, using only nodes in IN, has so far been determined. For every node z outside IN, we keep track of the shortest distance $d[z]$ from x to that node, using a path whose only non-IN node is z. We also keep track of the node adjacent to z on this path, $s[z]$.

How do we let IN grow; that is, which node should be moved into IN next? We pick the non-IN node with the smallest distance d. Once we add that node, call it p, to IN, then we have to recompute d for all the remaining non-IN nodes, because there may be a shorter path from x going through p than there was before p belonged to IN. If there is a shorter path, we must also update $s[z]$ so that p is now shown to be the node adjacent to z on the current shortest path. As soon as y is moved into IN, IN stops growing. The current value of $d[y]$ is the distance for the shortest path, and its nodes are found by looking at y, $s[y]$, $s[s[y]]$, and so forth, until we have traced the path back to x.

A pseudocode form of the algorithm is given in the accompanying box (algorithm $ShortestPath$). The input is the adjacency matrix for a simple, connected graph G with positive weights and nodes x and y; the algorithm writes out the shortest path between x and y and the distance for that path. Here shortest path means minimum-weight path. We actually assume a modified adjacency matrix $\mathbf{A}$, where $\mathbf{A}[i, j]$ is the weight of the arc between i and j if one exists and $\mathbf{A}[i, j]$ has the value ∞ if no arc exists (here the symbol ∞ denotes a number larger than any weight in the graph).

ALGORITHM   ShortestPath

ShortestPath ($n \times n$ matrix **A**; nodes x, y)
//Dijkstra's algorithm. **A** is a modified adjacency matrix for a simple, connected
//graph with positive weights; x and y are nodes in the graph; writes out nodes in the
//shortest path from x to y, and the distance for that path

Local variables:
set of nodes *IN*   //set of nodes whose shortest path from x is known
nodes z, p //temporary nodes
array of integers d //for each node, the distance from x using nodes in *IN*
array of nodes s //for each node, the previous node in the shortest path
integer *OldDistance*   //distance to compare against

  //initialize set *IN* and arrays d and s
  $IN = \{x\}$
  $d[x] = 0$
  for all nodes z not in *IN* **do**
      $d[z] = \mathbf{A}[x, z]$
      $s[z] = x$
  end for

  //process nodes into *IN*
  while y not in *IN* **do**
      //add minimum-distance node not in *IN*
      p = node z not in *IN* with minimum $d[z]$
      $IN = IN \cup \{p\}$

      //recompute d for non-*IN* nodes, adjust s if necessary
      for all nodes z not in *IN* **do**
          $OldDistance = d[z]$
          $d[z] = \min(d[z], d[p] + \mathbf{A}[p, z])$
          if $d[z] \neq OldDistance$ **then**
              $s[z] = p$
          end if
      end for
  end while

  //write out path nodes
  write("In reverse order, the path is")
  write (y)
  $z = y$
  repeat
      write ($s[z]$)
      $z = s[z]$
  until $z = x$

  // write out path distance
  write("The path distance is", $d[y]$)
end ShortestPath

EXAMPLE 8   Consider the graph in Figure 6.7 and the corresponding modified adjacency matrix shown in Figure 6.8.

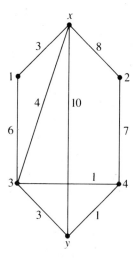

Figure 6.7

	x	1	2	3	4	y
x	∞	3	8	4	∞	10
1	3	∞	∞	6	∞	∞
2	8	∞	∞	∞	7	∞
3	4	6	∞	∞	1	3
4	∞	∞	7	1	∞	1
y	10	∞	∞	3	1	∞

Figure 6.8

We shall trace algorithm *ShortestPath* on this graph. At the end of the initialization phase, *IN* contains only x, and d contains all the direct distances from x to other nodes:

$$IN = \{x\}$$

	x	1	2	3	4	y
d	0	3	8	4	∞	10
s	—	x	x	x	x	x

In Figure 6.9, circled nodes are those in set *IN*, heavy lines show the current shortest paths, and the *d*-value for each node is written along with the node label. Figure 6.9a is the picture after initialization.

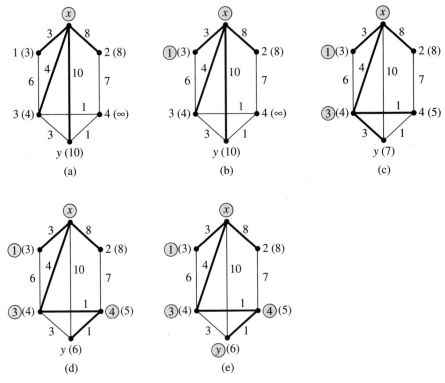

Figure 6.9

We now enter the **while** loop and search through the *d*-values for the node of minimum distance that is not in *IN*; this turns out to be node 1, with $d[1] = 3$. We throw node 1 into *IN*, and in the **for** loop we recompute all the *d*-values for the remaining nodes, 2, 3, 4, and *y*.

$p = 1$

$IN = \{x, 1\}$

$d[2] = \min(8, 3 + A[1, 2]) = \min(8, \infty) = 8$

$d[3] = \min(4, 3 + A[1, 3]) = \min(4, 9) = 4$

$d[4] = \min(\infty, 3 + A[1, 4]) = \min(\infty, \infty) = \infty$

$d[y] = \min(10, 3 + A[1, y]) = \min(10, \infty) = 10$

There were no changes in the *d*-values, so there were no changes in the *s*-values (there were no shorter paths from *x* by going through node 1 than by going directly from *x*). Figure 6.9b shows that 1 is now in *IN*.

The second pass through the **while** loop produces the following:

$p = 3$ (3 has the smallest d-value, namely 4, of 2, 3, 4, or y)

$IN = \{x, 1, 3\}$

$d[2] = \min(8, 4 + \mathbf{A}[3, 2]) = \min(8, 4 + \infty) = 8$

$d[4] = \min(\infty, 4 + \mathbf{A}[3, 4]) = \min(\infty, 4 + 1) = 5$ (a change, so update $s[4]$ to 3)

$d[y] = \min(10, 4 + \mathbf{A}[3, y]) = \min(10, 4 + 3) = 7$ (a change, so update $s[y]$ to 3)

	x	1	2	3	4	y
d	0	3	8	4	5	7
s	–	x	x	x	3	3

Shorter paths from x to the two nodes 4 and y were found by going through 3. Figure 6.9c reflects this.

On the next pass,

$p = 4$ (d-value $= 5$)

$IN = \{x, 1, 3, 4\}$

$d[2] = \min(8, 5 + 7) = 8$

$d[y] = \min(7, 5 + 1) = 6$ (a change, update $s[y]$)

	x	1	2	3	4	y
d	0	3	8	4	5	6
s	–	x	x	x	3	4

See Figure 6.9d.

Processing the **while** loop again, we get

$p = y$

$IN = \{x, 1, 3, 4, y\}$

$d[2] = \min(8, 6 + \infty) = 8$

	x	1	2	3	4	y
d	0	3	8	4	5	6
s	–	x	x	x	3	4

See Figure 6.9e.

Now that y is part of IN, the **while** loop terminates. The path goes through y, $s[y] = 4$, $s[4] = 3$, and $s[3] = x$. Thus the path uses nodes x, 3, 4, and y. (The algorithm gives us these nodes in reverse order.) The distance for the path is $d[y] = 6$. By looking at the graph in Figure 6.7 and checking all the possibilities, we can see that this is the shortest path from x to y.

Algorithm *ShortestPath* terminates when *y* is put into *IN*, even though there may be other nodes in the graph not yet in *IN* (such as node 2 in Example 8). How do we know that a still shorter path cannot be found through one of these excluded nodes? If we continue processing until all nodes have been included in *IN*, the *d*-values then represent the shortest path from *x* to any node, using all the values in *IN*, that is, the shortest path using any nodes of the graph. But new nodes are brought into *IN* in order of increasing *d*-values. A node *z* that is brought into *IN* later than *y* must have as its shortest path from *x* one whose distance is at least as great as the *d*-value of *y* when *y* was brought into *IN*. Therefore there cannot be a shorter path from *x* to *y* via *z* because there is not even a shorter path just between *x* and *z*.

PRACTICE 12   Trace algorithm *ShortestPath* on the graph shown in Figure 6.10. Show the values for *p* and *IN* and the *d*-values and *s*-values for each pass through the **while** loop. Write out the nodes of the shortest path and its distance.

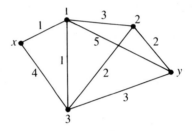

Figure 6.10

When looking for the next node to bring into *IN* in algorithm *ShortestPath*, more than one node *p* may have a minimum *d*-value, in which case *p* can be selected arbitrarily. There may also be more than one shortest path between *x* and *y* in a graph.

Algorithm *ShortestPath* also works for directed graphs if the adjacency matrix is in the appropriate form. It also works for unconnected graphs; if *x* and *y* are not in the same component, then *d*[*y*] will remain ∞ throughout. After *y* has been brought into *IN*, the algorithm will terminate, and this value of ∞ for *d*[*y*] will indicate that no path exists between *x* and *y*.

We may think of algorithm *ShortestPath* as being a "nearsighted" algorithm. It cannot see the entire graph at once to pick out overall shortest paths; it only picks out shortest paths relative to the set *IN* at each step. Such an algorithm is called a **greedy algorithm**—it does what seems best based on its limited immediate knowledge. In this case, what seems best at the time turns out to be best overall.

How efficient is the shortest-path algorithm? Most of the work seems to take place within the **for** loop that modifies the *d* and *s* arrays. Here the algorithm checks all *n* nodes to determine which nodes *z* are not in *IN* and recomputes *d*[*z*] for those nodes, possibly also changing *s*[*z*]. The necessary quantities *d*[*z*], *d*[p], and **A**[*p*, *z*] for a given *z* are directly available. Therefore the **for** loop requires $\Theta(n)$ operations. In addition, determining the node *p* to add to *IN* can also be done in $\Theta(n)$ operations by checking all *n* nodes. With the additional small amount of work to add *p* to *IN*, each execution of the **while** loop takes $\Theta(n)$ operations. In the worst case, *y* is the last node brought into *IN*, and the **while** loop will be executed *n* − 1 times. Therefore the total number of operations involved in the **while** loop is $\Theta(n(n - 1)) = \Theta(n^2)$.

Initialization and writing the output together take $\Theta(n)$ operations, so the algorithm requires $\Theta(n + n^2) = \Theta(n^2)$ operations in the worst case.

What if we keep *IN* (or rather the complement of *IN*) as some sort of linked list, so that all the nodes of the graph do not have to be examined to see which are not in *IN*? Surely this would make the algorithm more efficient. Note that the number of nodes not in *IN* is initially $n - 1$, and that number decreases by 1 for each pass through the **while** loop. Within the **while** loop the algorithm thus has to perform on the order of $n - 1$ operations on the first pass, then $n - 2$, then $n - 3$, and so on. But, as proof by induction will show,

$$(n - 1) + (n - 2) + \cdots + 1 = (n - 1)n / 2 = \Theta(n^2)$$

Thus the worst-case situation still requires $\Theta(n^2)$ operations.

Minimal Spanning Tree Problem

A problem encountered in designing networks is how to connect all the nodes efficiently, where nodes can be computers, telephones, warehouses, and so on. A minimal spanning tree may provide an economical solution, one that requires the least cable, pipeline, or whatever the connecting medium is. For reliability, however, the minimal spanning tree usually would be supplemented with additional arcs so that if one connection were broken for some reason, an alternate route could be found.

> **Definition: Spanning Tree**
> A **spanning tree** for a connected graph is a nonrooted tree whose set of nodes coincides with the set of nodes for the graph and whose arcs are (some of) the arcs of the graph.

A spanning tree thus connects all the nodes of a graph with no excess arcs (no cycles). There are algorithms for constructing a **minimal spanning tree**, a spanning tree with minimal weight, for a given simple, weighted, connected graph.

One of these algorithms, called Prim's algorithm, proceeds very much like the shortest-path algorithm. There is a set *IN*, which initially contains one arbitrary node. For every node z not in *IN*, we keep track of the shortest distance $d[z]$ between z and any node in *IN*. We successively add nodes to *IN*, where the next node added is one that is not in *IN* and whose distance $d[z]$ is minimal. The arc having this minimal distance is then made part of the spanning tree. Because there may be ties between minimal distances, the minimal spanning tree of a graph may not be unique. The algorithm terminates when all nodes of the graph are in *IN*.

The key difference in the implementation of the two algorithms comes in the computations of new distances for the nodes not yet in *IN*. In Dijkstra's algorithm, if p is the node that has just been added to *IN*, distances for non-*IN* nodes are recalculated by

$$d[z] = \min(d[z], d[p] + \mathbf{A}[p, z])$$

that is, by comparing the current distance of z from x with the distance of p from x plus the distance of z from p. In Prim's algorithm, if p is the node that has just been added to *IN*, distances for non-*IN* nodes are recalculated by

$$d[z] = \min(d[z], \mathbf{A}[p, z]))$$

that is, by comparing the current distance of z from *IN* with the distance of z from p.

We won't write out the algorithm (which, like the shortest-path algorithm, requires $\Theta(n^2)$ operations in the worst case and is a greedy algorithm); we shall simply illustrate it with an example.

EXAMPLE 9   We shall find a minimal spanning tree for the graph of Figure 6.7. We let node 1 be the arbitrary initial node in *IN*. Next we consider all the nodes adjacent to any node in *IN*, that is, all nodes adjacent to 1, and select the closest one, which is node *x*. Now $IN = \{1, x\}$, and the arc between 1 and *x* is part of the minimal spanning tree. Next we consider all nodes not in *IN* that are adjacent to either 1 or *x*. The closest such node is 3, which is 4 units away from *x*. The arc between 3 and *x* is part of the minimal spanning tree. For $IN = \{1, x, 3\}$, the next closest node is node 4, 1 unit away from 3. The remaining nodes are added in the order *y* and then 2. Figure 6.11 shows the minimal spanning tree.

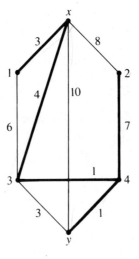

Figure 6.11

PRACTICE 13   Find a minimal spanning tree for the graph of Figure 6.10.

Section 6.3 Review

Techniques

- Find a shortest path from *x* to *y* in a graph (using Dijkstra's algorithm).
- Find a minimal spanning tree for a graph (using Prim's algorithm).

Main Idea

Algorithms that are $\Theta(n^2)$ in the worst case can find a shortest path between two nodes or a minimal spanning tree in a simple, positively weighted, connected graph with *n* nodes.

Exercises 6.3

For Exercises 1–4, use the accompanying graph. Apply algorithm *ShortestPath* (Dijkstra's algorithm) for the pairs of nodes given; show the values for *p* and *IN* and the *d*-values and *s*-values for each pass through the **while** loop. Write out the nodes in the shortest path and its distance.

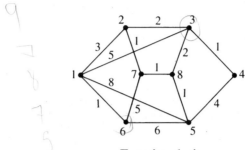

Exercises 1–4

★1. From 2 to 5   2. From 3 to 6   3. From 1 to 5   4. From 4 to 7

For Exercises 5 and 6, use the directed graph of the accompanying figure. Apply algorithm *ShortestPath* (Dijkstra's algorithm) to the nodes given; show the values for *p* and *IN* and the *d*-values and *s*-values for each pass through the **while** loop. Write out the nodes in the shortest path and its distance.

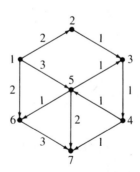

Exercises 5–6

★5. From 1 to 7   6. From 3 to 1

7. **a.** Modify algorithm *ShortestPath* so that it finds the shortest paths from *x* to all other nodes in the graph.
   b. Does this change the worst-case order of magnitude of the algorithm?

8. Give an example to show that algorithm *ShortestPath* does not work when negative weights are allowed.

Another algorithm for finding shortest paths from a single source node to all other nodes in the graph is the *Bellman–Ford algorithm*. In contrast to Dijkstra's algorithm, which keeps a set of nodes whose shortest path of whatever length (that is, number of hops) has been determined, the Bellman–Ford algorithm performs a series of computations that seeks to find successively shorter paths of length 1, then of length 2, then of length 3, and so on, up to a maximum of length $n - 1$ (if a path exists at all, then there is a path of length no greater than $n - 1$). A pseudocode description of the Bellman–Ford algorithm is given in the accompanying box (algorithm *AnotherShortestPath*); when using this algorithm, the adjacency matrix **A** must have $A[i, i] = 0$ for all i.

ALGORITHM   AnotherShortestPath

AnotherShortestPath ($n \times n$ matrix **A**; node x; array of integers d; array of nodes s)
//Bellman–Ford algorithm. **A** is a modified adjacency matrix for a simple,
//weighted, connected graph; x is a node in the graph; when procedure terminates,
//the nodes in the shortest path from x to a node y are y, $s[y]$, $s[s[y]]$, ... , x; the
//distance for that path is $d[y]$.

Local variables:
nodes z, p //temporary nodes
array of integers t //temporary distance array created at each iteration

    //initialize arrays d and s; this establishes the shortest 1-length paths from x
    $d[x] = 0$
    for all nodes z not equal to x **do**
        $d[z] = A[x, z]$
        $s[z] = x$
    end for

    //find shortest paths of length 2, 3, etc.
    for $i = 2$ to $n - 1$ **do**
        $t = d$ //copy current array d into array t

        //modify t to hold shortest paths of length i
        for all nodes z not equal to x **do**
            //find the shortest path with one more link
            $p = $ node in G for which ($d[p] + A[p, z]$) is minimum
            $t[z] = d[p] + A[p, z]$
            if $p \neq z$ **then**
                $s[z] = p$
            end if
        end for
        $d = t$; //copy array t back into d
    end for
end AnotherShortestPath

For Exercises 9–12 use algorithm *AnotherShortestPath* (the Bellman–Ford algorithm) to find the shortest path from the source node to any other node. Show the successive *d*-values and *s*-values.

★9. Graph for Exercises 1–4, source node = 2 (compare your answer to Exercise 1)

10. Graph for Exercises 1–4, source node = 1 (compare your answer to Exercise 3)

11. Graph for Exercises 5 and 6, source node = 1 (compare your answer to Exercise 5)

12. Accompanying graph, source node = 1 (compare your answer to Exercise 8)

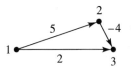

Exercise 12

To compute the distance for the shortest path between any two nodes in a graph, algorithm *ShortestPath* could be used repeatedly, with each node in turn as the source node. A different algorithm, very similar to Warshall's algorithm, can also be used to solve this "all pairs" shortest-path problem. A description follows, where **A** is the adjacency matrix of the graph with $A[i, i] = 0$ for all i.

ALGORITHM   **AllPairsShortestPath**

AllPairsShortestPath ($n \times n$ matrix **A**)
//Floyd's algorithm—computes the shortest path between any two nodes; initially, **A**
//is the adjacency matrix; upon termination, **A** will contain all the shortest-path
//distances
    for $k = 1$ to n **do**
        for $i = 1$ to n **do**
            for $j = 1$ to n **do**
                if $A[i, k] + A[k, j] < A[i, j]$ **then**
                    $A[i, j] = A[i, k] + A[k, j]$
                end if
            end for
        end for
    end for
end AllPairsShortestPath

For Exercises 13 and 14, use algorithm *AllPairsShortestPath* (Floyd's algorithm) to find the distances for all the shortest paths. Show the successive values of the **A** matrix for each pass through the outer loop.

★13. Figure 6.10   14. Graph for Exercises 1–4

For Exercises 15–18, find a minimal spanning tree for the graph in the specified figure.

15. Graph for Exercises 1–4

16.

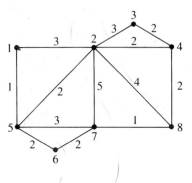

★17.

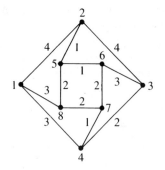

18.

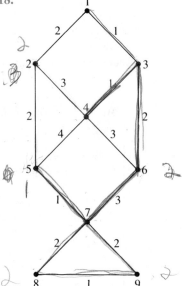

Kruskal's algorithm is another algorithm for finding a minimal spanning tree in a connected graph. Whereas Prim's algorithm "grows" the tree from an arbitrary starting point by attaching adjacent short arcs, Kruskal's algorithm adds arcs in order by increasing distance wherever they may be in the graph. Ties are resolved arbitrarily. The only restriction is that an arc is not added if adding it would create a cycle. The algorithm terminates when all nodes have been incorporated into a connected structure. A (very informal) pseudocode description follows:

ALGORITHM AnotherMST

AnotherMST ($n \times n$ matrix **A**; collection of arcs *T*)
//Kruskal's algorithm to find minimal spanning tree; *T* is initially empty; at
//termination, *T* = minimal spanning tree
   order arcs in *G* by increasing distance
   repeat
      if next arc in order does not complete a cycle **then**
         add that arc into *T*
      end if
   until *T* is connected and contains all nodes of *G*
end AnotherMST

For Exercises 19–22 use algorithm *AnotherMST* (Kruskal's algorithm) to find the minimal spanning tree.

★**19.** Graph for Exercises 1–4   **20.** Graph for Exercise 16

21. Graph for Exercise 17   **22.** Graph for Exercise 18

23. Give an example to show that adding the node closest to *IN* at each step, as is done in Prim's minimal spanning tree algorithm, will not guarantee a shortest path.

★**24.** Assume that arc weights represent distance. Then adding new nodes and arcs to a graph may result in a spanning tree for the new graph that has less weight than a spanning tree for the original graph. (The new spanning tree could represent a minimal-cost network for communications between a group of cities obtained by adding a switch in a location outside any of the cities.)

   a. Find a spanning tree of minimum weight for the labeled graph of the figure. What is its weight?

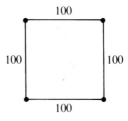

Exercise 24

   b. Put a node in the center of the square. Add new arcs from the center to the corners. Find a spanning tree for the new graph, and compute its (approximate) weight.

25. At the beginning of this chapter, you received the following assignment:

You are the network administrator for a wide-area backbone network that serves your company's many offices across the country. Messages travel through the network by being routed from point to point until they reach their destination. Each node in the network therefore acts as a switching station to forward messages to other nodes according to a routing table maintained at each node. Some connections in the network carry heavy traffic, while others are less used. Traffic may vary with the time of day; in addition, new nodes occasionally come on line and existing nodes may go off line. Therefore you must periodically provide each node with updated information so that it can forward messages along the most efficient (that is, the least heavily traveled) route.

How can you compute the routing table for each node?

You realize that you can represent the network as a weighted graph, where the arcs are the connections between nodes and the weights of the arcs represent traffic on the connections. The routing problem then becomes one of finding the shortest path in the graph from any node to any other node. Dijkstra's algorithm can be used to give the shortest path from any one node to all other nodes, so you could use the algorithm repeatedly with different start nodes. Or, you could use algorithm *AllPairsShortestPath* (preceding Exercise 13), which looks much shorter and neater. Analyze the order of magnitude of each approach.

Section 6.4   Traversal Algorithms

So far this chapter has considered various path questions about a graph G. Is there a path in G from node x to node y? Is there a path through G that uses each arc once? Is there a path through G that ends where we started and uses each node once? What is the minimum-weight path between x and y? In this section we deal with a simpler problem—we only want to write down all the nodes of a simple, connected graph G in some orderly way. This means we must find a path that visits each node at least once, but we can visit it more than once if we don't write it down again. We can also retrace arcs on the graph if necessary, and clearly this would in general be necessary if we were to visit each node in a tree. This process is called **graph traversal**. We already have several mechanisms for tree traversal (Section 5.2). The two algorithms in this section generalize traversal to apply to any graph.

Depth-First Search

In the **depth-first search** algorithm for graph traversal, we begin at an arbitrary node a of the graph, mark it visited, and write it down. We then strike out on a path away from a, visiting and writing down nodes, proceeding as far as possible until there are no more unvisited nodes on that path. We then back up the path, at each node exploring any new side paths, until finally we retreat back to a. We then explore any new paths remaining from a. Figure 6.12 shows a graph after the first few nodes (marked by circles) have been visited using depth-first search.

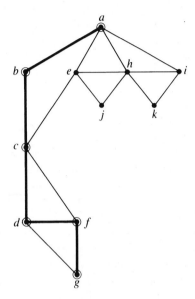

Figure 6.12

For a more formal description of the depth-first search algorithm we will use recursion, where the algorithm invokes itself in the course of its execution. In the following algorithm, the input is a simple, connected graph G and a specified node a; the output is a list of all nodes in G in depth-first order from a.

ALGORITHM   DepthFirst

DepthFirst(graph G; node a)
//Writes nodes in graph G in depth-first order from node a

    mark a visited
    write(a)
    for each node n adjacent to a **do**
        if n not visited **then**
            DepthFirst(G, n)
        end if
    end for

end DepthFirst

In the recursive step, the algorithm is invoked with a new node specified as the starting point. We have not indicated here how to mark visited nodes or how to find those nodes *n* that are adjacent to *a*.

EXAMPLE 10

We shall apply depth-first search to the graph of Figure 6.12, where *a* is the initial node. We first mark that we have visited *a* (it's helpful in tracing the execution of the algorithm to circle a visited node), and then we write out *a*. Next we search the nodes adjacent to *a* for an unvisited node. We have a choice here (*b*, *e*, *h*, and *i*); let us select node *b*. (Just so we all get the same answers, let's agree to choose the node that is alphabetically first when we have a choice; in practice, the choice would be determined by how the vertices were stored in the graph representation.) Then we invoke the depth-first search algorithm beginning with node *b*.

This means we go back to the beginning of the algorithm, where the specified node is now *b* rather than *a*. Thus we first mark *b* visited and write it out. Then we search through nodes adjacent to *b* to find an unmarked node. Node *a* is adjacent to *b*, but it is marked. Node *c* will do, and we invoke the depth-first search algorithm beginning with node *c*.

Node *c* is marked and written out, and we look for unmarked nodes adjacent to *c*. By our alphabetical convention, we select node *d*. Continuing in this fashion, we next visit node *f* and then node *g*. When we get to node *g*, we have reached a dead end because there are no unvisited adjacent nodes. Thus the **for** loop of the instance of the algorithm invoked with node *g* is complete. (The graph at this point looks like Figure 6.12.)

We are therefore done with the algorithm for node *g*, but node *g* was (one of) the unmarked nodes adjacent to node *f*, and we are still in the **for** loop for the instance of the algorithm invoked with node *f*. As it happens, *g* is the only unvisited node when we are processing *f*; therefore we complete the **for** loop and thus the algorithm for node *f*. Similarly, backing up to node *d*, the algorithm finds no other adjacent unmarked nodes, and it backs up again to the instance of the algorithm invoked with node *c*. Thus, after processing node *d* and everything that came after it until the dead end, we are still in the **for** loop for the algorithm applied to node *c*. We look for other unmarked nodes adjacent to *c* and find one—node *e*. Therefore we apply depth-first search to node *e*, which leads to nodes *h*, *i*, and *k* before another dead end is reached. Backing up, we have a final new path to try from node *h*, which leads to node *j*. The complete list of the nodes, in the order in which they would be written out, is

a, b, c, d, f, g, e, h, i, k, j

Example 10 makes the depth-first search process sound very complex, but it is much easier to carry out than to write down, as you shall see in Practice 14.

Write the nodes in a depth-first search of the graph in Figure 6.13. Begin with node *a*.

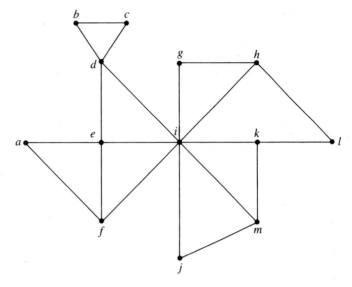

Figure 6.13

Breadth-First Search

In **breadth-first search,** beginning at an arbitrary node a, we first fan out from node a to visit nodes that are adjacent to a, then we fan out from those nodes, and so on, almost like the concentric circles of ripples in a pond. Figure 6.14 shows the first few nodes visited in the same graph as Figure 6.12, this time using breadth-first search.

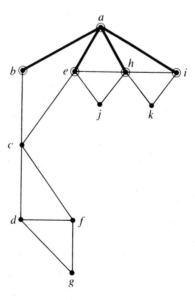

Figure 6.14

To write the breadth-first search algorithm in an elegant fashion, we shall use a **queue** structure. A queue is simply a line in which new arrivals are added at the back and departures take place at the front. A checkout line in a grocery store is an example of a queue of customers—a new customer joins the line at the back and departures take place from the front of the line as customers are checked through. The addition of an entry at the back of a queue is called an **enqueue** operation, and a departure from the front of the queue is called a **dequeue** operation. Thus the notation *enqueue*(*a*, *Q*) denotes adding *a* to the end of a queue called *Q*, and *dequeue*(*Q*) denotes removal of the entry currently at the front of *Q*. We'll also use a function *front*(*Q*), which returns the value of the entry currently at the front of *Q* but does not remove that entry. In the algorithm in the accompanying box (algorithm *BreadthFirst*) the input is a simple, connected graph *G* and a specified node *a*; the output is a list of all nodes in *G* in breadth-first order from *a*.

ALGORITHM   BreadthFirst

BreadthFirst(graph *G*; node *a*);
//writes nodes in graph *G* in breadth-first order from node *a*

Local Variable:
queue of nodes *Q*

   initialize *Q* to be empty
   mark *a* visited
   write(*a*)
   enqueue(*a*, *Q*)
   while *Q* is not empty **do**
      for each node *n* adjacent to front(*Q*) **do**
         if *n* not visited **then**
            mark *n* visited
            write(*n*)
            enqueue(*n*, *Q*)
         end if
      end for
      dequeue(*Q*)
   end while
end BreadthFirst

EXAMPLE 11

Let's walk through the algorithm for a breadth-first search of the graph of Figure 6.14 beginning at node *a* (this is the same graph on which we did the depth-first search in Example 10). We begin by initializing an empty queue *Q*, marking node *a* as visited, writing it out, and adding it to the queue. When we first reach the **while** loop, the queue is not empty and *a* is the entry at the front of the queue. In the **for** loop, we look for unvisited nodes adjacent to *a* to visit, write them out, and add them to the back of the queue. We may have a choice of nodes to visit here; as before, and purely as a convention, we shall agree to visit them in alphabetical order. Thus the first time we complete the **for** loop, we have visited and written out *b*, *e*, *h*, and *i*, in that order, and added them to the queue. The graph at this point looks like Figure 6.14. We then remove *a* from the front of the queue, which as a result contains (from front to back)

b, e, h, i

In the next iteration of the **while** loop, *b* is the front element in the queue, and the **for** loop searches for unvisited nodes adjacent to *b*. The only previously unvisited node here is *c*, which gets written out and added to the queue. After removing *b*, the queue contains

e, h, i, c

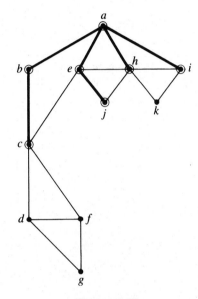

Figure 6.15

Performing the **while** loop again, *e* is at the front of the queue. A search of the nodes adjacent to node *e* produces one new node, node *j*. The graph now looks like Figure 6.15, and after removing *e* the queue contains

h, i, c, j

When searching for nodes adjacent to *h*, we pick up one new node, node *k*. When searching for nodes adjacent to *i*, no new nodes are added to the queue. When *c* becomes the first element in the queue, a search for nodes adjacent to *c* turns up two new nodes, *d* and *f*. After adding these to the queue (and removing *c*), the queue contains

j, k, d, f

Looking for nodes adjacent to *j* and then to *k* adds no new nodes to the queue. When the front of the queue is *d*, a new node *g* is found, and the queue (after removing *d*) is

f, g

Processing f and then g yields no new nodes. After g is removed from the queue, the queue is empty. The **while** loop—as well as the algorithm—terminates. The list of nodes written out by this process, that is, the nodes in breadth-first order from a, are

$a, b, e, h, i, c, j, k, d, f, g$ ●

Like the depth-first search, the breadth-first search is not difficult to trace; one must just keep track of the nodes that have been visited and the current contents of the queue.

PRACTICE 15   Write the nodes in a breadth-first search of the graph in Figure 6.13, beginning with node a.   ●

Analysis

How much work do depth-first search and breadth-first search perform? Both algorithms look for all unvisited nodes adjacent to a given node. Suppose the graph contains n nodes and m arcs. One of the advantages of representing a graph as an adjacency list rather than an adjacency matrix is that this particular operation is more efficient; to find nodes adjacent to node i requires traversing i's adjacency list, which may be short, rather than row i of the adjacency matrix, which must contain n entries. Therefore we shall assume an adjacency list representation of the graph.

In breadth-first search, the algorithm searches all at one time the entire adjacency list of the node at the front of the queue, marking, writing out, and enqueuing the unvisited nodes found. In depth-first search, the algorithm may be interrupted many times while traversing the adjacency list of a given node to go off (by virtue of the recursion) and process sections of the adjacency lists of other nodes. Eventually, however, every adjacency list is completely covered.

Traversing the adjacency lists of the graph drives the amount of work done in either search. There are n adjacency lists, so the amount of work is at least $\Theta(n)$ because each adjacency list must be checked, even if it turns out to be empty. Because there are m arcs, the work in traversing the total length of all the adjacency lists is at least $\Theta(m)$. Therefore both depth-first search and breadth-first search are $\Theta(\max(n, m))$ algorithms. If there are more arcs than nodes (the usual case), then $\Theta(\max(n, m)) = \Theta(m)$.

Applications

Depth-first search and breadth-first search can be used as the basis for performing other graph-related tasks, some of which we have solved before. A nonrooted tree structure that is a subgraph of the original graph can be associated with each search. When traversing node i's adjacency list, if node j is adjacent to i and is previously unvisited, then the i–j arc is added to this subgraph. Because no arc to a previously visited node is used, cycles are avoided and the subgraph is a nonrooted tree. Because all nodes ultimately are visited (for the first time), these trees are spanning trees for the graph. Each tree has $n - 1$ arcs, the minimal number of arcs to connect n nodes. Here we are assuming that arcs are unweighted, but if we consider them to be weighted arcs, each with weight 1, then these trees are minimal spanning trees.

The dark lines in Figure 6.12 are part of the depth-first search tree associated with the search of Example 10, and the dark lines in Figures 6.14 and 6.15 are part of the breadth-first search tree associated with the search of Example 11.

PRACTICE 16   **a.** Complete the depth-first search tree for Example 10.
b. Complete the breadth-first search tree for Example 11.   ●

The depth-first search and breadth-first search algorithms apply equally well to directed graphs and in the process yield a new algorithm for reachability. To determine if node j is reachable from node i, do a depth-first (or breadth-first) search beginning at node i; when the algorithm terminates, check whether node j has been visited. "All pairs" reachability, that is, which nodes are reachable from which nodes, can thus be determined by running depth-first or breadth-first searches using each node in turn as the source node. This would require $\Theta(n * \max(n, m))$ work. If the graph is very sparse, in which case we have $\max(n, m) = n$, we would have an $\Theta(n^2)$ algorithm for reachability. Recall that Warshall's algorithm (Section 6.1) was an $\Theta(n^3)$ algorithm. The improvement comes about because in a sparse graph, most adjacency lists will be short or empty, whereas Warshall's algorithm processes entries in the adjacency matrix even if those entries are 0s. But if the graph is not sparse, the number of arcs can be $\Theta(n^2)$, in which case $\Theta(n * \max(n, m)) = \Theta(n^3)$, the same as Warshall's algorithm. In addition, Warshall's algorithm has the advantage of succinct implementation.

In Section 4.2 we defined a topological sort as a way to extend a partial ordering on a finite set to a total ordering. Let the partially ordered set be represented by a directed graph. The topological sort will be achieved by counting the nodes, so let the initial value of the count be 0. Pick a node as a source node and perform a depth-first search from this node. Whenever the search backs up from a node for the final time, assign that node the next counting number. When the depth-first search algorithm terminates, pick an unvisited node (if one exists) to be the source for another depth-first search, and continue to increment the counting number. Continue this process until there are no unvisited nodes left in the graph. A topological sort results by ordering the nodes in the reverse order of their counting number. This process for topological sorting works because we assign the counting number when we back up from a node for the final time. Its counting number will then be higher than the numbers of all the nodes reachable from it, that is, all the nodes of which it is a predecessor in the partial ordering,

EXAMPLE 12   Figure 6.16a is a directed graph that represents a partial ordering. Choosing d (arbitrarily) as the source node and performing a depth-first search, we visit e and f, at which point we must back up. Node f is assigned the counting number 1, but we are not yet done with e, because we can go on to visit g. Backing up from g, g is assigned the counting number 2. At this point we back up from e for the final time and assign e the number 3 and then d the number 4. Choose a as the source node for another search. We visit node c and then must back up, so c and a are assigned the numbers 5 and 6, respectively. Beginning with b as a source node, there is nowhere to go, and b is assigned the number 7. There are no unvisited nodes left in the graph, so the process stops. The numbering scheme is shown in Figure 6.16b.

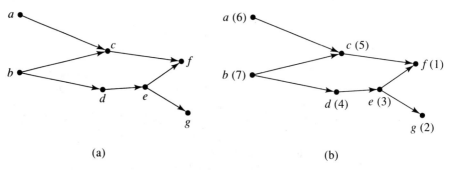

(a)   (b)

Figure 6.16

In reverse order of the counting numbers, we get

7	6	5	4	3	2	1
b	a	c	d	e	g	f

which is a topological ordering.

PRACTICE 17   Use the depth-first search algorithm to do a topological sort on the graph in Figure 6.17. Indicate the counting numbers on the graph.

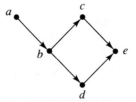

Figure 6.17

Now consider a graph G (undirected) that may not be a connected graph. A **connected component** of G is a subgraph of G that is both connected and not a subgraph of a larger connected subgraph. In Figure 6.18 there are three connected components. Of course, if the original graph is connected, then it has only one connected component.

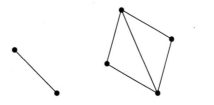

Figure 6.18

Depth-first or breadth-first search can be used to find the connected components of a graph. We pick an arbitrary node as a source node and then conduct a search. When the algorithm terminates, all visited nodes belong to one component. We then find an unvisited node in the graph to serve as a source for another search, which will produce a second component. We continue this process until there are no unvisited nodes in the graph.

Although we defined reachability only for directed graphs, the concept also makes sense for undirected, unconnected graphs. Let us consider only simple undirected, unconnected graphs but impose the convention that, even though there are no loops, each node is reachable from itself. Reachability then becomes an equivalence relation on the set of nodes of the graph; our convention imposes the reflexive property, and symmetry and transitivity follow because the graph is undirected. This equivalence relation partitions the nodes of the graph into equivalence classes, and each class consists of the nodes in one component of the graph. Warshall's algorithm can be applied to undirected graphs as well as directed graphs. Using Warshall's algorithm results in a matrix from which the nodes making up various components of the graph can be determined, but this requires more work than using depth-first search. The use of depth-first search to identify *biconnected components* in a graph will be discussed in Section 6.5.

As a final remark about depth-first search, we saw in Section 1.5 that the programming language Prolog, when processing a query based on a recursive definition, pursues a depth-first search strategy (Example 38).

Section 6.4 Review

Techniques

- Conduct a depth-first search of a graph.
- Conduct a breadth-first search of a graph.

Main Ideas

Algorithms exist to visit the nodes of a graph systematically.

Depth-first and breadth-first searches can serve as a basis for other tasks.

Exercises 6.4

For Exercises 1–6, write the nodes in a depth-first search of the graph in the accompanying figure, beginning with the node specified.

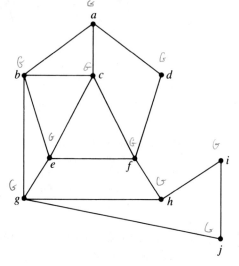

Exercises 1–6

★1. *a*   2. *c*   3. *d*   4. *g*   ★5. *e*   6. *h*

For Exercises 7–10, write the nodes in a depth-first search of the graph in the accompanying figure, beginning with the node specified.

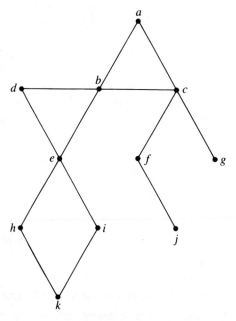

Exercises 7–10

★7. *a*   8. *e*

★9. *f*   10. *h*

For Exercises 11–16, write the nodes in a breadth-first search of the graph in the figure for Exercises 1–6, beginning with the node specified.

★11. *a*   12. *c*   13. *d*

14. *g*   15. *e*   16. *h*

For Exercises 17–20, write the nodes in a breadth-first search of the graph in the figure for Exercises 7–10, beginning with the node specified.

★17. *a*   18. *e*

19. *f*   20. *h*

For Exercises 21–23, write the nodes in a depth-first search of the graph in the accompanying figure, beginning with the node specified.

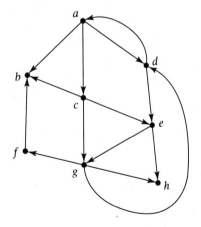

Exercises 21–26

★21. *a*   22. *g*   23. *f*

For Exercises 24–26, write the nodes in a breadth-first search of the graph in the figure accompanying Exercises 21–23, beginning with the node specified.

★24. *a*   25. *g*   26. *f*

★27. Use the depth-first search algorithm to do a topological sort on the graph in the accompanying figure. Indicate the counting numbers on the graph. Also state the starting node or nodes for the search.

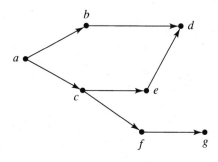

Exercise 27

28. Use the depth-first search algorithm to do a topological sort on the graph in the accompanying figure. Indicate the counting numbers on the graph. Also state the starting node or nodes for the search.

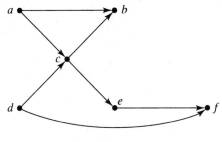

Exercise 28

29. Find a way to traverse a tree in level order, that is, so that all nodes at the same depth are listed from left to right for increasing depth. (*Hint:* We already have a way to do this.)

Section 6.5   Articulation Points and Computer Networks

Problem Statement

In a graph that represents a computer or communications network, the nodes denote the communicating entities (computers, telephones, etc.) and the arcs denote the communications medium (coaxial cable, telephone lines, etc.). Such a graph should be a connected graph so that there is a path between every pair of nodes. To minimize the length of cable or wire required, we would choose a minimum spanning tree. However, if an arc in a minimum spanning tree is removed (i.e., that section of cable or wire gets damaged or broken), then the graph is no longer connected. Each arc becomes a single point of failure for the network. That is why such a network usually contains more arcs than just those of a minimal spanning tree. However, even in a graph sufficiently rich in arcs to withstand the loss of a single arc, a node may be a single point of failure. If such a node fails (and thus is logically removed), the arcs of which that node is an endpoint are disabled and this may result in a disconnected graph.

Definition: Articulation Point
A node in a simple, connected graph is an **articulation point** if its removal (along with its attached arcs) causes the remaining graph to be disconnected.

EXAMPLE 13   Node d in the graph of Figure 6.19a is an articulation point. Removing d results in the disconnected graph of Figure 6.19b.

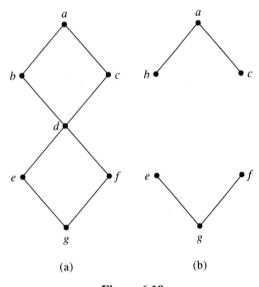

(a)   (b)

Figure 6.19

Definition: Biconnected Graph
A simple, connected graph is **biconnected** if it has no articulation points.

The presence of articulation points is clearly an undesirable feature of a network. Although it is easy to spot an articulation point in a graph as small as that of Figure 6.19a, we shall develop an algorithm that will detect such points no matter how large the graph is. Articulation points separate the graph into **biconnected components**, subgraphs that are biconnected and are not subgraphs of larger biconnected sub-graphs. In Figure 6.19, a–b–d–c and d–e–g–f are biconnected components.

Idea behind the Algorithm

The key to this algorithm is depth-first search. We know from the previous section that a depth-first search determines a nonrooted tree. An arc is added to the tree whenever the search progresses to a previously unvisited node. Arcs of the graph belonging to this tree are called **tree arcs**. The remaining arcs in the graph are called **back arcs**.

EXAMPLE 14

In Figure 6.19 a depth-first search from node a visits nodes in the order a, b, d, c, e, g, and f. In Figure 6.20 the tree arcs are dark and the back arcs are light.

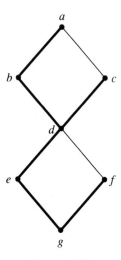

Figure 6.20

The depth-first search tree passes through all nodes. To detect nodes that are artic-ulation points, we examine their relative positions in the tree. First consider the single node that is the starting point of the depth-first search tree. If only one tree arc emanates from the starting node, then as the tree continues, all other nodes in the graph can be reached from the node at the other end of that tree arc. Therefore, removing the starting node will not disconnect the graph. However, if two or more tree arcs emanate from the starting node, then the only way to get from one subtree to another is to pass back through the starting node. In this case removing the starting node disconnects the graph.

Thus in Figure 6.20 node a is the starting node of the depth-first search tree, and there is a single tree arc emanating from a. Removing node a (and its two arcs) does not disconnect the graph. Had we begun a depth-first search at node d, however, the tree would have looked like Figure 6.21. There would be two tree arcs coming from node d, showing that d is an articulation point.

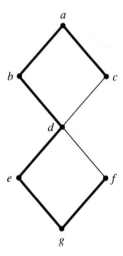

Figure 6.21

Consider any node n that is a leaf of the depth-first search tree (attached to the end of a single tree arc). Such a leaf may be a leaf of the graph itself, that is, a node of degree 1, in which case it is clearly not an articulation point. If not, then the other arcs emanating from n were not used in the depth-first search, so the nodes adjacent to n are reachable through alternate paths that do not go through n. Because n is not needed on a path to any other node, its removal will not disconnect the graph. Therefore no leaves of the depth-first search tree are articulation points. In Figure 6.20 node c, for example, is a leaf of the depth-first search tree; the arc from c to a is a back arc, so node a is accessible through another route that does not require node c. Node c can be removed without disconnecting the graph.

Now consider a node n that is not a leaf in the depth-first search tree and is not the starting node. Because n is not a leaf, there are one or more subtrees below n. Suppose there is a single subtree; let x be a node on this subtree. If x has a back arc to some node that precedes n in the depth-first search (an "ancestor" of n), then this arc provides part of an alternate path for x—and all other nodes in the subtree—to be connected with the rest of the graph without using node n. In this case n is not an articulation point. (See Figure 6.22a, where removing n and its attached arcs does not disconnect the graph.) If there is more than one sub-tree below n, then n will not be an articulation point if and only if each subtree has such an "escape route" allowing it to connect with the rest of the graph—including the other subtrees—without going through n. (See Figure 6.22b; note that the back arc from y to z does not help because it does not reach back to an ancestor of n.)

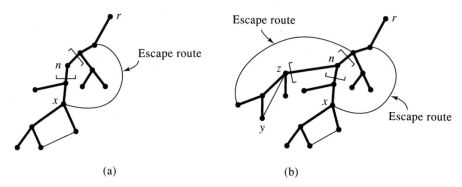

(a)   (b)

Figure 6.22

The Algorithm Itself

The key to the algorithm, as we may guess from the preceding discussion, is keeping track of the destinations of the back arcs. We will assign a tree number to each node that corresponds to the order in which that node is visited in the depth-first search. Thus the starting node in a depth-first search has tree number 1, the next node visited has tree number 2, and so on. In addition, we shall maintain a "back number" for each node x. The back number will be the minimum tree number of a node (the furthest back node) reachable from x using either back arcs from x or from descendants of x in the subtree. To incorporate information about back arcs of descendants of x, the back number of x is adjusted when the depth-first search backs up to node x from further down in the tree. Node n is an articulation point whenever a subtree of n has no back arc to an ancestor of n, and this is detected when the search backs up to n from the subtree. Suppose the search is backing up from x to n. If the back number of x at this point is not smaller than the tree number of n, then n is an articulation point.

The accompanying algorithm (algorithm *ArtPoint*) carries out the depth-first search and builds the depth-first search tree. It correctly handles both leaves and nonstarting-node nonleaves of the depth-first search tree, leaving only the starting node as a special case.

ALGORITHM   ArtPoint

ArtPoint(graph *G*; node *n*; integer *TreeNumber*)
//Detects articulation points by depth-first search from *n*;
//*TreeNumber* = 0 when first invoked

    mark *n* visited
    //first encountering *n*—give it tree number and back number
    TreeNumber = *TreeNumber* + 1
    TreeNumber[*n*] = *TreeNumber*
    BackNumber[*n*] = *TreeNumber*[*n*]
    for each node *x* adjacent to *n* by a nontree edge **do**
        if *x* not visited **then**
            make *n*–*x* a tree edge
            ArtPoint(*G*, *x*, *TreeNumber*)

            //depth-first search now returning to *n* from a subtree rooted at *x*
            //Is *n* an articulation point—does subtree rooted at *x* fail to have
            //back edge to ancestor of *n*?
            if *BackNumber*[*x*] >= *TreeNumber*[*n*] **then**   //line A
                write(*n*, "is an articulation point")
            else
                //adjust back number of *n*
                BackNumber[*n*] = min(*BackNumber*[*n*], *BackNumber*[*x*])   //line B
            end if
        else
            //arc *n*–*x* is a back edge, adjust *BackNumber*[*n*]
            BackNumber[*n*] = min(*BackNumber*[*n*], *TreeNumber*[*x*])   //line C
        end if
    end for
  end ArtPoint

EXAMPLE 15

We shall trace algorithm *ArtPoint* on the graph of Figure 6.19a, where *a* is the starting node. Node *a* is marked visited and numbered with *TreeNumber* = 1 and *BackNumber* = 1. The tree begins with arc *a*–*b*, then *ArtPoint* is recursively invoked with starting node *b*; *b* is marked visited and numbered with *TreeNumber* = *BackNumber* = 2. Moving on to nodes *d* and then *c*, each new node is numbered with a consecutive *TreeNumber*, and its *BackNumber* is set equal to its *TreeNumber*. (See Figure 6.23a, where the numbers in parentheses are the *TreeNumber* and the *BackNumber*, respectively.) While processing node *c*, the back edge to *a* is discovered, and *BackNumber* of *c* is adjusted down to 1, the *TreeNumber* of *a* (line C in the *ArtPoint* algorithm). This completes processing of node *c*, and the depth-first search backs up to *d*. *BackNumber* of *c* is less than *TreeNumber* of *d*, so *BackNumber* of *d* is adjusted to equal that of *c* (line B). The situation at this point is shown in Figure 6.23b.

    The depth-first search moves on to nodes *e*, *g*, and *f* (Figure 6.23c). At *f* the back arc to *d* is found, and *BackNumber* of *f* is set equal to *TreeNumber* of *d* (line C).

Backing up from f to g, *BackNumber* of g is adjusted to equal *BackNumber* of f (line B), and similarly for e (line B again). (See Figure 6.23d.)

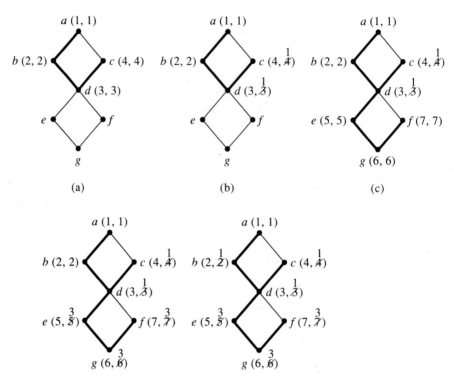

d = articulation point

Figure 6.23

Finally, in backing up from e to d, *BackNumber* of e is greater than or equal to *TreeNumber* of d, so d is declared an articulation point (line A). The recursion backs up to node b, adjusting *BackNumber* of b, and then node a, at which point line A would seem to apply (Figure 6.23e). But a is the starting node of the search and so is not an articulation point, because there is only one tree arc from a.   ●

PRACTICE 18   In Figure 6.24, the depth-first search begins at node a. Explain why each node is marked as it is and how it is concluded that c is an articulation point.

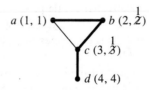

c = articulation point   **Figure 6.24**   ●

Section 6.5 Review

Technique

● Find articulation points in a simple, connected graph (using algorithm *ArtPoint*).

Main Idea

Articulation points represent single points of failure in a computer or communications network, but an algorithm exists to detect their presence.

Exercises 6.5

For Exercises 1–6, draw the depth-first search trees, where node *a* is the starting node of the depth-first search. Identify the back arcs.

★1.

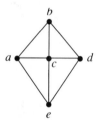

2.

3.

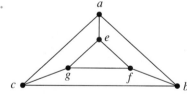

★4.

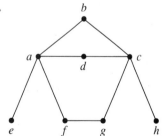

5.

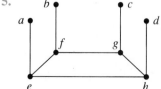

6.

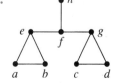

For Exercises 7–12, use algorithm *ArtPoint* to find the articulation points. Label *TreeNumber* and *BackNumber* for each node, both as first assigned and as changed. Draw the biconnected components of the graph

★7.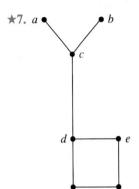

8.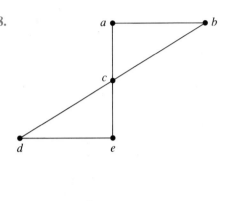

9.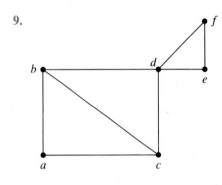

★10.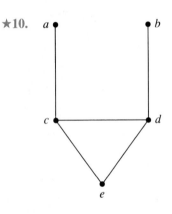

11.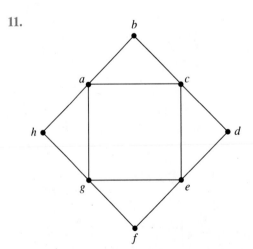

12.

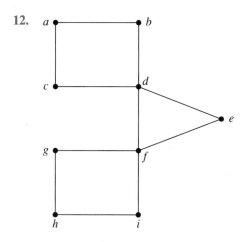

Chapter 6 Review

Terminology

adjacency relation
(p. 406)
articulation point (p. 455)
back arc (p. 456)
biconnected component
(p. 456)
biconnected graph
(p. 455)
breadth-first search
(p. 445)
connected component
(p. 450)

depth-first search
(p. 442)
dequeue (p. 446)
enqueue (p. 446)
Euler path (p. 421)
even node (p. 421)
graph traversal (p. 442)
greedy algorithm (p. 434)
Hamiltonian circuit
(p. 425)
minimal spanning tree
(p. 435)

odd node (p. 421)
queue (p. 446)
reachable node (p. 408)
reachability matrix
(p. 410)
spanning tree (p. 435)
tree arc (p. 456)

Self-Test

Answer the following true–false questions.

Section 6.1

1. Any binary relation on a set N has an associated adjacency matrix.

2. Transitive closure is the adjacency relation equivalent of reachability.

3. The reachability matrix **R** for a directed graph G is computed by taking the powers of the adjacency matrix up to n^2.

4. Warshall's algorithm proceeds by computing, in turn, the number of paths of length 1, then length 2, and so on, between nodes.

5. Warshall's algorithm computes symmetric closure in the case of a symmetric adjacency relation.

Section 6.2

6. A graph with four odd nodes can still be a connected graph.

7. An Euler path exists in any graph with an even number of odd nodes.

8. An $\Theta(n^2)$ algorithm exists to test the existence of an Euler path in a graph with n nodes.

9. A Hamiltonian circuit uses each arc and node of the graph exactly once except for the starting and ending node.

10. No algorithm to solve the Hamiltonian circuit problem is known.

Section 6.3

11. Dijkstra's algorithm for the shortest path in a graph maintains a set *IN* and adds at each step the node closest to a node in *IN*.

12. A greedy algorithm is one that recursively divides the problem into as many sub-problems as possible.

13. The minimal spanning tree for a graph may not be unique.

14. Using a linked-list representation for nodes not in *IN* does not improve the order of magnitude of the worst-case work done by Dijkstra's algorithm.

15. The collection of all arcs that are not in a minimal spanning tree for a graph will also form a spanning tree, but it may not be minimal.

Section 6.4

16. Depth-first search visits nodes at the bottom of the graph first.

17. In a breadth-first search beginning with node i, all of the nodes adjacent to i are visited in order.

18. An analysis of depth-first search and breadth-first search shows them to be algorithms of the same order of magnitude.

19. Preorder traversal is the tree equivalent of breadth-first search, using the root as the starting node.

20. Topological sorting can be done by a succession of breadth-first searches on a directed graph.

Section 6.5

21. If node n is an articulation point in a connected graph, then any path between any two nodes in the graph must pass through n.

22. A biconnected graph is a simple, connected graph with no articulation points.

23. When a node n is first reached during a depth-first search, any other arcs from n to previously visited nodes are back arcs.

24. A node n where every subtree of n in the depth-first search tree has a back arc to a predecessor of n is not an articulation point.

25. The root of a depth-first search is always an articulation point in the graph because any node's *BackNumber* will be greater than or equal to its *TreeNumber*.

On the Computer

For Exercises 1–5, write a computer program that produces the desired output from the given input.

1. *Input*: Adjacency matrix **A** for a directed graph
   Output: Reachability matrix **R** for the graph, computed from the formula

   $$\mathbf{R} = \mathbf{A} \vee \mathbf{A}^{(2)} \vee \cdots \vee \mathbf{A}^{(n)}$$

2. *Input*: Adjacency matrix **A** for a directed graph
   Output: Reachability matrix **R** for the graph, computed by using Warshall's algorithm

3. *Input*: Adjacency matrix **A** for a graph
   Output: Message indicating whether the graph has an Euler path

4. *Input*: Adjacency matrix **A** for a simple, weighted graph or directed graph and two nodes in the graph
   Output: Distance for the shortest path between the two nodes or a message that no path exists; vertices in the shortest path if one exists (*Hint:* You will need to find some way of denoting which vertices are currently in *IN*.)

5. *Input*: Adjacency matrix **A** for a simple, weighted, connected graph
   Output: Arcs (as ordered pairs) in a minimal spanning tree

For Exercises 6–8, first write a function that collects information from the user about a graph and builds an adjacency list representation of the graph; incorporate this function in the programs requested.

6. *Input*: Information about a graph (see instructions above) and a node in the graph
   Output: Nodes in a depth-first search of the graph beginning with the given node

7. *Input*: Information about a graph (see instructions above) and a node in the graph
   Output: Nodes in a breadth-first search of the graph beginning with the given node

8. *Input*: Information about a graph (see instructions above)
   Output: Articulation points in the graph

Boolean Algebra and Computer Logic

7

Chapter Objectives

After studying this chapter, you will be able to:

- Determine whether a given mathematical structure is a Boolean algebra.
- Prove properties about Boolean algebras.
- Understand what it means for an isomorphism function to preserve the effects of a binary operation or other property.
- Draw a logic network to represent a Boolean expression.
- Write a Boolean expression to represent a logic network.
- Write the truth function for a Boolean expression or logic network.
- Write a Boolean expression in canonical sum-of-products form for a given truth function.
- Use NAND and NOR gates as well as AND, OR, and NOT gates to build logic networks.
- Write a truth function from a description of a logical control device.
- Simplify Boolean expressions and logic networks using Karnaugh maps.
- Simplify Boolean expressions and logic networks using the Quine–McCluskey method.

You have been hired by Rats R Us to build the control logic for the production facilities for a new anticancer chemical compound being tested on rats. The control logic must manage the opening and closing of two valves, A and B, downstream of the mixing vat. Valve A is to open whenever the pressure in the vat exceeds 50 psi (pounds per square inch) and the salinity of the mixture exceeds 45 g/L (grams per liter). Valve B is to open whenever valve A is closed and the temperature exceeds 53°C and the acidity falls below 7.0 pH (lower pH values mean more acidity).

Question: How many and what type of logic gates will be needed in the circuit?

The answer to this electronics problem lies, surprisingly, in a branch of mathematics developed around 1850 by George Boole, an English mathematician. Boole was interested in developing rules of "algebra" for logical thinking, similar to the rules of algebra for numerical thinking. Derided at the time as useless, if harmless, Boole's work is the foundation for the electronics in computers today.

In Section 7.1 we define Boolean algebra as a mathematical model of both propositional logic and set theory. The definition requires every Boolean algebra to have certain properties, from which many additional properties can be derived. This section also discusses what it means for two instances of a Boolean algebra to be isomorphic.

Section 7.2 establishes a relationship between the Boolean algebra structure and the wiring diagrams for the electronic circuits in computers, calculators, industrial control devices, telephone systems, and so forth. Indeed, we shall see that truth functions, expressions made up of variables and the operations of Boolean algebra, and these wiring diagrams are all related. As a result, we can effectively pass from one formulation to another and still preserve characteristic behavior with respect to truth values. We shall also find that we can simplify wiring diagrams by using properties of Boolean algebras. In Section 7.3 we shall look at two other procedures for simplifying wiring diagrams.

Section 7.1   Boolean Algebra Structure

Let us revisit the wffs of propositional logic and associate with them a certain type of function. Suppose a propositional wff P has n statement letters. Then each row of the truth table for that wff associates a value of T or F with an n-tuple of T–F values. The entire truth table defines a function f such that $f: \{T, F\}^n \to \{T, F\}$ (see Example 31 in Chapter 4). The function associated with a tautology maps $\{T, F\}^n \to \{T\}$, while the function associated with a contradiction maps $\{T, F\}^n \to \{F\}$. Even a single statement letter A can be considered as defining such a function with n arguments by writing the truth table using n statement letters and then ignoring the truth value of all statement letters except for A.

EXAMPLE 1

The statement letter A defines a function $f: \{T, F\}^2 \to \{T, F\}$ by the following truthtable:

A	B	A
T	T	T
T	F	T
F	T	F
F	F	F

Here we've used two statement letters in the truth table, but the resulting truth values are those of A alone. Thus $f(T, T) = T$ and $f(T, F) = T$.   ●

467

Suppose we agree, for any propositional wff P with n statement letters, to let the symbol P denote not only the wff but also the corresponding function defined by the truth table. If P and Q are equivalent wffs, then they have the same truth tables and therefore define the same function. Then we can write $P = Q$ rather than $P \Leftrightarrow Q$. This simply confirms that a given function has multiple names, even though a given wff defines a unique function.

With this agreement, the short list of tautological equivalences from Section 1.1 can be written as follows, where $\vee$ and $\wedge$ denote disjunction and conjunction, respectively, A' denotes the negation of a statement A, 0 stands for any contradiction, and 1 stands for any tautology.

1a. $A \vee B = B \vee A$	1b. $A \wedge B = B \wedge A$	(commutative properties)
2a. $(A \vee B) \vee C =$	2b. $(A \wedge B) \wedge C =$	(associative properties)
$\quad A \vee (B \vee C)$	$\quad A \wedge (B \wedge C)$	
3a. $A \vee (B \wedge C) =$	3b. $A \wedge (B \vee C) =$	(distributive properties)
$\quad (A \vee B) \wedge (A \vee C)$	$\quad (A \wedge B) \vee (A \wedge C)$	
4a. $A \vee 0 = A$	4b. $A \wedge 1 = A$	(identity properties)
5a. $A \vee A' = 1$	5b. $A \wedge A' = 0$	(complement properties)

Switching gears a bit, in Section 3.1 we studied set identities among the subsets of a set S. We found the following list of set identities, where $\cup$ and $\cap$ denote the union and intersection of sets, respectively, A' is the complement of a set A, and $\varnothing$ is the empty set.

1a. $A \cup B = B \cup A$	1b. $A \cap B = B \cap A$	(commutative properties)
2a. $(A \cup B) \cup C =$	2b. $(A \cap B) \cap C =$	(associative properties)
$\quad A \cup (B \cup C)$	$\quad A \cap (B \cap C)$	
3a. $A \cup (B \cap C) =$	3b. $A \cap (B \cup C) =$	(distributive properties)
$\quad (A \cup B) \cap (A \cup C)$	$\quad (A \cap B) \cup (A \cap C)$	
4a. $A \cup \varnothing = A$	4b. $A \cap S = A$	(identity properties)
5a. $A \cup A' = S$	5b. $A \cap A' = \varnothing$	(complement properties)

These two lists of properties are similar. The disjunction of statements and the union of sets seem to play the same roles in their respective environments. So do the conjunction of statements and the intersection of sets. A contradiction seems to correspond to the empty set and a tautology to S. What should we make of this resemblance?

Models or Abstractions

We seem to have found two different examples—propositional logic and set theory—that share some common properties. One of the hallmarks of scientific thought is to look for patterns or similarities among various observed phenomena. Are these similarities manifestations of some underlying general principle? Can the principle itself be identified and studied? Could this shed light on the behavior of various instances of this principle? Sometimes, as seems to be the case with propositional logic and set theory, similar mathematical properties or behavior can be seen in different contexts. A mathematical structure is a formal model that serves to embody or explain this commonality, just as in physics the law of gravity is a formal model of why apples fall, the ocean has tides, and planets revolve around the sun.

Mathematical principles are models or abstractions intended to capture properties that may be common to different instances or manifestations. These principles are sometimes expressed as *mathematical structures*—abstract sets of objects, together with operations on or relationships among those objects that obey certain rules. (This may give you a clue about why this book is titled as it is.)

We can liken a mathematical structure to a human skeleton. We can think of the skeleton as the basic structure of the human body. People may be thin or fat, short or tall, black or white, and so on, but stripped down to skeletons they all look pretty much alike. Although the outward appearances differ, the inward structure, the shape and arrangement of the bones, is the same. Similarly, mathematical structures represent the underlying sameness in situations that may appear outwardly different.

It appears reasonable to abstract the common properties (tautological equivalences and set identities) for propositional wffs and set theory. Thus we shall soon define a mathematical structure called a Boolean algebra that incorporates these properties. First, however, we note that modeling or abstracting is not an entirely new idea to us:

1. We used predicate logic to model reasoning and formally defined an interpretation as a specific instance of predicate logic (Section 1.3).
2. We defined the abstract ideas of partial ordering and equivalence relation, and considered a number of specific instances that could be modeled as posets or sets on which an equivalence relation is defined (Section 4.1).
3. We noted that the graph and tree structures can model a great variety of instances (Sections 5.1 and 5.2).

Boolean algebra is just another model or abstraction for which we already have two instances.

Definition and Properties

Now suppose we try to characterize formally the similarities between propositional logic and set theory. In each case we are talking about items from a set: a set of wffs or a set of subsets of a set S. In each case we have two binary operations and one unary operation on the members of the set: disjunction/conjunction/negation or union/intersection/complementation. In each case there are two distinguished elements of the set: 0/1 or $\emptyset$/S. Finally, there are the 10 properties that hold in each case. Whenever all these features are present, we say that we have a Boolean algebra.

Definition: Boolean Algebra

A **Boolean algebra** is a set B on which are defined two binary operations $+$ and $\cdot$ and one unary operation $'$ and in which there are two distinct elements 0 and 1 such that the following properties hold for all $x, y, z \in B$:

1a. $x + y = y + x$	1b. $x \cdot y = y \cdot x$	(commutative properties)
2a. $(x + y) + z =$ $x + (y + z)$	2b. $(x \cdot y) \cdot z =$ $x \cdot (y \cdot z)$	(associative properties)
3a. $x + (y \cdot z) =$ $(x + y) \cdot (x + z)$	3b. $x \cdot (y + z) =$ $(x \cdot y) + (x \cdot z)$	(distributive properties)
4a. $x + 0 = x$	4b. $x \cdot 1 = x$	(identity properties)
5a. $x + x' = 1$	5b. $x \cdot x' = 0$	(complement properties)

What, then, is the Boolean algebra structure? It is a formalization that abstracts, or models, the two cases we have considered (and perhaps others as well). There is a subtle philosophical distinction between the formalization itself, the *idea* of the Boolean algebra structure, and any instance of the formalization, such as these two cases. Nevertheless, we shall often use the term *Boolean algebra* to describe both the idea and its occurrences. This usage should not be confusing. We often have a mental idea ("chair," for example), and whenever we encounter a concrete example of the idea, we also call it by our word for the idea (this object is a "chair").

The formalization helps us focus on the essential features common to all examples of Boolean algebras, and we can use these features—these facts from the definition of a Boolean algebra—to prove other facts about Boolean algebras. Then these new facts, once proved in general, hold in any particular instance of a Boolean algebra. To use our analogy, if we ascertain that in a typical human skeleton "the thighbone is connected to the kneebone," then we don't need to reconfirm this in every person we meet.

We denote a Boolean algebra by $[B, +, \cdot, ', 0, 1]$.

EXAMPLE 2   Let $B = \{0, 1\}$ (the set of integers 0 and 1) and define binary operations $+$ and $\cdot$ on B by $x + y = \max(x, y)$, $x \cdot y = \min(x, y)$. Then we can illustrate the operations of $+$ and $\cdot$ by the tables below.

+	0	1		$\cdot$	0	1
0	0	1		0	0	0
1	1	1		1	0	1

A unary operation $'$ can be defined by means of a table, as follows, instead of by a verbal description.

$'$	
0	1
1	0

Thus $0' = 1$ and $1' = 0$. Then $[B, +, \cdot, ', 0, 1]$ is a Boolean algebra. We can verify the 10 properties by checking all possible cases. Thus, for property 2b, the associativity of $\cdot$, we show that

$$(0 \cdot 0) \cdot 0 = 0 \cdot (0 \cdot 0) = 0$$
$$(0 \cdot 0) \cdot 1 = 0 \cdot (0 \cdot 1) = 0$$
$$(0 \cdot 1) \cdot 0 = 0 \cdot (1 \cdot 0) = 0$$
$$(0 \cdot 1) \cdot 1 = 0 \cdot (1 \cdot 1) = 0$$
$$(1 \cdot 0) \cdot 0 = 1 \cdot (0 \cdot 0) = 0$$
$$(1 \cdot 0) \cdot 1 = 1 \cdot (0 \cdot 1) = 0$$
$$(1 \cdot 1) \cdot 0 = 1 \cdot (1 \cdot 0) = 0$$
$$(1 \cdot 1) \cdot 1 = 1 \cdot (1 \cdot 1) = 1$$

For property 4a, we show that

$$0 + 0 = 0$$
$$1 + 0 = 1$$

PRACTICE 1   Verify property 4b for the Boolean algebra of Example 2.

There are many other properties that hold in any Boolean algebra. We can prove these additional properties by using the properties in the definition.

EXAMPLE 3   The **idempotent** (pronounced eye′-dem-po-tent) **property**

$$x + x = x$$

holds in any Boolean algebra because

$x + x = (x + x) \cdot 1$	(4b)
$= (x + x) \cdot (x + x')$	(5a)
$= x + (x \cdot x')$	(3a)
$= x + 0$	(5b)
$= x$	(4a)

Although ordinary arithmetic of integers has many of the properties of a Boolean algebra, the idempotent property should convince you that arithmetic is not a Boolean algebra. The property $x + x = x$ does not hold for ordinary numbers and ordinary addition unless x is zero.

In the proof of Example 3, we used property 5a to replace 1 with $x + x'$. The properties of Boolean algebra are equalities, and either side of an equal sign can be replaced with the other side. The Boolean algebra properties (rules) are like the equivalence rules in logic; in order to apply the rule, your situation must match exactly the pattern of the rule. For example, it is legal to replace

> **REMINDER:**
> A Boolean algebra property may be applied only when your expression exactly matches the pattern of one side of the property.

$$(y \cdot z) + x$$

with

$$x + (y \cdot z)$$

using property 1a, because $(y \cdot z) + x$ matches the right side of 1a where y is the Boolean algebra element $y \cdot z$, and $x + (y \cdot z)$ matches the left side of 1a under the same interpretation of y. We cannot say

$$x + (y \cdot z) = (x \cdot y) + (x \cdot z)$$

using either property 3a or 3b because we have mixed up the two properties. And, strictly speaking, we cannot replace

$$(y \cdot z) + x$$

with

$$(y + x) \cdot (z + x)$$

and claim that we are using property 3a because in property 3a the addition is to the left of the multiplication. We must reason as follows:

$$(y \cdot z) + x = x + (y \cdot z) \qquad \text{(1a)}$$
$$= (x + y) \cdot (x + z) \qquad \text{(3a)}$$
$$= (y + x) \cdot (z + x) \qquad \text{(1a twice)}$$

However, we will sometimes make implicit use of the associative property and write

$$x + y + z$$

with no parentheses.

Each property in the definition of a Boolean algebra has its dual as part of the definition, where the **dual** is obtained by interchanging $+$ and $\cdot$, and 1 and 0. Therefore, every time a new property P about Boolean algebras is proved, each step in that proof can be replaced by the dual of that step. The result is a proof of the dual of P. Thus, once we have proved P, we know that the dual of P also holds.

EXAMPLE 4   The dual of the property in Example 3, $x \cdot x = x$, is true in any Boolean algebra. ●

PRACTICE 2   **a.** What does the idempotent property of Example 3 become in the context of propositional logic?
b. What does it become in the context of set theory?   ●

Once a property about Boolean algebra is proved, we can use it to prove new properties.

PRACTICE 3   **a.** Prove that the property $x + 1 = 1$ holds in any Boolean algebra. Give a reason for each step.
b. What is the dual property?   ●

More properties of Boolean algebras appear in the exercises at the end of this section. Table 7.1 suggests hints that may help when trying to prove a Boolean algebra property of the form

some expression = some other expression

Hints for Proving Boolean Algebra Equalities
Usually the best approach is to start with the more complicated expression and try to show that it reduces to the simpler expression.
Think of adding some form of 0 (like $x \cdot x'$) or multiplying by some form of 1 (like $x + x'$).
Remember property 3a, the distributive property of addition over multiplication— easy to forget because it doesn't look like arithmetic.
Remember the idempotent properties $x + x = x$ and $x \cdot x = x$.

Table 7.1

For x an element of a Boolean algebra B, the element x' is called the **complement** of x. The complement of x satisfies

$$x + x' = 1 \quad \text{and} \quad x \cdot x' = 0$$

Indeed, x' is the unique element with these two properties. To prove this, suppose x_1 is an element of B with

$$x + x_1 = 1 \quad \text{and} \quad x \cdot x_1 = 0$$

Then

$$
\begin{aligned}
x_1 &= x_1 \cdot 1 & \text{(4b)} \\
&= x_1 \cdot (x + x') & (x + x' = 1) \\
&= (x_1 \cdot x) + (x_1 \cdot x') & \text{(3b)} \\
&= (x \cdot x_1) + (x' \cdot x_1) & \text{(1b)} \\
&= 0 + (x' \cdot x_1) & (x \cdot x_1 = 0) \\
&= (x \cdot x') + (x' \cdot x_1) & \text{(5b)} \\
&= (x' \cdot x) + (x' \cdot x_1) & \text{(1b)} \\
&= x' \cdot (x + x_1) & \text{(3b)} \\
&= x' \cdot 1 & (x + x_1 = 1) \\
&= x' & \text{(4b)}
\end{aligned}
$$

Thus $x_1 = x'$, and x' is unique. (Uniqueness in the context of propositional logic means that the truth table is unique, but there can be many different wffs associated with any particular truth table.)

The following theorem summarizes our observations.

Theorem on the Uniqueness of Complements
For any x in a Boolean algebra, if an element x_1 exists such that

$$x + x_1 = 1 \quad \text{and} \quad x \cdot x_1 = 0$$

then $x_1 = x'$.

PRACTICE 4

Prove that $0' = 1$ and $1' = 0$. (*Hint:* $1' = 0$ will follow by duality from $0' = 1$. To show $0' = 1$, use the theorem on the uniqueness of complements.)

There are many ways to define a Boolean algebra. Indeed, in our definition of Boolean algebra, we could have omitted the associative properties, since these can be derived from the remaining properties of the definition. It is much more convenient, however, to include them.

Isomorphic Boolean Algebras

What Is Isomorphism?

Two instances of a structure are **isomorphic** if there is a bijection (called an **isomorphism**) that maps the elements of one instance onto the elements of the other so that important properties are preserved. (Isomorphic graphs were discussed in

Section 5.1.) If two instances of a structure are isomorphic, each is a mirror image of the other, with the elements simply relabeled. The two instances are essentially the same. Therefore, we can use the idea of isomorphism to classify instances of a structure, lumping together those that are isomorphic.

EXAMPLE 5   Consider the two partially ordered sets

$$S_1 = \{1, 2, 3, 5, 6, 10, 15, 30\}; \ x \, \rho \, y \leftrightarrow x \text{ divides } y$$
$$S_2 = \wp(\{1, 2, 3\}); \ A \, \sigma \, B \leftrightarrow A \subseteq B$$

The Hasse diagram of each partially ordered set appears in Figure 7.1. These two diagrams certainly appear to be mirror images of each other; just by looking at the diagrams, an obvious relabeling of the nodes, as shown in Figure 7.2, suggests itself. The important properties of a partially ordered set are which elements are related, and the Hasse diagram displays this information. For example, Figure 7.1a shows that 1, because of its position at the bottom of the graph, is related to every element in S_1. Is this property preserved under the relabeling of Figure 7.2? Yes, because $\varnothing$ is the image of 1 under that relabeling, and $\varnothing$ is related to every element in S_2. Similarly, all the other "is related to" properties are preserved under the relabeling.

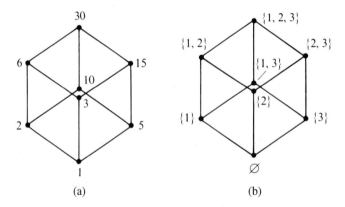

(a)   (b)

Figure 7.1

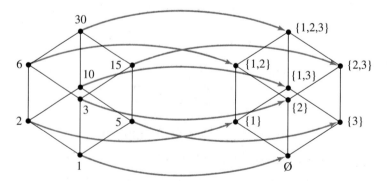

Figure 7.2

More formally, the relabeling is accomplished by the following bijection f from the set of nodes in Figure 7.1a onto the set of nodes in Figure 7.1b.

$$f(1) = \varnothing \qquad f(2) = \{1\} \qquad f(3) = \{2\} \qquad f(5) = \{3\}$$
$$f(6) = \{1, 2\} \qquad f(10) = \{1, 3\} \qquad f(15) = \{2, 3\} \qquad f(30) = \{1, 2, 3\}$$

The bijection f is an isomorphism from poset (S_1, ρ) to poset (S_2, σ). Because this isomorphism exists, the posets (S_1, ρ) and (S_2, σ) are isomorphic. (The function f^{-1} would be an isomorphism from (S_2, σ) to (S_1, ρ)).

In Example 5 it was relatively easy to find an isomorphism because of the visual representation that captured the important properties (which elements are related). Suppose that instead of a partially ordered set, we have a structure (like a Boolean algebra) where binary or unary operations are defined on a set. Then the important properties pertain to how these operations act. An isomorphism must preserve the effects of performing these operations. Each instance of two such structures that are isomorphic must be the mirror image of the other in the sense that "operate and then map" must equal "map and then operate."

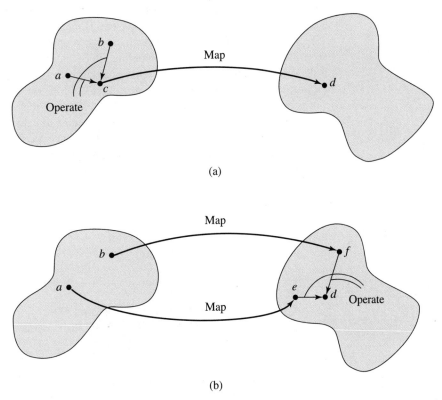

Figure 7.3

Figure 7.3 illustrates this general idea for a binary operation. In Figure 7.3a, the binary operation is performed on a and b, resulting in c, then c is mapped to d. In Figure 7.3b, a and b are mapped to e and f, on which a binary operation is performed, resulting in the same element d as before. Remember,

operate and map = map and operate

Still another view of this little equation appears in the commutative diagram of Figure 7.4.

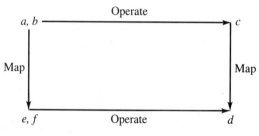

Figure 7.4

Isomorphism as Applied to Boolean Algebra

Now let's determine specifically what is involved when two instances of a Boolean algebra are isomorphic. Suppose we have two Boolean algebras, $[B, +, \cdot, ', 0, 1]$ and $[b, \&, *, ", \phi, +]$. This notation means that, for example, if x is in B, x' is the result of performing on x the unary operation defined in B, and if z is an element of b, z'' is the result of performing on z the unary operation defined in b. How would we define isomorphism between these two Boolean algebras? First, we would need a bijection f from B onto b. Then f must preserve in b the effects of the various operations in B. There are three operations, so we use three equations to express these preservations. To preserve the operation $+$, we want to be able to operate using $+$ on two elements in B and then map the result to b, or to map the two elements to b and operate using the corresponding operation $\&$ on the results there. (Think of the commutative diagram of Figure 7.4.) Thus, for x and y in B, we require

$$f(x + y) = f(x) \,\&\, f(y)$$

PRACTICE 5

a. Write the equation requiring f to preserve the effect of the binary operation $\cdot$.
b. Write the equation requiring f to preserve the effect of the unary operation $'$. ●

Here is the definition of an isomorphism for Boolean algebras.

Definition: Isomorphism for Boolean Algebras

Let $[B, +, \cdot, ', 0, 1]$ and $[b, \&, *, ", \phi, +]$ be Boolean algebras. A function $B \to b$ is an isomorphism from $[B, +, \cdot, ', 0, 1]$ to $[b, \&, *, ", \phi, +]$ if

1. f is a bijection
2. $f(x + y) = f(x) \,\&\, f(y)$
3. $f(x \cdot y) = f(x) * f(y)$
4. $f(x') = (f(x))''$

PRACTICE 6   Illustrate properties 2, 3, and 4 in the definition by commutative diagrams. ●

We already know (it was one of our original inspirations) that for any set S, $\wp(S)$ under the operations of union, intersection, and complementation constitutes a Boolean algebra. If we pick $S = \{1, 2\}$, then the elements of $\wp(S)$ are $\emptyset$, $\{1\}$, $\{2\}$, and $\{1, 2\}$. The operations are given by the following tables:

$\cup$	$\emptyset$	$\{1, 2\}$	$\{1\}$	$\{2\}$
$\emptyset$	$\emptyset$	$\{1, 2\}$	$\{1\}$	$\{2\}$
$\{1, 2\}$	$\{1, 2\}$	$\{1, 2\}$	$\{1, 2\}$	$\{1, 2\}$
$\{1\}$	$\{1\}$	$\{1, 2\}$	$\{1\}$	$\{1, 2\}$
$\{2\}$	$\{2\}$	$\{1, 2\}$	$\{1, 2\}$	$\{2\}$

$\cap$	$\emptyset$	$\{1, 2\}$	$\{1\}$	$\{2\}$	$'$	
$\emptyset$	$\emptyset$	$\emptyset$	$\emptyset$	$\emptyset$	$\emptyset$	$\{1, 2\}$
$\{1, 2\}$	$\emptyset$	$\{1, 2\}$	$\{1\}$	$\{2\}$	$\{1, 2\}$	$\emptyset$
$\{1\}$	$\emptyset$	$\{1\}$	$\{1\}$	$\emptyset$	$\{1\}$	$\{2\}$
$\{2\}$	$\emptyset$	$\{2\}$	$\emptyset$	$\{2\}$	$\{2\}$	$\{1\}$

A Boolean algebra can be defined on the set $B = \{0, 1, a, a'\}$ where the tables defining the operations of $+$, $\cdot$, and $'$ look like the following (see Exercise 1):

$+$	0	1	a	a'
0	0	1	a	a'
1	1	1	1	1
a	a	1	a	1
a'	a'	1	1	a'

$\cdot$	0	1	a	a'
0	0	0	0	0
1	0	1	a	a'
a	0	a	a	0
a'	0	a'	0	a'

$'$	
0	1
1	0
a	a'
a'	a

We claim that the mapping $f: B \rightarrow \wp(S)$ given by

$f(0) = \emptyset$
$f(1) = \{1, 2\}$
$f(a) = \{1\}$
$f(a') = \{2\}$

is an isomorphism. Certainly it is a bijection. For $x, y \in B$, we can verify each of the equations

$$f(x + y) = f(x) \cup f(y)$$
$$f(x \cdot y) = f(x) \cap f(y)$$
$$f(x') = (f(x))'$$

by examining all possible cases. Thus, for example,

$$f(a \cdot 1) = f(a) = \{1\} = \{1\} \cap \{1, 2\} = f(a) \cap f(1)$$

PRACTICE 7   Verify the following equations:

a. $f(0 + a) = f(0) \cup f(a)$
b. $f(a + a') = f(a) \cup f(a')$
c. $f(a \cdot a') = f(a) \cap f(a')$
d. $f(1') = (f(1))'$

The remaining cases also hold. Even without testing all cases, it is pretty clear here that f is going to work because it merely relabels the entries in the tables for B so that they resemble the tables for $\wp(S)$. In general, however, it may not be so easy to decide whether a given f is an isomorphism between two instances of a structure. Even harder to answer is the question of whether two given instances of a structure are isomorphic; we must either think up a function that works or show that no such function exists. One case where no such function exists is when the sets involved are not the same size; we cannot have a four-element Boolean algebra isomorphic to an eight-element Boolean algebra.

We just showed that a particular four-element Boolean algebra is isomorphic to $\wp(\{1, 2\})$. It turns out that any finite Boolean algebra is isomorphic to the Boolean algebra of a power set. Although we state this as a theorem, we shall not prove it.

Theorem on Finite Boolean Algebras
Let B be any Boolean algebra with n elements. Then $n = 2^m$ for some m, and B is isomorphic to $\wp(\{1, 2, \dots, m\})$.

The theorem above gives us two pieces of information. The number of elements in a finite Boolean algebra must be a power of 2. Also we learn that finite Boolean algebras that are power sets are—in our lumping together of isomorphic things—really the only kinds of finite Boolean algebras. In a sense we have come full circle. We defined a Boolean algebra to represent many kinds of situations; now we find that (for the finite case) the situations, except for the labels of objects, are the same anyway!

Section 7.1 Review

Techniques

- Decide whether something is a Boolean algebra.
- Prove properties about Boolean algebras.
- Write the equation meaning that a function f preserves an operation from one instance of a structure to another, and verify or disprove such an equation.

Main Ideas

Mathematical structures serve as models or abstractions of common properties found in diverse situations.

If there is an isomorphism (a bijection that preserves properties) from A to B, where A and B are instances of a structure, then except for labels, A and B are the same.

All finite Boolean algebras are isomorphic to Boolean algebras that are power sets.

Exercises 7.1

★**1.** Let $B = \{0, 1, a, a'\}$, and let $+$ and $\cdot$ be binary operations on B. The unary operation $'$ is defined by the table

$'$	
0	1
1	0
a	a'
a'	a

Suppose you know that $[B, +, \cdot, ', 0, 1]$ is a Boolean algebra. Making use of the properties that must hold in any Boolean algebra, fill in the following tables defining the binary operations $+$ and $\cdot$:

$+$	0	1	a	a'
0				
1				
a				
a'				

$\cdot$	0	1	a	a'
0				
1				
a				
a'				

2. Define two binary operations $+$ and $\cdot$ on the set $\mathbb{Z}$ of integers by $x + y = \max(x, y)$ and $x \cdot y = \min(x, y)$.

- **a.** Show that the commutative, associative, and distributive properties of a Boolean algebra hold for these two operations on $\mathbb{Z}$.
- **b.** Show that no matter what element of $\mathbb{Z}$ is chosen to be 0, the property $x + 0 = x$ of a Boolean algebra fails to hold.

3. Let S be the set $\{0, 1\}$. Then S^2 is the set of all ordered pairs of 0s and 1s; $S^2 = \{(0, 0), (0, 1), (1, 0), (1, 1)\}$. Consider the set B of all functions mapping S^2 to S. For example, one such function, $f(x, y)$, is given by

$$f(0, 0) = 0$$
$$f(0, 1) = 1$$
$$f(1, 0) = 1$$
$$f(1, 1) = 1$$

a. How many elements are in B?
b. For f_1 and f_2 members of B and $(x, y) \in S^2$, define

$$(f_1 + f_2)(x, y) = \max(f_1(x, y), f_2(x, y))$$
$$(f_1 \cdot f_2)(x, y) = \min(f_1(x, y), f_2(x, y))$$
$$f_1'(x, y) = \begin{cases} 1 \text{ if } f_1(x, y) = 0 \\ 0 \text{ if } f_1(x, y) = 1 \end{cases}$$

Suppose

$f_1(0, 0) = 1$	$f_2(0, 0) = 1$
$f_1(0, 1) = 0$	$f_2(0, 1) = 1$
$f_1(1, 0) = 1$	$f_2(1, 0) = 0$
$f_1(1, 1) = 0$	$f_2(1, 1) = 0$

What are the functions $f_1 + f_2, f_1 \cdot f_2$, and f_1'?
c. Prove that $[B, +, \cdot, ', 0, 1]$ is a Boolean algebra where the functions 0 and 1 are defined by

$0(0, 0) = 0$	$1(0, 0) = 1$
$0(0, 1) = 0$	$1(0, 1) = 1$
$0(1, 0) = 0$	$1(1, 0) = 1$
$0(1, 1) = 0$	$1(1, 1) = 1$

4. Prove the following properties of Boolean algebras. Give a reason for each step. (*Hint:* Remember the uniqueness of the complement.)
★a. $(x')' = x$ (double negation)
b. $(x + y)' = x' \cdot y', (x \cdot y)' = x' + y'$ (De Morgan's laws)

5. Prove the following properties of Boolean algebras. Give a reason for each step.
★a. $x + (x \cdot y) = x, x \cdot (x + y) = x$ (absorption properties)
b. $x \cdot [y + (x \cdot z)] = (x \cdot y) + (x \cdot z)$ (modular properties)
    $x + [y \cdot (x + z)] = (x + y) \cdot (x + z)$
c. $(x + y) \cdot (x' + y) = y$
   $(x \cdot y) + (x' \cdot y) = y$

d. $(x + (y \cdot z))' = x' \cdot y' + x' \cdot z'$
$(x \cdot (y + z))' = (x' + y') \cdot (x' + z')$
e. $(x + y) \cdot (x + 1) = x + (x \cdot y) + y$
$(x \cdot y) + (x \cdot 1) = x \cdot (x + y) \cdot y$
f. $(x + y) + (y \cdot x') = x + y$
$(x \cdot y) \cdot (y + x') = x \cdot y$

6. Prove the following properties of Boolean algebras. Give a reason for each step.

  ★**a.** $x + y' = x + (x' \cdot y + x \cdot y)'$
  b. $[(x \cdot y) \cdot z] + (y \cdot z) = y \cdot z$
  c. $(y' \cdot x) + x + (y + x) \cdot y' = x + (y' \cdot x)$
  d. $(x + y') \cdot z = [(x' + z') \cdot (y + z')]'$
  e. $(x \cdot y) + (x' \cdot z) + (x' \cdot y \cdot z') = y + (x' \cdot z)$
  f. $(x \cdot y') + (y \cdot z') + (x' \cdot z) = (x' \cdot y) + (y' \cdot z) + (x \cdot z')$

7. Prove that in any Boolean algebra, $x \cdot y' = 0$ if and only if $x \cdot y = x$.

★8. Prove that in any Boolean algebra, $x \cdot y' + x' \cdot y = y$ if and only if $x = 0$.

9. A new binary operation $\oplus$ in a Boolean algebra is defined by

$$x \oplus y = x \cdot y' + y \cdot x'$$

Prove that

  ★**a.** $x \oplus y = y \oplus x$ **b.** $x \oplus x = 0$ **c.** $0 \oplus x = x$ **d.** $1 \oplus x = x'$

10. Prove the following for any Boolean algebra:

  a. If $x + y = 0$, then $x = 0$ and $y = 0$.
  b. $x = y$ if and only if $x \cdot y' + y \cdot x' = 0$.

11. **a.** Find an example of a Boolean algebra with elements x, y, and z for which $x + y = x + z$ but $y \neq z$. (Here is further evidence that ordinary arithmetic of integers is not a Boolean algebra.)
  b. Prove that in any Boolean algebra, if $x + y = x + z$ and $x' + y = x' + z$, then $y = z$.

12. Prove that the 0 element in any Boolean algebra is unique; prove that the 1 element in any Boolean algebra is unique.

★13. Let $(S, \leqslant)$ and $(S', \leqslant')$ be two partially ordered sets. $(S, \leqslant)$ is isomorphic to $(S', \leqslant')$ if there is a bijection $f\colon S \to S'$ such that for x, y in S, $x < y \to f(x) <' f(y)$ and $f(x) <' f(y) \to x < y$.

  a. Show that there are exactly two nonisomorphic, partially ordered sets with two elements (use diagrams).
  b. Show that there are exactly five nonisomorphic, partially ordered sets with three elements.
  c. How many nonisomorphic, partially ordered sets with four elements are there?

14. Find an example of two partially ordered sets $(S, \preccurlyeq)$ and $(S', \preccurlyeq')$ and a bijection $f: S \to S'$ where, for x, y in S, $x < y \to f(x) <' f(y)$ but $f(x) <' f(y) \not\to x < y$.

15. Let $S = \{0, 1\}$ and let a binary operation $\cdot$ be defined on S by

$\cdot$	0	1
0	1	0
1	0	1

Let $T = \{5, 7\}$, and let a binary operation $+$ be defined on T by

$+$	5	7
5	7	5
7	5	7

Consider $[S, \cdot]$ and $[T, +]$ as mathematical structures.

a. If a function f is an isomorphism from $[S, \cdot]$ to $[T, +]$, what two properties must f satisfy?

b. Define a function $f: S \to T$ and prove it is an isomorphism from $[S, \cdot]$ to $[T, +]$.

★16. Let $\mathbb{R}$ denote the real numbers and $\mathbb{R}^+$ the positive real numbers. Addition is a binary operation on $\mathbb{R}$, and multiplication is a binary operation on $\mathbb{R}^+$. Consider $[\mathbb{R}, +]$ and $[\mathbb{R}^+, \cdot]$ as mathematical structures.

a. Prove that the function f defined by $f(x) = 2^x$ is a bijection from $\mathbb{R}$ to $\mathbb{R}^+$.

b. Write the equation that an isomorphism from $[\mathbb{R}, +]$ to $[\mathbb{R}^+, \cdot]$ must satisfy.

c. Prove that the function f of part (a) is an isomorphism from $[\mathbb{R}, +]$ to $[\mathbb{R}^+, \cdot]$

d. What is f^{-1} for this function?

e. Prove that f^{-1} is an isomorphism from $[\mathbb{R}^+, \cdot]$ to $[\mathbb{R}, +]$

17. An isomorphism from the Boolean algebra with set $B = \{0, 1, a, a'\}$ to the Boolean algebra with set $\wp(\{1, 2\})$ was defined in this section. Because the two Boolean algebras are essentially the same, an operation in one can be simulated by mapping to the other, operating there, and mapping back.

a. Use the Boolean algebra on $\wp(\{1, 2\})$ to simulate the computation $1 \cdot a'$ in the Boolean algebra on B.

b. Use the Boolean algebra on $\wp(\{1, 2\})$ to simulate the computation $(a)'$ in the Boolean algebra on B.

c. Use the Boolean algebra on B to simulate the computation $\{1\} \cup \{2\}$ in the Boolean algebra on $\wp(\{1, 2\})$.

d. Use the Boolean algebra on B to simulate the computation $\{1\} \cap \{1, 2\}$ in the Boolean algebra on $\wp(\{1, 2\})$.

18. Consider the set B of all functions mapping $\{0, 1\}^2$ to $\{0, 1\}$. We can define operations of $+$, $\cdot$, and $'$ on B by the following:

$$(f_1 + f_2)(x, y) = \max(f_1(x, y), f_2(x, y))$$
$$(f_1 \cdot f_2)(x, y) = \min(f_1(x, y), f_2(x, y))$$
$$f'_1(x, y) = \begin{cases} 1 \text{ if } f_1(x, y) = 0 \\ 0 \text{ if } f_1(x, y) = 1 \end{cases}$$

Then $[B, +, \cdot, ', 0, 1]$ is a Boolean algebra of 16 elements (see Exercise 3). The following table assigns names to these 16 functions:

(x, y)	0	1	f_1	f_2	f_3	f_4	f_5	f_6	f_7	f_8	f_9	f_{10}	f_{11}	f_{12}	f_{13}	f_{14}
$(0, 0)$	0	1	1	1	1	1	1	1	0	0	0	1	0	0	0	0
$(0, 1)$	0	1	0	1	1	0	1	0	1	1	1	0	0	1	0	0
$(1, 0)$	0	1	1	0	1	0	0	1	1	1	0	0	1	0	1	0
$(1, 1)$	0	1	0	0	0	0	1	1	1	0	1	1	1	0	0	1

According to the theorem on finite Boolean algebras, this Boolean algebra is isomorphic to $[\wp(\{1, 2, 3, 4\}), \cup, \cap, ', \varnothing, \{1, 2, 3, 4\}]$. Complete the following definition of an isomorphism from B to $\wp(\{1, 2, 3, 4\})$:

$0 \to \varnothing$

$1 \to \{1, 2, 3, 4\}$

$f_4 \to \{1\}$

$f_{12} \to \{2\}$

$f_{13} \to \{3\}$

$f_{14} \to \{4\}$

★19. Suppose that $[B, +, \cdot, ', 0, 1]$ and $[b, \&, *, '', \phi, +]$ are isomorphic Boolean algebras and that f is an isomorphism from B to b.

a. Prove that $f(0) = \phi$.

b. Prove that $f(1) = +$.

20. A Boolean algebra may also be defined as a partially ordered set with certain additional properties. Let $(B, \preccurlyeq)$ be a partially ordered set. For any $x, y \in B$, we define the *least upper bound* of x and y as an element z such that $x \preccurlyeq z, y \preccurlyeq z$,

and if there is any element z^* with $x \leqslant z^*$ and $y \leqslant z^*$, then $z \leqslant z^*$. The *greatest lower bound* of x and y is an element w such that $w \leqslant x$, $w \leqslant y$, and if there is any element w^* with $w^* \leqslant x$ and $w^* \leqslant y$, then $w^* \leqslant w$. A *lattice* is a partially ordered set in which every two elements x and y have a least upper bound, denoted by $x + y$, and a greatest lower bound, denoted by $x \cdot y$.

★**a.** Prove that in any lattice
  i. $x \cdot y = x$ if and only if $x \leqslant y$
  ii. $x + y = y$ if and only if $x \leqslant y$
b. Prove that in any lattice
  i. $x + y = y + x$
  ii. $x \cdot y = y \cdot x$
  iii. $(x + y) + z = x + (y + z)$
  iv. $(x \cdot y) \cdot z = x \cdot (y \cdot z)$
c. A lattice L is *complemented* if there exists a least element 0 and a greatest element 1, and for every $x \in L$ there exists $x' \in L$ such that $x + x' = 1$ and $x \cdot x' = 0$. Prove that in a complemented lattice L,

$$x + 0 = x \qquad \text{and} \qquad x \cdot 1 = x$$

for all $x \in L$.
d. A lattice L is *distributive* if

$$x + (y \cdot z) = (x + y) \cdot (x + z)$$

and

$$x \cdot (y + z) = (x \cdot y) + (x \cdot z)$$

for every $x, y, z \in L$. By parts (b) and (c), a complemented, distributive lattice is a Boolean algebra. Which of the Hasse diagrams of partially ordered sets in the accompanying figure do not represent Boolean algebras? Why? (*Hint:* In a Boolean algebra, the complement of an element is unique.)

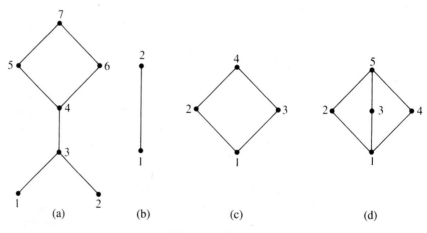

(a)   (b)   (c)   (d)

Exercise 20

21. According to the theorem on finite Boolean algebras, which we did not prove, any finite Boolean algebra must have 2^m elements for some m. Prove the weaker statement that no Boolean algebra can have an odd number of elements. (Note that in the definition of a Boolean algebra, 0 and 1 are distinct elements of B, so B has at least two elements. Arrange the remaining elements of B so that each element is paired with its complement.)

Section 7.2 Logic Networks

Combinational Networks

Basic Logic Elements

In 1938 the American mathematician Claude Shannon perceived the parallel between propositional logic and circuit logic and realized that Boolean algebra could play a part in systematizing this new realm of electronics.

Let us imagine that the electrical voltages carried along wires fall into one of two ranges, high or low, which we shall represent by 1 and 0, respectively. Voltage fluctuations within these ranges are ignored, so we are forcing a discrete, indeed binary, mask on an analogue phenomenon. We also suppose that switches can be wired so that a signal of 1 causes the switch to be closed and a signal of 0 causes the switch to be open (see Figure 7.5). Now we combine two such switches, controlled by lines x_1 and x_2, in parallel. Values of $x_1 = 0$ and $x_2 = 0$ will cause both switches to be open and thus break the circuit, so that the voltage level on the output line will be 0. If either or both lines carry a 1-value, however, one or both of the switches will be closed, and the output line will have a value of 1. Figure 7.6 illustrates the various cases.

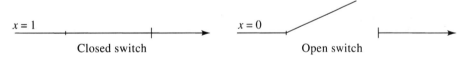

Closed switch   Open switch

Figure 7.5

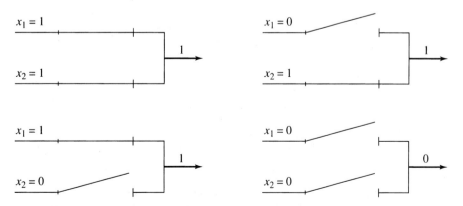

Figure 7.6

x_1	x_2	Output
1	1	1
1	0	1
0	1	1
0	0	0

Table 7.2

Table 7.2 summarizes the behavior of the circuit. Substituting T for 1 and F for 0 in the table results in the truth table for the logical connective of disjunction. Disjunction is an example of the Boolean algebra operation + in the realm of propositional logic. Thus we may think of the circuit more abstractly as an electronic device that performs the Boolean operation +. Other devices perform the Boolean operations · and '. Switches connected in series would serve to implement · , for example, where both switches must be closed ($x_1 = 1$ and $x_2 = 1$) in order to have an output of 1. However, we'll ignore the details of implementing the devices; suffice it to say that technology has progressed from mechanical switches through vacuum tubes and then transistors to integrated circuits. We shall simply represent these devices by their standard symbols.

The **OR gate**, Figure 7.7a, behaves like the Boolean operation +. The **AND gate**, Figure 7.7b, represents the Boolean operation · . Figure 7.7c shows an **inverter**, corresponding to the unary Boolean operation '. Because of the associativity property for + and ·, the OR and AND gates can have more than two inputs.

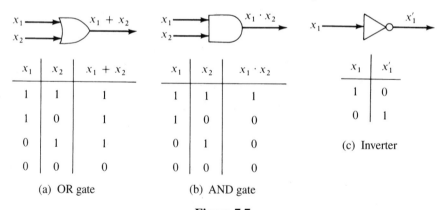

x_1	x_2	$x_1 + x_2$
1	1	1
1	0	1
0	1	1
0	0	0

(a) OR gate

x_1	x_2	$x_1 \cdot x_2$
1	1	1
1	0	0
0	1	0
0	0	0

(b) AND gate

x_1	x_1'
1	0
0	1

(c) Inverter

Figure 7.7

Boolean Expressions

Definition: Boolean Expression
A **Boolean expression** in n variables, $x_1, x_2, \ldots, x_n$, is any finite string of symbols formed by applying the following rules:

1. $x_1, x_2, \ldots, x_n$ are Boolean expressions.
2. If P and Q are Boolean expressions, so are $(P + Q)$, $(P \cdot Q)$, and (P').

(The definition of a Boolean expression is another example of a recursive definition; rule 1 is the basis step and rule 2 the inductive step.) When there is no chance of confusion, we can omit the parentheses introduced by rule 2. In addition, we define $\cdot$ to take precedence over $+$ and $'$ to take precedence over $+$ or $\cdot$, so that $x_1 + x_2 \cdot x_3$ stands for $x_1 + (x_2 \cdot x_3)$ and $x_1 + x_2'$ stands for $x_1 + (x_2')$; this convention also allows us to remove some parentheses. Finally, we shall generally omit the symbol $\cdot$ and use juxtaposition, so that $x_1 \cdot x_2$ is written $x_1 x_2$.

EXAMPLE 6 x_3, $(x_1 + x_2)'x_3$, $(x_1 x_3 + x_4')x_2$, and $(x_1'x_2)'x_1$ are all Boolean expressions.

In propositional logic, the logical connectives $\vee$, $\wedge$, and $'$ are instances of the operations of a Boolean algebra. Using only these connectives, the recursive definition for a propositional wff (Example 33 in Chapter 2) is an instance of the definition of a Boolean expression.

Truth Functions

> **Definition: Truth Function**
> A **truth function** is a function f such that $f\colon \{0, 1\}^n \to \{0, 1\}$ for some integer $n \geq 1$. The notation $\{0, 1\}^n$ denotes the set of all n-tuples of 0s and 1s. A truth function thus associates a value of 0 or 1 with each such n-tuple.

EXAMPLE 7   The truth table for the Boolean operation $+$ describes a truth function f with $n = 2$. The domain of f is $\{(1, 1), (1, 0), (0, 1), (0, 0)\}$, and $f(1, 1) = 1$, $f(1, 0) = 1$, $f(0, 1) = 1$, and $f(0, 0) = 0$. Similarly, the Boolean operation $\cdot$ describes a different truth function with $n = 2$, and the Boolean operation $'$ describes a truth function for $n = 1$.

PRACTICE 8
    a. If we are writing a truth function $f\colon \{0, 1\}^n \to \{0, 1\}$ in tabular form (like a truth table), how many rows will the table have?
    b. How many different truth functions are there that take $\{0, 1\}^2 \to \{0, 1\}$?
    c. How many different truth functions are there that take $\{0, 1\}^n \to \{0, 1\}$?

Any Boolean expression defines a unique truth function, just as do the simple Boolean expressions $x_1 + x_2$, $x_1 x_2$, and x_1'.

EXAMPLE 8   The Boolean expression $x_1 x_2' + x_3$ defines the truth function given in Table 7.3. (This is just like doing the truth tables of Section 1.1.)

x_1	x_2	x_3	$x_1 x_2' + x_3$
1	1	1	1
1	1	0	0
1	0	1	1
1	0	0	1
0	1	1	1
0	1	0	0
0	0	1	1
0	0	0	0

Table 7.3

Networks and Expressions

By combining AND gates, OR gates, and inverters, we can construct a logic network representing a given Boolean expression that produces the same truth function as that expression.

EXAMPLE 9   The logic network for the Boolean expression $x_1 x_2' + x_3$ is shown in Figure 7.8.

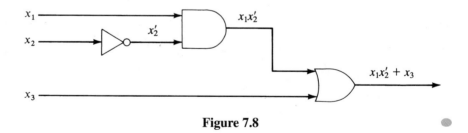

Figure 7.8

PRACTICE 9   Design the logic network for the following Boolean expressions:
a. $x_1 + x_2'$
b. $x_1(x_2 + x_3)'$

Conversely, if we have a logic network, we can write a Boolean expression with the same truth function.

EXAMPLE 10   A Boolean expression for the logic network in Figure 7.9 is $(x_1x_2 + x_3)' + x_3$.

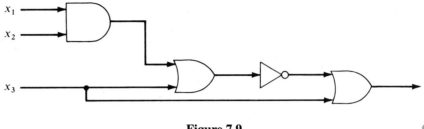

Figure 7.9

PRACTICE 10   **a.** Write a Boolean expression for the logic network in Figure 7.10.

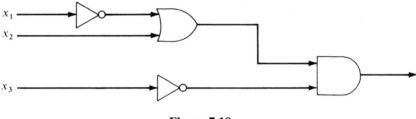

Figure 7.10

b. Write the truth function (in table form) for the network (and expression) of part (a).

Logic networks constructed of AND gates, OR gates, and inverters are also called **combinational networks**. They have several features that we should note. First, input or output lines are not tied together except by passing through gates. Lines can be split, however, to serve as input to more than one device. There are no loops where the output of an element is part of the input to that same element. Finally, the output of a network is an instantaneous function of the input; there are no delay elements that capture and remember input signals. Notice also that the picture of any network is, in effect, a directed graph.

Canonical Form

Here is the situation so far (arrows indicate a procedure that we can carry out):

truth function ← Boolean expression ↔ logic network

We can write a unique truth function from either a network or an expression. Given an expression, we can find a network with the same truth function, and conversely. The last part of the puzzle concerns how to get from an arbitrary truth function to an expression (and hence a network) having that truth function. An algorithm to solve this problem is explained in the next example.

EXAMPLE 11   Suppose we want to find a Boolean expression for the truth function f of Table 7.4. There are four rows in the table (rows 1, 3, 4, and 7) for which f is 1. The basic form of our expression will be a sum of four terms

$$(\ ) + (\ ) + (\ ) + (\ )$$

such that the first term has the value 1 for the input values of row 1 and for no others, the second term has the value 1 for the input values of row 3 and for no others, and so on. Thus, the entire expression has the value 1 for these inputs and for no others—precisely what we want. (Other inputs cause each term in the sum, and hence the sum itself, to be 0.)

x_1	x_2	x_3	$f(x_1, x_2, x_3)$
1	1	1	1
1	1	0	0
1	0	1	1
1	0	0	1
0	1	1	0
0	1	0	0
0	0	1	1
0	0	0	0

Table 7.4

Each term in the sum will be a product of the form $\alpha\beta\gamma$ where α is either x_1 or x_1', β is either x_2 or x_2', and γ is either x_3 or x_3' . If the input value of x_i, $i = 1, 2, 3$, in the row we are working on is 1, then x_i itself is used; if the input value of x_i in the row we are working on is 0, then x_i' is used. These values will force $\alpha\beta\gamma$ to be 1 for that row and 0 for all other rows. Thus, we have

row 1: $x_1x_2x_3$
row 3: $x_1x_2'x_3$
row 4: $x_1x_2'x_3'$
row 7: $x_1'x_2'x_3$

The final expression is

$$(x_1x_2x_3) + (x_1x_2'x_3) + (x_1x_2'x_3') + (x_1'x_2'x_3)$$ ●

The procedure described in Example 11 always leads to an expression that is a sum of products, called the **canonical sum-of-products form**, or the **disjunctive normal form**, for the given truth function. The only case not covered by this procedure is when the function has a value of 0 everywhere. Then we use an expression such as

$$x_1x_1'$$

which is also a sum (one term) of products. Therefore, we can find a sum-of-products expression to represent any truth function. A pseudocode description of the algorithm is given in the accompanying box. For this algorithm, the input is a truth table representing a truth function on n variables $x_1, x_2, \ldots, x_n$; the output is a Boolean expression in disjunctive normal form with the same truth function.

ALGORITHM   Sum-of-Products

Sum-of-Products (truth table; integer n)
//the truth table represents a truth function with n arguments;
//result is the canonical sum-of-products expression for this truth function

Local variables:
sum   //sum-of-products expression
product   //single term in sum, a product
i   //index for the columns of the table
row   //index for the rows of the table

   sum = empty
   for *row* = 1 to 2^n **do**
      if truth value for *row* is 1 **then**
         initialize *product*;
         for *i* = 1 **to** *n* **do**
            if x_i = 1 **then**
               put x_i in *product*
            else
               put x_i' in *product*
            end if
         end for
         sum = *sum* + *product*
      end if
   end for
   if *sum* is empty **then**
      sum = $x_1 x_1'$
   end if
   write ("The canonical sum-of-products expression for this truth function is ",
   sum)

end Sum-of-Products

Because any expression has a corresponding network, any truth function has a logic network representation. Furthermore, the AND gate, OR gate, and inverter are the only devices needed to construct the network. Thus, we can build a network for any truth function with only three kinds of parts—and lots of wire! Later we shall see that it is only necessary to stock one kind of part.

Given a truth function, the canonical sum-of-products form just described is one expression having this truth function, but it is not the only possible one. A method for obtaining a different expression for any truth function is given in Exercise 16 on page 505.

EXAMPLE 12   The network for the canonical sum-of-products form of Example 11 is shown in Figure 7.11. We have drawn the inputs to each AND gate separately because it looks neater, but actually a single x_1, x_2, or x_3 input can be split as needed.

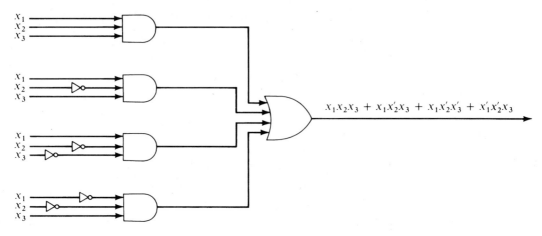

$$x_1 x_2 x_3 + x_1 x_2' x_3 + x_1 x_2' x_3' + x_1' x_2' x_3$$

Figure 7.11

PRACTICE 11   **a.** Find the canonical sum-of-products form for the truth function of Table 7.5.

x_1	x_2	x_3	$f(x_1, x_2, x_3)$
1	1	1	1
1	1	0	0
1	0	1	1
1	0	0	1
0	1	1	0
0	1	0	0
0	0	1	1
0	0	0	1

Table 7.5

b. Draw the network for the expression of part (a).

Minimization

As already noted, a given truth function may be represented by more than one Boolean expression and hence by more than one logic network composed of AND gates, OR gates, and inverters.

EXAMPLE 13

The Boolean expression

$$x_1 x_3 + x_2'$$

has the truth function of Table 7.5. The logic network corresponding to this expression is given by Figure 7.12. Compare this with your network in Practice 11(b)!

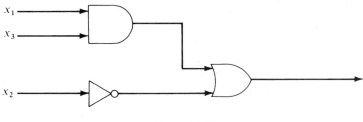

Figure 7.12

Definition: Equivalent Boolean Expressions
Two Boolean expressions are **equivalent** if they have the same truth functions.

We know that

$$x_1 x_2 x_3 + x_1 x_2' x_3 + x_1 x_2' x_3' + x_1' x_2' x_3 + x_1' x_2' x_3'$$

and

$$x_1 x_3 + x_2'$$

for example, are equivalent Boolean expressions.

Clearly, equivalence of Boolean expressions is an equivalence relation on the set of all Boolean expressions in n variables. Each equivalence class is associated with a distinct truth function. Given a truth function, algorithm *Sum-of-Products* produces one particular member of the class associated with that function, namely, the canonical sum-of-products form. However, if we are trying to design the logic network for that function, we want to find a member of the class that is as simple as possible. We would rather build the network of Figure 7.12 than the one for Practice 11(b).

How can we reduce a Boolean expression to an equivalent, simpler expression? We can use the properties of a Boolean algebra because they express the equivalence of Boolean expressions. If P is a Boolean expression containing the subexpression $x_1 + (x_2 x_3)$, for example, and Q is the expression obtained from P by replacing $x_1 + (x_2 x_3)$ with the equivalent expression $(x_1 + x_2)(x_1 + x_3)$, then P and Q are equivalent.

EXAMPLE 14

Using the properties of Boolean algebra, we can reduce

$$x_1 x_2 x_3 + x_1 x_2' x_3 + x_1 x_2' x_3' + x_1' x_2' x_3 + x_1' x_2' x_3'$$

to

$$x_1 x_3 + x_2'$$

as follows:

$$x_1x_2x_3 + x_1x_2'x_3 + x_1x_2'x_3' + x_1'x_2'x_3 + x_1'x_2'x_3'$$

$$= x_1x_2x_3 + x_1x_2'x_3 + x_1x_2'x_3 + x_1x_2'x_3' + x_1'x_2'x_3 + x_1'x_2'x_3' \qquad \text{(idempotent property)}$$

$$= x_1x_3x_2 + x_1x_3x_2' + x_1x_2'x_3 + x_1x_2'x_3' + x_1'x_2'x_3 + x_1'x_2'x_3' \qquad \text{(1b)}$$

$$= x_1x_3(x_2 + x_2') + x_1x_2'(x_3 + x_3') + x_1'x_2'(x_3 + x_3') \qquad \text{(3b)}$$

$$= x_1x_3 \cdot 1 + x_1x_2' \cdot 1 + x_1'x_2' \cdot 1 \qquad \text{(5a)}$$

$$= x_1x_3 + x_1x_2' + x_1'x_2' \qquad \text{(4b)}$$

$$= x_1x_3 + x_2'x_1 + x_2'x_1' \qquad \text{(1b)}$$

$$= x_1x_3 + x_2'(x_1 + x_1') \qquad \text{(3b)}$$

$$= x_1x_3 + x_2' \cdot 1 \qquad \text{(5a)}$$

$$= x_1x_3 + x_2' \qquad \text{(4b)} \quad \bullet$$

Unfortunately, one must be fairly clever to apply Boolean algebra properties to simplify an expression. In Section 7.3 we shall discuss more systematic approaches to this minimization problem that require less ingenuity. For now, we should say a bit more about why we want to minimize. When logic networks were built from separate gates and inverters, the cost of these elements was a considerable factor in the design, and it was desirable to have as few elements as possible. Now, however, most networks are built using integrated circuit technology, a development that began in the early 1960s. An integrated circuit is itself a logic network representing a certain truth function or functions, just as if some gates and inverters had been combined in the appropriate arrangement inside a package. These integrated circuits are then combined as needed to produce the desired result. Because the integrated circuits are extremely small and relatively inexpensive, it might seem pointless to bother minimizing a network. However, minimization is still important because the reliability of the final network is inversely related to the number of connections between the integrated circuit packages.

Moreover, the designers of integrated circuits are highly interested in the minimization problem. The square silicon chips in which integrated circuits are embedded may be no more than one-quarter inch on each side, yet they can contain the equivalent of several million transistors for implementing truth functions. The wiring channels required to connect components on the chip may be so numerous that the wiring takes up more of the chip's "floor space" than the components themselves. Minimizing the number of components and the amount of wiring required to realize a desired truth function makes the chip less crowded and easier to design. Minimization also makes it possible to embed more functions in a single chip.

Programmable Logic Arrays

Instead of designing a custom chip to implement particular truth functions, a **PLA** (*programmable logic array*) can be used. A PLA is a chip that is already implanted with an array of AND gates and an array of OR gates, together with a rectangular grid of wiring channels and some inverters. Once Boolean expressions in sum-of-products form have been determined for the truth functions, the required components in the PLA are activated. Although this chip is not very efficient and is practical only for smaller-scale circuit logic, the PLA can be mass-produced, and only a small amount of time (i.e., money) is then required to "program" it for the desired functions.

EXAMPLE 15   Figure 7.13a shows a PLA for the three inputs x_1, x_2, and x_3. There are four output lines, so four functions can be programmed in this PLA. When the PLA is programmed, the horizontal line going into an AND gate will pick up certain inputs, and the AND gate will form the product of these inputs. The vertical line going into an OR gate will, when programmed, allow the OR gate to form the sum of certain inputs. Figure 7.13b shows the same PLA programmed to produce the truth functions f_1 from Example 11 ($x_1x_2x_3 + x_1x_2'x_3 + x_1x_2'x_3' + x_1'x_2'x_3$) and f_2 from Practice 11 ($x_1x_2x_3 + x_1x_2'x_3 + x_1x_2'x_3' + x_1'x_2'x_3 + x_1'x_2'x_3'$). The dots represent activation points.

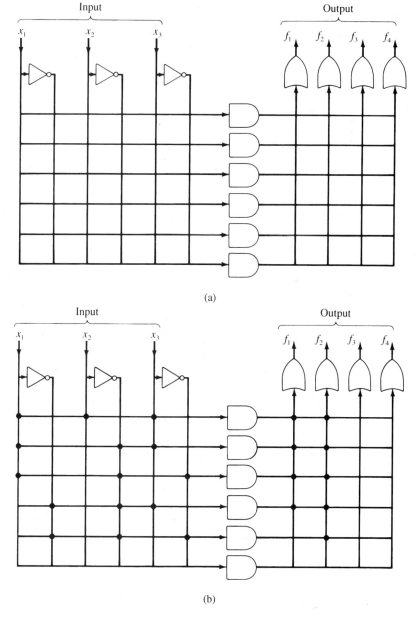

(a)

(b)

Figure 7.13

A Useful Network

We can design a network that adds binary numbers, a basic operation that a computer must be able to perform. The rules for adding two binary digits are summarized in Table 7.6.

x_1	x_2	Sum
1	1	10
1	0	1
0	1	1
0	0	0

Table 7.6

x_1	x_2	s
1	1	0
1	0	1
0	1	1
0	0	0

Table 7.7

x_1	x_2	c
1	1	1
1	0	0
0	1	0
0	0	0

Table 7.8

We can express the sum as a single sum digit s (the right-hand digit of the actual sum) together with a single carry digit c; this gives us the two truth functions of Tables 7.7 and 7.8, respectively. The canonical sum-of-products form for each truth function is

$$s = x_1'x_2 + x_1x_2'$$
$$c = x_1x_2$$

An equivalent Boolean expression for s is

$$s = (x_1 + x_2)(x_1x_2)'$$

Figure 7.14a shows a network with inputs x_1 and x_2 and outputs s and c. This device, for reasons which will be clear shortly, is called a **half-adder**.

To add two n-digit binary numbers, we add column by column from the low-order to the high-order digits. The ith column (except for the very first column) has as input its two binary digits x_1 and x_2 plus the carry digit from the addition of column $i - 1$ to its right. Thus we need a device incorporating the previous carry digit as input. This can be accomplished by adding x_1 and x_2 with a half-adder and then adding the previous carry digit c_{i-1} (using another half-adder) to the result. Again, a sum digit s and final carry digit c_i are output, where c_i is 1 if either half-adder produces a 1 as its carry digit. The **full-adder** is shown in Figure 7.14b. The full-adder is thus composed of two half-adders and an additional OR gate.

To add two n-digit binary numbers, the two low-order digits, where there is no input carry digit, can be added with a half-adder. Then the carry signal must be propagated through $n - 1$ full-adders. Although we have assumed that gates output instantaneously, there is in fact a small time delay that can be appreciable for large n. Circuitry that speeds up the addition process is available for today's computers, although more time can be saved by a clever representation of the numbers to be added.

Figure 7.15 shows the modules required to add two 3-digit binary numbers $z_1y_1x_1$ and $z_2y_2x_2$. The resulting sum is found by reading the output digits from top to bottom (low order to high order).

PRACTICE 12   Trace the operation of the circuit in Figure 7.15 as it adds 101 and 111.   ●

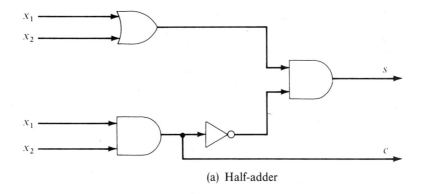

(a) Half-adder

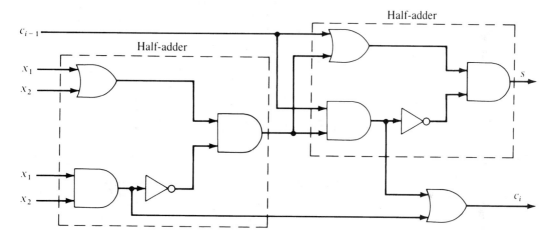

(b) Full-adder

Figure 7.14

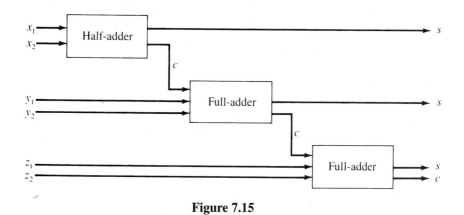

Figure 7.15

Other Logic Elements

The basic elements used in integrated circuits are not really AND and OR gates and inverters, but NAND and NOR gates. Figure 7.16 shows the standard symbol for the **NAND gate** (the NOT AND gate) and its truth function. The NAND gate alone is sufficient to realize any truth function because networks using only NAND gates can do the job of inverters, OR gates, and AND gates. Figure 7.17 shows these networks.

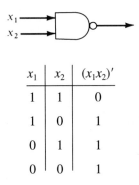

x_1	x_2	$(x_1 x_2)'$
1	1	0
1	0	1
0	1	1
0	0	1

Figure 7.16

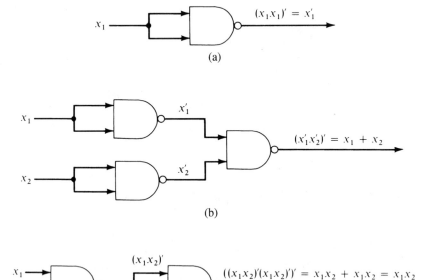

(a)

(b)

(c)

Figure 7.17

The **NOR gate** (the NOT OR gate) and its truth function appear in Figure 7.18. An exercise at the end of this section asks you to construct networks using only NOR gates for inverters, OR gates, and AND gates.

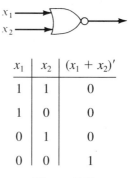

x_1	x_2	$(x_1 + x_2)'$
1	1	0
1	0	0
0	1	0
0	0	1

Figure 7.18

Although we can construct a NAND network for a truth function by replacing AND gates, OR gates, and inverters in the canonical form or a minimized form with the appropriate NAND networks, we can often obtain a simpler network by using the properties of NAND elements directly.

PRACTICE 13
    a. Rewrite the network of Figure 7.12 with NAND elements by directly replacing the AND gate, OR gate, and inverter, as in Figure 7.17.
    b. Rewrite the Boolean expression $x_1x_3 + x_2'$ for Figure 7.12 using De Morgan's laws, and then construct a network using only two NAND elements.   ●

Constructing Truth Functions

We know how to write a Boolean expression and construct a network from a given truth function. Often the truth function itself must first be deduced from the description of the actual problem.

EXAMPLE 16
At a mail-order cosmetics firm, an automatic control device is used to supervise the packaging of orders. The firm sells lipstick, perfume, makeup, and nail polish. As a bonus item, shampoo is included with any order that includes perfume or any order that includes lipstick, makeup, and nail polish. How can we design the logic network that controls whether shampoo is packaged with an order?

    The inputs to the network will represent the four items that can be ordered. We label these

x_1 = lipstick
x_2 = perfume
x_3 = makeup
x_4 = nail polish

The value of x_i will be 1 when that item is included in the order and 0 otherwise. The output from the network should be 1 if shampoo is to be packaged with the order and 0 otherwise. The truth table for the circuit appears in Table 7.9. The canonical sum-of-products form for this truth function is lengthy, but the expression $x_1x_3x_4 + x_2$ also represents the function. Figure 7.19 shows the logic network for this expression.

x_1	x_2	x_3	x_4	$f(x_1, x_2, x_3, x_4)$
1	1	1	1	1
1	1	1	0	1
1	1	0	1	1
1	1	0	0	1
1	0	1	1	1
1	0	1	0	0
1	0	0	1	0
1	0	0	0	0
0	1	1	1	1
0	1	1	0	1
0	1	0	1	1
0	1	0	0	1
0	0	1	1	0
0	0	1	0	0
0	0	0	1	0
0	0	0	0	0

Table 7.9

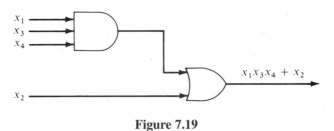

Figure 7.19

PRACTICE 14   A hall light is controlled by two light switches, one at each end. Find (a) a truth function, (b) a Boolean expression, and (c) a logic network that allows the light to be switched on or off by either switch.

In some problems the corresponding truth functions have certain undefined values because certain combinations of input cannot occur (see Exercise 25 at the end of this section). Under these "don't-care" conditions, any value may be assigned to the output.

In a programming language where the Boolean operators AND, OR, and NOT are available, designing the logic of a computer program may consist in part of choosing appropriate truth functions and their corresponding Boolean expressions (see Exercise 16 of Section 1.1).

Section 7.2 Review

Techniques

- Find the truth function corresponding to a given Boolean expression or logic network.
- Construct a logic network with the same truth function as a given Boolean expression.
- Write a Boolean expression with the same truth function as a given logic network.
- Write the Boolean expression in canonical sum-of-products form for a given truth function.
- Find a network composed only of NAND gates that has the same truth function as a given network with AND gates, OR gates, and inverters.
- Find a truth function satisfying the description of a particular problem.

Main Ideas

We can effectively convert information from any of the following three forms to any other form:

truth function $\leftrightarrow$ Boolean expression $\leftrightarrow$ logic network

A Boolean expression can sometimes be converted to a simpler, equivalent expression using the properties of Boolean algebra, thus producing a simpler network for a given truth function.

Exercises 7.2

1. Construct logic networks for the following Boolean expressions, using AND gates, OR gates, and inverters:
    ★**a.** $(x_1' + x_2)x_3$
    b. $(x_1 + x_2)' + x_1'x_3$
    c. $x_1'x_2 + (x_1x_2)'$
    d. $(x_1 + x_2)'x_3 + x_3'$

For Exercises 2–5, write a Boolean expression and a truth function for each of the logic networks shown.

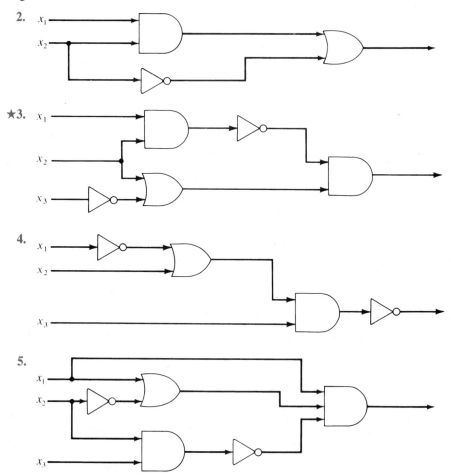

6. a. Write the truth function for the Boolean operation

$$x \oplus y = xy' + yx'$$

b. Draw the logic network for $x \oplus y$.

c. Show that the network of the accompanying figure also represents $x \oplus y$. Explain why the network illustrates that $\oplus$ is the *exclusive OR* operation.

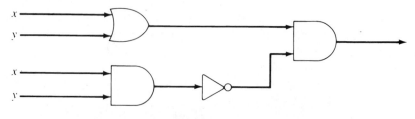

Exercise 6

For Exercises 7–12, find the canonical sum-of-products form for the truth functions in the given tables.

★7.

x_1	x_2	$f(x_1, x_2)$
1	1	0
1	0	0
0	1	0
0	0	1

8.

x_1	x_2	$f(x_1, x_2)$
1	1	1
1	0	0
0	1	1
0	0	0

9.

x_1	x_2	x_3	$f(x_1, x_2, x_3)$
1	1	1	0
1	1	0	1
1	0	1	1
1	0	0	0
0	1	1	1
0	1	0	0
0	0	1	0
0	0	0	1

10.

x_1	x_2	x_3	$f(x_1, x_2, x_3)$
1	1	1	0
1	1	0	0
1	0	1	1
1	0	0	1
0	1	1	0
0	1	0	1
0	0	1	0
0	0	0	0

11.

x_1	x_2	x_3	x_4	$f(x_1, x_2, x_3, x_4)$
1	1	1	1	1
1	1	1	0	0
1	1	0	1	1
1	1	0	0	0
1	0	1	1	1
1	0	1	0	0
1	0	0	1	1
1	0	0	0	0
0	1	1	1	0
0	1	1	0	0
0	1	0	1	0
0	1	0	0	0
0	0	1	1	1
0	0	1	0	1
0	0	0	1	0
0	0	0	0	0

12.

x_1	x_2	x_3	x_4	$f(x_1, x_2, x_3, x_4)$
1	1	1	1	1
1	1	1	0	0
1	1	0	1	1
1	1	0	0	0
1	0	1	1	1
1	0	1	0	1
1	0	0	1	0
1	0	0	0	0
0	1	1	1	1
0	1	1	0	0
0	1	0	1	1
0	1	0	0	0
0	0	1	1	0
0	0	1	0	0
0	0	0	1	0
0	0	0	0	0

★13.

x_1	x_2	x_3	$f(x_1, x_2, x_3)$
1	1	1	0
1	1	0	1
1	0	1	0
1	0	0	1
0	1	1	0
0	1	0	0
0	0	1	0
0	0	0	0

a. Find the canonical sum-of-products form for the truth function in the accompanying table.
b. Draw the logic network for the expression of part (a).
c. Use properties of a Boolean algebra to reduce the expression of part (a) to an equivalent expression whose network requires only two logic elements.

14.

x_1	x_2	x_3	$f(x_1, x_2, x_3)$
1	1	1	1
1	1	0	0
1	0	1	0
1	0	0	0
0	1	1	1
0	1	0	1
0	0	1	0
0	0	0	0

a. Find the canonical sum-of-products form for the truth function in the accompanying table.
b. Draw the logic network for the expression of part (a).
c. Use properties of a Boolean algebra to reduce the expression of part (a) to an equivalent expression whose network requires only three logic elements. Draw the network.

15. **a.** Show that the two Boolean expressions

$$(x_1 + x_2)(x_1' + x_3)(x_2 + x_3) \qquad \text{and} \qquad (x_1 x_3) + (x_1' x_2)$$

are equivalent by writing the truth table for each.
b. Write the canonical sum-of-products form equivalent to the two expressions of part (a).
c. Use properties of a Boolean algebra to reduce one of the expressions of part (a) to the other.

★16. There is also a *canonical product-of-sums form (conjunctive normal form)* for any truth function. This expression has the form

$$( )( ) \cdots ( )$$

with each factor a sum of the form

$$\alpha + \beta + \cdots + \omega$$

where $\alpha = x_1$ or x'_1, $\beta = x_2$ or x'_2, and so on. Each factor is constructed to have a value of 0 for the input values of exactly one of the rows of the truth function having value 0. Thus, the entire expression has value 0 for these inputs and no others. Find the canonical product-of-sums form for the truth functions of Exercises 7–10.

17. The accompanying figure shows an unprogrammed PLA for three inputs, x_1, x_2, and x_3. Program this PLA to generate the truth functions f_1 and f_3 represented by

$$f_1: x_1 x_2 x_3 + x'_1 x_2 x'_3 + x'_1 x'_2 x_3$$
$$f_3: x_1 x'_2 x'_3 + x'_1 x'_2 x_3 + x'_1 x'_2 x'_3$$

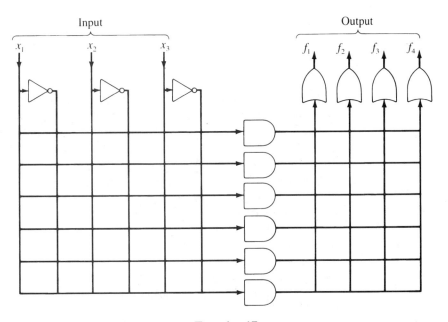

Exercise 17

18. The *2's complement* of an n-digit binary number p is an n-digit binary number q such that $p + q$ equals an n-digit representation of zero (any carry digit to column $n + 1$ is ignored). Thus 01110 is the 2's complement of 10010 because

$$\begin{array}{r} 10010 \\ + \ 01110 \\ \hline (1)00000 \end{array}$$

The 2's complement idea can be used to represent negative integers in binary form. After all, the negative of p is by definition a number that, when added to p, results in zero.

Given a binary number p, the 2's complement of p is found by scanning p from low-order to high-order digits (right to left). As long as digit i of p is 0, digit i of q is 0. When the first 1 of p is encountered, say at digit j, then digit j of q is 1, but for the remaining digits, $j < i \leq n$, $q_i = p_i'$. For $p = 10010$, for instance, the right-most 0 digit of p stays a 0 digit in q, and the first 1 digit stays a 1 digit. The remaining digits of q, however, are the reverse of the digits in p (see the accompanying figure).

$$\text{First 1}$$
$$p = 1\,0\,0 \mid 1\,0$$
$$q = 0\,1\,1 \mid 1\,0$$
$$q_i = p_i' \mid q_i = p_i$$

Exercise 18

For each binary number p, find the 2's complement of p, namely q, and then calculate $p + q$.

a. 1100   **b.** 1001   **c.** 001

19. For any digit x_i in a binary number p, let r_i be the corresponding digit in q, the 2's complement of p (see Exercise 18). The value of r_i depends on the value of x_i and also on the position of x_i relative to the first 1 digit in p. For the ith digit, let c_{i-1} denote a 0 if the digits p_j, $1 \leq j \leq i - 1$, are 0 and a 1 otherwise. A value c_i must be computed to move on to the next digit.

   a. Give a truth function for r_i with inputs x_i and c_{i-1}. Give a truth function for c_i with inputs x_i and c_{i-1}.
   b. Write Boolean expressions for the truth functions of part (a). Simplify as much as possible.
   c. Design a circuit module to output r_i and c_i from inputs x_i and c_{i-1}.
   d. Using the modules of part (c), design a circuit to find the 2's complement of a three-digit binary number zyx. Trace the operation of the circuit in computing the 2's complement of 110.

20. **a.** Construct a network for the following expression using only NAND elements. Replace the AND and OR gates and inverters with the appropriate NAND networks.

   $$x_3'x_1 + x_2'x_1 + x_3'$$

   b. Use the properties of a Boolean algebra to reduce the expression of part (a) to one whose network would require only three NAND gates. Draw the network.

★**21.** Replace the network of the accompanying figure with an equivalent network using one AND gate, one OR gate, and one inverter.

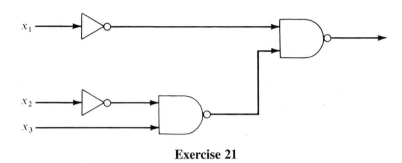

Exercise 21

22. Using only NOR elements, construct networks that can replace (a) an inverter, (b) an OR gate, and (c) an AND gate.

★**23.** Explain why Exercises 25 and 26 of Section 1.1 prove that the NAND gate and the NOR gate, respectively, are sufficient to realize any truth function.

24. Find an equivalent network for the half-adder module that uses exactly five NAND gates.

★**25.** You have just been hired at Mercenary Motors. Your job is to design a logic network so that a car can be started only when the automatic transmission is in neutral or park and the driver's seat belt is fastened. Find a truth function, a Boolean expression, and a logic network. (There is a don't-care condition to the truth function, since the car cannot be in both neutral and park.)

26. Mercenary Motors has expanded into the calculator business. You need to design the circuitry for the display readout on a new calculator. This design involves a two-step process.
   a. Any digit 0, 1, ... , 9 put into the calculator is first converted to binary form. Part (a) of the accompanying figure illustrates this conversion, which involves four separate networks, one each for x_1 to x_4. Each network has 10 inputs, but only 1 input can be on at any given moment. Write a Boolean expression and then draw a network for x_2.
   b. The binary form of the digit is then converted into a visual display by activating a pattern of seven outputs arranged as shown in part (b) of the figure. To display the digit 3, for example, y_1, y_2, y_3, y_5, and y_7 must be on, as in part (c) of the figure. Thus, the second step of the process can be represented by part (d), which involves seven separate networks, one each for y_1 to y_7, each with four inputs, x_1 to x_4. Write a truth function, a Boolean expression, and a network for y_5 and for y_6.

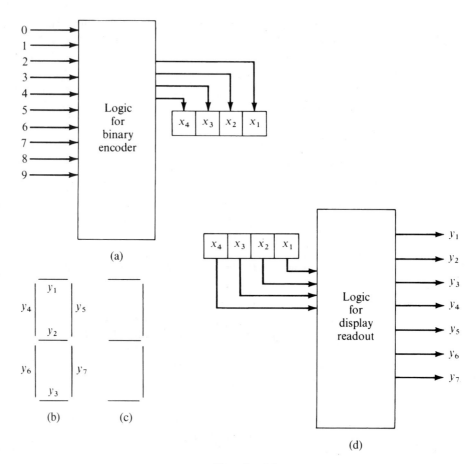

Exercise 26

27. At the beginning of this chapter, you were

... hired by Rats R Us to build the control logic for the production facilities for a new anticancer chemical compound being tested on rats. The control logic must manage the opening and closing of two valves, A and B, downstream of the mixing vat. Valve A is to open whenever the pressure in the vat exceeds 50 psi (pounds per square inch) and the salinity of the mixture exceeds 45 g/L (grams per liter). Valve B is to open whenever valve A is closed and the temperature exceeds 53°C and the acidity falls below 7.0 pH (lower pH values mean more acidity).

How many and what type of logic gates will be needed in the circuit?

Answer this question by finding the canonical sum-of-products form for the logic circuits to control A and B.

Section 7.3   Minimization

Minimization Process

Remember from Section 7.2 that a given truth function is associated with an equivalence class of Boolean expressions. If we want to design a logic network for the function, the ideal would be to have a procedure that chooses the simplest Boolean expression from the class. What we consider simple will depend on the technology employed in building the network, what kind of logic elements are available, and so on. At any rate, we probably want to minimize the total number of connections that must be made and the total number of logic elements used. (As we discuss minimization procedures, keep in mind that other factors may influence the economics of the situation. If a network is to be built only once, the time spent on minimization is costlier than building the network. But if the network is to be mass-produced, then the cost of minimization time may be worthwhile.)

We have had some experience in simplifying Boolean expressions by applying the properties of Boolean algebra. However, we had no procedure to use. We simply had to guess, attacking each problem individually. What we want now is a mechanical procedure that we can use without having to be clever or insightful. Unfortunately, we won't develop the ideal procedure. However, we already know how to select the canonical sum-of-products form from the equivalence class of expressions for a given truth function. In this section we shall discuss two procedures to reduce a canonical sum-of-products form to a minimal sum-of-products form. Therefore, we can minimize within the framework of a sum-of-products form and reduce, if not completely minimize, the number of elements and connections required.

EXAMPLE 17   The Boolean expression

$$x_1 x_2 x_3 + x_1' x_2 x_3 + x_1' x_2 x_3'$$

is in sum-of-products form. An equivalent minimal sum-of-products form is

$$x_2 x_3 + x_2 x_1'$$

Implementing a network for this form would require two AND gates, one OR gate, and an inverter. Using one of the distributive laws of Boolean algebra, this expression reduces to

$$x_2(x_3 + x_1')$$

which requires only one AND gate, one OR gate, and an inverter, but it is no longer in sum-of-products form. Thus, a minimal sum-of-products form may not be minimal in an absolute sense.   ●

There are two extremely useful equivalences in minimizing a sum-of-products form. They are

$$x_1 x_2 + x_1' x_2 = x_2$$

and

$$x_1 + x_1' x_2 = x_1 + x_2$$

PRACTICE 15

Use properties of Boolean algebra to reduce the following:

a. $x_1x_2 + x_1'x_2$ to x_2
b. $x_1 + x_1'x_2$ to $x_1 + x_2$

The equivalence $x_1x_2 + x_1'x_2 = x_2$ means, for example, that the expression $x_1'x_2x_3'x_4 + x_1'x_2'x_3'x_4$ reduces to $x_1'x_3'x_4$ Thus, when we have a sum of two products that differ in only one factor, we can eliminate that factor. However, the canonical sum-of-products form for a truth function of, say, four variables might be quite long and require some searching to locate two product terms differing by only one factor. To help us in this search, we can use the *Karnaugh map*. The Karnaugh map is a visual representation of the truth function so that terms in the canonical sum-of-products form that differ by only one factor can be matched quickly.

Karnaugh Map

In the canonical sum-of-products form for a truth function, we are interested in values of the input variables that produce outputs of 1. The Karnaugh map records the 1s of the function in an array that forces products of inputs differing by only one factor to be adjacent. The array form for a two-variable function is given in Figure 7.20. Notice that the square corresponding to x_1x_2, the upper left-hand square, is adjacent to squares $x_1'x_2$ and x_1x_2', which differ in one factor from x_1x_2; however, it is not adjacent to the $x_1'x_2'$ square, which differs in two factors from x_1x_2.

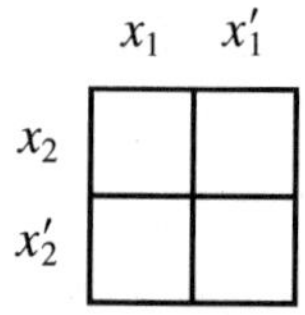

$$
\begin{array}{cc}
 & x_1 \quad x_1' \\
x_2 & \boxed{} \\
x_2' & \\
\end{array}
$$

Figure 7.20

EXAMPLE 18

The truth function of Table 7.10 is represented by the Karnaugh map of Figure 7.21. At once we can observe 1s in two adjacent squares, so there are two terms in the canonical sum-of-products form differing by one variable; again from the map, we see that the variable that changes is x_1. It can be eliminated. We conclude that the function can be represented by x_2. Indeed, the canonical sum-of-products form for the function is $x_1x_2 + x_1'x_2$, which, by our basic reduction rule, reduces to x_2. However, we did not have to write the canonical form—we only had to look at the map.

x_1	x_2	$f(x_1, x_2)$
1	1	1
1	0	0
0	1	1
0	0	0

Table 7.10

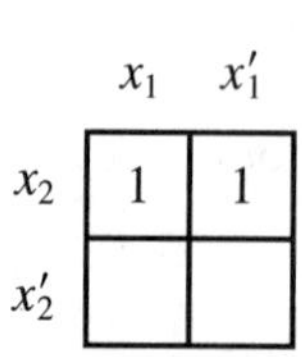

Figure 7.21

PRACTICE 16 Draw the Karnaugh map and use it to find a reduced expression for the function in Table 7.11.

x_1	x_2	$f(x_1, x_2)$
1	1	0
1	0	0
0	1	1
0	0	1

Table 7.11

Maps for Three and Four Variables

The array forms for functions of three and four variables are shown in Figure 7.22. In these arrays, adjacent squares also differ by only one variable. However, in Figure 7.22a, the leftmost and rightmost squares in a row also differ by one variable, so we consider them adjacent. (They would in fact be adjacent if we wrapped the map around a cylinder and glued the left and right edges together.) In Figure 7.22b, the leftmost and rightmost squares in a row are adjacent (differ by exactly one variable), and also the top and bottom squares in a column are adjacent.

REMINDER: It is crucial to label the Karnaugh map so that adjacent squares differ by one variable.

	x_1x_2	x_1x_2'	$x_1'x_2'$	$x_1'x_2$
x_3				
x_3'				

(a)

	x_1x_2	x_1x_2'	$x_1'x_2'$	$x_1'x_2$
x_3x_4				
x_3x_4'				
$x_3'x_4'$				
$x_3'x_4$				

(b)

Figure 7.22

In three-variable maps, when two adjacent squares are marked with 1, one variable can be eliminated; when four adjacent squares are marked with 1 (either in a single row or arranged in a square), two variables can be eliminated.

EXAMPLE 19 In the map of Figure 7.23, the squares that combine for a reduction are shown as a block. These four adjacent squares reduce to x_3 (eliminate the changing variables x_1 and x_2). The reduction uses our basic reduction rule more than once:

$$\begin{aligned} x_1x_2x_3 + x_1x_2'x_3 + x_1'x_2'x_3 + x_1'x_2x_3 &= x_1x_3(x_2 + x_2') + x_1'x_3(x_2' + x_2) \\ &= x_1x_3 + x_1'x_3 \\ &= x_3(x_1 + x_1') \\ &= x_3 \end{aligned}$$

	x_1x_2	x_1x_2'	$x_1'x_2'$	$x_1'x_2$
x_3	1	1	1	1
x_3'				

Figure 7.23

In four-variable maps, when two adjacent squares are marked with 1, one variable can be eliminated; when four adjacent squares are marked with 1, two variables can be eliminated; when eight adjacent squares are marked with 1, three variables can be eliminated.

Figure 7.24 illustrates some instances of two adjacent marked squares, Figure 7.25 illustrates some instances of four adjacent marked squares, and Figure 7.26 shows instances of eight.

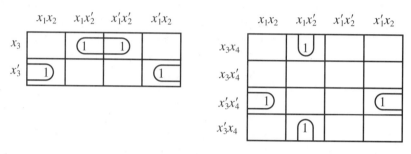

Figure 7.24

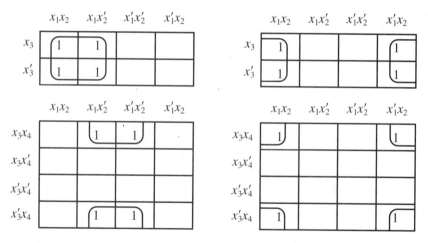

Figure 7.25

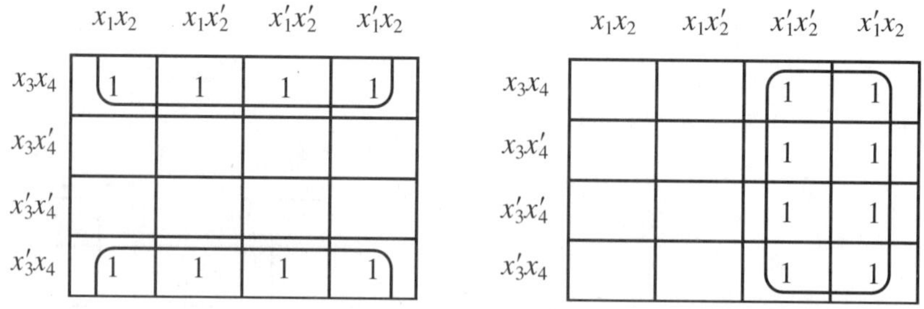

Figure 7.26

EXAMPLE 20   In the map of Figure 7.27, the four outside corners reduce to $x_2 x_4$ and the inside square reduces to $x_2' x_4'$.

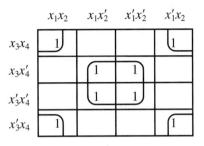

Figure 7.27

PRACTICE 17   Find the two terms represented by the map in Figure 7.28.

	$x_1 x_2$	$x_1 x_2'$	$x_1' x_2'$	$x_1' x_2$
$x_3 x_4$	1	1		
$x_3 x_4'$	1	1		
$x_3' x_4'$				1
$x_3' x_4$				1

Figure 7.28

Using the Karnaugh Map

How do we find a minimal sum-of-products form from a Karnaugh map (or from a truth function or a canonical sum-of-products form)? We must use every marked square of the map, and we want to include every marked square in the largest combination of marked squares possible, since doing so will reduce the expression as much as possible. However, we cannot begin by simply looking for the largest blocks of marked squares on the map.

EXAMPLE 21   In the Karnaugh map of Figure 7.29, if we simply looked for the largest block of marked squares, we would use the column of 1s and reduce it to $x_1' x_2'$. However, we would still have four marked squares unaccounted for. Each of these marked squares can be combined into a two-square block in only one way (see Figure 7.30), and each of these blocks has to be included. But when this is done, every square in the column of 1s is used, and the term $x_1' x_2'$ is superfluous. The minimal sum-of-products form for this map becomes

$$x_2' x_3 x_4 + x_1' x_3 x_4' + x_2' x_3' x_4' + x_1' x_3' x_4$$

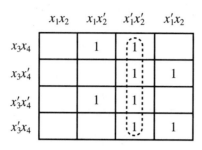

Figure 7.29

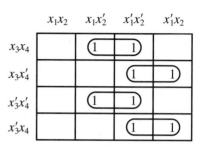

Figure 7.30

To avoid the redundancy illustrated by Example 21, we analyze the map as follows. First, we form terms for those marked squares that cannot be combined with anything. Then we use the remaining marked squares to find those that can be combined only into two-square blocks and in only one way. Then among the unused marked squares—that is, those not already assigned to a block—we find those that can be combined only into four-square blocks and in only one way; then we look for any unused squares that go uniquely into eight-square blocks. At each step, if an unused marked square can go into more than one block, we do nothing with it. Finally, we take any unused marked squares that are left (for which there was a choice of blocks) and select blocks that include them in the most efficient manner.

Table 7.12 shows the steps involved. Note, however, that this procedure for handling Karnaugh maps is not, strictly speaking, an algorithm because it doesn't always produce the correct result. If there are many 1s in the map, thus allowing many different blockings, even this procedure may not lead to a minimal form (see Example 26).

Steps in Using Karnaugh Maps
1. Set up the grid, using correct labeling for the number of Boolean variables.
2. Insert 1s in the table for the terms in the canonical sum-of-products expression.
3. Form terms for any isolated marked squares.
4. Combine squares uniquely into two-square blocks, if possible.
5. Combine squares uniquely into four-square blocks, if possible.
6. Combine squares uniquely into eight-square blocks, if possible.
7. Combine any remaining unused marked squares into blocks as efficiently as possible.

Table 7.12

EXAMPLE 22 In Figure 7.31a we have shown the only square that cannot be combined into a larger block. In Figure 7.31b, we have formed the unique two-square block for the $x_1 x_2' x_3'$ square and the unique two-square block for the $x_1' x_2' x_3$ square. All marked squares are covered. The minimal sum-of-products expression is

$$x_1 x_2 x_3 + x_2' x_3' + x_1' x_2'$$

Formally, the last two terms are obtained by expanding $x_1' x_2' x_3'$ into $x_1' x_2' x_3' + x_1' x_2' x_3'$ and then combining it with each of its neighbors.

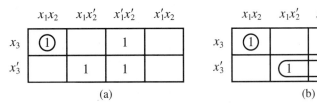

(a)   (b)

Figure 7.31

EXAMPLE 23 Figure 7.32a shows the unique two-square blocks for the $x_1' x_2' x_3 x_4$ square and the $x_1 x_2' x_3' x_4'$ square. In Figure 7.32b the two unused squares have been combined into a unique four-square block. The minimal sum-of-products expression is

$$x_1 x_3 + x_2' x_3 x_4 + x_1 x_2' x_4'$$

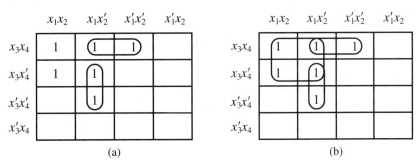

(a)   (b)

Figure 7.32

EXAMPLE 24 Figure 7.33a shows the unique two-square blocks. We can assign the remaining unused marked square to either of two different two-square blocks; these blocks are shown in Figure 7.33b. There are two minimal sum-of-products forms,

$$x_1 x_2' x_4' + x_1' x_2 x_3 + x_2' x_3 x_4'$$

and

$$x_1 x_2' x_4' + x_1' x_2 x_3 + x_1' x_3 x_4'$$

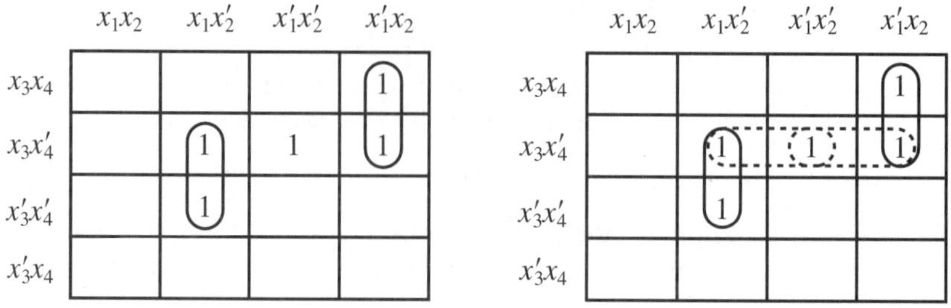

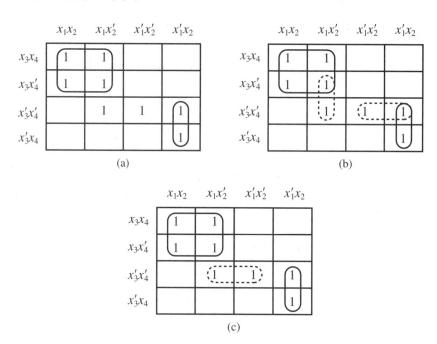

Figure 7.33

EXAMPLE 25   Figure 7.34a shows the unique two-square and four-square blocks. The remaining two unused marked squares can be assigned to two-square blocks in two different ways, as shown in parts (b) and (c). Assigning them together to a single two-square block is more efficient, since it produces a sum-of-products form with three terms rather than four. The minimal sum-of-products expression is

$$x_1 x_3 + x_1' x_2 x_3' + x_2' x_3' x_4'$$

Figure 7.34

EXAMPLE 26   Consider the map of Figure 7.35a. Here the two unique four-square blocks determined by the squares with $*$ have been chosen. In Figure 7.35b, the remaining unmarked squares, for which there was a choice of blocks, are combined into blocks as efficiently as possible. The resulting sum-of-products form is

$$x_1 x_3 + x_1' x_3' + x_3 x_4 + x_1' x_2 + x_1 x_2' x_4'$$

Yet in Figure 7.35c, choosing a different four-square block at the top leads to a simpler sum-of-products form,

$$x_2x_3 + x_1'x_3' + x_3x_4 + x_1x_2'x_4'$$

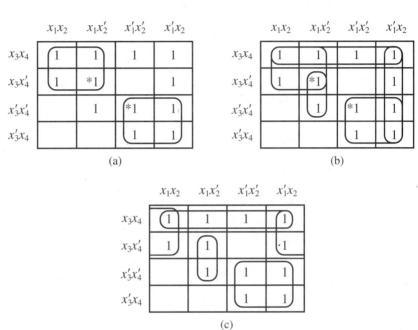

(a)

(b)

(c)

Figure 7.35

PRACTICE 18   Write the minimal sum-of-products expression for the map shown in Figure 7.36.

Figure 7.36

We have used Karnaugh maps for functions of two, three, and four variables. By using three-dimensional drawings or overlapping transparency sheets, Karnaugh maps for functions of five, six, or even more variables can be constructed, but the visualization gets too complicated to be worthwhile.

If the Karnaugh map corresponds to a function with don't-care conditions, then the don't-care squares on the map can be left blank or assigned the value 1, whichever aids the minimization process.

Quine–McCluskey Procedure

Remember that the key to reducing the canonical sum-of-products form for a truth function lies in recognizing terms of the sum that differ in only one factor. In the Karnaugh map, we see where such terms occur. A second method of reduction, the *Quine–McCluskey procedure*, organizes information from the canonical sum-of-products form into a table to simplify the search for terms differing by only one factor.

The procedure is a two-step process paralleling the use of the Karnaugh map. First we find groupings of terms (just as we looped together marked squares in the Karnaugh map); then we eliminate redundant groupings and make choices for terms that can belong to several groups.

EXAMPLE 27   Let's illustrate the Quine–McCluskey procedure by using the truth function for Example 21. We did not write the actual truth function there, but the information is contained in the Karnaugh map. The truth function is shown in Table 7.13. The eight 4-tuples of 0s and 1s producing a function value of 1 are listed in Table 7.14, which is separated into four groupings according to the number of 1s. Note that terms of the canonical sum-of-products form differing by only one factor must be in adjacent groupings, which simplifies the search for such terms.

x_1	x_2	x_3	x_4	$f(x_1, x_2, x_3, x_4)$
1	1	1	1	0
1	1	1	0	0
1	1	0	1	0
1	1	0	0	0
1	0	1	1	1
1	0	1	0	0
1	0	0	1	0
1	0	0	0	1
0	1	1	1	0
0	1	1	0	1
0	1	0	1	1
0	1	0	0	0
0	0	1	1	1
0	0	1	0	1
0	0	0	1	1
0	0	0	0	1

Table 7.13

Number of 1s	x_1	x_2	x_3	x_4
Three	1	0	1	1
Two	0	1	1	0
	0	1	0	1
	0	0	1	1
One	1	0	0	0
	0	0	1	0
	0	0	0	1
None	0	0	0	0

Table 7.14

We compare the first term, 1011, with each of the three terms of the second group, 0110, 0101, and 0011, to locate terms differing by only one factor. Such a term is 0011. The combination 1011 and 0011 reduces to –011 when the changing variable x_1 is eliminated. We shall write this reduced term with a dash in the x_1 position in the first row of a new table. The new table is Table 7.15b, where we've just seen how the first row was obtained. We've rewritten the original Table 7.14 as Table 7.15a, but we have also marked the two terms 1011 and 0011 in this table with a superscript 1. This superscript 1 is a pointer that indicates the row number of the reduced term in Table 7.15b that is formed from these two terms (numbering terms corresponds to putting loops in the Karnaugh map).

Number of 1s	x_1	x_2	x_3	x_4	
Three	1	0	1	1	1
Two	0	1	1	0	2
	0	1	0	1	3
	0	0	1	1	1,4,5
One	1	0	0	0	6
	0	0	1	0	2,4,7
	0	0	0	1	3,5,8
None	0	0	0	0	6,7,8

(a)

Number of 1s	x_1	x_2	x_3	x_4
Two	–	0	1	1
One	0	–	1	0
	0	–	0	1
	0	0	1	–
	0	0	–	1
None	–	0	0	0
	0	0	–	0
	0	0	0	–

(b)

Table 7.15

We continue this process with all the terms in Table 7.15a. A numbered term may still be used in other combinations, just as a marked square in a Karnaugh map can be in more than one loop. When we are done, the result is the completed Table 7.15b shown, where the terms in this table are again grouped by the number of 1s.

We now build still another table by processing the terms in Table 7.15b. Here not only the groupings but also the dashes help organize the search process, since terms differing by only one variable must have dashes in the same location. Tables 7.16a and 7.16b are the same as Tables 7.15a and 7.15b, and Table 7.16c is the new table. Again, numbers on terms in Table 7.16b that combine serve as pointers to the reduced terms in Table 7.16c. When we have processed all the terms in Table 7.16b, the reduction process cannot be continued. The unnumbered terms are irreducible, so they represent the possible maximum-sized loops on a Karnaugh map.

(a)

Number of 1s	x_1	x_2	x_3	x_4	
Three	1	0	1	1	1
Two	0	1	1	0	2
	0	1	0	1	3
	0	0	1	1	1,4,5
One	1	0	0	0	6
	0	0	1	0	2,4,7
	0	0	0	1	3,5,8
None	0	0	0	0	6,7,8

(b)

Number of 1s	x_1	x_2	x_3	x_4	
Two	–	0	1	1	
One	0	–	1	0	
	0	–	0	1	
	0	0	1	–	1
	0	0	–	1	1
None	–	0	0	0	
	0	0	–	0	1
	0	0	0	–	1

(c)

Number of 1s	x_1	x_2	x_3	x_4
None	0	0	–	–

Table 7.16

For the second step of the process, we compare the original terms with the irreducible terms. We form a table with the original terms as column headers and the irreducible terms (the unnumbered terms in the reduction tables just constructed) as row labels. A check in the comparison table (Table 7.17) indicates that the original term in that column eventually led to the irreducible term in that row, which can be determined by following the pointers.

	1011	0110	0101	0011	1000	0010	0001	0000
–011	✓			✓				
0–10		✓				✓		
0–01			✓				✓	
–000				✓				✓
00––				✓		✓	✓	✓

Table 7.17

If a column in the comparison table has a check in only one row, the irreducible term for that row is the only one covering the original term, so it is an essential term and must appear in the final sum-of-products form. Thus, we see from Table 7.17 that the terms –011, 0–10, 0–01, and –000 are essential and must be in the final expression. We also note that all columns with a check in row 5 also have checks in another row and so are covered by an essential reduced term already in the expression. Thus, 00–– is redundant. As in Example 21, the minimal sum-of-products form is

$$x_2' x_3 x_4 + x_1' x_3 x_4' + x_1' x_3' x_4 + x_2' x_3' x_4'$$

In situations where there is more than one minimal sum-of-products form, the comparison table will have nonessential, nonredundant reduced terms. A selection must be made from these reduced terms to cover all columns not covered by essential terms.

EXAMPLE 28   We shall use the Quine–McCluskey procedure on the problem presented in Example 24. The reduction tables are given in Table 7.18, and the comparison table appears in Table 7.19.

Number of 1s	x_1	x_2	x_3	x_4	
Three	0	1	1	1	1
Two	1	0	1	0	2,3
	0	1	1	0	1,4
One	0	0	1	0	2,4
	1	0	0	0	3

(a)

Number of 1s	x_1	x_2	x_3	x_4
Two	0	1	1	–
One	–	0	1	0
	1	0	–	0
	0	–	1	0

(b)

Table 7.18

	0111	1010	0110	0010	1000
011–	✓		✓		
–010		✓		✓	
10–0		✓			✓
0–10			✓	✓	

Table 7.19

We see from the comparison table that 011– and 10–0 are essential reduced terms and that there are no redundant terms. The only original term not covered by essential terms is 0010, column 4, and the choice of the reduced term for row 2 or for row 4 will cover it. Thus, the minimal sum-of-products form, as before, is

$$x_1'x_2x_3 + x_1x_2'x_4' + x_2'x_3x_4'$$

or

$$x_1'x_2x_3 + x_1x_2'x_4' + x_1'x_3x_4'$$

PRACTICE 19   Use the Quine–McCluskey procedure to find a minimal sum-of-products form for the truth function in Table 7.20.

x_1	x_2	x_3	$f(x_1, x_2, x_3)$
1	1	1	1
1	1	0	1
1	0	1	0
1	0	0	1
0	1	1	0
0	1	0	0
0	0	1	1
0	0	0	1

Table 7.20

The Quine–McCluskey procedure applies to truth functions with any number of input variables, but for a large number of variables, the procedure is extremely tedious to do by hand. However, it is exactly the kind of systematic, mechanical process that lends itself to a computerized solution. In contrast, Karnaugh maps make use of the human ability to quickly recognize visual patterns.

If the truth function f has few 0-values and a large number of 1-values, it may be simpler to implement the Quine–McCluskey procedure for the complement of the function, f', which will have 1-values where f has 0-values, and vice versa. Once a minimal sum-of-products expression is obtained for f', it can be complemented to obtain an expression for f, although the new expression will not be in sum-of-products form. (In fact, by De Morgan's laws, it will be equivalent to a product-of-sums form.) We can obtain the network for f from the sum-of-products network for f' by tacking an inverter on the end.

The whole object of minimizing a network is to simplify the internal configuration while preserving the external behavior. In Chapter 8 we shall attempt the same sort of minimization on finite-state machine structures.

Section 7.3 Review

Techniques

- Minimize the canonical sum-of-products form for a truth function by using a Karnaugh map.
- Minimize the canonical sum-of-products form for a truth function by using the Quine–McCluskey procedure.

Main Idea

Algorithms exist for reducing a canonical sum-of-products form to a minimized sum-of-products form.

Exercises 7.3

For Exercises 1–5, write the minimal sum-of-products form for the Karnaugh maps of the given figures.

★1.

	x_1x_2	x_1x_2'	$x_1'x_2'$	$x_1'x_2$
x_3			1	1
x_3'	1	1		1

2.

	x_1x_2	x_1x_2'	$x_1'x_2'$	$x_1'x_2$
x_3	1			1
x_3'		1		

★3.

	x_1x_2	x_1x_2'	$x_1'x_2'$	$x_1'x_2$
x_3	1	1	1	1
x_3'	1			1

4.

	x_1x_2	x_1x_2'	$x_1'x_2'$	$x_1'x_2$
x_3x_4		1		
x_3x_4'		1	1	1
$x_3'x_4'$	1	1	1	
$x_3'x_4$		1		

5.

	x_1x_2	x_1x_2'	$x_1'x_2'$	$x_1'x_2$
x_3x_4				1
x_3x_4'	1	1		
$x_3'x_4'$		1	1	
$x_3'x_4$			1	

For Exercises 6 and 7, use a Karnaugh map to find the minimal sum-of-products form for the truth functions shown.

★6.

x_1	x_2	x_3	$f(x_1, x_2, x_3)$
1	1	1	1
1	1	0	1
1	0	1	0
1	0	0	0
0	1	1	1
0	1	0	0
0	0	1	0
0	0	0	0

7.

x_1	x_2	x_3	x_4	$f(x_1, x_2, x_3, x_4)$
1	1	1	1	1
1	1	1	0	1
1	1	0	1	1
1	1	0	0	1
1	0	1	1	0
1	0	1	0	1
1	0	0	1	0
1	0	0	0	1
0	1	1	1	1
0	1	1	0	1
0	1	0	1	1
0	1	0	0	1
0	0	1	1	0
0	0	1	0	0
0	0	0	1	0
0	0	0	0	0

8. Use a Karnaugh map to find the minimal sum-of-products form for the truth function of Exercise 11, Section 7.2.

9. Use a Karnaugh map to find the minimal sum-of-products form for the truth function of Exercise 12, Section 7.2.

10. Use a Karnaugh map to find the minimal sum-of-products form for the following Boolean expressions:

   a. $x_1'x_2'x_3x_4 + x_1x_2x_3'x_4 + x_1'x_2'x_3'x_4 + x_1x_2'x_3x_4' + x_1'x_2x_3x_4 + x_1'x_2x_3'x_4 + x_1'x_2'x_3x_4'$

   b. $x_1'x_2'x_3'x_4' + x_1x_2x_3'x_4 + x_1'x_2'x_3'x_4 + x_1x_2x_3'x_4' + x_1'x_2x_3x_4 + x_1x_2'x_3'x_4'$

11. Use a Karnaugh map to find a minimal sum-of-products expression for the network of three variables shown in the accompanying figure. Sketch the new network.

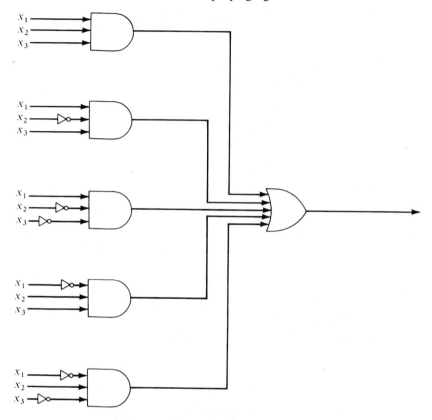

Exercise 11

★12. Use a Karnaugh map to find a minimal sum-of-products form for the truth function shown in two columns in the accompanying table. Don't-care conditions are shown by dashes.

x_1	x_2	x_3	x_4	$f(x_1, x_2, x_3, x_4)$	x_1	x_2	x_3	x_4	$f(x_1, x_2, x_3, x_4)$
1	1	1	1	0	0	1	1	1	0
1	1	1	0	1	0	1	1	0	1
1	1	0	1	0	0	1	0	1	0
1	1	0	0	–	0	1	0	0	1
1	0	1	1	0	0	0	1	1	1
1	0	1	0	–	0	0	1	0	0
1	0	0	1	0	0	0	0	1	–
1	0	0	0	0	0	0	0	0	0

Exercise 12

13. *At Rats R Us, you found a standard sum-of-products form for the logic to control valves A and B (Exercise 27, Section 7.2).*

Now earn yourself a raise by using Karnaugh maps to minimize these expressions.

★14. Use the Quine–McCluskey procedure to find a minimal sum-of-products form for the truth function illustrated by the map for Exercise 3.

15. Use the Quine–McCluskey procedure to find a minimal sum-of-products form for the network in the accompanying figure. Sketch the new network.

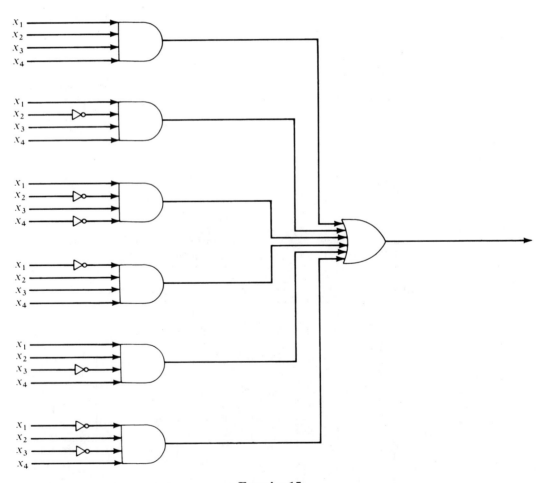

Exercise 15

For Exercises 16 and 17, Use the Quine–McCluskey procedure to find the minimal sum-of-products form for the truth functions in the given tables.

16.

x_1	x_2	x_3	x_4	$f(x_1, x_2, x_3, x_4)$
1	1	1	1	0
1	1	1	0	1
1	1	0	1	0
1	1	0	0	0
1	0	1	1	0
1	0	1	0	1
1	0	0	1	1
1	0	0	0	1
0	1	1	1	0
0	1	1	0	0
0	1	0	1	0
0	1	0	0	1
0	0	1	1	1
0	0	1	0	1
0	0	0	1	0
0	0	0	0	1

17.

x_1	x_2	x_3	x_4	$f(x_1, x_2, x_3, x_4)$
1	1	1	1	1
1	1	1	0	0
1	1	0	1	1
1	1	0	0	0
1	0	1	1	1
1	0	1	0	0
1	0	0	1	1
1	0	0	0	1
0	1	1	1	1
0	1	1	0	0
0	1	0	1	1
0	1	0	0	1
0	0	1	1	1
0	0	1	0	0
0	0	0	1	1
0	0	0	0	1

18. Use the Quine–McCluskey procedure to find the minimal sum-of-products form for the following Boolean expressions:

★**a.** $x_1 x_2' x_3 x_4' + x_1' x_2' x_3 x_4 + x_1' x_2 x_3 x_4 + x_1' x_2 x_3' x_4' + x_1' x_2 x_3 x_4' + x_1' x_2' x_3' x_4$

b. $x_1 x_2 x_3 x_4 + x_1 x_2' x_3 x_4 + x_1 x_2 x_3 x_4' + x_1 x_2' x_3 x_4' + x_1' x_2 x_3 x_4' + x_1 x_2 x_3' x_4' + x_1' x_2 x_3' x_4' + x_1 x_2 x_3' x_4 + x_1' x_2 x_3' x_4$

c. $x_1 x_2 x_3 x_4 + x_1 x_2 x_3 x_4' + x_1' x_2 x_3 x_4' + x_1 x_2 x_3' x_4' + x_1' x_2' x_3' x_4' + x_1' x_2 x_3' x_4' + x_1 x_2' x_3' x_4 + x_1' x_2' x_3' x_4 + x_1 x_2 x_3' x_4$

d. $x_1' x_2 x_3' x_4 x_5' + x_1' x_2 x_3' x_4' x_5 + x_1 x_2 x_3 x_4 x_5 + x_1' x_2' x_3 x_4' x_5 + x_1 x_2' x_3 x_4 x_5 + x_1' x_2' x_3' x_4' x_5 + x_1 x_2' x_3 x_4' x_5 + x_1 x_2 x_3 x_4' x_5 + x_1 x_2' x_3' x_4 x_5 + x_1' x_2' x_3' x_4 x_5'$

19. Use the Quine–McCluskey procedure to find a minimal sum-of-products form for the truth function illustrated by the map in Figure 7.34.

Chapter 7 Review

Terminology

AND gate (p. 486)
Boolean algebra (p. 469)
Boolean expression
  (p. 486)
canonical sum-of-products
  form (p. 490)
combinational network
  (p. 489)
complement (of a Boolean
  algebra element)
  (p. 473)

disjunctive normal form
  (p. 490)
dual (of a Boolean algebra
  property) (p. 472)
equivalent Boolean expressions (p. 493)
full-adder (p. 496)
half-adder (p. 496)
idempotent property
  (of a Boolean algebra)
  (p. 471)

inverter (p. 486)
isomorphic instances of a
  structure (p. 473)
isomorphism (p. 473)
isomorphism for Boolean
  algebras (p. 476)
NAND gate (p. 498)
NOR gate (p. 499)
OR gate (p. 486)
PLA (p. 494)
truth function (p. 487)

Self-Test

Answer the following true–false questions.

Section 7.1

1. In any Boolean algebra, $x + x' = 0$.

2. Set theory is an instance of a Boolean algebra in which $+$ is set union and $\cdot$ is set intersection.

3. In any Boolean algebra, $x + (y + x \cdot z) = x + y$.

4. The dual of the equation in the previous statement is $x \cdot [y \cdot (x + z)] = x \cdot y$.

5. Any two Boolean algebras with 16 elements are isomorphic.

Section 7.2

6. A logic network for the Boolean expression $(x + y)'$ could be built using one AND gate and two inverters.

7. The canonical sum-of-products form for a truth function $f: \{0, 1\}^n \to \{0, 1\}$ has n terms.

8. Two single-digit binary numbers can be added using a network consisting of two half-adders.

9. The following two logic networks represent the same truth function:

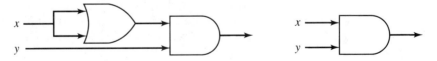

10. The most efficient way to construct a logic network for a given truth function using only NAND gates is to construct the logic network using AND, OR, and NOT gates and then replace each of these elements with its equivalent form in NAND gates.

Section 7.3

11. A Karnaugh map is a device to help change a canonical sum-of-products form to a reduced sum-of-products form.

12. The 1s in a Karnaugh map correspond to the 1-values of the truth function.

13. When using a Karnaugh map to reduce a Boolean expression, the largest possible blocks should be determined first because they provide the greatest reduction.

14. In the Quine–McCluskey procedure, terms that combine must have dashes in the same locations.

15. In the Quine–McCluskey procedure, a check in some row of the table indicates that the term for that row is an essential term that must appear in the reduced expression.

On the Computer

1. *Input*: n and tables defining two binary operations and one unary operation on a set of n objects
   Output: Indication of whether the structure is a Boolean algebra
   Algorithm: Testing the 10 properties for all cases

2. *Input*: n, tables defining a binary operation on each of two sets of n elements, and a table defining a bijection from one set to the other
   Output: Indication of whether the function is an isomorphism
   Algorithm: Testing all possible cases

3. *Input*: n and a table representing a truth function with n arguments
   Output: Canonical sum-of-products Boolean expression for the truth function

4. *Input*: n and a table representing a truth function with n arguments
   Output: Minimal sum-of-products Boolean expression for the truth function
   Algorithm: Using the Quine–McCluskey procedure

Modeling Arithmetic, Computation, and Languages

8

Chapter Objectives

After studying this chapter, you will be able to:

- See how algebraic structures, finite-state machines, and Turing machines are all models of various kinds of computation, and how formal languages attempt to model natural languages.
- Recognize certain well-known group structures.
- Prove some properties about groups.
- Understand what it means for groups to be isomorphic.
- Trace the operation of a given finite-state machine on an input string.
- Construct finite-state machines to recognize certain sets.
- For a given finite-state machine, find an equivalent machine with fewer states if one exists.
- Trace the operation of a given Turing machine on an input tape.
- Construct Turing machines to perform certain recognition or computation tasks.
- Understand the Church-Turing thesis and what it implies for the Turing machine as a model of computation.
- Be aware of the $P = NP$ question regarding computational complexity.
- Given a grammar G, construct the derivation of strings in $L(G)$.
- Understand the relationship between different classes of formal languages and different computational devices.

*Your team at Babel, Inc., is writing a compiler for a new programming language, currently codenamed ScrubOak after a tree outside your office window. During the first phase of compilation (called the lexical analysis phase) the compiler must break down statements into individual units called tokens. In particular, the compiler must be able to recognize identifiers in the language, which are strings of letters, and also recognize the two keywords in the language, which are **if** and **in**.*

Question: how can the compiler recognize the individual tokens in a statement?

A mathematical structure, as discussed in Chapter 7, is a formal model intended to capture common properties or behavior found in different contexts. A structure consists of an abstract set of objects, together with operations on or relationships among those objects that obey certain rules. The Boolean algebra structure of Chapter 7 is a model of the properties and behavior common to both propositional logic and set theory. As a formal model, it is an abstract entity, an idea; propositional logic and set theory are two instances, or realizations, of this idea.

In this chapter we shall study other structures. In Section 8.1, algebraic structures will be defined that model various types of arithmetic such as addition of integers and multiplication of positive real numbers. These arithmetics represent a limited form of computation, but we shall look for models of much broader forms of computation in Sections 8.2 and 8.3.

Our initial choice for such a model, the finite-state machine, is a useful device for certain tasks, such as the lexical analysis task facing your team at Babel. But the finite-state machine is ultimately too limited to model computation in the general sense. For a model that captures the notion of computation in all its generality, we shall turn to the Turing machine. Using the Turing machine as a model of computation will reveal that some well-defined tasks are not computable at all.

Finally, Section 8.4 discusses formal grammars and languages, which were developed as attempts to model natural languages such as English. While less than completely successful in this regard, formal grammars and languages do serve to model many constructs in programming languages and play an important role in compiler theory.

Section 8.1   Algebraic Structures

Definitions and Examples

Let us begin by analyzing a simple form of arithmetic, namely the addition of integers. There is a set $\mathbb{Z}$ of objects (the integers) and a binary operation on those objects (addition). Recall from Section 3.1 that a binary operation on a set must be *well defined* (giving a unique answer whenever it is applied to any two members of the set) and that the set must be *closed* under the operation (the answer must be a member of the set). The notation $[\mathbb{Z}, +]$ will denote the set together with the binary operation on that set.

In $[\mathbb{Z}, +]$, an equation such as

$$2 + (3 + 5) = (2 + 3) + 5$$

is true. On each side of the equation the integers remain in the same order, but the grouping of those integers, which indicates the order in which the additions are performed, changes. Changing the grouping has no effect on the answer. Another type of equation that holds in $[\mathbb{Z}, +]$ is

$$2 + 3 = 3 + 2$$

Changing the order of the integers being added has no effect on the answer.

Equations such as

$$2 + 0 = 2$$
$$0 + 3 = 3$$
$$-125 + 0 = -125$$

are also true. Adding zero to any integer does not change the value of that integer.

Finally, equations such as

$$2 + (-2) = 0$$
$$5 + (-5) = 0$$
$$-20 + 20 = 0$$

are true; adding the negative of an integer to the integer gives 0 as a result.

These equations represent four properties that occur so often that they each have a name.

Definitions: Properties of Binary Operations

Let S be a set and let $\cdot$ denote a binary operation on S. (Here $\cdot$ does *not* necessarily denote multiplication but simply any binary operation.)

1. The operation $\cdot$ is **associative** if

$$(\forall x)(\forall y)(\forall z)[x \cdot (y \cdot z) = (x \cdot y) \cdot z]$$

   Associativity allows us to write $x \cdot y \cdot z$ without using parentheses because grouping does not matter.

2. The operation $\cdot$ is **commutative** if

$$(\forall x)(\forall y)(x \cdot y = y \cdot x)$$

3. $[S, \cdot]$ has an **identity element** if

$$(\exists i)(\forall x)(x \cdot i = i \cdot x = x)$$

4. If $[S, \cdot]$ has an identity element i, then each element in S has an **inverse** with respect to $\cdot$ if

$$(\forall x)(\exists x^{-1})(x \cdot x^{-1} = x^{-1} \cdot x = i)$$

In the statements of the properties, the universal quantifiers range over the set S; if the associative property holds, the equation $x \cdot (y \cdot z) = (x \cdot y) \cdot z$ is true for any x, y, $z \in S$, and similarly for the commutative property. The existential quantifier also applies to the set S, so an identity element i, if it exists, must be an element of S, and an inverse element x^{-1}, if it exists, must be an element of S. Note the order of the quantifiers: In the definition of an identity, the existential quantifier comes first—there must be one identity element i that satisfies the equation $x \cdot i = i \cdot x = x$ for every x in S, just like the integer 0 in $[\mathbb{Z}, +]$. In the definition of the inverse element, the existential quantifier comes second—for each x, there is an x^{-1}, and if x is changed, then x^{-1} can change, just like the inverse of 2 in $[\mathbb{Z}, +]$ is -2 and the inverse of 5 is -5. If there is no identity element, then it does not make sense to talk about inverse elements.

Definitions: Group, Commutative Group

$[S, \cdot]$ is a **group** if S is a nonempty set and $\cdot$ is a binary operation on S such that

1. $\cdot$ is associative.
2. an identity element exists (in S).
3. each element in S has an inverse (in S) with respect to $\cdot$.

A group in which the operation $\cdot$ is commutative is called a **commutative group**.

Once again, the dot in the definitions is a generic symbol representing a binary operation. In any specific case, the particular binary operation has to be defined. If the operation is addition, for example, then the $+$ sign replaces the generic symbol, as in $[\mathbb{Z}, +]$. As an analogy with programming, we can think of the generic symbol as a formal parameter to be replaced by an actual argument—the specific operation—when its value becomes known. If it is clear what the binary operation is, we may refer to "the group S" rather than "the group $[S, \cdot]$."

From the discussion, it should be clear that $[\mathbb{Z}, +]$ is a commutative group, with an identity element of 0. The idea of a group would not be useful if there were not a number of other instances.

EXAMPLE 1   Let $\mathbb{R}^+$ denote the positive real numbers, and let $\cdot$ denote real-number multiplication, which is a binary operation on $\mathbb{R}^+$. Then $[\mathbb{R}^+, \cdot]$ is a commutative group. Multiplication is associative and commutative. The positive real number 1 serves as an identity because

$$x \cdot 1 = 1 \cdot x = x$$

for every positive real number x. Every positive real number x has an inverse with respect to multiplication, namely the positive real number $1/x$, because

$$x \cdot 1/x = 1/x \cdot x = 1$$

●

PRACTICE 1   The set in Example 1 is limited to the positive real numbers. Is $[\mathbb{R}, \cdot]$ a commutative group? Why or why not?

●

EXAMPLE 2

Let $M_2(\mathbb{Z})$ denote the set of 2×2 matrices with integer entries, and let $+$ denote matrix addition. Then $+$ is a binary operation on $M_2(\mathbb{Z})$ (note that closure holds). This is a commutative group because the integers are a commutative group, so each corner of the matrix behaves properly. For example, matrix addition is commutative because

$$\begin{bmatrix} a_{1,1} & a_{1,2} \\ a_{2,1} & a_{2,2} \end{bmatrix} + \begin{bmatrix} b_{1,1} & b_{1,2} \\ b_{2,1} & b_{2,2} \end{bmatrix} = \begin{bmatrix} a_{1,1} + b_{1,1} & a_{1,2} + b_{1,2} \\ a_{2,1} + b_{2,1} & a_{2,2} + b_{2,2} \end{bmatrix}$$

$$= \begin{bmatrix} b_{1,1} + a_{1,1} & b_{1,2} + a_{1,2} \\ b_{2,1} + a_{2,1} & b_{2,2} + a_{2,2} \end{bmatrix}$$

$$= \begin{bmatrix} b_{1,1} & b_{1,2} \\ b_{2,1} & b_{2,2} \end{bmatrix} + \begin{bmatrix} a_{1,1} & a_{1,2} \\ a_{2,1} & a_{2,2} \end{bmatrix}$$

The matrix

$$\begin{bmatrix} 0 & 0 \\ 0 & 0 \end{bmatrix}$$

is an identity. The matrix

$$\begin{bmatrix} 1 & -4 \\ 2 & 5 \end{bmatrix}$$

is an inverse of the matrix

$$\begin{bmatrix} -1 & 4 \\ -2 & -5 \end{bmatrix}$$

REMINDER:
It's important to understand all this new terminology. You can't prove that $P \rightarrow Q$ if you don't know what you're starting with or where you want to go.

A structure called a **monoid** results from dropping the inverse property in the definition of a group; thus a monoid has an associative operation and an identity element, but in a monoid that is not also a group, at least one element has no inverse. A **semigroup** results from dropping the identity property and the inverse property in the definition of a group; thus a semigroup has an associative operation, but in a semigroup that is not also a monoid, no identity element exists. Many familiar forms of arithmetic are instances of semigroups, monoids, and groups.

EXAMPLE 3

Consider $[M_2(\mathbb{Z}), \cdot]$ where $\cdot$ denotes matrix multiplication. Closure holds. It can be shown (Exercise 4) that matrix multiplication is associative. The matrix

$$\begin{bmatrix} 1 & 0 \\ 0 & 1 \end{bmatrix}$$

serves as an identity because

$$\begin{bmatrix} 1 & 0 \\ 0 & 1 \end{bmatrix} \cdot \begin{bmatrix} a & b \\ c & d \end{bmatrix} = \begin{bmatrix} a & b \\ c & d \end{bmatrix} \cdot \begin{bmatrix} 1 & 0 \\ 0 & 1 \end{bmatrix} = \begin{bmatrix} a & b \\ c & d \end{bmatrix}$$

Thus $[M_2(\mathbb{Z}), \cdot]$ is at least a monoid.

PRACTICE 2   Prove that $[M_2(\mathbb{Z}), \cdot]$ is not a commutative monoid.

PRACTICE 3   Prove that $[M_2(\mathbb{Z}), \cdot]$ is not a group.

Although the requirements for a structure to be a semigroup are relatively modest, not every arithmetic structure qualifies.

PRACTICE 4   Prove that $[\mathbb{Z}, -]$ is not a semigroup, where $-$ denotes integer subtraction.

PRACTICE 5   Let S be the set of noninteger rational numbers, and let $\cdot$ denote multiplication. Is $[S, \cdot]$ a semigroup?

EXAMPLE 4   Each of the following is an instance of a commutative semigroup. You should be able to verify closure, associativity, and commutativity for each:

$[\mathbb{N}, +], [\mathbb{N}, \cdot], [\mathbb{Q}, \cdot], [\mathbb{R}^+, +], [\mathbb{R}, +]$

EXAMPLE 5   For any Boolean algebra $[B, +, \cdot, ', 0, 1]$, $[B, +]$, and $[B, \cdot]$ are commutative semigroups. Therefore for any set S, $[\wp(S), \cup]$ and $[\wp(S), \cap]$ are commutative semigroups.

Because the requirements that must be satisfied in going from semigroup to monoid to group keep getting stiffer, we expect some examples to drop out, but those remaining should have richer and more interesting personalities.

PRACTICE 6   Which of the following semigroups are monoids? Name the identities.

$[\mathbb{N}, +], [\mathbb{N}, \cdot], [\mathbb{Q}, \cdot], [\mathbb{R}^+, +], [\mathbb{R}, +], [\wp(S), \cup], [\wp(S), \cap]$

PRACTICE 7   Which of the monoids from the list in Practice 6 are groups?

Now we shall look at a selection of other examples of semigroups, monoids, and groups where the elements are not just simple numbers or where the operations are less familiar.

EXAMPLE 6   An expression of the form

$$a_n x^n + a_{n-1} x^{n-1} + \cdots + a_0$$

where $a_i \in \mathbb{R}$, $i = 0, 1, \ldots, n$, and $n \in \mathbb{N}$ is a **polynomial in x with real-number coefficients** (or a **polynomial in x over $\mathbb{R}$.**) For each i, a_i is the **coefficient** of x^i. If i is the largest integer greater than 0 for which $a_i \neq 0$, the polynomial is of **degree** i; if no such i exists, the polynomial is of **zero degree**. Terms with zero coefficients are generally not written. Thus, $\pi x^4 - 2/3 x^2 + 5$ is a polynomial of degree 4, and the constant polynomial 6 is of zero degree. The set of all polynomials in x over $\mathbb{R}$ is denoted by $\mathbb{R}[x]$.

We define binary operations of $+$ and $\cdot$ in $\mathbb{R}[x]$ to be the familiar operations of polynomial addition and multiplication. For polynomials $f(x)$ and $g(x)$ members of $\mathbb{R}[x]$, the products $f(x) \cdot g(x)$ and $g(x) \cdot f(x)$ are equal because the coefficients are real numbers, and we can use all the properties of real numbers under multiplication and addition (properties such as commutativity and associativity). Similarly, for $f(x)$, $g(x)$, and $h(x)$ members of $\mathbb{R}[x]$, $(f(x) \cdot g(x)) \cdot h(x) = f(x) \cdot (g(x) \cdot h(x))$. The constant polynomial 1 is an identity because $1 \cdot f(x) = f(x) \cdot 1 = f(x)$ for every $f(x) \in \mathbb{R}[x]$. Thus, $[\mathbb{R}[x], \cdot]$ is a commutative monoid. It fails to be a group because only the nonzero constant polynomials have inverses. For example, there is no polynomial $g(x)$ such that $g(x) \cdot x = x \cdot g(x) = 1$, so the polynomial x has no inverse. However, $[\mathbb{R}[x], +]$ is a commutative group.

PRACTICE 8

a. For $f(x)$, $g(x)$, $h(x) \in \mathbb{R}[x]$, write the equations saying that $\mathbb{R}[x]$ under $+$ is commutative and associative.
b. What is an identity element in $[\mathbb{R}[x], +]$?
c. What is an inverse of $7x^4 - 2x^3 + 4$ in $[\mathbb{R}[x], +]$?

Polynomials play a special part in the history of group theory (the study of groups) because much research in group theory was prompted by the very practical problem of solving polynomial equations of the form $f(x) = 0, f(x) \in \mathbb{R}[x]$. The quadratic formula provides an algorithm for finding solutions for every $f(x)$ of degree 2, and the algorithm uses only the algebraic operations of addition, subtraction, multiplication, division, and taking roots. Other such algorithms exist for polynomials of degrees 3 and 4. One of the highlights of abstract algebra is the proof that no algorithm using only these operations exists for every $f(x)$ of degree 5. (Notice that this statement is much stronger than simply saying that no algorithm has yet been found; it says to stop looking for one.)

The next example uses modular arithmetic. You may recall from Section 4.1 (Example 15 and the subsequent discussion) that each computer has some limit on the size of the integers that it can store. Although we would like a computer to be able to exhibit the behavior of $[\mathbb{Z}, +]$, the best we can obtain is some finite approximation. The approximation is achieved by performing addition modulo n. The "answer" to the computation $x + y$ for $x, y \in \mathbb{Z}$ is then either the actual value $x + y$ if this value falls within the limit that can be stored or a remainder value obtained by doing modular arithmetic, which is equivalent to $x + y$ under the equivalence relation of congruence modulo n.

EXAMPLE 7

Let $\mathbb{Z}_5 = \{0, 1, 2, 3, 4\}$ and define *addition modulo 5*, denoted by $+_5$, on $\mathbb{Z}_5$ by $x +_5 y = r$, where r is the remainder when $x + y$ is divided by 5. For example, $1 +_5 2 = 3$ and $3 +_5 4 = 2$. *Multiplication modulo 5* is defined by $x \cdot_5 y = r$, where r is the remainder when $x \cdot y$ is divided by 5. Thus, $2 \cdot_5 3 = 1$ and $3 \cdot_5 4 = 2$. Then $[\mathbb{Z}_5, +_5]$ is a commutative group, and $[\mathbb{Z}_5, \cdot_5]$ is a commutative monoid.

PRACTICE 9   **a.** Complete the following tables defining $+_5$ and $\cdot_5$ on $\mathbb{Z}_5$:

$+_5$	0	1	2	3	4
0					
1			3		
2					
3				2	
4					

$\cdot_5$	0	1	2	3	4
0					
1					
2				1	
3					2
4					

b. What is an identity in $[\mathbb{Z}_5, +_5]$? In $[\mathbb{Z}_5, \cdot_5]$?
c. What is an inverse of 2 in $[\mathbb{Z}_5, +_5]$?
d. Which elements in $[\mathbb{Z}_5, \cdot_5]$ have inverses?

As we did on $\mathbb{Z}_5$, we can define operations of **addition modulo n** and **multiplication modulo n** on the set $\mathbb{Z}_n = \{0, 1, \dots, n-1\}$ where n is any positive integer. Again $[\mathbb{Z}_n, +_n]$ is a commutative group and $[\mathbb{Z}_n, \cdot_n]$ is a commutative monoid.

PRACTICE 10   **a.** Give the table for $\cdot_6$ on $\mathbb{Z}_6$.
b. Which elements in $[\mathbb{Z}_6, \cdot_6]$ have inverses?

Notice that when we use a table to define an operation on a finite set, it is easy to check for commutativity by looking for symmetry around the main diagonal. It is also easy to find an identity element because its row looks like the top of the table and its column looks like the side. And it is easy to locate an inverse of an element. Look along the row until you find a column where the identity appears; then check to see that changing the order of the elements still gives the identity. However, associativity (or the lack of it) is not immediately apparent from the table.

The next two examples give us algebraic structures where the elements are *functions*, mappings from a domain to a codomain.

EXAMPLE 8   Let A be a set and consider the set S of all functions f such that $f: A \to A$. The binary operation is function composition, denoted by $\circ$. Note that S is closed under $\circ$ and that function composition is associative (see Practice 11). Thus $[S, \circ]$ is a semigroup, called the **semigroup of transformations on A**. Actually $[S, \circ]$ is a monoid because the identity function i_A that takes each member of A to itself has the property that for any $f \in S$,

$$f \circ i_A = i_A \circ f = f$$

PRACTICE 11   Prove that function composition on the set S just defined is associative.

EXAMPLE 9

Again let A be a set and consider the set S_A of all bijections f such that $f: A \rightarrow A$ (permutations of A). Bijectiveness is preserved under function composition, function composition is associative, the identity function i_A is a permutation, and for any $f \in S_A$, the inverse function f^{-1} exists and is a permutation. Furthermore,

$$f \circ f^{-1} = f^{-1} \circ f = i_A$$

Thus, $[S_A, \circ]$ is a group, called the **group of permutations** on A.

If $A = \{1, 2, \ldots, n\}$ for some positive integer n, then S_A is called the **symmetric group of degree n** and denoted by S_n. Thus, S_3, for example, is the set of all permutations on $\{1, 2, 3\}$. There are six such permutations, which we shall name as follows (using the cycle notation of Section 4.4):

$$\alpha_1 = i \qquad \alpha_2 = (1, 2) \qquad \alpha_3 = (1, 3)$$
$$\alpha_4 = (2, 3) \qquad \alpha_5 = (1, 2, 3) \qquad \alpha_6 = (1, 3, 2)$$

Recall that the notation $(1, 2)$, for example, means that 1 maps to 2, 2 maps to 1, and unnamed elements map to themselves. The composition $(1, 2) \circ (1, 3)$ is done from right to left, so

By $(1, 3)$ By $(1, 2)$
$$1 \rightarrow 3 \rightarrow 3$$
$$2 \rightarrow 2 \rightarrow 1$$
$$3 \rightarrow 1 \rightarrow 2$$

resulting in $(1, 3, 2)$. Thus $\alpha_2 \circ \alpha_3 = (1, 2) \circ (1, 3) = (1, 3, 2) = \alpha_6$.

PRACTICE 12

a. Complete the group table for $[S_3, \circ]$.

$\circ$	α_1	α_2	α_3	α_4	α_5	α_6
α_1						
α_2			α_6			
α_3						
α_4						
α_5						
α_6						

b. Is $[S_3, \circ]$ a commutative group?

$[S_3, \circ]$ is our first example of a noncommutative group (although $[M_2(\mathbb{Z}), \cdot\,]$ was a noncommutative monoid).

The next example is very simple but particularly appropriate because it appears in several areas of computer science, including formal language theory and automata theory.

EXAMPLE 10   Let A be a finite set; its elements are called **symbols** and A itself is called an **alphabet**. A^* denotes the set of all finite-length **strings**, or **words**, over A. A^* can be defined recursively (as in Example 34 in Chapter 2), where $\cdot$ denotes **concatenation** of strings:

1. The **empty string** λ (the string with no symbols) belongs to A^*.
2. Any single member of A belongs to A^*.
3. If x and y are strings in A^*, so is $x \cdot y$.

Thus, if $A = \{a, b\}$, then *abbaa*, *bbbbba*, and *a* are all strings over A, and *abbaa* $\cdot$ *a* gives the string *abbaaa*. From the recursive definition, any string over A contains only a finite number of symbols. The number of symbols in a string is called its **length**. The empty string λ is the only zero-length string.

The empty string λ should not be confused with the empty set $\varnothing$; even if A itself is $\varnothing$, then $A^* = \{\lambda\}$. If A is nonempty, then whatever the size of A, A^* is a denumerable (countably infinite) set. If A contains only one element, say $A = \{a\}$, then λ, a, aa, aaa, ... , is an enumeration of A^*. If A contains more than one element, then a lexicographical (alphabetical) ordering can be imposed on the elements of A. An enumeration of A^* is then obtained by counting the empty string first, then lexicographically ordering all strings of length 1 (there is a finite number of these), then lexicographically ordering all strings of length 2 (there is a finite number of these), and so forth. Note also that if A is nonempty, strings of arbitrary length can be found in A^*.

Concatenation is a binary operation on A^*, and it is associative. The empty string λ is an identity because for any string $x \in A^*$,

$$x \cdot \lambda = \lambda \cdot x = x$$

Therefore, $[A^*, \cdot]$ is a monoid, called the **free monoid generated by** A.   ●

PRACTICE 13   For $A = \{a, b\}$,

    a. Is $[A^*, \cdot]$ a commutative monoid?
    b. Is $[A^*, \cdot]$ a group?   ●

Basic Results about Groups

We shall now prove some basic theorems about groups. There are hundreds of theorems about groups and many books devoted exclusively to group theory, so we are barely scratching the surface here. The results we shall prove follow almost immediately from the definitions involved.

REMINDER:
To prove that something is unique ...

By definition, a group $[G, \cdot]$ (or a monoid) has an identity element, and we have tried to be careful to refer to *an* identity element rather than *the* identity element. However, it is legal to say *the* identity because there is only one. To prove that the identity element is unique, suppose that i_1 and i_2 are both identity elements. Then

$$i_1 = i_1 \cdot i_2 = i_2$$

PRACTICE 14   Justify the foregoing equality signs.   ●

Because $i_1 = i_2$, the identity element is unique. Thus, we have proved the following theorem.

Theorem on the Uniqueness of the Identity in a Group
In any group (or monoid) $[G, \cdot]$, the identity element i is unique.

Each element x in a group $[G, \cdot]$ has an inverse element x^{-1}. Therefore, G contains many different inverse elements, but for each x, the inverse is unique.

Theorem on the Uniqueness of Inverses in a Group
For each x in a group $[G, \cdot]$, x^{-1} is unique.

PRACTICE 15

Prove the preceding theorem. (*Hint:* Assume two inverses for x, namely y and z, and let i be the identity. Then $y = y \cdot i = y \cdot (x \cdot z) = \cdots$.)   ●

If x and y belong to a group $[G, \cdot]$, then $x \cdot y$ belongs to G and must have an inverse element in G. Naturally, we expect that inverse to have some connection with x^{-1} and y^{-1}, which we know exist in G. We can show that $(x \cdot y)^{-1} = y^{-1} \cdot x^{-1}$; thus the inverse of a product is the product of the inverses in reverse order.

Theorem on the Inverse of a Product
For x and y members of a group $[G, \cdot]$, $(x \cdot y)^{-1} = y^{-1} \cdot x^{-1}$.

REMINDER:
If it walks like a duck ...

Proof: We shall show that $y^{-1} \cdot x^{-1}$ has the two properties required of $(x \cdot y)^{-1}$. Then, because inverses are unique, $(y^{-1} \cdot x^{-1})$ must be $(x \cdot y)^{-1}$.

$$(x \cdot y) \cdot (y^{-1} \cdot x^{-1}) = x \cdot (y \cdot y^{-1}) \cdot x^{-1}$$
$$= x \cdot i \cdot x^{-1}$$
$$= x \cdot x^{-1}$$
$$= i$$

Similarly, $(y^{-1} \cdot x^{-1}) \cdot (x \cdot y) = i$. Notice how associativity and the meaning of i and inverses all come into play in this proof.

PRACTICE 16

Write 10 as $7 +_{12} 3$ and use the theorem on the inverse of a product to find $(10)^{-1}$ in the group $[\mathbb{Z}_{12}, +_{12}]$.   ●

We know that many familiar number systems such as $[\mathbb{Z}, +]$ and $[\mathbb{R}, +]$ are groups. We make use of group properties when we do arithmetic or algebra in these systems. In $[\mathbb{Z}, +]$, for example, if we see the equation $x + 5 = y + 5$, we conclude that $x = y$. We are making use of the right cancellation law, which, we shall soon see, holds in any group.

> **Definition: Cancellation Laws**
> A set S with a binary operation $\cdot$ satisfies the **right cancellation law** if for x, y, $z \in S$, $x \cdot z = y \cdot z$ implies $x = y$. It satisfies the **left cancellation law** if $z \cdot x = z \cdot y$ implies $x = y$.

Now suppose that x, y, and z are members of a group $[G, \cdot]$ and that $x \cdot z = y \cdot z$. To conclude that $x = y$, we take advantage of z^{-1}. Thus,

$$x \cdot z = y \cdot z$$

implies

$$(x \cdot z) \cdot z^{-1} = (y \cdot z) \cdot z^{-1}$$
$$x \cdot (z \cdot z^{-1}) = y \cdot (z \cdot z^{-1})$$
$$x \cdot i = y \cdot i$$
$$x = y$$

Hence, G satisfies the right cancellation law.

PRACTICE 17   Show that any group $[G, \cdot]$ satisfies the left cancellation law.

We have proved the following theorem.

> **Theorem on Cancellation in a Group**
> Any group $[G, \cdot]$ satisfies the left and right cancellation laws.

EXAMPLE 11   We know that $[\mathbb{Z}_6, \cdot_6]$ is not a group. Here the equation

$$4 \cdot_6 2 = 1 \cdot_6 2$$

holds, but of course $4 \neq 1$.

Again, working in $[\mathbb{Z}, +]$, we would solve the equation $x + 6 = 13$ by adding -6 to both sides, producing a unique answer of $x = 13 + (-6) = 7$. The property of being able to solve linear equations for unique solutions holds in all groups. Consider the equation $a \cdot x = b$ in the group $[G, \cdot]$ where a and b belong to G and x is to be found. Then $x = a^{-1} \cdot b$ is an element of G satisfying the equation. Should x_1 and x_2 both be solutions to the equation $ax = b$, then $a \cdot x_1 = a \cdot x_2$ and, by left cancellation, $x_1 = x_2$. Similarly, the unique solution to $x \cdot a = b$ is $x = b \cdot a^{-1}$.

> **Theorem on Solving Linear Equations in a Group**
> Let a and b be any members of a group $[G, \cdot]$. Then the linear equations $a \cdot x = b$ and $x \cdot a = b$ have unique solutions in G.

PRACTICE 18   Solve the equation $x +_8 3 = 1$ in $[\mathbb{Z}_8, +_8]$.

The theorem on solving linear equations tells us something about tables for finite groups. As we look along row a of the group operation table, does element b appear twice? If so, then the table says that there are two distinct elements x_1 and x_2 of the group such that $a \cdot x_1 = b$ and $a \cdot x_2 = b$. But by the theorem on solving linear equations, this double occurrence can't happen. Thus, a given element of a finite group appears at most once in a given row of the group table. However, to complete the row, each element must appear at least once. A similar result holds for columns. Therefore, in a group table, each element appears exactly once in each row and each column. This property alone, however, is not sufficient to insure that a table represents a group; the operation must also be associative (see Exercise 24 at the end of this section).

PRACTICE 19   Assume that ∘ is an associative binary operation on $\{1, a, b, c, d\}$. Complete the following table to define a group with identity 1:

∘	1	a	b	c	d
1	1				
a				c	d	1
b		c	d		
c		d		a	
d				b	c

If $[G, \cdot]$ is a group where G is finite with n elements, then n is said to be the **order of the group**, denoted by $|G|$. If G is an infinite set, the group is of infinite order.

PRACTICE 20   **a.** Name a commutative group of order 18.
b. Name a noncommutative group of order 6.

More properties of groups appear in the Exercises at the end of this section.

Subgroups

We know what groups are and we know what subsets are, so it should not be hard to guess what a subgroup is. However, we shall look at an example before we give the definition. We know that $[\mathbb{Z}, +]$ is a group. Now let A be any nonempty subset of $\mathbb{Z}$. For any x and y in A, x and y are also in $\mathbb{Z}$, so that $x + y$ exists and is unique. The set A "inherits" a well-defined operation, $+$, from $[\mathbb{Z}, +]$. The associativity property is also inherited, because for any $x, y, z \in A$, it is also true that $x, y, z \in \mathbb{Z}$ and the equation

$$(x + y) + z = x + (y + z)$$

holds. Perhaps A under the inherited operation has all of the structure of $[\mathbb{Z}, +]$ and is itself a group. This depends on A.

Suppose that $A = E$, the set of even integers. E is closed under addition, E contains 0 (the identity element), and the inverse of every even integer (its negative) is an even integer. $[E, +]$ is thus a group. But suppose that $A = O$, the set of odd integers. $[O, +]$ fails to be a group for several reasons. For one thing, it is not closed—adding two odd integers produces an even integer. (Closure depends on the set as well as the operation, so it is not an inherited property). For another, a subgroup must have an identity with respect to addition; 0 is the only integer that will serve, and 0 is not an odd integer.

> **Definition: Subgroup**
> Let $[G, \cdot]$ be a group and $A \subseteq G$. Then $[A, \cdot]$ is a **subgroup** of $[G, \cdot]$ if $[A, \cdot]$ is itself a group.

In order for $[A, \cdot]$ to be a group, it must have an identity element, which we'll denote by i_A. Of course G also has an identity element, which we'll denote by i_G. It turns out that $i_A = i_G$, but this equation does not follow from the uniqueness of a group identity, because the element i_A, as far as we know, may not be an identity for all of G, and we cannot yet say that i_G is an element of A. However, $i_A = i_A \cdot i_A$ because i_A is the identity for $[A, \cdot]$, and $i_A = i_A \cdot i_G$ because i_G is the identity for $[G, \cdot]$. Because of the left cancellation law holding in the group $[G, \cdot]$, it follows that $i_A = i_G$.

To test whether $[A, \cdot]$ is a subgroup of $[G, \cdot]$, we can assume the inherited properties of a well-defined operation and associativity and check for the three remaining properties required.

> **Theorem on Subgroups**
> For $[G, \cdot]$ a group with identity i and $A \subseteq G$, $[A, \cdot]$ is a subgroup of $[G, \cdot]$ if it meets the following three tests:
>
> 1. A is closed under $\cdot$
> 2. $i \in A$.
> 3. Every $x \in A$ has an inverse element in A.

PRACTICE 21   The definition of a group requires that the set be nonempty. In the theorem on subgroups, why isn't there a specific test that $A \neq \varnothing$?

EXAMPLE 12   **a.** $[\mathbb{Z}, +]$ is a subgroup of the group $[\mathbb{R}, +]$.
b. $[\{1,4\}, \cdot_5]$ is a subgroup of the group $[\{1, 2, 3, 4\}, \cdot_5]$ (closure holds; $1 \in \{1, 4\}$, $1^{-1} = 1$, $4^{-1} = 4$).

PRACTICE 22   **a.** Show that $[\{0, 2, 4, 6\}, +_8]$ is a subgroup of the group $[\mathbb{Z}_8, +_8]$.
b. Show that $[\{1, 2, 4\}, \cdot_7]$ is a subgroup of the group

$$[\{1, 2, 3, 4, 5, 6\}, \cdot_7]$$

If $[G, \cdot]$ is a group with identity i, then it is true that $[\{i\}, \cdot]$ and $[G, \cdot]$ are subgroups of $[G, \cdot]$. These somewhat trivial subgroups of $[G, \cdot]$ are called **improper subgroups**. Any other subgroups of $[G, \cdot]$ are **proper subgroups**.

PRACTICE 23   Find all the proper subgroups of S_3, the symmetric group of degree 3. (You can find them by looking at the group table; see Practice 12.)   ●

One point of confusing terminology: The set of *all* bijections on a set A into itself under function composition (like S_3) is called *the group of permutations on A*, and any subgroup of this set (such as those in Practice 23) is called a **permutation group**. The distinction is that *the* group of permutations on a set A includes all bijections on A into itself, but *a* permutation group may not. Permutation groups are of particular importance, not only because they were the first groups to be studied, but also because they are the only groups if we consider isomorphic structures to be the same. We shall see this result shortly.

There is an interesting subgroup we can always find in the symmetric group S_n for $n > 1$. We know that every member of S_n can be written as a composition of cycles, but it is also true that each cycle can be written as the composition of cycles of length 2, called **transpositions**. In S_7, for example, $(5, 1, 7, 2, 3, 6) = (5, 6) \circ (5, 3) \circ (5, 2) \circ (5, 7) \circ (5, 1)$. We can verify this by computing $(5, 6) \circ (5, 3) \circ (5, 2) \circ (5, 7) \circ (5, 1)$. Working from right to left,

$$1 \rightarrow 5 \rightarrow 7 \rightarrow 7 \rightarrow 7 \rightarrow 7$$

so 1 maps to 7. Similarly,

$$7 \rightarrow 7 \rightarrow 5 \rightarrow 2 \rightarrow 2 \rightarrow 2$$

so 7 maps to 2, and so on, resulting in $(5, 1, 7, 2, 3, 6)$. It is also true that $(5, 1, 7, 2, 3, 6) = (1, 5) \circ (1, 6) \circ (1, 3) \circ (1, 2) \circ (2, 4) \circ (1, 7) \circ (4, 2)$.

For any $n > 1$, the identity permutation i in S_n can be written as $i = (a, b) \circ (a, b)$ for any two elements a and b in the set $\{1, 2, \ldots, n\}$. This equation also shows that the inverse of the transposition (a, b) in S_n is (a, b). Now we borrow (without proof) one more fact: Even though there are various ways to write a cycle as the composition of transpositions, for a given cycle the number of transpositions will either always be even or always be odd. Consequently, we classify any permutation in S_n, $n > 1$, as **even** or **odd** according to the number of transpositions in any representation of that permutation. For example, in S_7, $(5, 1, 7, 2, 3, 6)$ is odd. If we denote by A_n the set of all even permutations in S_n, then A_n determines a subgroup of $[S, \circ]$. The composition of even permutations produces an even permutation, and $i \in A_n$. If $\alpha \in A_n$, and α as a product of transpositions is $\alpha = \alpha_1 \circ \alpha_2 \circ \cdots \circ \alpha_k$, then $\alpha^{-1} = \alpha_k^{-1} \circ \alpha_{k-1}^{-1} \circ \cdots \circ \alpha_1^{-1}$. Each inverse of a transposition is a transposition, so α^{-1} is also even.

The order of the group $[S_n, \circ]$ (the number of elements) is $n!$. What is the order of the subgroup $[A, \circ]$? We might expect half the permutations in S_n to be even and half to be odd. Indeed, this is the case. If we let O_n denote the set of odd permutations in S_n (which is not closed under function composition), then the mapping $f: A_n \rightarrow O_n$ defined by $f(\alpha) = \alpha \circ (1, 2)$ is a bijection.

PRACTICE 24   Prove that $f: A_n \rightarrow O_n$, given by $f(\alpha) = \alpha \circ (1, 2)$ is one-to-one and onto.   ●

Because there is a bijection from A_n onto O_n, each set has the same number of elements. But $A_n \cap O_n = \varnothing$ and $A_n \cup O_n = S_n$, so $|A_n| = |S_n|/2 = n!/2$.

Theorem on Alternating Groups

For $n \in \mathbb{N}$, $n > 1$, the set A_n of even permutations determines a subgroup called the **alternating group**, of $[S_n, \circ]$ of order $n!/2$.

We have now seen several examples of subgroups of finite groups. In Example 12b and Practice 22, there were three such examples, and the orders of the groups and subgroups were

Group of order 4, subgroup of order 2
Group of order 8, subgroup of order 4
Group of order 6, subgroup of order 3

The theorem on alternating groups says that a particular group of order $n!$ has a subgroup of order $n!/2$.

Based on these examples, one might conclude that subgroups are always half the size of the parent group. This is not always true, but there is a relationship between the size of a group and the size of a subgroup. This relationship is stated in Lagrange's theorem, proved by the great French mathematician Joseph-Louis Lagrange in 1771 (we shall omit the proof here).

Lagrange's Theorem

The order of a subgroup of a finite group divides the order of the group.

Lagrange's theorem helps us narrow down the possibilities for subgroups of a finite group. If $|G| = 12$, for example, we would not look for any subgroups of order 7, since 7 does not divide 12. Also, the fact that 6 divides 12 does not imply the existence of a subgroup of G of order 6. In fact, A_4 is a group of order $4!/2 = 12$, but it can be shown that A_4 has no subgroups of order 6. The converse to Lagrange's theorem does not always hold. In certain cases the converse can be shown to be true—for example, in finite commutative groups (note that A_4 is not commutative).

Finally, we consider subgroups of the group $[\mathbb{Z}, +]$. For n any fixed element of $\mathbb{N}$, the set $n\mathbb{Z}$ is defined as the set of all integral multiples of n; $n\mathbb{Z} = \{nz \mid z \in \mathbb{Z}\}$. Thus, for example, $3\mathbb{Z} = \{0, \pm 3, \pm 6, \pm 9, \dots \}$.

PRACTICE 25 Show that for any $n \in \mathbb{N}$, $[n\mathbb{Z}, +]$ is a subgroup of $[\mathbb{Z}, +]$. ●

Not only is $[n\mathbb{Z}, +]$ a subgroup of $[\mathbb{Z}, +]$ for any fixed n, but sets of the form $n\mathbb{Z}$ are the only subgroups of $[\mathbb{Z}, +]$. To illustrate, let $[S, +]$ be any subgroup of $[\mathbb{Z}, +]$. If $S = \{0\}$, then $S = 0\mathbb{Z}$. If $S \neq \{0\}$, let m be a member of S, $m \neq 0$. Either m is positive or, if m is negative, $-m \in S$ and $-m$ is positive. The subgroup S, therefore, contains at least one positive integer. Let n be the smallest positive integer in S (which exists by the Principle of Well-Ordering). We shall now see that $S = n\mathbb{Z}$.

First, since 0 and $-n$ are members of S and S is closed under $+$, $n\mathbb{Z} \subseteq S$. To obtain inclusion in the other direction, let $s \in S$. Now we divide the integer s by the integer n to get an integer quotient q and an integer remainder r with $0 \leq r < n$. Thus, $s = nq + r$. Solving for r, $r = s + (-nq)$. But $nq \in S$, therefore $-nq \in S$, and $s \in S$, so by closure of S under $+$, $r \in S$. If r is positive, we have a contradiction of the definition of n as the smallest positive number in S. Therefore, $r = 0$ and $s = nq + r = nq$. We now have $S \subseteq n\mathbb{Z}$, and thus $S = n\mathbb{Z}$, which completes the proof of the following theorem.

Theorem on Subgroups of $[\mathbb{Z}, +]$

Subgroups of the form $[n\mathbb{Z}, +]$ for $n \in \mathbb{N}$ are the only subgroups of $[\mathbb{Z}, +]$.

Isomorphic Groups

Suppose that $[S, \cdot]$ and $[T, +]$ are isomorphic groups; what would this mean? From the discussion of isomorphism in Section 7.1, isomorphic structures are the same except for relabeling. There must be a bijection from S to T that accomplishes the relabeling. This bijection must also preserve the effects of the binary operation; that is, it must be true that "operate and map" yields the same result as "map and operate." The following definition is more precise.

Definition: Group Isomorphism

Let $[S, \cdot]$ and $[T, +]$ be groups. A mapping $f: S \rightarrow T$ is an **isomorphism** from $[S, \cdot]$ to $[T, +]$ if

1. the function f is a bijection.
2. for all $x, y \in S, f(x \cdot y) = f(x) + f(y)$.

Property (2) is expressed by saying that f is a **homomorphism**.

PRACTICE 26   Illustrate the homomorphism property of the definition of group isomorphism by a commutative diagram.   ●

To further emphasize that isomorphic groups are the same except for relabeling, we shall show that under an isomorphism, the identity of one group maps to the identity of the other, and inverses map to inverses. Also, if one group is commutative, so is the other.

Suppose, then, that f is an isomorphism from the group $[S, \cdot]$ to the group $[T, +]$ and that i_S and i_T are the identities in the respective groups. Under the function f, i_S maps to an element $f(i_S)$ in T. Let t be any element in T. Then, because f is an onto function, $t = f(s)$ for some $s \in S$. It follows that

$$f(i_S) + t = f(i_S) + f(s)$$
$$= f(i_S \cdot s) \qquad \text{(because } f \text{ is a homomorphism)}$$
$$= f(s) \qquad \text{(because } i_S \text{ is the identity in } S)$$
$$= t$$

Therefore

$$f(i_S) + t = t$$

Similarly,

$$t + f(i_S) = t$$

The element $f(i_S)$ acts like an identity element in $[T, +]$, and because the identity is unique, $f(i_S) = i_T$.

PRACTICE 27   Prove that if f is an isomorphism from the group $[S, \cdot]$ to the group $[T, +]$, then for any $s \in S, f(s^{-1}) = -f(s)$ (inverses map to inverses). (*Hint:* Show that $f(s^{-1})$ acts like the inverse of $f(s)$.)   ●

PRACTICE 28   Prove that if f is an isomorphism from the commutative group $[S, \cdot]$ to the group $[T, +]$, then $[T, +]$ is a commutative group.   ●

EXAMPLE 13 $[\mathbb{R}^+, \cdot]$ and $[\mathbb{R}, +]$ are both groups. Let b be a positive real number, $b \neq 1$, and let f be the function from $\mathbb{R}^+$ to $\mathbb{R}$ defined by

$$f(x) = \log_b x$$

Then f is an isomorphism. To prove this, we must show that f is a bijection (one-to-one and onto) and that f is a homomorphism (preserves the operation). We can show that f is onto: For $r \in \mathbb{R}$, $b^r \in \mathbb{R}^+$ and $f(b^r) = \log_b b^r = r$. Also, f is one-to-one: If $f(x_1) = f(x_2)$, then $\log_b x_1 = \log_b x_2$. Let $p = \log_b x_1 = \log_b x_2$. Then $b^p = x_1$ and $b^p = x_2$, so $x_1 = x_2$. Finally f is a homomorphism: For $x_1, x_2 \in \mathbb{R}^+$, $f(x_1 \cdot x_2) = \log_b(x_1 \cdot x_2) = \log_b x_1 + \log_b x_2 = f(x_1) + f(x_2)$. Note that $\log_b 1 = 0$, so f maps 1, the identity of $[\mathbb{R}^+, \cdot]$, to 0, the identity of $[\mathbb{R}, +]$. Also note that

$$\log_b(1/x) = \log_b 1 - \log_b x = 0 - \log_b x = -\log_b x = -f(x)$$

so f maps the inverse of x in $[\mathbb{R}^+, \cdot]$ to the inverse of $f(x)$ in $[\mathbb{R}, +]$. Finally, both groups are commutative.   ●

Because the two groups in Example 13 are isomorphic, each is the mirror image of the other, and each can be used to simulate a computation in the other. Suppose, for example, that $b = 2$. Then $[\mathbb{R}, +]$ can be used to simulate the computation $64 \cdot 512$ in $[\mathbb{R}^+, \cdot]$. First, map from $\mathbb{R}^+$ to $\mathbb{R}$:

$$f(64) = \log_2 64 = 6$$
$$f(512) = \log_2 512 = 9$$

Now in $[\mathbb{R}, +]$ perform the computation

$$6 + 9 = 15$$

Finally, use f^{-1} to map back to $\mathbb{R}^+$:

$$f^{-1}(15) = 2^{15} = 32{,}768$$

(In the age B.C.—before calculators and computers—large numbers were multiplied by using tables of common logarithms, where $b = 10$, to convert a multiplication problem to an addition problem.) Either of two isomorphic groups can always simulate computations in the other, just as in Example 13.

According to the definition, in order for f to be an isomorphism, it must be both a bijection and a homomorphism.

EXAMPLE 14   Consider the following functions from $\mathbb{Z}$ to $\mathbb{Z}$:

$$f(x) = 0$$
$$g(x) = x + 1$$

The function f is a homomorphism from the group $[\mathbb{Z}, +]$ to the group $[\mathbb{Z}, +]$ because $f(x + y) = 0 = 0 + 0 = f(x) + f(y)$. However, f is not a bijection, so it is not an isomorphism. The function g is a bijection because $g(x) = g(y)$ implies $x + 1 = y + 1$, or $x = y$, so g is one-to-one; g is also onto because for any $z \in \mathbb{Z}$, $z - 1 \in \mathbb{Z}$ and $g(z - 1) = z$. But g is not a homomorphism because $g(x + y) = (x + y) + 1 \neq (x + 1) + (y + 1) = g(x) + g(y)$. Hence g is not an isomorphism. ●

EXAMPLE 15   Let $f: M_2(\mathbb{Z}) \to M_2(\mathbb{Z})$ be given by

$$f\left(\begin{bmatrix} a & b \\ c & d \end{bmatrix}\right) = \begin{bmatrix} a & c \\ b & d \end{bmatrix}$$

To show that f is one-to-one, let

$$f\left(\begin{bmatrix} a & b \\ c & d \end{bmatrix}\right) = f\left(\begin{bmatrix} e & f \\ g & h \end{bmatrix}\right)$$

Then

$$\begin{bmatrix} a & c \\ b & d \end{bmatrix} = \begin{bmatrix} e & g \\ f & h \end{bmatrix}$$

so $a = e$, $c = g$, $b = f$, and $d = h$, or

$$\begin{bmatrix} a & b \\ c & d \end{bmatrix} = \begin{bmatrix} e & f \\ g & h \end{bmatrix}$$

To show that f is onto, let

$$\begin{bmatrix} a & b \\ c & d \end{bmatrix} \in M_2(\mathbb{Z})$$

Then

$$\begin{bmatrix} a & c \\ b & d \end{bmatrix} \in M_2(\mathbb{Z}) \quad \text{and} \quad f\left(\begin{bmatrix} a & c \\ b & d \end{bmatrix}\right) = \begin{bmatrix} a & b \\ c & d \end{bmatrix}$$

Also, f is a homomorphism from $[M_2(\mathbb{Z}), +]$ to $[M_2(\mathbb{Z}), +]$ because

$$f\left(\begin{bmatrix} a & b \\ c & d \end{bmatrix} + \begin{bmatrix} e & f \\ g & h \end{bmatrix}\right) = f\left(\begin{bmatrix} a+e & b+f \\ c+g & d+h \end{bmatrix}\right) = \begin{bmatrix} a+e & c+g \\ b+f & d+h \end{bmatrix}$$

$$= \begin{bmatrix} a & c \\ b & d \end{bmatrix} + \begin{bmatrix} e & g \\ f & h \end{bmatrix} = f\left(\begin{bmatrix} a & b \\ c & d \end{bmatrix}\right) + f\left(\begin{bmatrix} e & f \\ g & h \end{bmatrix}\right)$$

The function f is therefore an isomorphism from $[M_2(\mathbb{Z}), +]$ to $[M_2(\mathbb{Z}), +]$. ●

PRACTICE 29   Let $5\mathbb{Z} = \{5z \mid z \in \mathbb{Z}\}$. Then $[5\mathbb{Z}, +]$ is a group. Show that $f\colon \mathbb{Z} \to 5\mathbb{Z}$ given by $f(x) = 5x$ is an isomorphism from $[\mathbb{Z}, +]$ to $[5\mathbb{Z}, +]$.   ●

If f is an isomorphism from $[S, \cdot]$ to $[T, +]$, then f^{-1} exists and is a bijection. Further, f^{-1} is also a homomorphism, this time from T to S. To see this, let t_1 and t_2 belong to T and consider $f^{-1}(t_1 + t_2)$. Because $t_1, t_2 \in T$ and f is onto, $t_1 = f(s_1)$ and $t_2 = f(s_2)$ for some s_1 and s_2 in S. Thus,

$$\begin{aligned} f^{-1}(t_1 + t_2) &= f^{-1}(f(s_1) + f(s_2)) \\ &= f^{-1}\,(f(s_1 \cdot s_2)) \\ &= (f^{-1} \circ f)(s_1 \cdot s_2) \\ &= s_1 \cdot s_2 \\ &= f^{-1}(t_1) \cdot f^{-1}(t_2) \end{aligned}$$

This is why we can speak of S and T as being simply isomorphic, denoted by $S \simeq T$, without having to specify that the isomorphism is from S to T or vice versa.

Checking whether a given function is an isomorphism from S to T, as in Practice 29, is not hard. Deciding whether S and T are isomorphic may be harder. To prove that they are isomorphic, we must produce a function. To prove that they are not isomorphic, we must show that no such function exists. Since we can't try all possible functions, we use ideas such as the following: There is no one-to-one correspondence between S and T, S is commutative but T is not, and so on.

We have noted that isomorphic groups are alike except for relabeling and that each can be used to simulate the computations in the other. Isomorphism of groups is really an equivalence relation, as Practice 30 shows; thus isomorphic groups belong to the same equivalence class. Thinking of isomorphic groups as "alike except for labeling" is consistent with the idea that elements in an equivalence class represent different names for the same thing.

PRACTICE 30   **a.** Let $f\colon S \to T$ be an isomorphism from the group $[S, \cdot]$ onto the group $[T, +]$ and $g\colon T \to U$ be an isomorphism from $[T, +]$ onto the group $[U, *]$. Show that $g \circ f$ is an isomorphism from S onto U.
b. Let $\mathcal{T}$ be a collection of groups and define a binary relation ρ on $\mathcal{T}$ by $S \, \rho \, T \leftrightarrow S \simeq T$. Show that ρ is an equivalence relation on $\mathcal{T}$.   ●

We shall finish this section by looking at some equivalence classes of groups under isomorphism. Often we pick out one member of an equivalence class and note that it is the typical member of that class and that all other groups in the class look just like it (with different names).

A result concerning the nature of very small groups follows immediately from Exercise 18 at the end of this section.

Theorem on Small Groups

Every group of order 2 is isomorphic to the group whose group table is

$\cdot$	1	a
1	1	a
a	a	1

Every group of order 3 is isomorphic to the group whose group table is

$\cdot$	1	a	b
1	1	a	b
a	a	b	1
b	b	1	a

Every group of order 4 is isomorphic to one of the two groups whose group tables are

$\cdot$	1	a	b	c
1	1	a	b	c
a	a	1	c	b
b	b	c	1	a
c	c	b	a	1

$\cdot$	1	a	b	c
1	1	a	b	c
a	a	b	c	1
b	b	c	1	a
c	c	1	a	b

We can also prove that any group is essentially a permutation group. Suppose $[G, \cdot]$ is a group. We want to establish an isomorphism from G to a permutation group; each element g of G must be associated with a permutation α_g on some set. In fact, the set will be G itself; for any $x \in G$, we define $\alpha_g(x)$ to be $g \cdot x$. We must show that $\{\alpha_g \mid g \in G\}$ forms a permutation group and that this permutation group is isomorphic to G. First we need to show that for any $g \in G$, α_g is indeed a permutation on G. From the definition $\alpha_g(x) = g \cdot x$, it is clear that $\alpha_g \colon G \to G$, but it must be shown that α_g is a bijection.

PRACTICE 31   Show that α_g as defined above is a permutation on G.   ●

Now we consider $P = \{\alpha_g \mid g \in G\}$ and show that P is a group under function composition. P is nonempty because G is nonempty, and associativity always holds for function composition. We must show that P is closed and has an identity and that each $\alpha_g \in P$ has an inverse in P. To show closure, let α_g and $\alpha_h \in P$. For any $x \in G$, $(\alpha_g \circ \alpha_h)(x) = \alpha_g(\alpha_h(x)) = \alpha_g(h \cdot x) = g \cdot (h \cdot x) = (g \cdot h) \cdot x$. Thus, $\alpha_g \circ \alpha_h = \alpha_{g \cdot h}$ and $\alpha_{g \cdot h} \in P$.

PRACTICE 32   **a.** Let 1 denote the identity of G. Show that α_1 is an identity for P under function composition.
b. For $\alpha_g \in P$, $\alpha_{g^{-1}} \in P$; show that $\alpha_{g^{-1}} = (\alpha_g)^{-1}$.

We now know that $[P, \circ]$ is a permutation group, and it only remains to show that the function $f: G \rightarrow P$ given by $f(g) = \alpha_g$ is an isomorphism. Clearly, f is an onto function.

PRACTICE 33   Show that $f: G \rightarrow P$ defined by $f(g) = \alpha_g$ is

a. one-to-one
b. a homomorphism

We have now proved the following theorem, first stated and proved by the English mathematician Arthur Cayley in the mid-1800s.

Cayley's Theorem
Every group is isomorphic to a permutation group.

Section 8.1 Review

Techniques

- Test whether a given set and operation have the properties necessary to form a semigroup, monoid, or group structure.
- Test whether a given subset of a group is a subgroup.
- Test whether a given function from one group to another is an isomorphism.
- Decide whether two groups are isomorphic.

Main Ideas

Many elementary arithmetic systems are instances of a semigroup, monoid, or group structure.

In any group structure, the identity and inverse elements are unique, cancellation laws hold, and linear equations are solvable; these and other properties follow from the definitions involved.

A subset of a group may itself be a group under the inherited operation.

The order of a subgroup of a finite group divides the order of the group.

The only subgroups of the group $[\mathbb{Z}, +]$ are of the form $[n\mathbb{Z}, +]$, where $n\mathbb{Z}$ is the set of all integral multiples of a fixed $n \in \mathbb{N}$.

If f is an isomorphism from one group to another, f maps the identity to the identity and inverses to inverses, and it preserves commutativity.

If S and T are isomorphic groups, they are identical except for relabeling, and each simulates any computation in the other.

Isomorphism is an equivalence relation on groups.

To within an isomorphism, there is only one group of order 2, one group of order 3, and two groups of order 4.

Every group is essentially a permutation group.

Exercises 8.1

1. Each case below defines a binary operation, denoted by · , on a given set. Which are associative? Which are commutative?

   ★**a.** On $\mathbb{Z}$: $x \cdot y = \begin{cases} x \text{ if } x \text{ is even} \\ x + 1 \text{ if } x \text{ is odd} \end{cases}$

   ★**b.** On $\mathbb{N}$: $x \cdot y = (x + y)^2$

   c. On $\mathbb{R}^+$: $x \cdot y = x^4$

   d. On $\mathbb{Q}$: $x \cdot y = xy/2$

   e. On $\mathbb{R}^+$: $x \cdot y = 1/(x + y)$

★2. **a.** A binary operation · is defined on the set $\{a, b, c, d\}$ by the table on the left below. Is · commutative? Is · associative?

   b. Let $S = \{p, q, r, s\}$. An associative operation · is partly defined on S by the table on the right below. Complete the table to preserve associativity. Is · commutative?

·	a	b	c	d
a	a	c	d	a
b	b	c	a	d
c	c	a	b	d
d	d	b	a	c

·	p	q	r	s
p	p	q	r	s
q	q	r	s	p
r			p	
s	s		q	r

3. Define binary operations on the set $\mathbb{N}$ that are

   a. commutative but not associative

   b. associative but not commutative

   c. neither associative nor commutative

   d. both associative and commutative

4. Show that matrix multiplication on $M_2(\mathbb{Z})$ is associative.

5. Determine whether the following structures $[S, \cdot]$ are semigroups, monoids, groups, or none of these. Name the identity element in any monoid or group structure.

   ★**a.** $S = \mathbb{N}$; $x \cdot y = \min(x, y)$

   ★**b.** $S = \mathbb{R}$; $x \cdot y = (x + y)^2$

★**c.** $S = \{a\sqrt{2} \mid a \in \mathbb{N}\}; \cdot = $ multiplication
★**d.** $S = \{a + b\sqrt{2} \mid a, b \in \mathbb{Z}\}; \cdot = $ multiplication
★**e.** $S = \{a + b\sqrt{2} \mid a, b \in \mathbb{Q}, a$ and b not both $0\}; \cdot = $ multiplication
★**f.** $S = \{1, -1, i, -i\}; \cdot = $ multiplication (where $i^2 = -1$)
★**g.** $S = \{1, 2, 4\}; \cdot = \cdot_6$
 h. $S = \{1, 2, 3, 5, 6, 10, 15, 30\}; x \cdot y = $ least common multiple of x and y
 i. $S = \mathbb{N} \times \mathbb{N}; (x_1, y_1) \cdot (x_2, y_2) = (x_1, y_2)$
 j. $S = \mathbb{N} \times \mathbb{N}; (x_1, y_1) \cdot (x_2, y_2) = (x_1 + x_2, y_1 y_2)$
 k. $S = $ set of even integers; $\cdot = $ addition
 l. $S = $ set of odd integers; $\cdot = $ addition
 m. $S = $ set of all polynomials in $\mathbb{R}[x]$ of degree $\leq 3; \cdot = $ polynomial addition
 n. $S = $ set of all polynomials in $\mathbb{R}[x]$ of degree ≤ 3;
          $\cdot = $ polynomial multiplication
 o. $S = \left\{ \begin{bmatrix} 1 & z \\ 0 & 1 \end{bmatrix} \middle| z \in \mathbb{Z} \right\}; \cdot = $ matrix multiplication
 p. $S = \{1, 2, 3, 4\}; \cdot = \cdot_5$
 q. $S = \mathbb{R} - \{-1\}; x \cdot y = x + y + xy$
 r. $S = \{f \mid f: \mathbb{N} \to \mathbb{N}\}; \cdot = $ function addition, that is, $(f + g)(x) = f(x) + g(x)$

★**6.** Let $A = \{1, 2\}$.

  a. Describe the elements and write the table for the semigroup of transformations on A.

  b. Describe the elements and write the table for the group of permutations on A.

★**7.** Given an equilateral triangle, six permutations can be performed on the triangle that will leave its image in the plane unchanged. Three of these permutations are clockwise rotations in the plane of 120°, 240°, and 360° about the center of the triangle; these permutations are denoted R_1, R_2, and R_3, respectively. The triangle can also be flipped about any of the axes 1, 2, and 3 (see the accompanying figure); these permutations are denoted F_1, F_2, and F_3, respectively. During any of these permutations, the axes remain fixed in the plane. Composition of permutations is a binary operation on the set D_3 of all six permutations. For example, $F_3 \circ R_2 = F_2$. The set D_3 under composition is a group, called the **group of symmetries of an equilateral triangle**. Complete the group table below for $[D_3, \circ]$. What is an identity element in $[D_3, \circ]$? What is an inverse element for F_1? For R_2?

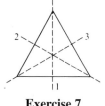

Exercise 7

$\circ$	R_1	R_2	R_3	F_1	F_2	F_3
R_1						
R_2						
R_3						
F_1						
F_2						
F_3		F_2				

8. The set S_3, the symmetric group of degree 3, is isomorphic to D_3, the group of symmetries of an equilateral triangle (see Exercise 7). Find a bijection from the elements of S_3 to the elements of D_3 that preserves the operation. (*Hint:* R_1 of D_3 may be considered a permutation in S_3 sending 1 to 2, 2 to 3, and 3 to 1.)

9. In each case, decide whether the structure on the left is a subgroup of the group on the right. If not, why not? (Note that here S^* denotes $S - \{0\}$.)

   ★**a.** $[\mathbb{Z}_5^*, \cdot_5]$; $[\mathbb{Z}_5, +_5]$
   ★**b.** $[P, +]$; $[\mathbb{R}[x], +]$ where P is the set of all polynomials in x over $\mathbb{R}$ of degree ≥ 3
   ★**c.** $[\mathbb{Z}^*, \cdot]$; $[\mathbb{Q}^*, \cdot]$
   d. $[A, \circ]$; $[S, \circ]$ where S is the set of all bijections on $\mathbb{N}$ and A is the set of all bijections on $\mathbb{N}$ mapping 3 to 3
   e. $[\mathbb{Z}, +]$; $[M_2(\mathbb{Z}), +]$
   f. $[K, +]$; $[\mathbb{R}[x], +]$ where K is the set of all polynomials in x over $\mathbb{R}$ of degree $\leq k$ for some fixed k
   g. $[\{0, 3, 6\}, +_8]$; $[\mathbb{Z}_8, +_8]$

★10. Find all the distinct subgroups of $[\mathbb{Z}_{12}, +_{12}]$.

11. **a.** Show that the subset

$$\alpha_1 = i \qquad\qquad \alpha_3 = (1, 4) \circ (2, 3)$$
$$\alpha_2 = (1, 2) \circ (3, 4) \qquad \alpha_4 = (1, 3) \circ (2, 4)$$

   forms a subgroup of the symmetric group S_4.
   b. Show that the subset

$$\alpha_1 = i \qquad\qquad \alpha_5 = (1, 2) \circ (3, 4)$$
$$\alpha_2 = (1, 2, 3, 4) \qquad \alpha_6 = (1, 4) \circ (2, 3)$$
$$\alpha_3 = (1, 3) \circ (2, 4) \qquad \alpha_7 = (2, 4)$$
$$\alpha_4 = (1, 4, 3, 2) \qquad \alpha_8 = (1, 3)$$

   forms a subgroup of the symmetric group S_4.

★12. Find the elements of the alternating group A_4.

13. In each case, decide whether the given function is a homomorphism from the group on the left to the one on the right. Are any of the homomorphisms also isomorphisms?

   ★**a.** $[\mathbb{Z}, +]$, $[\mathbb{Z}, +]$; $f(x) = 2$
   ★**b.** $[\mathbb{R}, +]$, $[\mathbb{R}, +]$; $f(x) = |x|$
   ★**c.** $[\mathbb{R}^*, \cdot]$, $[\mathbb{R}^*, \cdot]$ (where $\mathbb{R}^*$ denotes the set of nonzero real numbers); $f(x) = |x|$

d. $[\mathbb{R}[x], +], [\mathbb{R}, +]; f(a_n x^n + a_{n-1} x^{n-1} + \cdots + a_1 x + a_0) = a_n + a_{n-1} + \cdots + a_0$

e. $[S_3, \circ], [\mathbb{Z}_2, +_2]; f(\alpha) = \begin{cases} 1 \text{ if } \alpha \text{ is an even permutation} \\ 0 \text{ if } \alpha \text{ is an odd permutation} \end{cases}$

14. In each case, decide whether the given groups are isomorphic. If they are, produce an isomorphism function. If they are not, give a reason why they are not.

    a. $[\mathbb{Z}, +], [12\mathbb{Z}, +]$ (where $12\mathbb{Z} = \{12z \mid z \in \mathbb{Z}\}$)
    b. $[\mathbb{Z}_5, +_5], [5\mathbb{Z}, +]$
    c. $[5\mathbb{Z}, +], [12\mathbb{Z}, +]$
    d. $[S_3, \circ], [\mathbb{Z}_6, +_6]$
    e. $[\{ a_1 x + a_0 \mid a_1, a_0 \in \mathbb{R}\}, +], [\mathbb{C}, +]$
    f. $[\mathbb{Z}_6, +_6], [S_6, \circ]$
    g. $[\mathbb{Z}_2, +_2], [S_2, \circ]$

15. **a.** Let $S = \{1, -1\}$. Show that $[S, \cdot]$ is a group where $\cdot$ denotes ordinary integer multiplication.

    b. Let f be the function from the group $[S_n, \circ]$ to the group $[S, \cdot]$ given by

$$f(\alpha) = \begin{cases} 1 \text{ if } \alpha \text{ is even} \\ -1 \text{ if } \alpha \text{ is odd} \end{cases}$$

    Prove that f is a homomorphism.

★16. Let $M_2^0(\mathbb{Z})$ be the set of all 2×2 matrices of the form

$$\begin{bmatrix} 1 & z \\ 0 & 1 \end{bmatrix}$$

where $z \in \mathbb{Z}$.

    a. Show that $[M_2^0(\mathbb{Z}), \cdot]$ is a group, where $\cdot$ denotes matrix multiplication.
    b. Let a function $f: M_2^0(\mathbb{Z}) \to \mathbb{Z}$ be defined by

$$f\left(\begin{bmatrix} 1 & z \\ 0 & 1 \end{bmatrix} \right) = z$$

    Prove that f is an isomorphism from $[M_2^0(\mathbb{Z}), \cdot]$ to $[\mathbb{Z}, +]$.
    c. Use $[\mathbb{Z}, +]$ to simulate the computation

$$\begin{bmatrix} 1 & 7 \\ 0 & 1 \end{bmatrix} \cdot \begin{bmatrix} 1 & -3 \\ 0 & 1 \end{bmatrix}$$

    in $[M_2^0(\mathbb{Z}), \cdot]$.
    d. Use $[M_2^0(\mathbb{Z}), \cdot]$ to simulate the computation $2 + 3$ in $[\mathbb{Z}, +]$.

17. In any group $[G, \cdot]$, show that
    a. $i^{-1} = i$
    b. $(x^{-1})^{-1} = x$ for any $x \in G$

18. **a.** Show that any group of order 2 is commutative by constructing a group table on the set $\{1, a\}$ with 1 as the identity.

   b. Show that any group of order 3 is commutative by constructing a group table on the set $\{1, a, b\}$ with 1 as the identity. (You may assume associativity.)

   c. Show that any group of order 4 is commutative by constructing a group table on the set $\{1, a, b, c\}$ with 1 as the identity. (You may assume associativity.) There will be four such tables, but three of them are isomorphic because the elements have simply been relabeled from one to the other. Find these three groups and indicate the relabeling. Thus, there are two essentially different groups of order 4, and both of these are commutative.

★19. Let $[S, \cdot]$ be a semigroup. An element $i_L \in S$ is a *left identity element* if for all $x \in S$, $i_L \cdot x = x$. An element $i_R \in S$ is a *right identity element* if for all $x \in S$, $x \cdot i_R = x$.

   a. Show that if a semigroup $[S, \cdot]$ has both a left identity element and a right identity element, then $[S, \cdot]$ is a monoid.

   b. Give an example of a finite semigroup with two left identities and no right identity.

   c. Give an example of a finite semigroup with two right identities and no left identity.

   d. Give an example of a semigroup with neither a right nor a left identity.

20. Let $[S, \cdot]$ be a monoid with identity i, and let $x \in S$. An element x_L^{-1} in S is a *left inverse* of x if $x_L^{-1} \cdot x = i$. An element x_R^{-1} in S is a *right inverse* of x if $x \cdot x_R^{-1} = i$.

   a. Show that if every element in a monoid $[S, \cdot]$ has both a left inverse and a right inverse, then $[S, \cdot]$ is a group.

   b. Let S be the set of all functions f such that $f: \mathbb{N} \to \mathbb{N}$. Then S under function composition is a monoid. Define a function $f \in S$ by $f(x) = 2x$, $x \in \mathbb{N}$. Then define a function $g \in S$ by

$$g(x) = \begin{cases} x/2 & \text{if } x \in \mathbb{N}, x \text{ even} \\ 1 & \text{if } x \in \mathbb{N}, x \text{ odd} \end{cases}$$

   Show that g is a left inverse for f. Also show that f has no right inverse.

21. For x a member of a group $[G, \cdot]$, we can define x^n for any positive integer n by $x^1 = x$, $x^2 = x \cdot x$, and $x^n = x^{n-1} \cdot x$ for $n > 2$. Prove that in a finite group $[G, \cdot]$, for each $x \in G$ there is a positive integer k such that $x^k = i$.

★22. Let $[G, \cdot]$ be a group and let $x, y \in G$. Define a relation ρ on G by $x \, \rho \, y \leftrightarrow g \cdot x \cdot g^{-1} = y$ for some $g \in G$.

   a. Show that ρ is an equivalence relation on G.

   b. Prove that for each $x \in G$, $[x] = \{x\}$ if and only if G is commutative.

23. Let $[S, \cdot]$ be a semigroup having a left identity i_L (see Exercise 19) and the property that for every $x \in S$, x has a left inverse y such that $y \cdot x = i_L$. Prove that $[S, \cdot]$ is a group. (*Hint:* y also has a left inverse in S.)

24. Show that if $[S, \cdot]$ is a semigroup in which the linear equations $a \cdot x = b$ and $x \cdot a = b$ are solvable for any $a, b \in S$, then $[S, \cdot]$ is a group. (*Hint:* Use Exercise 23.)

25. Prove that a finite semigroup satisfying the left and right cancellation laws is a group. (*Hint:* Use Exercise 23.)

26. Show that a group $[G, \cdot]$ is commutative if and only if $(x \cdot y)^2 = x^2 \cdot y^2$ for each $x, y \in G$.

27. Show that a group $[G, \cdot]$ in which $x \cdot x = i$ for each $x \in G$ is commutative.

★28. **a.** Let $[G, \cdot]$ be a group and let $[S, \cdot]$ and $[T, \cdot]$ be subgroups of $[G, \cdot]$. Show that $[S \cap T, \cdot]$ is a subgroup of $[G, \cdot]$.
    b. Will $[S \cup T, \cdot]$ be a subgroup of $[G, \cdot]$? Prove or give a counterexample.

29. Let $[G, \cdot]$ be a commutative group with subgroups $[S, \cdot]$ and $[T, \cdot]$. Let $ST = \{s \cdot t \mid s \in S, t \in T\}$. Show that $[ST, \cdot]$ is a subgroup of $[G, \cdot]$.

★30. Let $[G, \cdot]$ be a commutative group with identity i. For a fixed positive integer k, let $B_k = \{x \mid x \in G, x^k = i \}$. Show that $[B_k, \cdot]$ is a subgroup of $[G, \cdot]$.

31. For any group $[G, \cdot]$, the *center* of the group is $A = \{x \in G \mid x \cdot g = g \cdot x$ for all $g \in G\}$.
    a. Prove that $[A, \cdot]$ is a subgroup of $[G, \cdot]$
    b. Find the center of the group of symmetries of an equilateral triangle, $[D_3, \circ]$ (see Exercise 7).
    c. Show that G is commutative if and only if $G = A$.
    d. Let x and y be members of G with $x \cdot y^{-1} \in A$. Show that $x \cdot y = y \cdot x$.

32. **a.** Let S_A denote the group of permutations on a set A, and let a be a fixed element of A. Show that the set H_a of all permutations in S_A leaving a fixed forms a subgroup of S_A.
    b. If A has n elements, what is $|H_a|$?

33. **a.** Let $[G, \cdot]$ be a group and $A \subseteq G, A \neq \emptyset$. Show that $[A, \cdot]$ is a subgroup of $[G, \cdot]$ if for each $x, y \in A, x \cdot y^{-1} \in A$. This subgroup test is sometimes more convenient to use than the theorem on subgroups.
    b. Use the test of part (a) to work Exercise 30.

34. **a.** Let $[G, \cdot]$ be any group with identity i. For a fixed $a \in G$, a^0 denotes i and a^{-n} means $(a^n)^{-1}$. Let $A = \{a^z \mid z \in \mathbb{Z}\}$. Show that $[A, \cdot]$ is a subgroup of G.
    b. The group $[G, \cdot]$ is a *cyclic group* if for some $a \in G, A = \{a^z \mid z \in \mathbb{Z}\}$ is the entire group G. In this case, a is a *generator* of $[G, \cdot]$. For example, 1 is a generator of the group $[\mathbb{Z}, +]$; remember that the operation is addition. Thus, $1^0 = 0, 1^1 = 1, 1^2 = 1 + 1 = 2, 1^3 = 1 + 1 + 1 = 3, ... ; 1^{-1} = (1)^{-1} = -1,$ $1^{-2} = (1^2)^{-1} = -2, 1^{-3} = (1^3)^{-1} = -3, ...$. Every integer can be written as an integral "power" of 1, and $[\mathbb{Z}, +]$ is cyclic with generator 1. Show that the group $[\mathbb{Z}_7, +_7]$ is cyclic with generator 2.
    c. Show that 5 is also a generator of the cyclic group $[\mathbb{Z}_7, +_7]$.
    d. Show that 3 is a generator of the cyclic group $[\mathbb{Z}_4, +_4]$.

⋆35. Let $[G, \cdot]$ be a cyclic group with generator a (see Exercise 34). Show that G is commutative.

36. Let $[G, \cdot]$ be a commutative group with identity i. Prove that the function f: $G \to G$ given by $f(x) = x^{-1}$ is an isomorphism.

⋆37. **a.** Let $[S, \cdot]$ be a semigroup. An isomorphism from S to S is called an *automorphism* on S. Let Aut(S) be the set of all automorphisms on S, and show that Aut(S) is a group under function composition.
   b. For the group $[\mathbb{Z}_4, +_4]$, find the set of automorphisms and show its group table under $\circ$.

38. Let f be a homomorphism from a group G onto a group H. Show that f is an isomorphism if and only if the only element of G that is mapped to the identity of H is the identity of G.

39. Let $[G, \cdot]$ be a group and g a fixed element of G. Define $f: G \to G$ by $f(x) = g \cdot x \cdot g^{-1}$ for any $x \in G$. Prove that f is an isomorphism from G to G.

Section 8.2   Finite-State Machines

The algebraic structures of the previous section served as models for various simple arithmetic systems. However, we would surely agree that computation should go beyond mere arithmetic. We would like a model that captures the general nature of computation. Perhaps looking at a simplified version of a modern digital computer would be a start.

A computer stores information internally in binary form. At any instant, the computer contains certain information, so its internal storage is set in certain patterns of binary digits, which we'll call the state of the computer at that instant. Because a computer contains a finite amount of storage, there is a finite (although large) number of different states that the computer can assume. An internal clock synchronizes the actions of the computer. On a clock pulse, input can be read, which can change some of the storage locations and thus change the state of the machine to a new state. What the new state is will depend on what the input was, as well as what the previous state was. If these two factors are known, the change is predictable and nonrandom. Because the contents of certain storage cells are available as output, the state of the machine determines its output. In this way, over a succession of clock pulses, the machine produces a sequence of outputs in response to a sequence of inputs.

Definition

The finite-state machine is a model that captures the characteristics of the computer described above. As you read the definition, look for the following properties in the behavior of our abstract machine:

1. Operations of the machine are *synchronized* by discrete clock pulses.
2. The machine proceeds in a *deterministic* fashion; that is, its actions in response to a given sequence of inputs are completely predictable.
3. The machine responds to *inputs*.
4. There is a *finite number of states* that the machine can attain. At any given moment, the machine is in exactly one of these states. Which state it will be in next is a function of both the present state and the present input. The present state, however, depends on the previous state and input, while the previous state depends on its previous state and input, and so on, going all the way back to the initial configuration. Thus, the state of the machine at any moment serves as a form of memory of past inputs.
5. The machine is capable of *output*. The nature of the output is a function of the present state of the machine, meaning that it also depends on past inputs.

Definition: Finite-State Machine

$M = [S, I, O, f_S, f_O]$ is a **finite-state machine** if S is a finite set of states, I is a finite set of input symbols (the **input alphabet**), O is a finite set of output symbols (the **output alphabet**), and f_S and f_O are functions where $f_S: S \times I \rightarrow S$ and $f_O: S \rightarrow O$. The machine is always initialized to begin in a fixed starting state s_0.

The function f_S is the **next-state function**. It maps a (state, input) pair to a state. Thus, the state at clock pulse t_{i+1}, state(t_{i+1}), is obtained by applying the next-state function to the state at time t_i and the input at time t_i:

$$\text{state}(t_{i+1}) = f_S(\text{state}(t_i), \text{input}(t_i))$$

The function f_O is the **output function**. When f_O is applied to a state at time t_i, we get the output at time t_i:

$$\text{output}(t_i) = f_O(\text{state}(t_i))$$

Notice that the effect of applying function f_O is available instantly, but the effect of applying function f_S is not available until the next clock pulse.

Examples of Finite-State Machines

To describe a particular finite-state machine, we have to define the three sets and two functions involved.

EXAMPLE 16   A finite-state machine M is described as follows: $S = \{s_0, s_1, s_2\}$, $I = \{0, 1\}$, $O = \{0, 1\}$. Because the two functions f_S and f_O act on finite domains, they can be defined by a **state table**, as in Table 8.1. The machine M begins in state s_0, which has an output of 0. If the first input symbol is a 0, the next state of the machine is then s_1, which has an output of 1.

Present State	Next State		Output
	Present Input		
	0	1	
s_0	s_1	s_0	0
s_1	s_2	s_1	1
s_2	s_2	s_0	1

Table 8.1

If the next input symbol is a 1, the machine stays in state s_1 with an output of 1. By continuing this analysis, we see that an input sequence consisting of the characters 01101 (read left to right) would produce the following effect:

Time	t_0	t_1	t_2	t_3	t_4	t_5
Input	0	1	1	0	1	–
State	s_0	s_1	s_1	s_1	s_2	s_0
Output	0	1	1	1	1	0

The initial 0 of the output string is spurious—it merely reflects the starting state, not the result of any input.

In a similar way, the input sequence 1010 produces an output of 00111.

Another way to define the functions f_S and f_O (in fact all of M) is by a directed graph called a **state graph**. Each state of M with its corresponding output is the label of a node of the graph. The next-state function is given by directed arcs of the graph, each arc showing the input symbol(s) that produces that particular state change. The state graph for M appears in Figure 8.1.

> **REMINDER:**
> Each input symbol must appear on one and only one transition arc from each state.

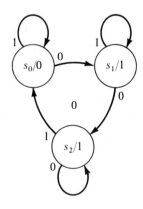

Figure 8.1

PRACTICE 34   For the machine M of Example 16, what output sequence is produced by the input sequence 11001?

PRACTICE 35   A machine *M* is given by the state graph of Figure 8.2. Give the state table for *M*.

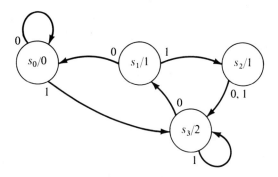

Figure 8.2

PRACTICE 36   A machine *M* is described by the state table shown in Table 8.2.

Present State	Next State			Output
	Present Input			
	0	1	2	
s_0	s_0	s_1	s_1	0
s_1	s_1	s_0	s_0	1

Table 8.2

a. Draw the state graph for *M*.
b. What output corresponds to an input sequence of 2110?

The machine of Example 16 is not particularly interesting. If finite-state machines model real-world computers, they should be able to do something. Let's try to build a finite-state machine that will add two binary numbers. The input will consist of a sequence of pairs of binary digits, each of the form 00, 01, 10, or 11. Each pair represents one column of digits of the two numbers to be added, least significant digits first. Thus to add the two numbers

011

101

the number pairs are 11, 10, and 01. The output gives the least significant digits of the answer first. Recall the basic facts of binary addition:

$$\frac{0}{0}\quad\frac{0}{1}\quad\frac{1}{0}\quad\frac{0}{1}$$
$$\begin{array}{cccc}0&0&1&0\\\underline{0}&\underline{1}&\underline{0}&\underline{1}\\0&1&1&10\end{array}$$

(Note that in the fourth addition a carry to the next column takes place.)

A moment's thought shows us that we can encounter four cases in adding the digits in any given column, and we shall use states of the machine to represent these cases.

- The output should be 0 with no carry—state s_0.
- The output should be 0 but with a carry to the next column—state s_1.
- The output should be 1 with no carry—state s_2.
- The output should be 1 with a carry to the next column—state s_3.

State s_0, as always, is the starting state. We have already indicated the output for each state, but we need to determine the next state based on the present state and the input. For example, suppose we are in state s_1 and the input is 11. The output for the present state is 0, but there is a carry, so in the next column we are adding $1 + 1 + 1$, which results in an output of 1 and a carry. The next state is s_3.

PRACTICE 37

In the binary adder under construction:

a. What is the next state if the present state is s_2 and the input is 11?
b. What is the next state if the present state is s_3 and the input is 10?   ●

After considering all possible cases, we have the complete state graph of Figure 8.3.

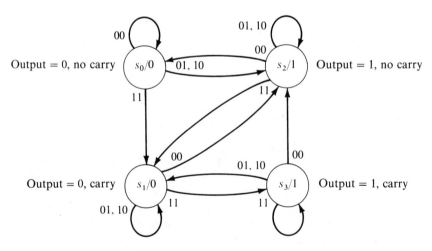

Figure 8.3

The operation of this machine in adding the two numbers 011 and 101 (low-order digits first) can be traced as follows:

Time	t_0	t_1	t_2	t_3	t_4
Input	11	10	01	00	–
State	s_0	s_1	s_1	s_1	s_2
Output	0	0	0	0	1

The output is 1000 when we ignore the initial 0, which does not reflect the action of any input. Converting this arithmetic to decimal form, we have computed $3 + 5 = 8$. Note the symmetry of this machine with respect to the inputs of 10 and 01, reflecting that binary addition is commutative.

PRACTICE 38   Compute the sum of 01110110 and 01010101 by using the binary adder machine of Figure 8.3.   ●

Recognition

We have already noted that a given input signal may affect the behavior of a finite-state machine for longer than just one clock pulse. Because of the (limited) memory of past inputs represented by the states of a machine, we can use these machines as *recognizers*. A machine can be built to recognize, say by producing an output of 1, when the input it has received matches a certain description. We shall soon discuss more fully the capabilities of finite-state machines as recognizers. Here we shall simply construct some examples.

EXAMPLE 17   The machine described in Figure 8.4 is a parity check machine. When the input received through time t_i contains an even number of 1s, then the output at time t_{i+1} is 1; otherwise, the output is 0.

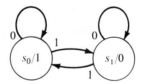

Figure 8.4   ●

EXAMPLE 18   Suppose we want to design a machine having an output of 1 exactly when the input string received to that point ends in 101. As a special case, an input sequence consisting of just 101 could be handled by progressing directly from state s_0 to states s_1, s_2, and s_3 with outputs of 0 except for s_3, which has an output of 1. This much of the design results in Figure 8.5a. This figure shows that we want to be in state s_2 whenever the input has been such that one more 1 takes us to s_3 (with an output of 1); thus we should be in s_2 whenever the two most recent input symbols were 10, regardless of what came before. In particular, a string of 1010 should put us in s_2; hence, the next-state function for s_3 with an input of 0 is s_2. Similarly, we can use s_1 to "remember" that the most recent input symbol received was 1, and that a 01 will take us to s_3. In particular, 1011 should put us in s_1; hence, the next-state function for s_3 with an input of 1 is s_1. The rest of the next-state function can be determined the same way; Figure 8.5b shows the complete state graph.

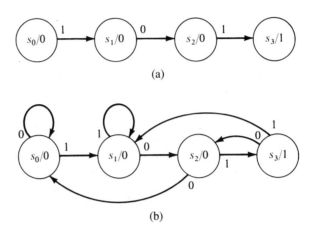

(a)

(b)

Figure 8.5

Notice that the machine is in state s_2 at the end of an input of 0110 and at the end of an input of 011010—in fact, at the end of any input ending in 10; yet s_2 cannot distinguish between these inputs. *Each state of M represents a class of indistinguishable input histories, s_3 being the state representing all inputs ending in 101.*

PRACTICE 39   Draw the state graph for a machine producing an output of 1 exactly when the input string received to that point ends in 00.

Now we want to see exactly what sets finite-state machines can recognize. Remember that recognition is possible because machine states have a limited memory of past inputs. Even though the machine is finite, a particular input signal can affect the behavior of a machine "forever." However, not every input signal can do so, and some classes of inputs require remembering so much information that no machine can detect them.

To avoid writing down outputs, we shall designate those states of a finite-state machine with an output of 1 as **final states** and denote them in the state graph with a double circle. Then we can give the following formal definition of recognition, where I^* denotes the set of finite-length strings over the input alphabet.

Definition: Finite-State Machine Recognition
A finite-state machine M with input alphabet I **recognizes** a subset S of I^* if M, beginning in state s_0 and processing an input string α, ends in a final state if and only if $\alpha \in S$.

PRACTICE 40   Describe the sets recognized by the machines in Figure 8.6.

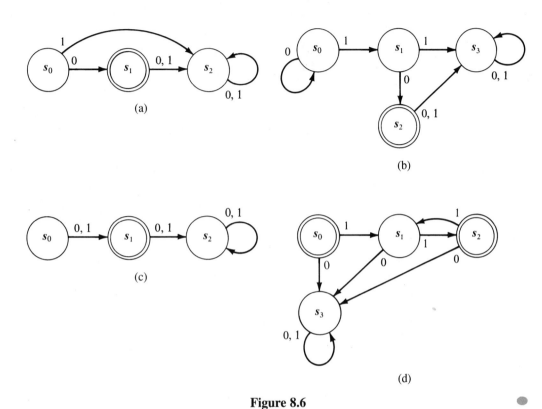

Figure 8.6

Regular Sets and Kleene's Theorem

We want a compact, symbolic way to describe sets such as those appearing in the answer to Practice 40. We shall describe such sets by using *regular expressions*; each regular expression describes a particular set. First, we shall define what regular expressions are; then we shall see how a regular expression describes a set. We assume here that I is some finite set of symbols; later I will be the input alphabet for a finite-state machine.

Definition: Regular Expressions over I
Regular expressions over I are

1. the symbol $\varnothing$ and the symbol λ.
2. the symbol i for any $i \in I$.
3. the expressions (AB), $(A \vee B)$, and $(A)^*$ if A and B are regular expressions.

(This definition of a regular expression over I is still another example of a recursive definition.)

> **Definition: Regular Set**
> Any set represented by a regular expression according to the following conventions
> is a **regular set**:
>
> 1. $\emptyset$ represents the empty set.
> 2. λ represents the set $\{\lambda\}$ containing the empty string.
> 3. i represents the set $\{i\}$.
> 4. For regular expressions A and B,
>   **a.** (AB) represents the set of all elements of the form $\alpha\beta$ where α belongs to the set represented by A and β belongs to the set represented by B.
>   **b.** $(A \vee B)$ represents the union of A's set and B's set.
>   **c.** $(A)^*$ represents the set of all concatenations of members of A's set.

In our discussion, we shall be a little sloppy and say things like "the regular set AB" instead of "the set represented by the regular expression AB." Informally, an element in AB is an item from A followed by an item from B. An element in $A \vee B$ is a single item chosen from either A or B. An element in $(A)^*$ is zero or more repetitions of elements from A. We note that λ, the empty string, is a member of the set represented by A^* for any A because it is the case of zero repetitions of elements from A. In writing regular expressions, we can eliminate parentheses when no ambiguity results. The regular expression $0^* \vee 10$ therefore consists of $\lambda, 0, 00, 000, 0000, \ldots , 10$.

EXAMPLE 19   Here are some regular expressions and a description of the set each one represents.

a. $1^*0(01)^*$	Any number (including none) of 1s, followed by a single 0, followed by any number (including none) of 01 pairs
b. $0 \vee 1^*$	A single 0 or any number (including none) of 1s
c. $(0 \vee 1)^*$	Any string of 0s or 1s, including λ
d. $11((10)^*11)^*(00^*)$	A nonempty string of pairs of 1s interspersed with any number (including none) of 10 pairs, followed by at least one 0

PRACTICE 41   Which strings belong to the set described by the regular expression?

a. 10100010; $(0^*10)^*$
b. 011100; $(0 \vee (11)^*)^*$
c. 000111100; $((011 \vee 11)^*(00)^*)^*$

PRACTICE 42   Write regular expressions for the sets recognized by the machines of Practice 40. ●

A regular set may be described by more than one regular expression. For example, the set of all strings of 0s and 1s, which we already know from Example 19(c) to be described by (0 ∨ 1)*, is also described by the regular expression [(0 ∨ 1*)* ∨ (01)*]*. We might, therefore, write the equation

$$(0 \vee 1)^* = [(0 \vee 1^*)^* \vee (01)^*]^*$$

Although we may be quite willing to accept this particular equation, it can be difficult to decide in general whether two regular expressions are equal, that is, whether they represent the same set. An efficient algorithm that will make this decision for any two regular expressions has not been found.

We have introduced regular sets because, as it turns out, these are exactly the sets finite-state machines are capable of recognizing. This result was first proved by the American mathematician Stephen Kleene in 1956. We state his theorem below without proof.

Kleene's Theorem

Any set recognized by a finite-state machine is regular, and any regular set can be recognized by some finite-state machine.

Kleene's theorem outlines the limitations as well as the capabilities of finite-state machines, because there are certainly many sets that are not regular. For example, consider $S = \{0^n1^n \mid n \geq 0\}$ where a^n stands for a string of n copies of a. Strings in S have some number of 0s followed by the same number of 1s. S is not regular. (Notice that 0*1* does not do the job.) By Kleene's theorem, there is no finite-state machine capable of recognizing S. Yet S seems like such a reasonable set, and surely we humans could count a string of 0s followed by 1s and see whether we had the same number of 1s as 0s. This lapse suggests some deficiency in our use of a finite-state machine as a model of computation. We shall investigate this further in Section 8.3.

Machine Minimization

Although we have treated finite-state machines as abstractions, circuits that act like finite-state machines can be built from electronic devices like the logic elements of Section 7.2 and others. If we wish to construct a physical machine, the number of internal states is a factor in the cost of construction. Minimization is the process of finding, for a given finite-state machine M, a machine M' with two properties:

1. If M and M' are both begun in their respective start states and are given the same sequence of input symbols, they will produce identical output sequences.
2. M' has, if possible, fewer states than M (if this is not possible, then M is already a minimal machine and cannot be further reduced).

Unreachable States

First, let's observe that we can remove any **unreachable states** of M, those states that cannot be attained from the starting state no matter what input sequence occurs.

EXAMPLE 20   Let M be given by the state table of Table 8.3. Although the state table contains the same information as the state graph (Figure 8.7), the graph shows us at a glance that state s_2 can never be reached from the starting state s_0. If we simply remove state s_2 and its associated arcs, we have the state graph of Figure 8.8 for a machine M' with one less state than M that behaves exactly like M; that is, it gives the same output as M for any input string.

Present State	Next State		Output
	Present Input		
	0	1	
s_0	s_1	s_3	0
s_1	s_3	s_0	0
s_2	s_1	s_3	1
s_3	s_0	s_1	1

Table 8.3

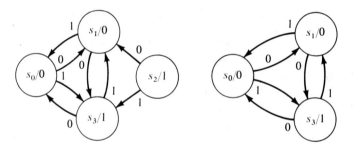

Figure 8.7   **Figure 8.8**

PRACTICE 43   What state(s) are unreachable from s_0 in the machine of Table 8.4? Try to get your answer directly from the state table.

Present State	Next State		Output
	Present Input		
	0	1	
s_0	s_1	s_4	0
s_1	s_4	s_1	1
s_2	s_2	s_2	1
s_3	s_3	s_1	0
s_4	s_0	s_0	1

Table 8.4

Because the state graph of a finite-state machine is a directed graph, it has an associated adjacency matrix. Warshall's algorithm (Section 6.1) can be used to detect unreachable states.

Minimization Procedure

Assuming now that all unreachable states have been removed from M, we shall continue to look for a reduced machine M'. The key to finding a reduced M', if one exists, lies in the notion of equivalent states.

> **Definition: Equivalent States**
> Two states s_i and s_j of M are **equivalent** if for any $\alpha \in I^*, f_O(s_i, \alpha) = f_O(s_j, \alpha)$, where I^* again denotes the set of finite-length strings over the input alphabet.

In this definition of equivalent states, the definition of the output function has been extended to denote the sequence of output symbols obtained by repeatedly applying f_O to a sequence α of input symbols. Thus equivalent states of a machine produce identical output strings for any input string.

PRACTICE 44   Prove that state equivalence is an equivalence relation on the states of a machine. ●

For the time being, we shall postpone the problem of how to identify equivalent states in a given machine M. Let's simply assume that we have somehow found which states are equivalent and have partitioned the states of M into the corresponding equivalence classes. These classes have two properties: (1) All states in the same class have the same output, and (2) for each input symbol, all states in the same class proceed under the next-state function to states that are all in the same class.

PRACTICE 45   Show that properties 1 and 2 are satisfied when M is partitioned into classes of equivalent states.   ●

We define a machine M' whose states are the equivalence classes of M. M' has the same input and output alphabet as M, and its start state is the class to which s_0, the start state of M, belongs. The output of a class is the output symbol common to all states of M in that class (property 1). The next state of class X under an input symbol is that class to which all states of M in X proceed under that input symbol (property 2). M' is a well-defined machine. M' produces the same output strings when processing a given input string as does M. Also, the number of states of M' (equivalence classes of M) will be no greater than the number of states of M.

The minimization problem for M thus boils down to finding the equivalent states of M. Perhaps we should note first that the obvious approach of directly trying to satisfy the definition of equivalent states will not work. Given two states s_i and s_j of M, we cannot actually compare the outputs corresponding to each possible input string. Fortunately, the problem is not as infinite as it sounds; we only need to identify k-equivalent states.

> **Definition: k-Equivalent States**
> Two states s_i and s_j of M are **k-equivalent** if for any $\alpha \in I^*$, where α has no more than k symbols, $f_O(s_i, \alpha) = f_O(s_j, \alpha)$.

It is not hard to see that k-equivalence is an equivalence relation on the states of M (check the reflexive, symmetric, and transitive properties). It is possible to test two states of M for k-equivalence directly, since we can actually produce the finite number of input strings having no more than k symbols. However, it turns out that we don't have to do this. We can begin by finding 0-equivalent states. These are states producing the same output for 0-length input strings, that is, states having the same associated output symbol. Thus, we can identify the 0-equivalence classes directly from the description of M.

EXAMPLE 21   Let M be defined by the state table of Table 8.5. (Here we've started writing 0, 1, 2 ... for states instead of $s_0, s_1, s_2, \ldots$.) The 0-equivalence classes of the states of M are

$$\{0, 2, 5\} \quad \text{and} \quad \{1, 3, 4, 6\}$$

Present State	Next State		Output
	Present Input		
	0	1	
0	2	3	0
1	3	2	1
2	0	4	0
3	1	5	1
4	6	5	1
5	2	0	0
6	4	0	1

Table 8.5

Our procedure to find k-equivalent states is a recursive one; we know how to find 0-equivalent states, and we shall show how to find k-equivalent states once we have identified states that are $(k - 1)$-equivalent. Suppose, then, that we already know which states are $(k - 1)$-equivalent. If states s_i and s_j are k-equivalent, they must produce the same output strings for any input string of length k or less, in particular, for any string of length no greater than $k - 1$. Thus, s_i and s_j must at least be $(k - 1)$-equivalent. But they also must produce the same output strings for any k-length input string.

An arbitrary k-length input string consists of a single arbitrary input symbol followed by an arbitrary $(k - 1)$-length input string. If we apply such a k-length string to states s_i and s_j (which themselves have the same output symbol), the single input symbol moves s_i and s_j to next states s_i' and s_j'; then s_i' and s_j' must produce identical output strings for the remaining, arbitrary $(k - 1)$-length string, which will surely happen if s_i' and s_j' are $(k - 1)$-equivalent. Therefore, to find k-equivalent states, look for $(k - 1)$-equivalent states whose next states under any input symbol are $(k - 1)$-equivalent.

EXAMPLE 22   Consider again the machine M of Example 21. We know the 0-equivalent states. To find 1-equivalent states, we look for 0-equivalent states with 0-equivalent next states. For example, the states 3 and 4 are 0-equivalent; under the input symbol 0, they proceed to states 1 and 6, respectively, which are 0-equivalent states, and under the input symbol 1 they both proceed to 5, which of course is 0-equivalent to itself. Therefore, states 3 and 4 are 1-equivalent. But states 0 and 5, themselves 0-equivalent, proceed under the input symbol 1 to states 3 and 0, respectively, which are not 0-equivalent states. So states 0 and 5 are not 1-equivalent; the input string 1 will produce an output string of 01 from state 0 and of 00 from state 5. The 1-equivalence classes for M are

$$\{0, 2\}, \{5\}, \{1, 3, 4, 6\}$$

To find 2-equivalent states, we look for 1-equivalent states with 1-equivalent next states. States 1 and 3, although 1-equivalent, proceed under input 1 to states 2 and 5, respectively, which are not 1-equivalent states. Therefore, states 1 and 3 are not 2-equivalent. The 2-equivalence classes for M are

$$\{0, 2\}, \{5\}, \{1, 6\}, \{3, 4\}$$

The 3-equivalence classes for M are the same as the 2-equivalence classes.   ●

Definition: Partition Refinement
Given two partitions π_1, and π_2 of a set S, π_1 is a **refinement** of π_2 if each block of π_1 is a subset of a block of π_2.

In Example 22 each successive partition of the states of M into equivalence classes is a refinement of the previous partition. This refinement will always happen; k-equivalent states must also be $(k - 1)$-equivalent, so the blocks of the $(k - 1)$-partition can only be further subdivided. However, the subdivision process cannot continue indefinitely (at worst it can go on only until each partition block contains only one state); there will eventually be a point where $(k - 1)$-equivalent states and k-equivalent states coincide. (In Example 22, 2-equivalent and 3-equivalent states coincide.) Once this happens, all next states for members of a partition block under any input symbol fall within a partition block. Thus, k-equivalent states are also $(k + 1)$-equivalent and $(k + 2)$-equivalent, and so on. Indeed, these states are equivalent.

The total procedure for finding equivalent states is to start with 0-equivalent states, then l-equivalent states, and so on, until the partition no longer subdivides. A pseudocode description of this algorithm is given.

ALGORITHM   Minimize

Minimize (finite-state machine table *M*)
//produces a minimized version of *M*

Local variable:
boolean flag   //flag for loop exit when nonequivalent states found

```
    find 0-equivalent states of M
    repeat
        while untested equivalence classes remain do
            select untested equivalence class
            while untested state pairs in current class remain do
                select untested state pair in current class
                flag = false
                while untried input symbols remain and not flag do
                    select untried input symbol
                    for both states in current pair, find next state
                        under current input symbol
                    if next states not equivalent then
                        flag = true
                    end if
                end while
                if flag then
                    mark current states for different classes;
                end if
            end while
            form new equivalence classes
        end while
    until set of new equivalence classes = set of old equivalence classes
end Minimize
```

EXAMPLE 23

For the machine *M* of Examples 21 and 22, the reduced machine *M'* will have states

$A = \{0, 2\}$
$B = \{5\}$
$C = \{1, 6\}$
$D = \{3, 4\}$

The state table for *M'* (Table 8.6) is obtained from that for *M*. Machine *M'* (starting state *A*) will reproduce *M*'s output for any input string, but it has four states instead of seven.

Present State	Next State		Output
	Present Input		
	0	1	
A	A	D	0
B	A	A	0
C	D	A	1
D	C	B	1

Table 8.6

EXAMPLE 24   We shall minimize M where M is given by the state table of Table 8.7.

Present State	Next State		Output
	Present Input		
	0	1	
0	3	1	1
1	4	1	0
2	3	0	1
3	2	3	0
4	1	0	1

Table 8.7

The 0-equivalence classes of M are

$\{0, 2, 4\}, \{1, 3\}$

The 1-equivalence classes of M are

$\{0\}, \{2, 4\}, \{1, 3\}$

No further refinement is possible. Let

$A = \{0\}$
$B = \{2, 4\}$
$C = \{1, 3\}$

The reduced machine is shown in Table 8.8.

Present State	Next State		Output
	Present Input		
	0	1	
A	C	C	1
B	C	A	1
C	B	C	0

Table 8.8

PRACTICE 46   Minimize the machines whose state tables are shown in Tables 8.9 and 8.10.

Present State	Next State		Output
	Present Input		
	0	1	
0	2	1	1
1	2	0	1
2	4	3	0
3	2	3	1
4	0	1	0

Table 8.9

Present State	Next State		Output
	Present Input		
	0	1	
0	1	3	1
1	2	0	0
2	1	3	0
3	2	1	0

Table 8.10

Section 8.2 Review

Techniques

- Compute the output string for a given finite-state machine and a given input string.
- Draw a state graph from a state table and vice versa.
- Construct a finite-state machine to act as a recognizer for a certain type of input.
- Find a regular expression given the description of a regular set.
- Decide whether a given string belongs to a given regular set.
- Minimize finite-state machines.

Main Ideas

Finite-state machines have a synchronous, deterministic mode of operation and limited memory capabilities.

The class of sets that finite-state machines can recognize is the class of all regular sets; hence, their recognition capabilities are limited.

Unreachable states can be removed from a machine.

After unreachable states have been removed from a machine, a minimized version of that machine can be found that produces the same output strings for all input strings.

Exercises 8.2

★**1.** For each input sequence and machine given, compute the corresponding output sequence (starting state is always s_0).

a. 011011010

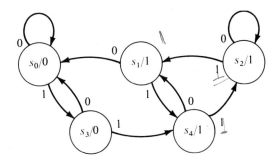

b. *abccaab*

Present State	Next State			Output
	Present Input			
	a	*b*	*c*	
s_0	s_2	s_0	s_3	*a*
s_1	s_0	s_2	s_3	*b*
s_2	s_2	s_0	s_1	*a*
s_3	s_1	s_2	s_0	*c*

c. 0100110

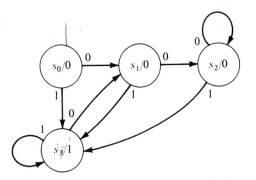

2. **a.** For the machine described in Exercise 1(a), find all input sequences yielding an output sequence of 0011110.

   b. For the machine described in Exercise 1(b), find all input sequences yielding an output sequence of *abaaca*.

   c. For the machine described in Exercise 1(c), what will be the output for an input sequence $a_1a_2a_3a_4a_5$ where $a_i \in \{0, 1\}$, $1 \le i \le 5$?

In Exercises 3–6, write the state table for the machine, and compute the output sequence for the given input sequence.

3. 00110

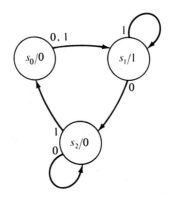

4. 1101100

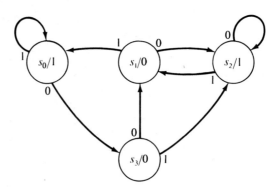

★5. 01011

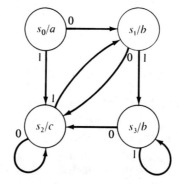

6. *acbabc*

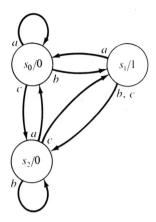

In Exercises 7–10, draw the state graph for the machine, and compute the output sequence for the given input sequence.

7. 10001

Present State	Next State		Output
	Present Input		
	0	1	
s_0	s_0	s_2	1
s_1	s_1	s_0	0
s_2	s_0	s_1	0

8. 0011

Present State	Next State		Output
	Present Input		
	0	1	
s_0	s_2	s_3	0
s_1	s_0	s_1	1
s_2	s_1	s_3	0
s_3	s_1	s_2	1

★9. *acbbca*

Present State	Next State			Output
	Present Input			
	a	b	c	
s_0	s_1	s_1	s_1	0
s_1	s_2	s_2	s_1	0
s_2	s_0	s_2	s_1	1

10. 21021

Present State	Next State			Output
	Present Input			
	0	1	2	
s_0	s_3	s_1	s_2	1
s_1	s_3	s_0	s_1	2
s_2	s_2	s_1	s_1	0
s_3	s_1	s_4	s_0	0
s_3	s_1	s_4	s_0	2

11. Construct a finite-state machine that will compute $x + 1$ where x is the input given in binary form, least significant digit first.

12. a. Construct a finite-state machine that will compute the 2's complement of p where p is a binary number input with the least significant digit first. (See Exercise 18, Section 7.2.)

b. Use the machine of part (a) to find the 2's complement of 1100 and of 1011.

★**13. a.** Construct a delay machine having input and output alphabet $\{0, 1\}$ that, for any input sequence $a_1a_2a_3 \cdots$ produces an output sequence of $00a_1a_2a_3 \cdots$.

b. Explain (intuitively) why a finite-state machine cannot be built that, for any input sequence $a_1a_2a_3 \cdots$, produces the output sequence $0a_10a_20a_3 \cdots$.

14. You are designing a windows-based, event-driven program to handle customers for a small business. You design the user interface with three screens. The opening screen contains an exit button to quit the program and displays a list box of customer names. Double-clicking on one of the entries in the list box brings up a second screen showing complete data for that customer. This screen contains a button to get back to the opening screen. The opening screen also contains a button that brings up a form to enter the data for a new customer. Construct a finite-state machine that describes the user interaction with the program.

15. Whenever a video tape is inserted into a VCR, the machine automatically turns on and plays the tape. At the end of the tape, the machine turns off. To program the VCR, you must manually turn it on and then select the menu function; when you are finished, you turn the machine off, but its timer is set. At the appropriate time, the machine records, then at the appropriate time turns itself completely off. Construct a finite-state machine to describe the behavior of the VCR.

★16. You have an account at First National Usury Trust (FNUT) and a card to operate their ATM (automated teller machine). Once you have inserted your card, the ATM will allow you to process a transaction only if you enter your correct code number, which is 417. Draw a state graph for a finite-state machine designed to recognize this code number. The output alphabet should have three symbols: "bingo" (correct code), "wait" (correct code so far), and "dead" (incorrect code). The input alphabet is {0, 1, 2, ... , 9}. To simplify notation, you may designate an arc by I-{3}, for example, meaning that the machine will take this path for an input symbol that is any digit except 3. (At FNUT, you get only one chance to enter the code correctly.)

17. Construct finite-state machines that act as recognizers for the input described by producing an output of 1 exactly when the input received to that point matches the description. (The input and output alphabet in each case is {0, 1}.)

  ★a. set of all strings consisting of an even number of 0s
  ★b. set of all strings consisting of two or more 1s followed by a 0
  ★c. set of all strings containing two consecutive 0s and the rest 1s
   d. set of all strings ending with one or more 0s
   e. set of all strings where the number of 0s is a multiple of 3
   f. set of all strings containing at least four 1s

18. Construct finite-state machines that act as recognizers for the input described by producing an output of 1 exactly when the input received to that point matches the description. (The input and output alphabet in each case is {0, 1}.)

   a. set of all strings containing exactly one 1
   b. set of all strings beginning with 000
   c. set of all strings where the second input is 0 and the fourth input is 1
   d. set of all strings consisting entirely of any number (including none) of 01 pairs or consisting entirely of two 1s followed by any number (including none) of 0s
   e. set of all strings ending in 110
   f. set of all strings containing 00

19. A paragraph of English text is to be scanned and the number of words beginning with "con" counted. Design a finite-state machine that will output a 1 each time such a word is encountered. The output alphabet is {0, 1}. The input alphabet is the 26 letters of the English alphabet, a finite number of punctuation symbols (period, comma, etc.), and a special character β for blank. To simplify your description, you may use I-{m}, for example, to denote any input symbol not equal to m.

20. **a.** In many computer languages, any decimal number N must be presented in one of the following forms:

$$sd* \qquad sd*.d* \qquad d* \qquad d*.d* \tag{1}$$

where s denotes the sign (i.e., $s \in \{ +, - \}$), d is a digit (i.e., $d \in \{0, 1, 2, ... , 9\}$), and $d*$ denotes a string of digits where the string may be of any length, including length zero (the empty string). Thus, the following would be examples of valid decimal numbers:

$$+2.74 \qquad -.58 \qquad 129 \qquad +$$

Design a finite-state machine that recognizes valid decimal numbers by producing an output of 1. The input symbols are $+$, $-$, . , and the 10 digits. To simplify notation, you may use d to denote any digit input symbol.
   b. Modify the machine of part (a) to recognize any sequence of decimal numbers as defined in part (a) separated by commas. For example, such a machine would recognize

$$+2.74, -.58, 129, +$$

The input alphabet should be the same as for the machine of part (a) with the addition of the symbol c for comma.
   c. Suppose a decimal number must be presented in a form similar to that for part (a) except that any decimal point that appears must have at least one digit before it and after it. Write an expression similar to expression (1) in part (a) to describe the valid form for a decimal number. How would you modify the machine of part (a) to recognize such a number?

★21. Let M be a finite-state machine with n states. The input alphabet is $\{0\}$. Show that for any input sequence that is long enough, the output of M must eventually be periodic. What is the maximum number of inputs before periodic output begins? What is the maximum length of a period?

22. At the beginning of the chapter, we learned that

*Your team at Babel, Inc., is writing a compiler for a new programming language, currently codenamed ScrubOak after a tree outside your office window. During the first phase of compilation (called the lexical analysis phase) the compiler must break down statements into individual units called tokens. In particular, the compiler must be able to recognize identifiers in the language, which are strings of letters, and also recognize the two keywords in the language, which are **if** and **in**.*

How can the compiler recognize the individual tokens in a statement?

Construct a finite-state machine that operates on a stream of characters and moves into one of two final states representing that a keyword has just been processed or that another legitimate identifier has just been processed. Use β to denote a separating blank between tokens.

23. Give a regular expression for the set recognized by each finite-state machine in the accompanying figure.

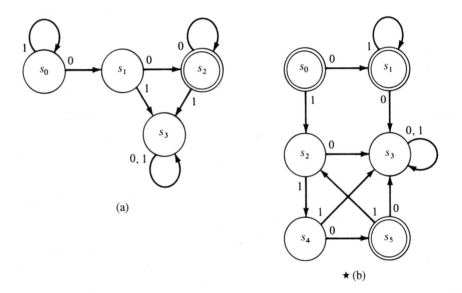

(a)

★ (b)

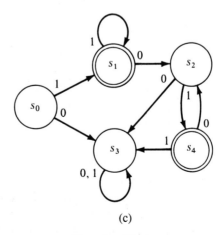

(c)

Exercise 23

24. Give a regular expression for the set recognized by each finite-state machine in the accompanying table.

Present State	Next State		Output
	Present Input		
	0	1	
s_0	s_3	s_1	0
s_1	s_1	s_2	0
s_2	s_3	s_3	1
s_3	s_3	s_3	0

(a)

Present State	Next State		Output
	Present Input		
	0	1	
s_0	s_3	s_1	1
s_1	s_1	s_2	1
s_2	s_2	s_2	0
s_3	s_0	s_2	0

(b)

Present State	Next State		Output
	Present Input		
	0	1	
s_0	s_2	s_1	1
s_1	s_3	s_1	1
s_2	s_3	s_4	0
s_3	s_3	s_3	0
s_4	s_5	s_3	0
s_5	s_2	s_3	1

(c)

Exercise 24

25. Give a regular expression for each of the following sets.
   a. set of all strings of 0s and 1s beginning with 0 and ending with 1
   b. set of all strings of 0s and 1s having an odd number of 0s
   ★**c.** {101, 1001, 10001, 100001, ...}
   d. set of all strings of 0s and 1s containing at least one 0
   e. set of all strings of a's and b's where each a is followed by two b's
   f. set of all strings of 0s and 1s containing exactly two 0s

★26. Does the given string belong to the given regular set?
   a. 01110111; (1*01)*(11 ∨ 0*)
   b. 11100111; [(1*0)* ∨ 0*11]*
   c. 011100101; 01*10*(11*0)*
   d. 1000011; (10* ∨ 11)*(0*1)*

27. Write a regular expression for the set of all alphanumeric strings beginning with a letter, the set of legal identifiers in some programming languages.

★28. Write a regular expression for the set of all arithmetic expressions indicating the addition or subtraction of two positive integers.

29. Write regular expressions for each of the strings described in Exercise 17.

30. Write regular expressions for each of the strings described in Exercise 18.

31. **a.** Prove that if A is a regular set, then the set A^R consisting of the reverse of all strings in A is also regular.
   b. For any string α, let α^R be the reverse string. Do you think the set $\{\alpha\alpha^R \mid \alpha \in I^*\}$ is regular?

32. Prove that if A is a regular set whose symbols come from the alphabet I, then $I^* - A$ is a regular set.

★33. Identify any unreachable states of each M in the accompanying table.

Present State	Next State		Output
	Present Input		
	0	1	
s_0	s_2	s_0	0
s_1	s_2	s_1	1
s_2	s_2	s_0	1

(a)

Present State	Next State			Output
	Present Input			
	a	b	c	
s_0	s_1	s_0	s_3	0
s_1	s_1	s_3	s_0	1
s_2	s_3	s_2	s_1	0
s_3	s_1	s_1	s_0	0

(b)

★34. Minimize *M*.

Present State	Next State		Output
	Present Input		
	0	1	
0	3	6	1
1	4	2	0
2	4	1	0
3	2	0	1
4	5	0	1
5	3	5	0
6	4	2	1

35. Minimize *M*.

Present State	Next State		Output
	Present Input		
	0	1	
0	5	3	1
1	5	2	0
2	1	3	0
3	2	4	1
4	2	0	1
5	1	4	0

36. Minimize *M*.

Present State	Next State		Output
	Present Input		
	0	1	
0	1	2	0
1	2	3	1
2	3	4	0
3	2	1	1
4	5	4	1
5	6	7	0
6	5	6	1
7	8	1	0
8	7	3	0

37. Minimize M.

Present State	Next State		Output
	Present Input		
	0	1	
0	7	1	1
1	0	3	1
2	5	1	0
3	7	6	1
4	5	6	0
5	2	3	0
6	3	0	1
7	4	0	0

38. Minimize M.

Present State	Next State		Output
	Present Input		
	0	1	
0	1	3	0
1	2	4	1
2	5	4	0
3	1	2	2
4	2	1	1
5	4	0	2

★39. Minimize M.

Present State	Next State			Output
	Present Input			
	a	b	c	
0	1	4	0	1
1	4	2	3	0
2	3	4	2	1
3	4	0	1	0
4	1	0	2	0

40. Minimize M.

Present State	Next State		Output
	Present Input		
	0	1	
0	1	3	0
1	2	0	0
2	0	3	0
3	2	1	0

41. Minimize M.

Present State	Next State		Output
	Present Input		
	0	1	
0	1	3	1
1	2	0	0
2	4	3	1
3	0	1	1
4	2	4	0

Section 8.3   Turing Machines

In Section 8.2, we noted that because $S = \{0^n1^n \mid n \geq 0\}$ is not a regular set, Kleene's theorem tells us that it is not recognizable by any finite-state machine. We didn't actually prove that S is not a regular set, however; we only noted that we were not able to come up with a regular expression for it. Let's take a slightly different approach.

Suppose that S is recognized by a finite-state machine M with m states. Then all strings from S, and only strings from S, lead M from its start state to a final state. Now let us run M a number of times on successive input strings of $\lambda, 0, 0^2, 0^3 \dots, 0^m$. At the end of processing each of these $m + 1$ strings, M will be in some state. Because M has only m distinct states, there must be two strings from this list, say 0^v and 0^w, $v \neq w$, each of which lead M from the start state to the same state. (This is actually a result of the Pigeonhole Principle of Chapter 3, where the items are the input strings and the bins into which we put the items are the states M is in after processing the strings.) Because M recognizes S, the input string 0^v1^v will cause M to end in a final state. But because M is in the same state after processing 0^w as after processing 0^v, the string 0^w1^v, which does not belong to S, will take M to the same final state. This contradiction proves that no finite-state machine can recognize S.

We probably consider ourselves to be finite-state machines and imagine that our brains, being composed of a large number of cells, can take on only a finite, although immensely large, number of configurations, or states. We feel sure, however, that if someone presented us with an arbitrarily long string of 0s followed by an arbitrarily long string of 1s, we could detect whether the number of 0s and 1s was the same. Let's think of some techniques we might use.

For small strings of 0s and 1s, we could just look at the strings and decide. Thus, we can tell without great effort that $000111 \in S$ and that $000110 \notin S$. However, for the string

$$00000000000000001111111111111111$$

we must devise another procedure, probably resorting to counting. We would count the number of 0s received, and when we got to the first 1, we would write the number of 0s down (or remember it) for future reference; then we would begin counting 1s. (This process is what we did mentally for smaller strings.)

However, we have now made use of some extra memory, because when we finished counting 1s, we would have to retrieve the number representing the total number of 0s to make a comparison. But such information retrieval is what the finite-state machine cannot do; its only capacity for remembering input is to have a given input symbol send it to a particular state. We have already seen that no finite-state machine can "remember" 0^n for arbitrarily large n because it runs out of distinct states. In fact, if we try to solve this problem on a real computer, we encounter the same difficulty. If we set a counter as we read in 0s, we might get an overflow because our counter can go only so high. To process $0^n 1^n$ for arbitrarily large n requires an unlimited auxiliary memory for storing the value of our counter, which in practice cannot exist.

Another way we humans might attack the problem of recognizing S is to wait until the entire string has been presented. Then we would go to one end of the string and cross out a 0, go to the other end and cross out a 1, and then continue this back-and-forth operation until we ran out of 0s or 1s. The string belongs to S if and only if we run out of both at the same time. Although this approach sounds rather different from the first one, it still requires remembering each of the inputs, since we must go back and read them once the string is complete. The finite-state machine, of course, cannot reread input.

We have come up with two computational procedures—algorithms—to decide, given a string of 0s and 1s, whether that string belongs to $S = \{0^n 1^n \mid n \geq 0\}$. Both require some form of additional memory unavailable in a finite-state machine. Evidently, the finite-state machine is not a model of the most general form of computational procedure.

Definition

To simulate more general computational procedures than the finite-state machine can handle, we use a *Turing machine*, proposed by the British mathematician Alan M. Turing in 1936. A Turing machine is essentially a finite-state machine with the added ability to reread its input and also to erase and write over its input. It also has unlimited auxiliary memory. Thus, the Turing machine overcomes the deficiencies we noted in finite-state machines. Unlimited auxiliary memory makes the Turing machine a hypothetical "machine"—a model—and not a real device.

A Turing machine consists of a finite-state machine and an unlimited tape divided into cells, each cell containing at most one symbol from an allowable finite alphabet. At any one instant, only a finite number of cells on the tape are nonblank. We use the special symbol b to denote a blank cell. The finite-state unit, through its read–write head, reads one cell of the tape at any given moment (see Figure 8.9).

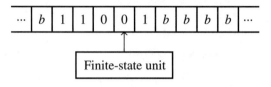

Figure 8.9

By the next clock pulse, depending on the present state of the unit and the symbol read, the unit either does nothing (halts) or completes three actions:

1. Print a symbol from the alphabet on the cell read (it might be the same symbol that's already there).
2. Go to the next state (it might be the same state as before).
3. Move the read–write head one cell left or right.

We can describe the actions of any particular Turing machine by a set of quintuples of the form (s, i, i', s', d), where s and i indicate the present state and the tape symbol being read, i' denotes the symbol printed, s' denotes the new state, and d denotes the direction in which the read–write head moves (R for right, L for left).

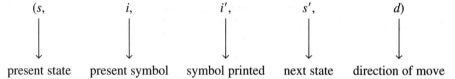

Thus, a machine in the configuration illustrated by Figure 8.10a, if acting according to the instructions contained in the quintuple $(2, 1, 0, 1, R)$, would move to the configuration illustrated in Figure 8.10b. The symbol 1 being read on the tape has been changed to a 0, the state of the unit has been changed from 2 to 1, and the head has moved one cell to the right.

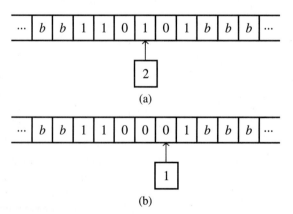

Figure 8.10

The term "Turing machine" is used both in the generic sense and also as the collection of quintuples that describe the actions of any particular machine. This is the same double usage for both the name of the abstraction and any instance of that abstraction that we mentioned for Boolean algebra in Chapter 7.

Definition: Turing Machine

Let S be a finite set of states and I a finite set of tape symbols (the **tape alphabet**) including a special symbol b. A **Turing machine** is a set of quintuples of the form (s, i, i', s', d) where $s, s' \in S$; $i, i' \in I$; and $d \in \{R, L\}$ and no two quintuples begin with the same s and i symbols.

The restriction that no two quintuples begin with the same s and i symbols ensures that the action of the Turing machine is deterministic and completely specified by its present state and symbol read. If a Turing machine gets into a configuration for which its present state and symbol read are not the first two symbols of any quintuple, the machine halts.

Just as in the case of ordinary finite-state machines, we specify a fixed starting state, denoted by 0, in which the machine begins any computation. We also assume an initial configuration for the read–write head, namely, a position over the farthest left nonblank symbol on the tape. (If the tape is initially all blank, the read–write head can be positioned anywhere to start.)

EXAMPLE 25   *A* Turing machine is defined by the set of quintuples

$(0, 0, 1, 0, R)$

$(0, 1, 0, 0, R)$

$(0, b, 1, 1, L)$

$(1, 0, 0, 1, R)$

$(1, 1, 0, 1, R)$

The action of this Turing machine when processing a particular initial tape is shown by the sequence of configurations in Figure 8.11, which also shows the quintuple that applies at each step. Again, which quintuple applies is determined by the present state and present symbol; as a result, the order in which quintuples are applied has nothing to do with the order in which they are presented in the machine's definition, quintuples can be used more than once, or may not be used at all. Since there are no quintuples defining the action to be taken when in state 1 reading b, the machine halts with final tape

| $\cdots$ | b | 1 | 0 | 0 | 0 | 0 | b | $\cdots$ |

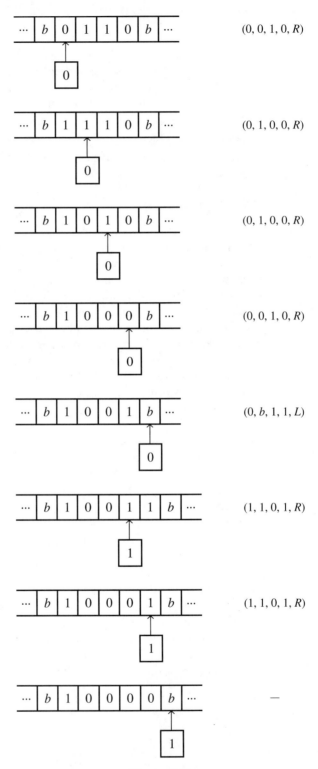

Figure 8.11

The tape serves as a memory medium for a Turing machine, and, in general, the machine can reread cells of the tape. Since it can also write on the tape, the nonblank portion of the tape can be as long as desired, although there are still only a finite number of nonblank cells at any time. Hence the machine has available an unbounded, though finite, amount of storage. Because Turing machines overcome the limitations of finite-state machines, Turing machines should have considerably higher capabilities. In fact, a finite-state machine is a very special case of a Turing machine, one that always prints the old symbol on the cell read, always moves to the right, and always halts on the symbol b.

PRACTICE 47   Consider the following Turing machine:

$(0, 0, 0, 1, R)$

$(0, 1, 0, 0, R)$

$(0, b, b, 0, R)$

$(1, 0, 1, 0, R)$

$(1, 1, 1, 0, L)$

a. What is the final tape, given the initial tape

...	b	1	0	b	...

(Since it is tedious to draw all the little squares, you don't need to do so; just write down the contents of the final tape.)

b. Describe the behavior of the machine when started on the tape

...	b	0	1	b	...

c. Describe the behavior of the machine when started on the tape

...	b	0	0	b	...

Parts (b) and (c) of Practice 47 illustrate two ways in which a Turing machine can fail to halt: by endlessly cycling or by moving forever along the tape.

Turing Machines as Set Recognizers

Although the Turing machine computations we have seen so far are not particularly meaningful, we shall use the Turing machine to do two kinds of jobs. First, we'll use it as a recognizer, much as we considered finite-state machines as recognizers in the previous section. We can even give a very similar definition, provided we first define a final state for a Turing machine. A **final state** in a Turing machine is one that is not the first symbol in any quintuple. Thus, on entering a final state, whatever the symbol read, the Turing machine halts.

Definition: Turing Machine Recognition (Acceptance)

A Turing machine T with tape alphabet I **recognizes** (**accepts**) a subset S of I^* if T, beginning in standard initial configuration on a tape containing a string α of tape symbols, halts in a final state if and only $\alpha \in S$.

Note that our definition of acceptance leaves open two possible behaviors for T when applied to a string α of tape symbols not in S. T may halt in a nonfinal state, or T may fail to halt at all.

We can now build a Turing machine to recognize our old friend $S = \{0^n 1^n \mid n \geq 0\}$. The machine is based on our second approach to this recognition problem, sweeping back and forth across the input and crossing out 0–1 pairs.

EXAMPLE 26   We want to build a Turing machine that will recognize $S = \{0^n 1^n \mid n \geq 0\}$. We shall use one additional special symbol, call it X, to mark out the 0s and 1s already examined. Thus the tape alphabet is $I = \{0, 1, b, X\}$. State 8 is the only final state. The quintuples making up T are given below, together with a description of their function.

$(0, b, b, 8, R)$	Recognizes the empty tape, which is in S.
$(0, 0, X, 1, R)$	Erases the leftmost 0 and begins to move right.
$(1, 0, 0, 1, R)$	
$(1, 1, 1, 1, R)$	Moves right in state 1 until it reaches the end of the initial
$(1, b, b, 2, L)$	string; then moves left in state 2.
$(2, 1, X, 3, L)$	Erases the rightmost 1 and begins to move left.
$(3, 1, 1, 3, L)$	Moves left over 1s.
$(3, 0, 0, 4, L)$	Goes to state 4 if more 0s are left.
$(3, X, X, 7, R)$	Goes to state 7 if no more 0s in string.
$(4, 0, 0, 4, L)$	Moves left over 0s.
$(4, X, X, 5, R)$	Finds left end of binary string and begins sweep again.
$(5, 0, X, 6, R)$	Erases the leftmost 0 and begins to move right.
$(6, 0, 0, 6 R)$	
$(6, 1, 1, 6, R)$	Moves right in state 6 until it reaches the end of the binary
$(6, X, X, 2, L)$	string; then moves left in state 2.
$(7, X, X, 8, R)$	No more 1s in string; machine accepts.

Reading down the columns in Figure 8.12, we can see the key configurations in the machine's behavior on the tape

$\cdots$	b	0	0	0	1	1	1	b	$\cdots$

which, of course, it should accept.

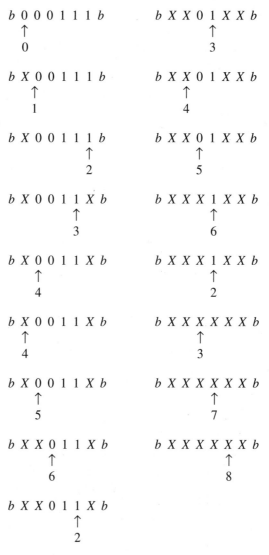

Figure 8.12

PRACTICE 48   For the Turing machine of Example 26, describe the final configuration after processing the following input tapes:

a.

...	b	0	0	1	1	1	b	...

b.

...	b	0	0	0	1	1	b	...

c.

...	b	0	0	0	0	1	1	b	...

REMINDER:
Give the states
of your Turing
machine big
enough jobs to
do so that the
machine will
work in general,
not just for
special cases.
Test using a
variety of input
tapes.

Notice how each state of the Turing machine in Example 26 is designed to accomplish a certain task, as indicated by the "comments." The job of state 1, for example, is to move right until the end of the input is found, then turn the computation over to state 2. A change of state should occur only when something significant happens. For example, a Turing machine cannot pass right over an indeterminate number of 1s by changing state at each move because its behavior is tied to a specific input tape. On the other hand, if the machine needs to count over a certain fixed number of 1s, then changing states at each move would accomplish this.

PRACTICE 49   Design a Turing machine to recognize the set of all strings of 0s and 1s ending in 00. (This set can be described by the regular expression $(0 \lor 1)*00$, so you should be able to use a Turing machine that changes no tape symbols and always moves to the right.)   ●

PRACTICE 50   Modify the Turing machine of Example 26 to recognize $\{0^n 1^{2n} \mid n \geq 0\}$.   ●

Turing Machines as Function Computers

The second job for which we shall use the Turing machine is to compute functions. Given a particular Turing machine T and a string α of tape symbols, we begin T in standard initial configuration on a tape containing α. If T eventually halts with a string β on the tape, we may consider β as the value of a function evaluated at α. Using function notation, $T(\alpha) = \beta$. The domain of the function T consists of all strings α for which T eventually halts. We can also think of T as computing **number-theoretic functions**, functions from a subset of $\mathbb{N}^k$ into $\mathbb{N}$ for any $k \geq 1$. We shall think of a string of 1s of length $n + 1$ as the unary representation of the nonnegative integer n; we'll denote this encoding of n by $\bar{n}$. (The extra 1 in the encoding enables us to distinguish 0 from a blank tape.) Then a tape containing the string $\bar{n}_1 * \bar{n}_2 * \cdots * \bar{n}_k$ can be thought of as the representation of the k-tuple $(n_1, n_2, \ldots, n_k)$ of nonnegative integers. If T begun in the standard initial configuration on such a tape eventually halts with a final tape that is the representation $\bar{m}$ of a nonnegative integer m, then T has acted as a k-variable function T^k, where $T^k(n_1, n_2, \ldots, n_k) = m$. If T begun in standard initial configuration on such a tape either fails to halt or halts with the final tape not a representation of a nonnegative integer, then the function T^k is undefined at $(n_1, n_2, \ldots, n_k)$.

There is thus an infinite sequence $T^1, T^2, \ldots, T^k, \ldots$ of number-theoretic functions computed by T associated with each Turing machine T. For each k, the function T^k is a **partial function** on $\mathbb{N}^k$, meaning that its domain may be a proper subset of $\mathbb{N}^k$. A special case of a partial function on $\mathbb{N}^k$ is a **total function** on $\mathbb{N}^k$, where the function is defined for all k-tuples of nonnegative integers.

EXAMPLE 27   Let a Turing machine T be given by the quintuples

$(0, 1, 1, 0, R)$

$(0, b, 1, 1, R)$

If T is begun in standard initial configuration on the tape

	b	1	1	1	b	
$\cdots$						$\cdots$

then T will halt with final configuration

$b\ 1\ 1\ 1\ 1\ b$
$\uparrow$
1

Therefore, T defines a one-variable function T^1 that maps $\overline{2}$ to $\overline{3}$. In general, T maps $\overline{n}$ to $\overline{n+1}$, so $T^1(n) = n + 1$, a total function of one variable. ●

In Example 27, we began with a Turing machine and observed a particular function it computed, but we can also begin with a number-theoretic function and try to find a Turing machine to compute it.

Definition: Turing-Computable Function
A **Turing-computable function** is a number-theoretic function computed by some Turing machine.

A Turing-computable function f can in fact be computed by an infinite number of Turing machines. Once a machine T is found to compute f, we can always include extraneous quintuples in T, producing other machines that also compute f.

EXAMPLE 28   We want to find a Turing machine that computes the function f defined as follows:

$$f(n_1, n_2) = \begin{cases} n_2 - 1 & \text{if } n_2 \neq 0 \\ \text{undefined} & \text{if } n_2 = 0 \end{cases}$$

Thus f is a partial function of two variables. Let's consider the Turing machine given by the following quintuples. The actions performed by various sets of quintuples are described.

$(0, 1, 1, 0, R)$
$(0, *, *, 1, R)$ $\left.\right\}$ Passes right over $\bar{n}_1$ to $\bar{n}_2$

$(1, 1, 1, 2, R)$ Counts first 1 in $\bar{n}_2$.

$(2, b, b, 3, R)$ $n_2 = 0$; halts.

$(2, 1, 1, 4, R)$
$(4, 1, 1, 4, R)$ $\left.\right\}$ Finds the right end of $\bar{n}_2$.
$(4, b, b, 5, L)$

$(5, 1, b, 6, L)$ Erases last 1 in $\bar{n}_2$.

$(6, 1, 1, 6, L)$
$(6, *, b, 7, L)$ $\left.\right\}$ Passes left to $\bar{n}_1$, erasing $*$.

$(7, 1, b, 7, L)$ Erases $\bar{n}_1$.

$(7, b, b, 8, L)$ $\bar{n}_1$ erased; halts with $\overline{n_2 - 1}$ on tape.

If T is begun on the tape

	b	1	1	$*$	1	1	1	1	b	
$\cdots$										$\cdots$

then T will halt with final configuration

$b\ b\ b\ b\ b\ 1\ 1\ 1\ b$
$\uparrow$
8

This configuration agrees with the requirement that $f(1, 3) = 2$. If T is begun on the tape

	b	1	1	$*$	1	b	
$\cdots$							$\cdots$

then T will halt with final configuration

$b\ 1\ 1\ *\ 1\ b\ b$
$\uparrow$
3

Because the final tape is not $\bar{m}$ for any nonnegative integer m, the function computed by T is undefined at $(1, 0)$—just as we want. It is easy to see that this Turing machine computes f and that f is therefore a Turing-computable function.   ●

PRACTICE 51   Design a Turing machine to compute the function

$$f(n) = \begin{cases} n - 2 & \text{if } n \geq 2 \\ 1 & \text{if } n < 2 \end{cases}$$

●

Church–Turing Thesis

In this chapter we have talked about models of "computation" or of "computational procedures." Although we have not defined the term, by a computational procedure we mean an algorithm. We have talked about algorithms often in this book and have given a number of algorithms for various tasks. Recall that our (somewhat intuitive) definition of an algorithm is a set of instructions that can be mechanically executed in a finite amount of time in order to solve some problem. This means that, given input appropriate to the task, the algorithm must eventually stop (halt) and produce the correct answer if an answer exists. (If no answer exists, let us agree that the algorithm can either halt and declare that no answer exists, or it can go on indefinitely searching for an answer.)

Now we ask, Is the Turing machine a better model of a computational procedure than the finite-state machine? We are quite likely to agree that any Turing-computable function f is a function whose values can be found by a computational procedure or algorithm. In fact, if f is computed by the Turing machine T, then the set of quintuples of T is itself the algorithm; as a list of instructions that can be carried out mechanically, it satisfies the various properties in our notion of an algorithm. Therefore, we are probably willing to accept the proposal illustrated by Figure 8.13. The figure shows "computable by algorithm" as a "cloudy," intuitive idea and "Turing computable" as a mathematically precise, well-defined idea. The arrow asserts that any Turing-computable function is computable by an algorithm.

Given the simplicity of the definition of a Turing machine, it is a little startling to contemplate Figure 8.14, which asserts that any function computable by any means we might consider to be an algorithm is also Turing computable. Combining Figures 8.13 and 8.14, we get the Church–Turing thesis (Figure 8.15), named after Turing and another well-known mathematician, Alonzo Church.

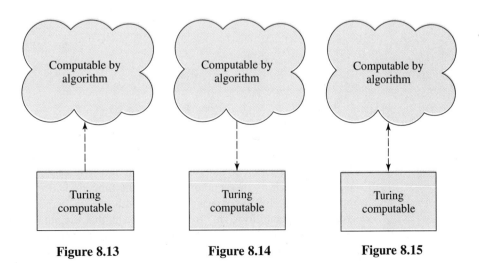

Figure 8.13   **Figure 8.14**   **Figure 8.15**

> **Church–Turing Thesis**
>
> A number-theoretic function is computable by an algorithm if and only if it is Turing computable.

Because the Church–Turing thesis equates an intuitive idea with a mathematical idea, it can never be formally proved and must remain a thesis, not a theorem. What, then, is its justification?

One piece of evidence is that whenever a procedure generally agreed to be an algorithm has been proposed to compute a function, someone has been able to design a Turing machine to compute that function. (Of course, there is always the nagging thought that someday this might not happen.)

Another piece of evidence is that other mathematicians, several of them at about the same time Turing developed the Turing machine, proposed other models of a computational procedure. On the surface, each proposed model seemed quite unrelated to any of the others. However, because all the models were formally defined, just as Turing computability is, it was possible to determine on a formal, mathematical basis whether any of them were equivalent. All the models, as well as Turing computability, were proved equivalent; that is, they all defined the same class of functions, which suggests that Turing computability embodies everyone's concept of an algorithm. Figure 8.16 illustrates what has been done; here solid lines represent mathematical proofs and dashed lines correspond to the Church–Turing thesis. The dates indicate when the various models were proposed.

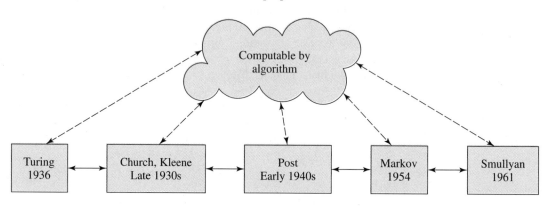

Figure 8.16

The Church–Turing thesis is now widely accepted as a working tool by researchers dealing with computational procedures. If, in a research paper, a method is set forth for computing a function and the method intuitively seems to be an algorithm, then the Church–Turing thesis is invoked and the function is declared to be Turing computable (or one of the names associated with one of the equivalent formulations of Turing computability). This invocation means that the author presumably could, if pressed, produce a Turing machine to compute the function, but, again, the Church–Turing thesis is so universally accepted that no one bothers with these details anymore.

Although the Church–Turing thesis is stated in terms of number-theoretic functions, it can be interpreted more broadly. Any algorithm in which a finite set of symbols is manipulated can be translated into a number-theoretic function by a suitable encoding of the symbols as nonnegative integers, much as input to a real computer is encoded and stored in binary form. Thus, by the Church–Turing thesis we can say that if there is an algorithm to do a symbol manipulation task, there is a Turing machine to do it.

By accepting the Church–Turing thesis, we have accepted the Turing machine as the ultimate model of a computational procedure. Turing machine capabilities exceed those of any actual computer, which, after all, does not have the unlimited tape storage of a Turing machine. It is remarkable that Turing proposed this concept in 1936, well before the advent of the modern computer.

Decision Problems

We have spent quite a bit of time discussing what Turing machines can do. By the Church–Turing thesis, they can do a great deal indeed, although not very efficiently. It is even more important, however, to consider what Turing machines *cannot* do. Because a Turing machine's abilities to perform tasks exceed those of an actual computer, if we find something no Turing machine can do, then a real computer cannot do it either. In fact, by invoking the Church–Turing thesis, no algorithm exists to do it, and the task is not computable. The type of task we have in mind here is generally that of determining the truth value of each of a number of related statements.

> **Definition: Decision Problem**
> A **decision problem** asks if an algorithm exists to decide whether individual statements from some large class of statements are true.

The solution to a decision problem answers the question of whether an algorithm exists. A **positive solution** consists of proving that an algorithm exists, and it is generally given by actually producing an algorithm that works. A **negative solution** consists of proving that no algorithm exists. Note that this statement is much stronger than simply saying that a lot of people have tried but no one has come up with an algorithm—this might simply mean that the algorithm is hard. It must be shown that it is impossible for anyone ever to come up with an algorithm. When a negative solution to a decision problem is found, the problem is said to be **unsolvable**, or **undecidable**. This terminology can be confusing because the decision problem itself—the question of whether an algorithm exists to do a task—has been solved; what must forever be unsolvable is the task itself.

Examples of Decision Problems

We shall look at some decision problems that have been answered.

EXAMPLE 29

Does an algorithm exist to decide, given integers a, b, and c, whether $a^2 = b^2 + c^2$? Clearly, this question is a solvable decision problem. The algorithm consists of multiplying b by itself, multiplying c by itself, adding the two results, and comparing the sum with the result of multiplying a by itself.

Obviously, Example 29 is a rather trivial decision problem. Historically, much of mathematics has concerned itself at least indirectly with finding positive solutions to decision problems, that is, producing algorithms. Negative solutions to decision problems arose only in the twentieth century.

EXAMPLE 30   One of the earliest decision problems to be formulated was Hilbert's Tenth Problem, tenth in a list of problems David Hilbert posed to the International Congress of Mathematicians in 1900. The problem is, Does an algorithm exist to decide for any polynomial equation $P(x_1, x_2, \dots, x_n) = 0$ with integral coefficients whether it has integral solutions? For polynomial equations of the form $ax + by + c = 0$, where a, b, and c are integers, it is known that integer solutions exist if and only if the greatest common divisor of a and b also divides c. Thus, for particular subclasses of polynomial equations, there might be algorithms to decide whether integer solutions exist, but the decision problem as stated applies to the whole class of polynomial equations with integer coefficients. When this problem was posed and for some time after, the general belief was that surely an algorithm existed and the fact that no one had found such an algorithm merely implied that it must be difficult. In the mid-1930s, results such as those in the next example began to cast doubt on this view. It was not until 1970, however, that this problem was finally shown to be unsolvable.   ●

EXAMPLE 31   An alternative formulation to the logic systems we discussed in Chapter 1 is to identify certain strings of symbols as *axioms* and to give *rules of inference* whereby a new string can be obtained from old strings. Any string that is the last one in a finite list of strings consisting of either axioms or strings obtainable by the rules of inference from earlier strings in the list is said to be a *theorem*. The decision problem for such a formal theory is, Does an algorithm exist to decide whether a given string in the formal theory is a theorem of the theory?

The work of Church and the famous twentieth-century logician Kurt Gödel showed that any formal theory that axiomatizes properties of arithmetic (making commutativity of addition an axiom, for example) and is not completely trivial (not everything is a theorem) is undecidable. Their work can be considered good news for mathematicians, because it means that ingenuity in answering questions in number theory will never be replaced by a mechanical procedure.   ●

EXAMPLE 32   A particular Turing machine T begun on a tape containing a string α will either eventually halt or never halt. The **halting problem** for Turing machines is a decision problem: Does an algorithm exist to decide, given a Turing machine T and string α, whether T begun on a tape containing α will eventually halt? Turing proved the unsolvability of the halting problem in the late 1930s.   ●

Halting Problem

We shall prove the unsolvability of the halting problem after two observations. First, it might occur to us that "run T on α" would constitute an algorithm to see whether T halted on α. If within 25 steps of T's computation T has halted, then we know T halts on α. But if within 25,000 steps T has not halted, what can we conclude? T may still eventually halt. How long should we wait? This so-called algorithm will not always give us the answer to our question.

A second observation is that the halting problem asks for one algorithm to be applied to a large class of statements. The halting problem asks, Does an algorithm exist to decide, for any given (T, α) pair, whether T halts when begun on a tape containing α? The algorithm comes first, and that single algorithm has to give the correct answer for all (T, α) pairs. In the notation of predicate logic, the halting problem asks about the truth value of a statement in the form

$(\exists$ algorithm$)(\forall(T, \alpha))(\dots)$

Consider the following statement, which seems very similar: Given a particular (T, α) pair, does an algorithm exist to decide whether T halts when begun on a tape containing α? Here, the (T, α) pair comes first and an algorithm is chosen based on the particular (T, α); for a different (T, α), there can be a different algorithm. The statement has become $(\forall(T, \alpha))(\exists$ algorithm$)(\dots)$. This problem is solvable. Suppose someone gives us a (T, α). Two algorithms are (1) "say yes" and (2) "say no." Since T acting on α either does or does not halt, one of these two algorithms correctly answers the question. This solution may seem trivial or even sneaky, but consider again the problem statement: Given a particular (T, α) pair, *does an algorithm exist* to decide, and so forth. Such an algorithm *does exist*; it is either to say yes or to say no—we are not required to choose which one is correct!

This turnabout of words changes the unsolvable halting problem into a trivially solvable problem. It also points out the character of a decision problem, asking whether a single algorithm exists to solve a large class of problems. An unsolvable problem has both a good side and a bad side. That no algorithm exists to solve a large class of problems guarantees jobs for creative thinkers who cannot be replaced by Turing machines. But that the class of problems considered is so large might make the result too general to be of interest.

We shall state the halting problem again and then prove its unsolvability.

Definition: Halting Problem
The **halting problem** asks, Does an algorithm exist to decide, given any Turing machine T and string α, whether T begun on a tape containing α will eventually halt?

Theorem on the Halting Problem
The halting problem is unsolvable.

Proof: We want to prove that something *does not exist*, a situation made to order for proof by contradiction. Therefore we assume that the halting problem is solvable and that a single algorithm exists that can act on any (T, α) pair as input and eventually decide whether T running on α halts. We are asking this algorithm to solve a task of symbol manipulation, since we can imagine the set of quintuples of T encoded as some unique string s_T of symbols; we'll use (s_T, α) to denote the string s_T concatenated with the string α. The task then becomes transforming the string (s_T, α) into a string representing a yes (the Turing machine with description s_T halts when begun on a tape containing α) or a no (the Turing machine with description s_T does not halt when begun on a tape containing α). By the Church–Turing thesis, because we have assumed the existence of an algorithm that performs this task, we can assume the existence of a single Turing machine X that performs this task. Thus X acts on a tape containing (s_T, α) for any T and α and eventually halts, at the same time telling us whether T on α halts. To be definite, suppose that X begun on (s_T, α) halts with a 1 left on the tape if and only if T begun on α halts, and X begun on (s_T, α) halts with a 0 left on the tape if and only if T begun on α fails to halt; these are the only two possibilities.

Now we add to X's quintuples to create a new machine Y. Machine Y modifies X's behavior so that whenever X halts with a 1 on its tape, Y goes to a state that moves Y endlessly to the right so that it never halts. If X halts with a 0 on its tape, so does Y. Finally, we modify Y to get a new machine Z that acts on any input β by first copying β (see, for example, Exercise 13) and then turning the computation over to Y so that Y acts on (β, β).

Now by the way Z is constructed, if Z acting on s_Z halts, it is because Y acting on (s_Z, s_Z) halts, and that happens because X acting on (s_Z, s_Z) halts with a 0 on the tape; but if this happens, it implies that Z begun on s_Z fails to halt! Therefore,

$$Z \text{ on } s_Z \text{ halts} \rightarrow Z \text{ on } s_Z \text{ fails to halt} \tag{1}$$

This implication is very strange; let's see what happens if Z on s_Z does not halt. By the way Z is constructed, if Z acting on s_Z does not halt, neither does Y acting on (s_Z, s_Z). Now Y acting on (s_Z, s_Z) fails to halt exactly when X acting on (s_Z, s_Z) halts with a 1 on the tape; but this result implies that Z begun on s_Z halts! Therefore

$$Z \text{ on } s_Z \text{ fails to halt} \rightarrow Z \text{ on } s_Z \text{ halts} \tag{2}$$

Together, implications (1) and (2) provide an airtight contradiction, so our assumption that the halting problem is solvable is incorrect. *End of Proof*

The proof of the unsolvability of the halting problem depends on two ideas. One is that of encoding a Turing machine into a string description, and the other is that of having a machine look at and act on its own description. Notice also that neither (1) nor (2) alone in the proof is sufficient to prove the result. Both are needed to contradict the original assumption of the solvability of the halting problem.

We have previously encountered another proof of this nature, where the observation that makes the proof work is self-contradictory. You might want to review here the proof of Cantor's theorem in Chapter 4.

Computational Complexity

As a model of computation, the Turing machine has provided us with a way to prove the existence of unsolvable (uncomputable) problems. Not only does the Turing machine help us find the limits of computability, but it can also help us classify problems that are computable—that have an algorithm for their solution—by the amount of work required to carry out the algorithm.

Finding the amount of work required to carry out an algorithm sounds like analysis of algorithms. We have analyzed a number of real algorithms in this book and classified them as $\Theta(\log n)$, $\Theta(n)$, $\Theta(n^2)$, or what have you. By the Church–Turing thesis, any algorithm can be expressed in Turing machine form. In this form, the amount of work is the number of Turing machine steps (one per clock pulse) required before the Turing machine halts. (We assume here that we are considering only tasks that "have answers" so that the Turing machine halts on all appropriate input.)

Turing machine computations are quite inefficient. Therefore if algorithms A and A' both solve the same problem, but A is expressed as a description of a Turing machine and A' as pseudocode for instructions in a high-level programming language, then comparing the number of operations each algorithm performs is rather meaningless. Therefore we shall assume that all algorithms are expressed in Turing machine form so that we can readily compare the efficiency of different algorithms.

Rather than discuss whether a Turing machine algorithm is $\Theta(n)$ or $\Theta(n^2)$, let us simply note whether it is a polynomial-time algorithm. (Only quite trivial algorithms can be better than polynomial time, because it takes a Turing machine n steps just to examine its tape.) Problems for which no polynomial-time algorithms exist are called **intractable**. Such problems may be solvable, but only by inefficient algorithms.

Definition: *P*

P is the collection of all sets recognizable by Turing machines in polynomial time.

Consideration of set recognition in our definition of P is not as restrictive as it may seem. Because the Turing machine halts on all appropriate input, it actually decides, by halting in a final or nonfinal state, whether the initial string was or was not a member of the set. Many problems can be posed as set decision problems by suitably encoding the objects involved in the problem.

For example, consider the Hamiltonian circuit problem (Section 6.2) of whether a graph has a cycle that uses every node of the graph. We may define some encoding process to represent any graph as a string of symbols. Strings that are the representations of graphs become appropriate input, and we want to decide, given such a string, whether it belongs to the set of strings whose associated graphs have Hamiltonian circuits. If we can build a Turing machine to make this decision in polynomial time, then the Hamiltonian circuit problem belongs to P.

We noted in Section 6.2 that the Hamiltonian circuit problem is solvable by the brute-force approach of tracing all possible paths, but this is an exponential solution because of the number of paths. We said that there is no known efficient (polynomial) algorithm to solve the Hamiltonian circuit problem, so we have no proof that the Hamiltonian circuit problem belongs to P. But there is also no proof that the Hamiltonian circuit problem does not belong to P. Might a clever, efficient algorithm someday be found? To see why this is unlikely, we'll consider a new kind of Turing machine.

Ordinary Turing machines act deterministically, due to our restriction that no two quintuples begin with the same present state/present symbol pair. A relaxation of this requirement results in a **nondeterministic Turing machine**, which may have a choice of actions at any step. A nondeterministic Turing machine recognizes a string on its tape if some sequence of actions leads to halting in a final state.

Definition: *NP*

NP is the collection of all sets recognizable by nondeterministic Turing machines in polynomial time. (*NP* comes from *n*ondeterministic *p*olynomial time.)

While a set in P requires that a deterministic Turing machine be able to make a decision (in polynomial time) about whether some string on its tape does or does not belong to the set, a set in NP requires only that a nondeterministic Turing machine be able to verify (in polynomial time) by a fortuitous choice of actions that an input string is in the set. Given a graph that has a Hamiltonian circuit, for example, this fact can be confirmed in polynomial time by a nondeterministic Turing machine that picks the correct path, so the Hamiltonian circuit problem belongs to NP.

If a Turing machine can decide in polynomial time whether an arbitrary string belongs to a set, it can surely use the same process to verify a member of the set in polynomial time. Therefore $P \subseteq NP$. However, it is not known whether this inclusion is proper, that is, whether $P \subset NP$ so that there could be NP problems—including perhaps the Hamiltonian circuit problem—that are intractable.

The Hamiltonian circuit problem belongs to a third class of problems known as **NP-complete problems**, meaning that not only are they in NP, but if a polynomial-time decision algorithm were ever found for any one of them, that is, if any of them were ever found to be in P, then indeed we would have $P = NP$. A large number of problems from many different fields have been found to be NP-complete since this idea was formulated in 1971, and no polynomial-time decision algorithm has been found for any of them. Therefore it is now suspected that $P \subset NP$ and that all these problems are intractable, but to prove this remains a tantalizing goal in computer science research.

Section 8.3 Review

Techniques

- Describe the action of a given Turing machine on a given initial tape.
- Construct a Turing machine to recognize a given set.
- Construct a Turing machine to compute a given number-theoretic function.

Main Ideas

Turing machines have a deterministic mode of operation, the ability to reread and rewrite input, and an unbounded auxiliary memory.

A finite-state machine is a special case of a Turing machine.

Turing machines can be used as set recognizers and as function computers.

The Church–Turing thesis equates a function computable by an algorithm with a Turing-computable function. Because this thesis expresses a relationship between an intuitive idea and a formally defined one, it can never be proved but has nonetheless been widely accepted.

A decision problem asks if an algorithm exists to decide whether individual statements from a large class of statements are true; if no algorithm exists, the decision problem is unsolvable.

Unsolvable decision problems have arisen in a number of contexts in this century.

The halting problem is unsolvable.

$P \subseteq NP$, but it is unknown whether $P \subset NP$.

Exercises 8.3

★1. Consider the following Turing machine:

$(0, 0, 0, 0, L)$

$(0, 1, 0, 1, R)$

$(0, b, b, 0, L)$

$(1, 0, 0, 1, R)$

$(1, 1, 0, 1, R)$

a. What is its behavior when started on the tape

...	b	1	0	0	1	1	b	...

b. What is its behavior when started on the tape

...	b	0	0	1	1	1	b	...

2. Given the Turing machine

$(0, 1, 1, 0, R)$

$(0, 0, 0, 1, R)$

$(1, 1, 1, 1, R)$

$(1, b, 1, 2, L)$

$(2, 1, 1, 2, L)$

$(2, 0, 0, 2, L)$

$(2, b, 1, 0, R)$

a. What is its behavior when started on the tape

...	b	1	0	1	0	b	...

b. What is its behavior when started on the tape

...	b	1	0	1	b	...

3. Find a Turing machine that replaces every 0 in a string of 0s and 1s with a 1 and every 1 with a 0.

4. Find a Turing machine that recognizes the set of all strings of 0s and 1s containing at least one 1.

★5. Find a Turing machine that recognizes the set of all unary strings consisting of an even number of 1s (this includes the empty string).

6. Find a Turing machine that recognizes 0*10*1.

7. Find a Turing machine to accept the set of nonempty strings of well-balanced parentheses. (Note that (( )(( ))) is well balanced and (( )(( )) is not.)

★8. Find a Turing machine that recognizes $\{0^{2n}1^n2^{2n} \mid n \geq 0\}$.

9. Find a Turing machine that recognizes $\{w * w^R \mid w \in \{0, 1\}^*$ and w^R is the reverse of the string $w\}$.

10. Find a Turing machine that recognizes $\{w_1 * w_2 \mid w_1, w_2 \in \{0, 1\}^*$ and $w_1 \neq w_2\}$.

11. Find a Turing machine that recognizes the set of palindromes on $\{0, 1\}^*$, that is, the set of all strings in $\{0, 1\}^*$ that read the same forward and backward, such as 101.

★12. Find a Turing machine that converts a string of 0s and 1s representing a nonzero binary number into a string of that number of 1s. As an example, the machine should, when started on a tape containing

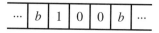

halt on a tape containing

$$\cdots \boxed{b} \boxed{1} \boxed{1} \boxed{1} \boxed{1} \boxed{b} \cdots$$

13. Find a Turing machine that, given an initial tape containing a nonempty string of 1s, marks the right end of the string with a * and puts a copy of the string to the right of the *. As an example, the machine should, when started on a tape containing

$$\cdots \boxed{b} \boxed{1} \boxed{1} \boxed{1} \boxed{b} \cdots$$

halt on a tape containing

$$\cdots \boxed{b} \boxed{1} \boxed{1} \boxed{1} \boxed{*} \boxed{1} \boxed{1} \boxed{1} \boxed{b} \cdots$$

★14. What number-theoretic function of three variables is computed by the following Turing machine?

$(0, 1, b, 0, R)$

$(0, *, b, 1, R)$

$(1, 1, 1, 2, R)$

$(2, *, *, 3, R)$

$(3, 1, 1, 2, L)$

$(2, 1, 1, 4, R)$

$(4, 1, 1, 4, R)$

$(4, *, 1, 5, R)$

$(5, 1, b, 5, R)$

$(5, b, b, 6, R)$

15. Find a Turing machine to compute the function

$$f(n) = \begin{cases} n & \text{if } n \text{ is even} \\ n + 1 & \text{if } n \text{ is odd} \end{cases}$$

★16. Find a Turing machine to compute the function

$$f(n) = \begin{cases} 1 & \text{if } n = 0 \\ 2 & \text{if } n \neq 0 \end{cases}$$

17. Find a Turing machine to compute the function

$$f(n) = 2n$$

18. Find a Turing machine to compute the function

$$f(n) = \begin{cases} n/3 & \text{if } 3 \text{ divides } n \\ \text{undefined} & \text{otherwise} \end{cases}$$

★19. Find a Turing machine to compute the function

$$f(n_1, n_2) = n_1 + n_2$$

20. Find a Turing machine to compute the function

$$f(n_1, n_2) = \begin{cases} n_1 - n_2 & \text{if } n_1 \geq n_2 \\ 0 & \text{otherwise} \end{cases}$$

21. Find a Turing machine to compute the function

$$f(n_1, n_2) = \max(n_1, n_2)$$

22. Do Exercise 17 again, this time making use of the machines T_1 and T_2 of Exercises 13 and 19, respectively, as "functions." (Formally, the states of these machines would have to be renumbered as the quintuples are inserted into the "main program," but you may omit this tiresome detail and merely "invoke T_1" or "invoke T_2.")

23. Describe verbally the actions of a Turing machine that computes the function $f(n_1, n_2) = n_1 \cdot n_2$; that is, design the algorithm but do not bother to create all the necessary quintuples. You may make use of Exercises 13 and 19.

Section 8.4   Formal Languages

Suppose we come upon the English language sentence "The walrus talks loudly." Although we might be surprised at the meaning, or *semantics*, of the sentence, we accept its form, or *syntax*, as valid in the language, meaning that the various parts of speech (noun, verb, etc.) are strung together in a reasonable way. In contrast, we reject "Loudly walrus the talks" as an illegal combination of parts of speech, or as syntactically incorrect and not part of the language. We must also worry about correct syntax in programming languages, but in these, unlike natural languages (English, French, etc.), legal combinations of symbols are specified in detail. Let's give a formal definition of *language*; the definition will be general enough to include both natural and programming languages.

> **Definitions: Alphabet, Vocabulary, Word, Language**
> An **alphabet**, or **vocabulary**, V is a finite, nonempty set of symbols. A **word** over V is a finite-length string of symbols from V. The set V^* is the set of all words over V. (See Example 34 in Chapter 2 for a recursive definition of V^*.) A **language** over V is any subset of V^*.

Viewing syntactically correct English language as a subset L of the set of all strings over the usual alphabet, we feel that "The walrus talks loudly" belongs to L while "Loudly walrus the talks" does not.

For any given language L, how can we describe L, that is, specify exactly those words belonging to L? If L is finite, we can just list its members, but if L is infinite, can we find a finite description of L? Not always—there are many more languages than possible finite descriptions. Although we shall consider only languages that can be finitely described, we can still think of two possibilities. We may be able to describe an algorithm to *decide* membership in L; that is, given any word in V^*, we could apply our algorithm and receive a yes or no answer as to whether the word belongs to L. Or we may be able to describe a procedure allowing us only to *generate* members of L, that is, crank out one at a time a list of all the members of L. We shall settle for languages for which this second option is possible and describe such a language by defining its generative process, or giving a *grammar* for the language.

Before we give a formal definition of what constitutes a grammar, let's look again at why "The walrus talks loudly" seems to be an acceptable sentence by seeing how it could be generated. Starting from the notion of sentence, we would agree that one legitimate form for a sentence is a noun-phrase followed by a verb-phrase. Symbolically,

> **sentence → noun-phrase verb-phrase**

A legitimate form of noun-phrase is an article followed by a noun,

> **noun-phrase → article noun**

and a legitimate form of verb-phrase is a verb followed by an adverb,

> **verb-phrase → verb adverb**

We would also agree with the substitutions

> **article →** the
>
> **noun →** walrus
>
> **verb →** talks
>
> **adverb →** loudly

Thus we can generate the sentence "The walrus talks loudly" by making successive substitutions:

> sentence ⇒ **noun-phrase verb-phrase**
>
> ⇒ **article noun verb-phrase**
>
> ⇒ the **noun verb-phrase**
>
> ⇒ the walrus **verb-phrase**
>
> ⇒ the walrus **verb adverb**
>
> ⇒ the walrus talks **adverb**
>
> ⇒ the walrus talks loudly

The foregoing boldface terms are those for which further substitutions can be made. The nonboldface terms stop or terminate the substitution process. These ideas are incorporated in the next definition.

Definition: Phrase-Structure (Type 0) Grammar

A **phrase-structure grammar (type 0 grammar)** G is a 4-tuple, $G = (V, V_T, S, P)$, where

V = vocabulary
V_T = nonempty subset of V called the set of **terminals**
S = element of $V - V_T$ called the **start symbol**
P = finite set of **productions** of the form $\alpha \to \beta$ where α is a word over V containing at least one nonterminal symbol and β is a word over V

EXAMPLE 33   Here is a very simple grammar: $G = (V, V_T, S, P)$ where $V = \{0, 1, S\}$, $V_T = \{0, 1\}$, and $P = \{S \to 0S, S \to 1 \}$.

The productions of a grammar allow us to transform some words over V into others; the productions can be called rewriting rules.

> **Definition: Generations (Derivations) in a Language**
>
> Let G be a grammar, $G = (V, V_T, S, P)$, and let w_1 and w_2 be words over V. Then w_1 **directly generates (directly derives)** w_2, written $w_1 \Rightarrow w_2$, if $\alpha \to \beta$ is a production of G, w_1 contains an instance of α, and w_2 is obtained from w_1 by replacing that instance of α with β. If $w_1, w_2, \ldots, w_n$ are words over V and $w_1 \Rightarrow w_2$, $w_2 \Rightarrow w_3$, $\ldots w_{n-1} \Rightarrow w_n$, then w_1 **generates (derives)** w_n, written $w_1 \stackrel{*}{\Rightarrow} w_n$. (By convention, $w_1 \stackrel{*}{\Rightarrow} w_1$.)

EXAMPLE 34

In the grammar of Example 33, $00S \Rightarrow 000S$ because the production $S \to 0S$ has been used to replace the S in $00S$ with $0S$. Also $00S \stackrel{*}{\Rightarrow} 00000S$.   ●

PRACTICE 52

Show that in the grammar of Example 33, $0S \stackrel{*}{\Rightarrow} 00001$.   ●

> **Definition: Language Generated by a Grammar**
>
> Given a grammar G, the **language L generated by G,** sometimes denoted $L(G)$, is the set
>
> $$L = \{w \in V_T^* \mid S \stackrel{*}{\Rightarrow} w\}$$
>
> In other words, L is the set of all strings of terminals generated from the start symbol.

REMINDER:
Productions may be used in any order. The only requirement is that the left side of the production must appear in the string you are processing.

Notice that once a string w of terminals has been obtained, no productions can be applied to w, and w cannot generate any other words.

The following procedure generates a list of the members of L: Begin with the start symbol S and systematically apply some sequence of productions until a string w_1 of terminals has been obtained; then $w_1 \in L$. Go back to S and repeat this procedure using a different sequence of productions to generate another word $w_2 \in L$, and so forth. Actually, this procedure doesn't quite work because we might get started on an infinite sequence of direct derivations that never leads to a string of terminals and thus never contributes a word to our list. Instead, we need to run a number of derivations from S simultaneously (parallel processing), checking on each one after each step and adding the final word to the list of members of L for any that terminate. That way we cannot get stuck waiting indefinitely while unable to do anything else.

PRACTICE 53

Describe the language generated by the grammar G of Example 33.   ●

Languages derived from grammars such as we have defined are called **formal languages.** If the grammar is defined first, the language will follow as an outcome of the definition. Alternatively, the language, as a well-defined set of strings, may be given first, and we then seek a grammar that generates it.

EXAMPLE 35   Let L be the set of all nonempty strings consisting of an even number of 1s. Then L is generated by the grammar $G = (V, V_T, S, P)$ where $V = \{1, S\}$, $V_T = \{1\}$, and $P = \{S \rightarrow SS, S \rightarrow 11\}$. A language can be generated by more than one grammar. L is also generated by the grammar $G' = (V', V'_T, S', P')$ where $V' = \{1, S\}$, $V'_T = \{1\}$, and $P' = \{S \rightarrow 1S1, S \rightarrow 11\}$.   ●

PRACTICE 54   a. Find a grammar that generates the language $L = \{0^n 10^n \mid n \geq 0\}$.
   b. Find a grammar that generates the language $L = \{0^n 10^n \mid n \geq 1\}$.   ●

Trying to describe concisely the language generated by a given grammar and defining a grammar to generate a given language can both be quite difficult. We'll look at another example where the grammar is a bit more complicated than any we've seen so far. Don't worry about how you might think up this grammar; just convince yourself that it works.

EXAMPLE 36   Let $L = \{a^n b^n c^n \mid n \geq 1\}$. A grammar generating L is $G = (V, V_T, S, P)$ where $V = \{a, b, c, S, B, C\}$, $V_T = \{a, b, c\}$, and P consists of the following productions:

1. $S \rightarrow aSBC$ 5. $bB \rightarrow bb$
2. $S \rightarrow aBC$ 6. $bC \rightarrow bc$
3. $CB \rightarrow BC$ 7. $cC \rightarrow cc$
4. $aB \rightarrow ab$

It is fairly easy to see how to generate any particular member of L using these productions. Thus, a derivation of the string $a^2 b^2 c^2$ is

$$S \Rightarrow aSBC$$
$$\Rightarrow aaBCBC$$
$$\Rightarrow aaBBCC$$
$$\Rightarrow aabBCC$$
$$\Rightarrow aabbC\overset{\curvearrowleft}{C}$$
$$\Rightarrow aabbcC$$
$$\Rightarrow aabbcc$$

In general, $L \subseteq L(G)$ where the outline of a derivation for any $a^n b^n c^n$ is given below; the numbers refer to the productions used.

$$S \overset{*}{\underset{1}{\Rightarrow}} a^{n-1} S (BC)^{n-1}$$
$$\underset{2}{\Rightarrow} a^n (BC)^n$$
$$\overset{*}{\underset{3}{\Rightarrow}} a^n B^n C^n$$
$$\underset{4}{\Rightarrow} a^n b B^{n-1} C^n$$
$$\overset{*}{\underset{5}{\Rightarrow}} a^n b^n C^n$$
$$\underset{6}{\Rightarrow} a^n b^n c C^{n-1}$$
$$\overset{*}{\underset{7}{\Rightarrow}} a^n b^n c^n$$

We must also show that $L(G) \subseteq L$, which involves arguing that some productions must be used before others and that the general derivation shown above is the only sort that will lead to a string of terminals.   ●

In trying to invent a grammar to generate the L of Example 36, we might first try to use productions of the form $B \to b$ and $C \to c$ instead of productions 4 through 7. Then we would indeed have $L \subseteq L(G)$, but $L(G)$ would also include words such as $a^n(bc)^n$.

Formal languages were developed in the 1950s by linguist Noam Chomsky in an attempt to model natural languages, such as English, with an eye toward automatic translation. However, since a natural language already exists and is quite complex, defining a formal grammar to generate a natural language is very difficult. Attempts to do this for English have been only partially successful.

EXAMPLE 37

We can describe a formal grammar that will generate a very restricted class of English sentences. The terminals in the grammar are the words "the," "a," "river," "walrus," "talks," "flows," "loudly," and "swiftly," and the nonterminals are the words **sentence**, **noun-phrase**, **verb-phrase**, **article**, **noun**, **verb**, and **adverb**. The start symbol is **sentence** and the productions are

> **sentence** → **noun-phrase verb-phrase**
>
> **noun-phrase** → **article noun**
>
> **verb-phrase** → **verb adverb**
>
> **article** → the
>
> **article** → a
>
> **noun** → river
>
> **noun** → walrus
>
> **verb** → talks
>
> **verb** → flows
>
> **adverb** → loudly
>
> **adverb** → swiftly

We know how to derive "The walrus talks loudly" in this grammar. Here is a derivation of "A river flows swiftly":

> **sentence** ⇒ **noun-phrase verb-phrase**
>
> ⇒ **article noun verb-phrase**
>
> ⇒ a **noun verb-phrase**
>
> ⇒ a river **verb-phrase**
>
> ⇒ a river **verb adverb**
>
> ⇒ a river flows **adverb**
>
> ⇒ a river flows swiftly

A few other sentences making various degrees of sense, such as "a walrus flows loudly," are also part of the language defined by this grammar. The difficulty of specifying a grammar for English as a whole becomes more apparent when we consider that a phrase such as "time flies" can be an instance of either a noun followed by a verb or of a verb followed by a noun! This situation is "ambiguous" (see Exercise 18 in this section). ●

Programming languages are less complex than natural languages, and their syntax often can be described successfully using formal language notation.

EXAMPLE 38   A section of formal grammar to generate identifiers in some programming language could be presented as follows:

> **identifier** → **letter**
> **identifier** → **identifier letter**
> **identifier** → **identifier digit**
> **letter** → a
> **letter** → b
> $\vdots$
> **letter** → z
> **digit** → 0
> **digit** → 1
> $\vdots$
> **digit** → 9

Here the set of terminals is $\{a, b, \dots, z, 0, 1, \dots, 9\}$ and **identifier** is the start symbol. ●

EXAMPLE 39   A shorthand that can avoid a long listing of productions is called **Backus–Naur form** (BNF). The productions of Example 38 can be given in BNF by three lines (as in Example 35 of Chapter 2):

> $<$identifier$> :: = <$letter$> \mid <$identifier$> <$letter$> \mid <$identifier$> <$digit$>$
> $<$letter$> ::= a \mid b \mid c \mid \dots \mid z$
> $<$digit$> ::= 0 \mid 1 \mid \dots \mid 9$

In BNF, nonterminals are identified by $< >$, the production arrow becomes $::=$, and | stands for "or," identifying various productions having the same left-hand symbol.

In modern times, BNF notation was originally used to define the programming language Algol (*Algo*rithmic *L*anguage) in the early 1960s. However, it appears that a similar notation was used between 400 B.C. and 200 B.C. to describe the rules of Sanskrit grammar.[1] ●

Classes of Grammars

Before we identify some types of grammars, let's look at one more example.

---

[1]*Ingerman, P. Z., "Panini-Backus form suggested,"* Communications of the ACM, *vol. 10, no. 3, 1967.*

EXAMPLE 40

Let L be the empty string λ together with the set of all strings consisting of an odd number n of 0s, $n \geq 3$. The grammar $G = (V, V_T, S, P)$ generates L where $V = \{0, A, B, E, F, W, X, Y, Z, S\}$, $V_T = \{0\}$, and the productions are

$S \rightarrow FA$	$0X \rightarrow X0$	$0Z \rightarrow Z0$
$S \rightarrow FBA$	$Y0 \rightarrow 0Y$	$WBZ \rightarrow EB$
$FB \rightarrow F0EB0$	$FX \rightarrow F0W$	$F \rightarrow \lambda$
$EB \rightarrow 0$	$YA \rightarrow Z0A$	$A \rightarrow \lambda$
$EB \rightarrow XBY$	$W0 \rightarrow 0W$	

The derivation $S \Rightarrow FA \overset{*}{\Rightarrow} \lambda\lambda = \lambda$ produces λ. The derivation

$$S \Rightarrow FBA$$
$$\Rightarrow F0EB0A$$
$$\Rightarrow F0XBY0A$$
$$\overset{*}{\Rightarrow} FX0B0YA$$
$$\overset{*}{\Rightarrow} F0W0B0Z0A$$
$$\overset{*}{\Rightarrow} F00WBZ00A$$
$$\Rightarrow F00EB00A$$
$$\Rightarrow F00000A$$
$$\overset{*}{\Rightarrow} 00000$$

produces five 0s. Notice how X and Y, and also W and Z, march back and forth across the strings of 0s, adding one more 0 on each side. This activity is highly reminiscent of a Turing machine read–write head sweeping back and forth across its tape and enlarging the printed portion.   ●

REMINDER: The erasing convention says how to do erasing IF erasing is to be done. It does not require the production $S \rightarrow \lambda$ in every grammar.

The preceding grammar allows the erasing productions $F \rightarrow \lambda$ and $A \rightarrow \lambda$. To generate any language containing λ, we have to be able to erase somewhere. In the following grammar types, we shall limit erasing, if it occurs at all, to a single production of the form $S \rightarrow \lambda$, where S is the start symbol; if this production occurs, we shall not allow S to appear on the right-hand side of any other productions. This restriction allows us to crank out λ from S as a special case and then get on with other derivations, none of which allow any erasing. Let's call this the **erasing convention**. The following definition defines three special types of grammars by further restricting the productions allowed.

> **Definitions: Context-Sensitive, Context-Free, and Regular Grammars; Chomsky Hierarchy**
>
> A grammar G is **context-sensitive (type 1)** if it obeys the erasing convention and if, for every production $\alpha \rightarrow \beta$ (except $S \rightarrow \lambda$), the word β is at least as long as the word α. A grammar G is **context-free (type 2)** if it obeys the erasing convention and for every production $\alpha \rightarrow \beta$, α is a single nonterminal. A grammar G is **regular (type 3)** if it obeys the erasing convention and for every production $\alpha \rightarrow \beta$ (except $S \rightarrow \lambda$), α is a single nonterminal and β is of the form t or tW, where t is a terminal symbol and W is a nonterminal symbol. This hierarchy of grammars, from type 0 to type 3, is called the **Chomsky hierarchy**.

In a context-free grammar, a single, nonterminal symbol on the left of a production can be replaced wherever it appears by the right side of the production. In a context-sensitive grammar, a given nonterminal symbol can perhaps be replaced only if it is part of a particular string, or context—hence the names *context-free* and *context-sensitive*. It is clear that any regular grammar is also context-free, and any context-free grammar is also context-sensitive. The grammar of Example 33 is regular (the two productions have the single nonterminal S on the left, and on the right either 1— a terminal—or $0S$—a terminal followed by a nonterminal). Both grammars of Example 35 are context-free but not regular (again the single nonterminal S appears on the left of all productions, but the right sides consist of three symbols or two non-terminals or two terminals, respectively). The grammar of Example 36 is context-sensitive but not context-free (the productions do not shrink any strings, but some left sides have multiple symbols). The grammars of Example 37 and Example 38 are context-free but not regular (for example, the first three productions of Example 38 violate the requirement for a regular grammar). Finally, the grammar of Example 40 is a type 0 grammar, but it is not context-sensitive (for example, the production $EB \rightarrow 0$ is a "shrinking" production; also, the erasing convention is violated).

Definition: Language Types
A language is **type 0** (**context-sensitive**, **context-free**, or **regular**) if it can be generated by a type 0 (context-sensitive, context-free, or regular) grammar.

Because of the relationships among the four grammar types, we can classify languages as shown in Figure 8.17. Thus, any regular language is also context-free because any regular grammar is also a context-free grammar, and so on. However, although it turns out to be true, we do not know from what we have done that these sets are properly contained in one another. For example, the language L described in Example 40 was generated in that example by a grammar that was type 0 but not context-sensitive, but that does not imply that L itself falls into that category. Different grammars can generate the same language.

Formal language hierarchy

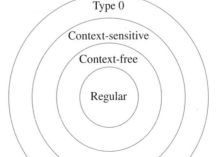

Figure 8.17

> **Definition: Equivalent Grammars**
> Two grammars are **equivalent** if they generate the same language.

EXAMPLE 41

Example 40 gave a grammar G to generate a language L, which is λ together with all odd-length strings of 0s of length at least 3. We shall now give three more grammars equivalent to G. (See if you agree that each of these grammars also generates L.)

$G_1 = (V, V_T, S, P)$ where $V = \{0, A, B, S\}$, $V_T = \{0\}$, and the productions are

$$S \to \lambda \qquad AB \to 00$$
$$S \to ABA \qquad 0A \to 000A$$
$$A \to 0$$

G_1 is context-sensitive but not context-free.

$G_2 = (V, V_T, S, P)$ where $V = \{0, A, S\}$, $V_T = \{0\}$, and the productions are

$$S \to \lambda \qquad A \to 00A$$
$$S \to 00A \qquad A \to 0$$

G_2 is context-free but not regular.

$G_3 = (V, V_T, S, P)$ where $V = \{0, A, B, C, S\}$, $V_T = \{0\}$, and the productions are

$$S \to \lambda \qquad B \to 0$$
$$S \to 0A \qquad B \to 0C$$
$$A \to 0B \qquad C \to 0B$$

G_3 is regular.

Thus, L is a regular language.   ●

PRACTICE 55   Give the derivation of 00000 in G_1, G_2, and G_3.   ●

Formal Languages and Computational Devices

The language L of Example 40 can be described by the regular expression $\lambda \vee (000)(00)^*$, so L is a regular set. From Example 41, L is also a regular language. It is not coincidental that a regular set turned out to be a regular language. It can be shown that for any finite-state machine, the set it recognizes is a regular language. It can also be shown that for any regular language, there is a finite-state machine that recognizes exactly that language. (In the proofs of these results, the productions of a regular grammar correspond to the state transitions of a finite-state machine.) Hence those sets recognized by finite-state machines—the regular sets—correspond to regular languages. Therefore the class of sets recognized by a computational device of limited capacity coincides with the most-restricted class of languages.

On the other end of the spectrum, the most general computational device is the Turing machine and the most general language is a type 0 language. As it happens, the sets recognized by Turing machines correspond to type 0 languages.

There are computational devices with capabilities midway between those of finite-state machines and those of Turing machines. These devices recognize exactly the context-free languages and the context-sensitive languages, respectively.

The type of device that recognizes the context-free languages is called a **pushdown automaton**, or **pda**. A pda consists of a finite-state unit that reads input from a tape and controls activity in a stack. Symbols from some alphabet can be pushed onto or popped off of the top of the stack. The finite-state unit in a pda, as a function of the input symbol read, the present state, and the top symbol on the stack, has a finite number of possible next moves. The moves are of the following types:

1. Go to a new state, pop the top symbol off the stack, and read the next input symbol.
2. Go to a new state, pop the top symbol off the stack, push a finite number of symbols onto the stack, and read the next input symbol.
3. Ignore the input symbol being read, manipulate the stack as above, but do not read the next input symbol.

A pda has a choice of next moves, and it recognizes the set of all inputs for which some sequence of moves exists that causes it to empty its stack. It can be shown that any set recognized by a pda is a context-free language, and conversely.

The type of device that recognizes the context-sensitive languages is called a **linear bounded automaton**, or **lba**. An lba is a Turing machine whose read–write head is restricted to that portion of the tape containing the original input; in addition, at each step it has a choice of possible next moves. An lba recognizes the set of all inputs for which some sequence of moves exists that causes it to halt in a final state. Any set recognized by an lba can be shown to be a context-sensitive language, and conversely.

Figure 8.18 shows the relationship between the hierarchy of languages and the hierarchy of computational devices.

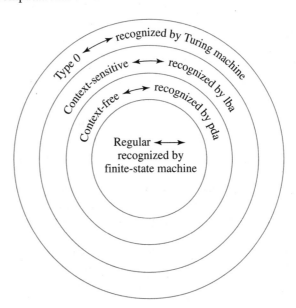

Figure 8.18

Context-Free Grammars

Context-free grammars are important for three reasons. Context-free grammars seem to be the easiest to work with, since they allow replacing only one symbol at a time. Furthermore, many programming languages are defined such that sections of syntax, if not the whole language, can be described by context-free grammars. Finally, a derivation in a context-free grammar has a nice graphical representation called a **parse tree**.

EXAMPLE 42   The grammar of Example 38 is context-free. The word $d2q$ can be derived as follows: **identifier** $\Rightarrow$ **identifier letter** $\Rightarrow$ **identifier digit letter** $\Rightarrow$ **letter digit letter** $\Rightarrow$ d **digit letter** $\Rightarrow$ $d2$ **letter** $\Rightarrow$ $d2q$. We can represent this derivation as a tree with the start symbol for the root. When a production is applied to a node, that node is replaced at the next lower level of the tree by the symbols in the right-hand side of the production used. A tree for the derivation appears in Figure 8.19.

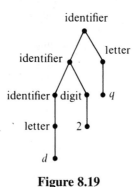

Figure 8.19

PRACTICE 56   Draw a parse tree for the word $m34s$ in the grammar of Example 38.

Suppose that a context-free grammar G describes a programming language. The programmer uses the rules of G to generate legitimate strings of symbols, that is, words in the language. Here we may think of a word as corresponding to a program instruction. Thus, a word consists of various subwords, for example, identifiers, operators, and key words for the language. The program instructions are fed into the compiler for the language so that the program can be translated into machine language code for the computer. The compiler must decide whether each program instruction is legitimate in the language. In a sense, the compiler has to undo the process that the programmer used to construct the statement; that is, the compiler must start with the statement and decompose it to see whether it follows the rules of the language. This really entails two questions: Are the subwords themselves legitimate strings? Is the program instruction a legitimate way of grouping the subwords together?

Usually the set of legitimate subwords of a language can be described by a regular expression, and then a finite-state machine can be used to detect the subwords; the lexical analysis or scanner portion of the compiler handles this phase of compilation. (See Exercise 22 of Section 8.2 for the lexical analysis of ScrubOak.) If all goes well, the scanner then passes the program instruction, in the form of a string of legitimate subwords, to the syntax analyzer. The syntax analyzer determines whether the string is correct by trying to parse it (construct its parse tree).

Various parsing techniques, which we won't go into, have been devised. Given a string to be tested, one approach is to construct a tree by beginning with the start symbol, applying productions (keeping an eye on the "goal," i.e., the given string) and ending with the goal string. This procedure is called **top-down parsing**. The alternative is to begin with the string, see what productions were used to create it, apply productions "backwards," and end with the start symbol. This process is called **bottom-up parsing**. The trick to either approach is to decide exactly which productions should be used.

EXAMPLE 43

Consider the context-free grammar G given by $G = (V, V_T, S, P)$ where $V = \{a, b, c, A, B, C, S\}$, $V_T = \{a, b, c\}$, and the productions are

$$S \rightarrow B \qquad B \rightarrow C \qquad A \rightarrow abc$$
$$S \rightarrow A \qquad B \rightarrow ab \qquad C \rightarrow c$$

Suppose we want to test the string abc. A derivation for abc is $S \Rightarrow A \Rightarrow abc$. If we try a top-down parse, we might begin with

S

B

Then we have to detect that this will not work and try something else. If we try a bottom-up parse, we might begin with

BC

abc

Then we have to detect that this will not work and try something else. Parsing techniques automate this process and attempt to minimize the amount of false starts and backtracking required.   ●

Notice the distinction between *generating* members of a set, which the programmer does, and *deciding* membership in a set, which the compiler does. Since we ask the compiler to decide membership in a set, a decision algorithm must exist for the set. It turns out that decision algorithms do exist for context-free languages, another point in their favor.

Section 8.4 Review

Techniques

- Describe $L(G)$ for a given grammar G.
- Define a grammar to generate a given language L.
- Construct parse trees in a context-free grammar.

Main Ideas

A grammar G is a generating mechanism for its language $L(G)$.

Formal languages were developed in an attempt to describe correct syntax for natural languages; although this attempt has largely failed because of the complexity of natural languages, it has been quite successful for high-level programming languages.

Special classes of grammars are defined by restricting the allowable productions.

The various types of formal languages correspond to the sets recognized by various automata; in particular, (1) regular languages are the sets recognized by finite-state machines, (2) context-free languages are the sets recognized by pushdown automata, (3) context-sensitive languages are the sets recognized by linear bounded automata, and (4) type 0 languages are the sets recognized by Turing machines.

Derivations in context-free grammars can be illustrated by parse trees.

A compiler for a context-free programming language checks correct syntax by parsing.

Exercises 8.4

★1. Describe $L(G)$ for each of the following grammars G.

    a. $G = (V, V_T, S, P)$ where $V = \{a, A, B, C, S\}$, $V_T = \{a\}$, and P consists of

$$S \to A \qquad B \to A$$
$$A \to BC \qquad aC \to \lambda$$
$$A \to a$$

    b. $G = (V, V_T, S, P)$ where $V = \{0, 1, A, B, S\}$, $V_T = \{0, 1\}$, and P consists of

$$S \to 0A \qquad A \to 1BB \qquad B \to 01$$
$$S \to 1A \qquad \qquad \qquad B \to 11$$

    c. $G = (V, V_T, S, P)$ where $V \{0, 1, A, B, S\}$, $V_T = \{0, 1\}$, and P consists of

$$S \to 0 \qquad A \to 1B \qquad B \to 0A$$
$$S \to 0A \qquad \qquad \qquad B \to 0$$

    d. $G = (V, V_T, S, P)$ where $V \{0, 1, A, S\}$, $V_T = \{0, 1\}$, and P consists of

$$S \to 0S \qquad A \to 1A$$
$$S \to 11A \qquad A \to 1$$

2. **a.** Which of the grammars of Exercise 1 are regular? Which are context-free?
    b. Find regular grammars to generate each of the languages of Exercise 1.

★3. Describe $L(G)$ for the grammar $G = (V, V_T, S, P)$ where $V = \{a, b, A, B, S\}$, $V_T = \{a, b\}$, and P consists of

$$S \to AB \qquad AB \to AAB$$
$$A \to a \qquad AB \to ABB$$
$$B \to b$$

What type of grammar is G? Find a regular grammar G' that generates $L(G)$.

4. Write the productions of the following grammars in BNF:
   a. G_3 in Example 41
   b. G in Exercise 1(b)
   c. G in Exercise 1(c)
   d. G in Exercise 1(d)

5. A grammar G is described in BNF as

$$<S> ::= 01 \,|\, 0<S> \,|\, <S>1$$

   a. Find a regular expression for $L(G)$.
   b. Find a regular grammar for $L(G)$.

★6. A grammar G is described in BNF as

$$<S> ::= 1 \,|\, 1<S> \,|\, <S>00$$

   a. Find a regular expression for $L(G)$.
   b. Find a regular grammar for $L(G)$.

7. Find a grammar that generates the set of all strings of well-balanced parentheses.

8. A word w in V^* is a palindrome if $w = w^R$, where w^R is the reverse of the string w. A language L is a palindrome language if L consists entirely of palindromes.
   a. Find a grammar that generates the set of all palindromes over the alphabet $\{a, b\}$.
   b. Let L be a palindrome language. Prove that $L^R = \{w^R \,|\, w \in L\}$ is a palindrome language.
   c. Let w be a palindrome. Prove that the language described by the regular expression w^* is a palindrome language.

★9. Find a grammar that generates the language $L = (0 \lor 1)^*01$.

10. Find a context-free grammar that generates the language $L = \{0^n1^n \,|\, n \geq 0\}$.

11. Find a grammar that generates the language $L = \{0^{2^i} \,|\, i \geq 0\}$.

12. Find a context-free grammar that generates the language L where L consists of the set of all nonempty strings of 0s and 1s with an equal number of 0s and 1s.

13. Find a context-free grammar that generates the language L where L consists of the set of all nonempty strings of 0s and 1s with twice as many 0s as 1s.

★14. Find a context-free grammar that generates the language $L = \{ww^R \,|\, w \in \{0, 1\}^*$ and w^R is the reverse of the string $w\}$.

15. Find a grammar that generates the language $L = \{ww \,|\, w \in \{0, 1\}^*\}$.

16. Find a grammar that generates the language $L = \{a^{n^2} \mid n \geq 1\}$. (By Exercise 20, L is not a context-free language, so your grammar cannot be too simple.)

17. Draw parse trees for the following words:

★**a.** 111111 in the grammar G of Example 35
   b. 111111 in the grammar G' of Example 35
   c. 011101 in the grammar of Exercise 1(b)
   d. 00111111 in the grammar of Exercise 1(d)

★18. Consider the context-free grammar $G = (V, V_T, S, P)$ where $V = \{0, 1, A, S\}$, $V_T = \{0, 1\}$, and P consists of

$$S \rightarrow A1A$$
$$A \rightarrow 0$$
$$A \rightarrow A1A$$

Draw two distinct parse trees for the word 01010 in G. A grammar in which a word has more than one parse tree is *ambiguous*.

19. Show that for any context-free grammar G, there exists a context-free grammar G' in which for every production $\alpha \rightarrow \beta$, β is a longer string than α, $L(G') \subseteq L(G)$, and $L(G) - L(G')$ is a finite set.

20. The following is the *pumping lemma* for context-free languages. Let L be any context-free language. Then there exists some constant k such that for any word w in L with $|w| \geq k$, w can be written as the string $w_1 w_2 w_3 w_4 w_5$ with $|w_2 w_3 w_4| \leq k$ and $|w_2 w_4| \geq 1$. Furthermore, the word $w_1 w_2^i w_3 w_4^i w_5 \in L$ for each $i \geq 0$.

   a. Use the pumping lemma to show that $L = \{a^n b^n c^n \mid n \geq 1\}$ is not context free.
   b. Use the pumping lemma to show that $L = \{a^{n^2} \mid n \geq 1\}$ is not context free.

Chapter 8 Review

Terminology

Self-Test

Answer the following true–false questions.

Section 8.1

1. A binary operation is associative if the order of the elements being operated on does not matter.

2. The identity i in a group $[G, \cdot]$ has the property that $x^{-1} \cdot i = i \cdot x^{-1} = x^{-1}$ for all x in G.

3. Every group is also a monoid.

4. A group of order 10 cannot have a subgroup of order 6.

5. If $[S, \cdot]$ and $[T, +]$ are two groups, then a function $f: S \to T$ for which $f(x \cdot y) = f(x) + f(y)$ is an isomorphism.

Section 8.2

6. The next state of a finite-state machine is determined by its present state and the present input symbol.

7. The set of all binary strings ending in two 0s is regular.

8. A finite-state machine cannot get to a state from which there is no exit.

9. According to Kleene's theorem, a set that cannot be described by a regular expression cannot be recognized by a finite-state machine.

10. In a finite-state machine, k-equivalent states are also $(k + 1)$-equivalent.

Section 8.3

11. A Turing machine halts if and only if it enters a final state.

12. A Turing machine that computes the function $f(n) = n + 1$, given input n, will halt with $(n + 1)$ 1s on its tape.

13. Church's thesis says that the halting problem is unsolvable.

14. The halting problem says that, given a Turing machine and its input, there is no algorithm to decide whether the Turing machine halts when run on that input.

15. A set in P is recognizable by a Turing machine in no more than a polynomial number of steps.

Section 8.4

16. The language generated by a type 0 grammar G is the set of all strings of terminals generated from the start symbol by applying G's productions a finite number of times.

17. Beginning at the start symbol and applying the productions of a grammar G eventually leads to a string of terminals.

18. A language generated by a grammar that is context-sensitive but not context-free is a context-sensitive but not context-free language.

19. Any regular set is a regular language.

20. A parse tree will have as many leaves as terminal symbols in the word being derived.

On the Computer

For Exercises 1–9, write a computer program that produces the desired output from the given input.

1. *Input*: Two words from an alphabet A
   Output: Their concatenation

2. *Input*: Positive integer n and finite alphabet A
Output: All words over A of length $\leq n$

3. *Input*: Positive integer n
Output: Tables for addition and multiplication modulo n

4. *Input*: Positive integer n
Output: The $n!$ elements of S_n expressed both in array form and in cycle notation, the group table for $[S_n, \circ]$, and the group table for $[A_n, \circ]$

5. *Input*: $n \times n$ array, $n \leq 10$, that purports to represent a binary operation on the finite set of integers from 1 to n
Output: Determination of whether the set under this operation is a commutative group

6. *Input*: Two $n \times n$ arrays, $n \leq 10$, that represent two groups and an array that represents a function from the first group to the second
Output: Determination of whether the function is an isomorphism

7. *Input*: Positive integer n, $n \leq 50$, representing the number of states of a finite state machine with input alphabet = output alphabet = $\{0, 1\}$ and an $n \times 3$ array representing the state table description of such a machine
Output: List of any states unreachable from the start state s_0

8. *Input*: Positive integer n, $n \leq 50$, representing the number of states of a finite state machine with input alphabet = output alphabet = $\{0, 1\}$ and an $n \times 3$ array representing the state table description of such a machine
Output: $m \times 3$ array representing the state table of a minimized version of M

9. *Input*: Set of terminals in a grammar and a description of the productions in a grammar; ability for the user to set a maximum number of steps for any derivation
Output: List of words in the language that can be derived within that maximum

10. Write a finite-state machine simulator. That is, given

a positive integer n, $n \leq 50$, representing the number of states of a finite state machine with input alphabet = output alphabet = $\{0, 1\}$

an $n \times 3$ array representing the state table description of such a machine

your program should request input strings and write the corresponding output strings as long as the user wishes.

11. Write a Turing machine simulator. That is, given a set of quintuples describing a Turing machine, your program should request the initial tape contents and write out a sequence of successive tape configurations. Assume that there are at most 100 quintuples and that the number of cells used on the tape is at most 70, and allow the user to set a maximum number of steps in case the computation does not halt before then.

Summation Notation

Summation notation is a shorthand way of writing certain expressions that are sums of terms. As an example, consider the sum of the integers from 1 to 5:

$$1 + 2 + 3 + 4 + 5$$

This can be thought of in the following way: Suppose we have some quantity i that initially has the value 1 and then takes on successive values of 2, 3, 4, and 5. The expression above is the sum of i at all its different values. The summation notation is

$$\sum_{i=1}^{5} i$$

The uppercase Greek letter sigma, Σ, denotes summation. Here the number 1 is the *lower limit of summation*, and the number 5 is the *upper limit of summation*. The variable i is called the *index of summation*. The index of summation is initially set equal to the lower limit and then keeps increasing its value until it reaches the upper limit. All the values that the index of summation takes on are added together. Thus

$$\sum_{i=1}^{5} i = 1 + 2 + 3 + 4 + 5 = 15$$

Similarly

$$\sum_{i=1}^{3} i = 1 + 2 + 3 = 6$$

and

$$\sum_{i=4}^{8} i = 4 + 5 + 6 + 7 + 8 = 30$$

In these examples, the expression after the summation sign is just i, the index of summation. However, what appears after the summation sign can be any expression, and the successive values of the index are simply substituted into the expression. Thus

$$\sum_{i=1}^{5} i^2 = 1^2 + 2^2 + 3^2 + 4^2 + 5^2 = 55$$

A way to symbolize summation in general is

$$\sum_{i=p}^{q} a_i$$

Here the lower limit, upper limit, and expression behind the summation are not specifically given but merely symbolized. The notation a_i is a reminder that the expression will be evaluated at the different values i takes on in going from the lower to the upper limit.

There are three special cases to consider:

1. $\sum_{i=p}^{q} 0 = 0$

Here the expression behind the summation is the constant 0, which has the value 0 no matter what the value of the index of summation. The sum of any number of 0s is 0.

2. $\sum_{i=1}^{n} 1 = n$

Here again the expression behind the summation is a constant, and the summation says to add n copies of 1, which results in n.

3. $\sum_{i=1}^{0} a_i = 0$

Here the upper limit is smaller than the lower limit; the usual interpretation of summation does not apply, but by convention the summation is assigned the value 0.

The index of summation is a *dummy variable*, meaning that it merely acts as a placeholder and that a different variable could be used without changing the value of the summation. Thus

$$\sum_{i=1}^{3} i = \sum_{j=1}^{3} j = 6$$

It may be convenient to change the limits on a summation, which is legitimate as long as the final value of the summation remains the same. For example,

$$\sum_{i=1}^{3} i = \sum_{i=0}^{2} (i + 1)$$

since both have the value

$$1 + 2 + 3 = 6$$

Finally, the following three rules hold, as we shall see shortly.

Rules of Summation

1. $\sum_{i=p}^{q} (a_i + b_i) = \sum_{i=p}^{q} a_i + \sum_{i=p}^{q} b_i$

2. $\sum_{i=p}^{q} (a_i - b_i) = \sum_{i=p}^{q} a_i - \sum_{i=p}^{q} b_i$

3. $\sum_{i=p}^{q} ca_i = c \sum_{i=p}^{q} a_i$ where c is a constant

To prove rule 1, note that

$$a_p + b_p + a_{p+1} + b_{p+1} + \cdots + a_q + b_q$$
$$= a_p + a_{p+1} + \cdots + a_q + b_p + b_{p+1} + \cdots + b_q$$

because of the commutative property of addition. The proof of rule 2 is similar.
To prove rule 3, note that

$$ca_p + ca_{p+1} + \cdots + ca_q = c(a_p + a_{p+1} + \cdots + a_q)$$

because of the distributive property. This rule allows a constant to be "moved through" a summation.

Sometimes a summation can be represented by an even shorter expression that does not involve adding separate terms. For example, according to Exercise 7 of Section 2.2,

$$\sum_{i=1}^{n} i^2 = \frac{n(n + 1)(2n + 1)}{6} \tag{1}$$

so that the value of $\displaystyle\sum_{i=1}^{5} i^2$ can be found by substituting the upper limit, 5, into the right side of (1), giving

$$\frac{5(5 + 1)(2 * 5 + 1)}{6} = 55$$

as before. Section 2.2 and its exercises give a number of other "closed form" expressions for certain summations, all of them provable by mathematical induction.

The Logarithm Function

The logarithm function is closely related to the *exponential function*

$$y = b^x$$

where b, the *base*, is a constant greater than 1. (Actually b can be any positive number, but the only interesting cases occur when $b > 1$.) Recall the following rules of exponents:

1. $b^n b^m = b^{n+m}$ (when you multiply, you add exponents)
2. $b^n / b^m = b^{n-m}$ (when you divide, you subtract exponents)
3. $(b^n)^m = b^{nm}$ (when you raise a power to a power, you multiply exponents)

If we select a specific base, $b = 2$ for example, we can plot $y = 2^x$ for various values of x and fill in the remaining values, getting the graph

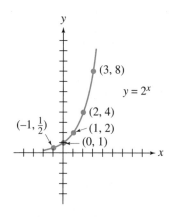

In this function, x can take on any real value, and y will always be positive. Another way to say this is that the domain of the function is the set $\mathbb{R}$ of real numbers, and the range is the set $\mathbb{R}^+$ of positive real numbers.

A related function (in fact, the inverse function) is the *logarithm function*, defined by

$$y = \log_b x \qquad \text{meaning} \qquad b^y = x$$

Therefore $\log_2 16 = 4$, for example, because $2^4 = 16$. These two equations are the logarithmic form and exponential form of the same fact. Similarly, $\log_2 8 = 3$ and $\log_2 2 = 1$.

A graph of $y = \log_2 x$ follows.

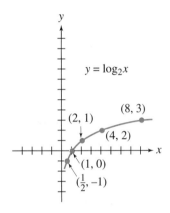

Because the logarithm function $y = \log_b x$ is the inverse of the exponential function, its domain (the values x can take on) is the set $\mathbb{R}^+$ of positive real numbers and the range (the values y can take on) is the set $\mathbb{R}$ of real numbers. The logarithm function for any base $b > 1$ has a domain, range, and shape similar to the case for $b = 2$.

Certain properties about the logarithm function are true either because of its definition or because of corresponding properties about the exponential function, We'll list all the properties of the logarithm function, and then prove them.

Properties of the Logarithm Function $y = \log_b x$

1. If $p < q$ then $\log_b p < \log_b q$ (the log function is strictly increasing)
2. If $\log_b p = \log_b q$ then $p = q$ (the log function is one-to-one)
3. $\log_b 1 = 0$
4. $\log_b b = 1$
5. $\log_b (b^p) = p$
6. $b^{\log_b p} = p$
7. $\log_b (pq) = \log_b p + \log_b q$ (the log of a product equals the sum of the logs)
8. $\log_b (p/q) = \log_b p - \log_b q$ (the log of a quotient equals the difference of the logs)

9. $\log_b (p^q) = q(\log_b p)$ (the log of something to a power equals the power times the log)

10. $\log_a p = \log_b p/\log_b a$ (change-of-base formula)

In proving the properties of the logarithm function, we first note that many of the properties involve the quantities $\log_b p$ and $\log_b q$. Let us name these quantities r and s, respectively, so that

$$\log_b p = r \quad \text{and} \quad \log_b q = s$$

This in turn means that

$$b^r = p \quad \text{and} \quad b^s = q$$

1. Because $b > 1$, the larger the power to which b is raised, the larger the result. Thus if $p < q$, then $b^r < b^s$, so $r < s$ and therefore $\log_b p < \log_b q$.

2. If $\log_b p = \log_b q$, then $r = s$, so $b^r = b^s$ and $p = q$.

3. $\log_b 1 = 0$ because $b^0 = 1$.

4. $\log_b b = 1$ because $b^1 = b$.

5. $\log_b (b^p) = p$ because (translating this equation to its exponential form) $b^p = b^p$.

6. $b^{\log_b p} = p$ because (translating this equation to its logarithmic form) $\log_b p = \log_b p$.

7. $\log_b (pq) = \log_b p + \log_b q = r + s$ because it is true that $b^{r+s} = b^r b^s = pq$, which is the exponential form of the equation we are trying to prove.

8. $\log_b (p/q) = \log_b p - \log_b q = r - s$ because it is true that $b^{r-s} = b^r/b^s = p/q$, which is the exponential form of the equation we are trying to prove.

9. $\log_b (p^q) = q(\log_b p) = qr$ because it is true that $b^{qr} = (b^r)^q = p^q$, which is the exponential form of the equation we are trying to prove.

10. $\log_a p = \log_b p/\log_b a$.

Let $\log_a p = w$. Then $a^w = p$. Now take the logarithm to the base b of both sides of this equation:

$$\log_b(a^w) = w(\log_b a) = \log_b p$$

or

$$w = \log_b p/\log_b a$$

which is the desired result.

The three most useful bases for logarithms are

$b = 10$ (common logarithm)
$b = e, e \sim 2.7183$ (natural logarithm)
$b = 2$ (what we use throughout this book)

Common logarithms were used as computational aids before calculators and computers became commonplace. Property 7 of the logarithm function says that to multiply two numbers, one can take the logarithm of each number, add the results, and then find the number with that logarithm value. Addition was easier than multiplication, and tables of common logarithms allowed one to look up a number and find its logarithm, or vice versa.

Natural logarithms are useful in calculus and are often written "ln p" rather than "$\log_e p$." Base 2 logarithms are sometimes denoted by "lg p" rather than "$\log_2 p$." In this book, all logarithms are base 2, so we use log p to denote $\log_2 p$.

A final inequality involving base 2 logarithms (used in Section 2.5) is

$$1 + \log n < n \text{ for } n \geq 3$$

To prove this, note that

$$n < 2^{n-1} \text{ for } n \geq 3$$

so by property 1 of logarithms,

$$\log n < \log 2^{n-1}$$

By property 5 of logarithms, $\log 2^{n-1} = n - 1$. Therefore

$$\log n < n - 1$$

or

$$1 + \log n < n \text{ for } n \geq 3$$

Answers to Practice Problems

Note to student: Finish all parts of a practice problem before turning to the answers.

Chapter 1

1. False, false, false

2.

A	B	$A \vee B$
T	T	T
T	F	T
F	T	T
F	F	F

3.

A	B	$A \rightarrow B$
T	T	T
T	F	F
F	T	T
F	F	T

4.

A	A'
T	F
F	T

5. a. Antecedent: The rain continues
   Consequent: The river will flood
   b. Antecedent: The central switch goes down
   Consequent: Network failure
   c. Antecedent: The avocados are ripe
   Consequent: They are dark and soft
   d. Antecedent: A healthy cat
   Consequent: A good diet

6. Answer (d). This is negation of $A \wedge B$, the same as the negation of "Peter is tall and thin."

7. a.

A	B	$A \to B$	$B \to A$	$(A \to B) \leftrightarrow (B \to A)$
T	T	T	T	T
T	F	F	T	F
F	T	T	F	F
F	F	T	T	T

b.

A	B	A'	B'	$A \vee A'$	$B \wedge B'$	$(A \vee A') \to (B \wedge B')$
T	T	F	F	T	F	F
T	F	F	T	T	F	F
F	T	T	F	T	F	F
F	F	T	T	T	F	F

c.

A	B	C	B'	$A \wedge B'$	C'	$(A \wedge B') \to C'$	$[(A \wedge B') \to C']'$
T	T	T	F	F	F	T	F
T	T	F	F	F	T	T	F
T	F	T	T	T	F	F	T
T	F	F	T	T	T	T	F
F	T	T	F	F	F	T	F
F	T	F	F	F	T	T	F
F	F	T	T	F	F	T	F
F	F	F	T	F	T	T	F

d.

A	B	A'	B'	$A \to B$	$B' \to A'$	$(A \to B) \leftrightarrow (B' \to A')$
T	T	F	F	T	T	T
T	F	F	T	F	F	T
F	T	T	F	T	T	T
F	F	T	T	T	T	T

8.

A	1	A'	$A \vee A'$	$A \vee A' \leftrightarrow 1$
T	T	F	T	T
F	T	T	T	T

9. To prove $(P \rightarrow Q) \leftrightarrow (P' \vee Q)$, just construct a truth table:

P	Q	$P \rightarrow Q$	P'	$P' \vee Q$	$(P \rightarrow Q) \leftrightarrow (P' \vee Q)$
T	T	T	F	T	T
T	F	F	F	F	T
F	F	T	T	T	T
F	F	T	T	T	T

10. $(A \wedge B')'$ 1, 2, mt

11. $[(A \vee B') \rightarrow C] \wedge (C \rightarrow D) \wedge A \rightarrow D$

1. $(A \vee B') \rightarrow C$ hyp
2. $C \rightarrow D$ hyp
3. A hyp
4. $A \vee B'$ 3, add
5. C 1, 4, mp
6. D 2, 5, mp

12. $(A \rightarrow B) \wedge (B \rightarrow C) \rightarrow (A \rightarrow C)$

1. $A \rightarrow B$ hyp
2. $B \rightarrow C$ hyp
3. A hyp
4. B 1, 3, mp
5. C 2, 4, mp

13. $(A \rightarrow B) \wedge (C' \vee A) \wedge C \rightarrow B$

1. $A \rightarrow B$ hyp
2. $C' \vee A$ hyp
3. C hyp
4. $C \rightarrow A$ 2, imp
5. $C \rightarrow B$ 1, 4, hs
6. B 3, 5, mp

14. The argument is $(S \rightarrow R) \wedge (S' \rightarrow B) \rightarrow (R' \rightarrow B)$. A proof sequence is

1. $S \rightarrow R$ hyp
2. $S' \rightarrow B$ hyp
3. R' hyp
4. S' 1, 3, mt
5. B 2, 4, mp

15. a. True (all buttercups are yellow)
  b. False (not true that all flowers are yellow)
  c. True (all flowers are plants)
  d. False (zero is neither positive nor negative)

16. For example:

a. The domain is the collection of licensed drivers in the United States; $P(x)$ is the property that x is older than 14.

b. The domain is the collection of all fish; $P(x)$ is the property that x weighs more than 3 pounds.

c. No; if all objects in the domain have property P, then (since the domain must contain objects) there is an object in the domain with property P.

d. The domain is all the people who live in Boston; $P(x)$ is the property that x is a male. (Not every person who lives in Boston is a male, but someone is.)

17. Let $x = 1$; then x is positive and any integer less than x is ≤ 0, so the truth value of the statement is true. For the second interpretation, let $A(x)$ be "x is even," $B(x, y)$ be "$x < y$," and $C(y)$ be "y is odd"; the statement is false because no even integer has the property that all larger integers are odd.

18. a. $(\forall x)[S(x) \to I(x)]$

b. $(\exists x)[I(x) \wedge S(x) \wedge M(x)]$

c. $(\forall x)(M(x) \to S(x) \wedge [I(x)]')$

d. $(\forall x)(M(x) \to S(x) \wedge I(x))$

19. Answer (d). If $L(x, y, t)$ means "x loves y at time t," the original statement is

$$(\forall x)(\exists y)(\exists t)L(x, y, t)$$

and the negation is

$$[(\forall x)(\exists y)(\exists t)L(x, y, t)]' \leftrightarrow (\exists x)[(\exists y)(\exists t)L(x, y, t)]'$$
$$\leftrightarrow (\exists x)(\forall y)[(\exists t)L(x, y, t)]'$$
$$\leftrightarrow (\exists x)(\forall y)(\forall t)[L(x, y, t)]'$$

or "There is some person who, for all persons and all times, dislikes those persons at those times," or "Somebody hates everybody all the time." Answers (a) and (c) can be eliminated because they begin with a universal quantifier instead of an existential quantifier; answer (b) is wrong because "loves" has not been negated.

20. Invalid. In the interpretation where the domain consists of the integers, $P(x)$ is "x is odd" and $Q(x)$ is "x is even," the antecedent is true (every integer is even or odd), but the consequent is false (it is not the case that every integer is even or that every integer is odd).

21. $(\forall x)[P(x) \to R(x)] \wedge [R(y)]' \to [P(y)]'$

1. $(\forall x)[P(x) \to R(x)]$	hyp
2. $[R(y)]'$	hyp
3. $P(y) \to R(y)$	1, ui
4. $[P(y)]'$	2, 3, mt

22. $(\forall x)[P(x) \wedge Q(x)] \to (\forall x)[Q(x) \wedge P(x)]$

1. $(\forall x)[P(x) \wedge Q(x)]$	hyp
2. $P(x) \wedge Q(x)$	1, ui
3. $Q(x) \wedge P(x)$	2, comm
4. $(\forall x)[Q(x) \wedge P(x)]$	3, ug

23. $(\forall y)[P(x) \to Q(x, y)] \to [P(x) \to (\forall y)Q(x, y)]$

1. $(\forall y)[P(x) \to Q(x, y)]$	hyp
2. $P(x)$	hyp
3. $P(x) \to Q(x, y)$	1, ui
4. $Q(x, y)$	2, 3, mp
5. $(\forall y)Q(x, y)$	4, ug

24. $(\forall x)[(B(x) \lor C(x)) \to A(x)] \to (\forall x)[B(x) \to A(x)]$

1. $(\forall x)[(B(x) \lor C(x)) \to A(x)]$	hyp
2. $(B(x) \lor C(x)) \to A(x)$	1, ui
3. $B(x)$	temporary hyp
4. $B(x) \lor C(x)$	3, add
5. $A(x)$	2, 4, mp
6. $B(x) \to A(x)$	temporary hyp discharged
7. $(\forall x)[B(x) \to A(x)]$	6, ug

25. $(\exists x)R(x) \land [(\exists x)[R(x) \land S(x)]]' \to (\exists x)[S(x)]'$
The argument is valid. If something has property R but nothing has both properties R and S, then something fails to have property S. A proof sequence is

1. $(\exists x)R(x)$	hyp
2. $[(\exists x)[R(x) \land S(x)]]'$	hyp
3. $(\forall x)[R(x) \land S(x)]'$	2, neg
4. $R(a)$	1, ei
5. $[R(a) \land S(a)]'$	3, ui
6. $[R(a)]' \lor [S(a)]'$	5, De Morgan
7. $[[R(a)]']'$	4, dn
8. $[S(a)]'$	6, 7, ds
9. $(\exists x)[S(x)]'$	8, eg

26. The argument is $(\forall x)[R(x) \to L(x)] \land (\exists x)R(x) \to (\exists x)L(x)$.

1. $(\forall x)[R(x) \to L(x)]$	hyp
2. $(\exists x)R(x)$	hyp
3. $R(a)$	2, ei
4. $R(a) \to L(a)$	1, ui
5. $L(a)$	3, 4, mp
6. $(\exists x)L(x)$	5, eg

27. deer (deer eat grass)

28. a. *predator*(x) **if** *eat*(x, y) **and** *animal*(y)
b. bear
   fish
   raccoon
   bear
   bear
   fox
   bear
   wildcat

29. Responses 7–9 result from *in-food-chain*(raccoon, *y*); responses 10 and 11 result from *in-food-chain*(fox, *y*); response 12 results from *in-food-chain*(deer, *y*).

30. $x - 2 = y$, or $x = y + 2$.

31. Working backward from the postcondition using the assignment rule,

$$\{x + 4 = 7\}$$
$$y = 4$$
$$\{x + y = 7\}$$
$$z = x + y$$
$$\{z = 7\}$$

The first assertion, $x + 4 = 7$, is equivalent to the precondition, $x = 3$. The assignment rule, applied twice, proves the program segment correct.

32. By the assignment rule

$$\{x = 4\}\ y = x - 1\ \{y = 3\}$$

is true and therefore

$$\{x = 4 \text{ and } x < 5\}\ y = x - 1\ \{y = 3\}$$

is true. Also,

$$\{x = 4 \text{ and } x \geq 5\}\ y = 7\ \{y = 3\}$$

is true because the antecedent is false. The program segment is correct by the conditional rule.

Chapter 2

1. Possible answers:

a. A whale

b. The integer 4. Four is less than 10, but it is not bigger than 5.

2. a. Show that the conjecture is true for all cases:

n	n^2	$10 + 5n$
1	1	15
2	4	20
3	9	25
4	16	30
5	25	35

b. For $n = 7$, n^2 is 49 but $10 + 5n$ is only 45.

3. Let x be divisible by 6. Then $x = 6k$ where k is an integer, and $2x = 2(6k) = 12k = 4(3k)$. Since $3k$ is an integer, $2x$ is divisible by 4.

4. a. If the river will not flood, then the rain will not continue.
 b. If there is not a network failure, then the central switch does not go down.
 c. If the avocados are not dark or not soft, then they are not ripe.
 d. If the diet is not good, the cat is not healthy.

5. a. If the river will flood, then the rain will continue.
 b. If there is network failure, then the central switch goes down.
 c. If the avocados are dark and soft, then they are ripe.
 d. If the diet is good, then the cat is healthy.

6. Let $x = 2m + 1$ and $y = 2n + 1$ where m and n are integers, and assume that xy is even. Then

$$xy = 2k \qquad \text{for some integer } k$$

or

$$(2m + 1)(2n + 1) = 2k$$

Multiplying out the left side,

$$4mn + 2m + 2n + 1 = 2k$$

Rearranging terms in the equation,

$$1 = 2k - 4mn - 2m - 2n$$

Factoring out 2 on the right side,

$$1 = 2(k - 2mn - m - n) \qquad \text{where } k - 2mn - m - n \text{ is an integer}$$

This is a contradiction since 1 is not even.

7. $P(1)$: $1 = 1(1 + 1)/2$, true
 Assume $P(k)$: $1 + 2 + \cdots + k = k(k + 1)/2$
 Show $P(k + 1)$: $1 + 2 + \cdots + (k + 1) \overset{?}{=} \dfrac{(k + 1)[(k + 1) + 1]}{2}$

$$1 + 2 + \cdots + (k + 1) = 1 + 2 + \cdots + k + (k + 1)$$
$$= \frac{k(k + 1)}{2} + (k + 1) = (k + 1)\left(\frac{k}{2} + 1\right)$$
$$= (k + 1)\left(\frac{k + 2}{2}\right) = \frac{(k + 1)[(k + 1) + 1]}{2}$$

8. The base case is $n = 2$.
 $P(2)$: $2^{2+1} < 3^2$, or $8 < 9$, true
 Assume $P(k)$: $2^{k+1} < 3^k$ and $k > 1$

Show $P(k + 1)$: $2^{k+2} \overset{?}{<} 3^{k+1}$

$$2^{k+2} = 2(2^{k+1})$$
$$< 2(3^k) \qquad \text{(by the inductive hypothesis)}$$
$$< 3(3^k) \qquad \text{(since } 2 < 3\text{)}$$
$$= 3^{k+1}$$

9. a. To verify $P(k + 1)$ in implication $2'$, we subtract 3 from $k + 1$. In order for the inductive hypothesis to hold, it must be the case that $(k + 1) - 3 \geq 8$, so $k + 1$ must be ≥ 11. Therefore implication $2'$ cannot be used to verify $P(9)$ or $P(10)$.

b. The truth of $P(k + 1)$ cannot be verified from the truth of $P(k)$.

10. $Q(0)$: $j_0 = x + i_0$ true since $j = x$, $i = 0$ before the loop is entered
Assume $Q(k)$: $j_k = x + i_k$

Show $Q(k + 1)$: $j_{k+1} \overset{?}{=} x + i_{k+1}$

$$
\begin{aligned}
j_{k+1} = j_k + 1 \qquad &\text{(by the assignment } j = j + 1) \\
= (x + i_k) + 1 \qquad &\text{(by inductive hypothesis)} \\
= x + (i_k + 1) \qquad & \\
= x + i_{k+1} \qquad &\text{(by the assignment } i = i + 1)
\end{aligned}
$$

Upon loop termination, $i = y$ and $j = x + y$.

11. 1, 4, 7, 10, 13

12. 1, 1, 2, 3, 5, 8, 13, 21

13. In proving the $k + 1$ case, the terms $F(k - 1)$ and $F(k)$ are used. If $k + 1 = 2$, then $F(k - 1)$ is undefined. Therefore in the inductive step, we must have $k + 1 \geq 3$ and the case $n = 2$ must be done separately. Put another way, the inductive step does not demonstrate the truth of the $n = 2$ case from the truth of the $n = 1$ case.

14. A, B, and C are wffs by rule 1. By rule 2, (B') is a wff, and so then is $[A \lor (B')]$ and $([A \lor (B')] \to C)$. This can be written as $(A \lor B') \to C$.

15. Every parent of an ancestor of James is an ancestor of James.

16. 1011001, 0011011, 00110111011

17. 1. 1, λ, 0, and 1 are binary palindromes.
   2. If x is a binary palindrome, so are $0x0$ and $1x1$.

18. 1. $x^1 = x$
   2. $x^n = x^{n-1}x$ for $n > 1$

19. **if** $n = 1$ **then**
        return 1
    else
        return $T(n - 1) + 3$
    end if

20. 10, 7, 8

21. $T(n) = T(n-1) + 3$

$$= [T(n-2) + 3] + 3 = T(n-2) + 2 * 3$$

$$= [T(n-3) + 3] + 2 * 3 = T(n-3) + 3 * 3$$

$$\vdots$$

In general, we guess that

$$T(n) = T(n-k) + k * 3$$

When $n - k = 1$, that is, $k = n - 1$,

$$T(n) = T(1) + (n-1) * 3 = 1 + (n-1) * 3$$

Now prove by induction that $T(n) = 1 + (n-1) * 3$.

$T(1)$: $T(1) = 1 + (1-1) * 3 = 1$, true

Assume $T(k)$: $T(k) = 1 + (k-1) * 3$

Show $T(k+1)$: $T(k+1) \overset{?}{=} 1 + k * 3$

$$T(k+1) = T(k) + 3 \qquad \text{by the recurrence relation}$$

$$= 1 + (k-1) * 3 + 3 \qquad \text{by the inductive hypothesis}$$

$$= 1 + k * 3$$

22. The recurrence relation matches equation (6) with $c = 1$ and $g(n) = 3$. From equation (8), the closed-form solution is

$$T(n) = 1^{n-1}(1) + \sum_{i=2}^{n} 1^{n-i}(3)$$

$$= 1 + \sum_{i=2}^{n} 3$$

$$= 1 + (n-1)3$$

23. This is in the form of equation (1) with $c = 2$ and $g(n) = 1$. By equation (6), the solution is

$$2^{\log n}(1) + \sum_{i=1}^{\log n} 2^{(\log n)-i}(1) = 2^{\log n} + 2^{(\log n)-1} + 2^{(\log n)-2} + \cdots + 2^0$$

$$= 2^{(\log n)+1} - 1$$

$$= (2)2^{\log n} - 1 = 2n - 1$$

Chapter 3

1. a. $\{4, 5, 6, 7\}$
   b. $\{\text{April, June, September, November}\}$
   c. $\{\text{Washington, D.C.}\}$

2. a. $\{x \mid x$ is one of the first four perfect squares$\}$
   b. $\{x \mid x$ is one of the Three Men in a Tub)
   c. $\{x \mid x$ is a prime number$\}$

3. a. $A = \{x \mid x \in N \text{ and } x \geq 5\}$
   b. $B = \{3, 4, 5\}$

4. $x \in B$

5. $A \subset B$ means $(\forall x)(x \in A \rightarrow x \in B) \land (\exists y)[y \in B \land (y \in A)']$

6. a, b, d, e, h, i, l

7. Let $x \in A$. Then $x \in \mathbb{R}$ and $x^2 - 4x + 3 = 0$, or $(x - 1)(x - 3) = 0$, which gives $x = 1$ or $x = 3$. In either case, $x \in \mathbb{N}$ and $1 \leq x \leq 4$, so $x \in B$. Therefore $A \subseteq B$. The value 4 belongs to B but not to A, so $A \subset B$.

8. $\wp(A) = \{\emptyset, \{1\}, \{2\}, \{3\}, \{1, 2\}, \{1, 3\}, \{2, 3\}, \{1, 2, 3\}\}$

9. 2^n

10. By the definition of equality for ordered pairs,

$$2x - y = 7 \quad \text{and} \quad x + y = -1$$

Solving the system of equations, $x = 2$, $y = -3$.

11. $(3, 3), (3, 4), (4, 3), (4, 4)$

12. a. S is not closed under division.
   c. 0^0 is not defined.
   f. $x^\#$ is not unique for, say, $x = 4$ ($2^2 = 4$ and $(-2)^2 = 4$).

13. Yes; if $x \in A \cap B$, then $x \in A$ (and $x \in B$, but we don't need this fact), so $x \in A \cup B$.

14. $A' = \{x \mid x \in S \text{ and } x \notin A\}$

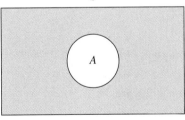

A'

15. $A - B = \{x \mid x \in A \text{ and } x \notin B\}$

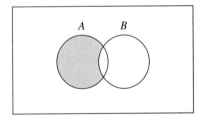

$A - B$

16. a. $\{1, 2, 3, 4, 5, 7, 8, 9, 10\}$
   b. $\{1, 2, 3\}$
   c. $\{1, 3, 5, 10\}$

17. a. $A \times B = \{(1, 3), (1, 4), (2, 3), (2, 4)\}$
   b. $B \times A = \{(3, 1), (3, 2), (4, 1), (4, 2)\}$
   c. $A^2 = \{(1, 1), (1, 2), (2, 1), (2, 2)\}$
   d. $A^3 = \{(1, 1, 1), (1, 1, 2), (1, 2, 1), (1, 2, 2), (2, 1, 1), (2, 1, 2), (2, 2, 1), (2, 2, 2)\}$

18. Show set inclusion in each direction. To show $A \cup \emptyset \subseteq A$, let $x \in A \cup \emptyset$. Then $x \in A$ or $x \in \emptyset$, but since $\emptyset$ has no elements, $x \in A$. To show $A \subseteq A \cup \emptyset$, let $x \in A$. Then $x \in A$ or $x \in \emptyset$, so $x \in A \cup \emptyset$.

19. a. $[C \cap (A \cup B)] \cup [(A \cup B) \cap C']$

$$= [(A \cup B) \cap C] \cup [(A \cup B) \cap C'] \qquad \text{(1b)}$$
$$= (A \cup B) \cap (C \cup C') \qquad \text{(3b)}$$
$$= (A \cup B) \cap S \qquad \text{(5a)}$$
$$= A \cup B \qquad \text{(4b)}$$

b. $[C \cup (A \cap B)] \cap [(A \cap B) \cup C'] = A \cap B$

20. An enumeration of the even positive integers is 2, 4, 6, 8, 10, 12, ...

21. 1/5, 5/1   **22.** $4(8)(5) = 160$ **23.** $7(5) + 9 = 44$

24. Although the problem consists of successive events—the five tosses—the number of outcomes of each event is not constant, but varies between one and two depending on the outcome of the preceding event.

25.

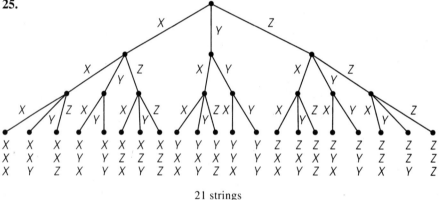

21 strings

26. $A \cup B$

27. Equation (2) gives the result of Example 28 because if A and B are disjoint, then $|A \cap B| = 0$.

28. The reasons for the equalities are

set union is associative
equation (2)
equation (2) and set identity 3b (distributive property)
equation (2)
rearranging terms.

29. 7   **30.** $P(20, 2) = \dfrac{20!}{18!} = 380$

31. $6! = 720$ **32.** $C(12, 3) = \dfrac{12!}{3!9!} = 220$

33. a. 18 **b.** 24 **34.** $\dfrac{9!}{3!2!}$ **35.** $C(8, 6)$

36. $(a + b)^3 = a^3 + 3a^2b + 3ab^2 + b^3$

Coefficients: 1 3 3 1, which is row $n = 3$ in Pascal's triangle

$(a + b)^4 = a^4 + 4a^3b + 6a^2b^2 + 4ab^3 + b^4$

Coefficients: 1 4 6 4 1, which is row $n = 4$ in Pascal's triangle

37. $(x + 1)^5 = C(5, 0)x^5 + C(5, 1)x^4 + C(5, 2)x^3 + C(5, 3)x^2 + C(5, 4)x + C(5, 5)$
$= x^5 + 5x^4 + 10x^3 + 10x^2 + 5x + 1$

38. $C(7, 4)x^3y^4$

Chapter 4

1. a. $(3, 2) \in \rho$ **c.** $(3, 4), (5, 6) \in \rho$
b. $(2, 4), (2, 6) \in \rho$ **d.** $(2, 1), (5, 2) \in \rho$

2. a. Many-to-one **b.** One-to-one **c.** Many-to-many

3. a. $x\,(\rho \cup \sigma)\,y \leftrightarrow x \le y$ **c.** $x\,\sigma'\,y \leftrightarrow x \ge y$
b. $x\,\rho'\,y \leftrightarrow x \ne y$ **d.** $\rho \cap \sigma = \varnothing$

4. a. $(1, 1), (2, 2), (3, 3)$
b. Knowing that a relation is symmetric does not by itself give information about any of the ordered pairs that might belong to ρ. If we know that a relation is symmetric and we know some ordered pairs that belong to the relation, then we know certain other pairs that must belong to the relation (see part (c)).
c. (b, a)
d. $a = b$
e. The transitive property says $(x, y) \in \rho \wedge (y, z) \in \rho \to (x, z) \in \rho$. In this case $(1, 2)$ is the only element of ρ and $(2, z) \notin \rho$ for any z in S. Therefore the antecedent of the implication is always false, and the implication is true; ρ is transitive.

5. a. Reflexive, symmetric, transitive
b. Reflexive, antisymmetric, transitive
c. Reflexive, symmetric, transitive
d. Antisymmetric
e. Reflexive, symmetric, antisymmetric, transitive
f. Antisymmetric (recall the truth table for implication), transitive
g. Reflexive, symmetric, transitive
h. Reflexive, symmetric, transitive

6. No. If the relation has the antisymmetry property, then it is its own antisymmetric closure. If the relation is not antisymmetric, there must be two ordered pairs (x, y) and (y, x) in the relation with $x \neq y$. Extending the relation by adding more ordered pairs will not change this situation, so no extension will be antisymmetric.

7. Reflexive closure: $\{(a, a), (b, b), (c, c), (a, c), (a, d), (b, d), (c, a), (d, a), (d, d)\}$
Symmetric closure: $\{(a, a), (b, b), (c, c), (a, c), (a, d), (b, d), (c, a), (d, a), (d, b)\}$
Transitive closure: $\{(a, a), (b, b), (c, c), (a, c), (a, d), (b, d), (c, a), (d, a), (d, d),$
$(d, c), (b, a), (b, c), (c, d)\}$

8. a. $(1, 1), (1, 2), (2, 2), (1, 3), (3, 3), (1, 6), (6, 6), (1, 12), (12, 12), (1, 18), (18, 18),$
$(2, 6), (2, 12), (2, 18), (3, 6), (3, 12), (3, 18), (6, 12), (6, 18)$
b. 1, 2, 3
c. 2, 3

9.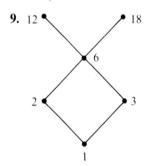

10. $y \in S$ is a greatest element if $x \leqslant y$ for all $x \in S$.
$y \in S$ is a maximal element if there is no $x \in S$ with $y < x$.

11.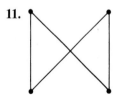

12. Let $q \in [x]$. Then $x \rho q$. Because $x \rho z$, by symmetry $z \rho x$. By transitivity $z \rho x$ together with $x \rho q$ gives $z \rho q$. Therefore, $q \in [z]$.

13. Reflexive: For any $x \in S$, x is in the same subset as itself, so $x \rho x$.
Symmetric: If $x \rho y$, then x is in the same subset as y, so y is in the same subset as x, or $y \rho x$ and ρ is symmetric.
Transitive: If $x \rho y$ and $y \rho z$, then x is in the same subset as y and y is in the same subset as z, so x is in the same subset as z, or $x \rho z$.

Therefore ρ is an equivalence relation.

14. a. The equivalence classes are sets consisting of lines in the plane with the same slope.
b. $[n] = \{n\}$; the equivalence classes are all the singleton sets of elements of $\mathbb{N}$.
c. $[1] = [2] = \{1, 2\}, [3] = \{3\}$

15. $[0] = \{ \dots, -15, -10, -5, 0, 5, 10, 15, \dots \}$
$[1] = \{ \dots, -14, -9, -4, 1, 6, 11, 16, \dots \}$
$[2] = \{ \dots, -13, -8, -3, 2, 7, 12, 17, \dots \}$
$[3] = \{ \dots, -12, -7, -2, 3, 8, 13, 18, \dots \}$
$[4] = \{ \dots, -11, -6, -1, 4, 9, 14, 19, \dots \}$

16. 2

17.

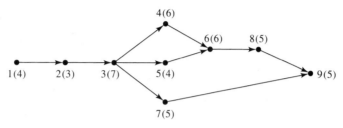

18. Minimum time to completion is 36 days. Critical path is 1, 2, 3, 4, 6, 8, 9.

19. For example: 1, 3, 2, 6, 7, 5, 4, 8, 9, 10, 11, 12

20. For example: 1, 2, 3, 7, 5, 4, 6, 8, 9

21.

Locale	
Name	State
Patrick, Tom	FL
Smith, Mary	IL
Collier, Jon	IL
Jones, Kate	OH
Smith, Bob	MA
White, Janet	GA
Garcia, Maria	NY

22. a. project(join(restrict Pet-Owner **where** *Pet-Type* = "Dog") **and** Person over *Name*) **over** *City* **giving** *Result*

b. SELECT *City*
FROM Person, Pet-Owner
WHERE Person.*Name* = Pet-Owner.*Name*
AND Pet-Owner.*Pet-Type* = "Dog"

c. Range of *x* is Person
Range of *y* is Pet-Owner
$\{x.City \mid \textbf{exists } y(y.Name = x.Name \textbf{ and } y.Pet\text{-}type = \text{"Dog"})\}$

23. a. Not a function; $2 \in S$ has two values associated with it
b. Function
c. Not a function; for values 0, 1, 2, 3 of the domain, the corresponding $h(x)$ values fall outside the codomain
d. Not a function; not every member of S has a Social Security number
e. Function (not every value in the codomain need be used)
f. Function
g. Function
h. Not a function; $5 \in N$ has two values associated with it

24. a. 16
b. ± 3

25. T, F

26.

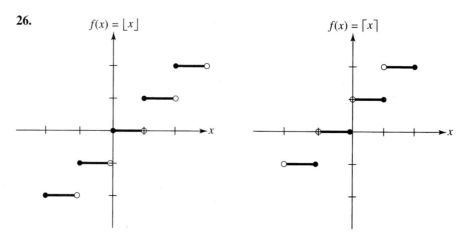

$f(x) = \lfloor x \rfloor$ $f(x) = \lceil x \rceil$

27. f and g have the same domain and codomain, so we must show that each function has the same effect on each member of the domain.

$$f(1) = 1 \qquad g(1) = \frac{\sum\limits_{k=1}^{1}(4k-2)}{2} = \frac{4 \cdot 1 - 2}{2} = \frac{2}{2} = 1$$

$$f(2) = 4 \qquad g(2) = \frac{\sum\limits_{k=1}^{2}(4k-2)}{2} = \frac{(4 \cdot 1 - 2) + (4 \cdot 2 - 2)}{2} = \frac{2+6}{2} = 4$$

$$f(3) = 9 \qquad g(3) = \frac{\sum\limits_{k=1}^{3}(4k-2)}{2} = \frac{(4 \cdot 1 - 2) + (4 \cdot 2 - 2) + (4 \cdot 3 - 2)}{2}$$

$$= \frac{2+6+10}{2} = 9$$

Therefore $f = g$.

28. b, f, g

29. If P is either a tautology or a contradiction.

30. e, g

31. $(g \circ f)(2.3) = g(f(2.3)) = g((2.3)^2) = g(5.29) = \lfloor 5.29 \rfloor = 5$
$(f \circ g)(2.3) = f(g(2.3)) = f(\lfloor 2.3 \rfloor) = f(2) = 2^2 = 4$

32. Let $(g \circ f)(s_1) = (g \circ f)(s_2)$. Then $g(f(s_1)) = g(f(s_2))$ and because g is one-to-one, $f(s_1) = f(s_2)$. Because f is one-to-one, $s_1 = s_2$.

33. Let $t \in T$. Then $(f \circ g)(t) = f(g(t)) = f(s) = t$.

34. $f^{-1}: \mathbb{R} \to \mathbb{R}, f^{-1}(x) = (x - 4)/3$

35. a. $(1, 4, 5) = (4, 5, 1) = (5, 1, 4)$

    b. $\begin{pmatrix} 1 & 2 & 3 & 4 & 5 \\ 1 & 4 & 2 & 5 & 3 \end{pmatrix}$

36. a. $g \circ f = (1, 3, 5, 2, 4) = (3, 5, 2, 4, 1) = \cdots$
          $f \circ g = (1, 5, 2, 3, 4) = (5, 2, 3, 4, 1) = \cdots$

    b. $g \circ f = \begin{pmatrix} 1 & 2 & 3 & 4 & 5 \\ 4 & 2 & 5 & 1 & 3 \end{pmatrix}$

         $f \circ g = \begin{pmatrix} 1 & 2 & 3 & 4 & 5 \\ 2 & 1 & 3 & 5 & 4 \end{pmatrix}$

    c. $g \circ f = f \circ g = \begin{pmatrix} 1 & 2 & 3 & 4 & 5 \\ 3 & 5 & 1 & 4 & 2 \end{pmatrix}$

37. $(1, 2, 4) \circ (3, 5)$ or $(3, 5) \circ (1, 2, 4)$

38.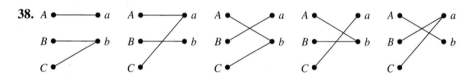

39. One possibility: $\{(0, 0), (1, 1), (-1, 2), (2, 3), (-2, 4), (3, 5), (-3, 6), \ldots\}$

40. a. $10.87 \leq 12 \leq 1087$
      $22.27 \leq 27 \leq 2227$
      $37.67 \leq 48 \leq 3767$
      $57.07 \leq 75 \leq 5707$

    b. No

    c. $n_0 = 1, c_1 = 1/200, c_2 = 1$

41. a. Let $f \rho g$. Then there are positive constants n_0, c_1, and c_2 with $c_1 g(x) \leq f(x) \leq c_2 g(x)$ for $x \geq n_0$. Then for $x \geq n_0$, it is true that $(1/c_2) f(x) \leq g(x) \leq (1/c_1) f(x)$, so $g \rho f$.

    b. Let $f \rho g$ and $g \rho h$. Then there are positive constants n_0, n_1, c_1, c_2, d_1, and d_2 with $c_1 g(x) \leq f(x) \leq c_2 g(x)$ for $x \geq n_0$ and $d_1 h(x) \leq g(x) \leq d_2 h(x)$ for $x \geq n_1$. Then for $x \geq \max(n_0, n_1)$, $c_1 d_1 h(x) \leq f(x) \leq c_2 d_2 h(x)$, so $f \rho h$.

42. $3x^2 = \Theta(x^2)$ using constants $n_0 = 1, c_1 = c_2 = 3$.
    $200x^2 + 140x + 7 = \Theta(x^2)$ using constants $n_0 = 2, c_1 = 1, c_2 = 300$.

43. $a_{23} = 1, a_{24} = -7, a_{13} = -6$

44.
$$2\mathbf{A} + \mathbf{B} = \begin{bmatrix} 6 & 14 \\ 3 & 10 \\ 9 & 16 \end{bmatrix}$$

45.

$$\mathbf{A} \cdot \mathbf{B} = \begin{bmatrix} 15 & 22 \\ 12 & 28 \end{bmatrix}$$

$$\mathbf{B} \cdot \mathbf{A} = \begin{bmatrix} 39 & 0 \\ 27 & 4 \end{bmatrix}$$

46.

$$\mathbf{I} \cdot \mathbf{A} = \begin{bmatrix} 1(a_{11}) + 0(a_{21}) & 1(a_{12}) + 0(a_{22}) \\ 0(a_{11}) + 1(a_{21}) & 0(a_{12}) + 1(a_{22}) \end{bmatrix} = \begin{bmatrix} a_{11} & a_{12} \\ a_{21} & a_{22} \end{bmatrix} = \mathbf{A}$$

Similarly, $\mathbf{A} \cdot \mathbf{I} = \mathbf{A}$.

47.

$$\mathbf{A} \cdot \mathbf{B} = \begin{bmatrix} -1 & 2 & -3 \\ 2 & 1 & 0 \\ 4 & -2 & 5 \end{bmatrix} \begin{bmatrix} -5 & 4 & -3 \\ 10 & -7 & 6 \\ 8 & -6 & 5 \end{bmatrix} = \begin{bmatrix} 1 & 0 & 0 \\ 0 & 1 & 0 \\ 0 & 0 & 1 \end{bmatrix}$$

$$\mathbf{B} \cdot \mathbf{A} = \begin{bmatrix} -5 & 4 & -3 \\ 10 & -7 & 6 \\ 8 & -6 & 5 \end{bmatrix} \begin{bmatrix} -1 & 2 & -3 \\ 2 & 1 & 0 \\ 4 & -2 & 5 \end{bmatrix} = \begin{bmatrix} 1 & 0 & 0 \\ 0 & 1 & 0 \\ 0 & 0 & 1 \end{bmatrix}$$

48.

x	y	$x \wedge y$
1	1	1
1	0	0
0	1	0
0	0	0

x	y	$x \vee y$
1	1	1
1	0	1
0	1	1
0	0	0

49.

$$\text{No, } \mathbf{A} \cdot \mathbf{B} = \begin{bmatrix} 2 & 1 & 1 \\ 1 & 1 & 1 \\ 0 & 0 & 1 \end{bmatrix}$$

50.

$$\mathbf{B} \times \mathbf{A} = \begin{bmatrix} 1 & 1 & 0 \\ 1 & 1 & 1 \\ 0 & 0 & 1 \end{bmatrix}$$

Chapter 5

1. One possible picture:

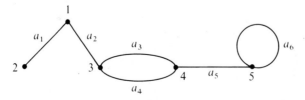

2. a.

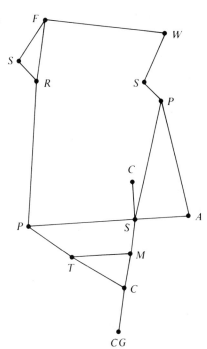

b.

3. Possible answers:

    a. 2 and 3

    b. 5

    c. a_6

    d. a_3 and a_4

    e. 3

    f. 2, a_1, 1, a_2, 3, a_3, 4, a_4, 3, a_3, 4

    g. 3, a_3, 4, a_4, 3

    h. No

    i. Yes

4.

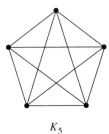

K_5

5.

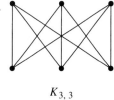

$K_{3,3}$

6. a. In a complete graph, any two distinct nodes are adjacent, so there is a path of length 1 from any node to any other node; hence the graph is connected.

    b. For example, the graph of Figure 5.10b.

7. $f_2: a_4 \rightarrow e_3$

    $a_5 \rightarrow e_8$

    $a_6 \rightarrow e_7$

    $a_7 \rightarrow e_5$ (or e_6)

    $a_8 \rightarrow e_6$ (or e_5)

8. $f: 1 \rightarrow d$

    $2 \rightarrow e$

    $3 \rightarrow f$

    $4 \rightarrow c$

    $5 \rightarrow b$

    $6 \rightarrow a$

9. The graph at the left in Figure 5.19 has two nodes of degree 2, but the graph at the right does not; or the graph at the left has parallel arcs, but the graph at the right does not.

10. K_4 can be represented as

11. Making 1–3 and 1–4 exterior arcs leads to the graph below, where it is still impossible to make 3 and 5 adjacent while preserving planarity.

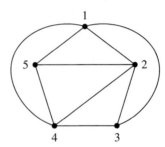

12. An attempt to construct $K_{3,3}$ as a planar graph leads to the graph below; there is no way to connect nodes 3 and 5. Any other construction leads to a similar difficulty.

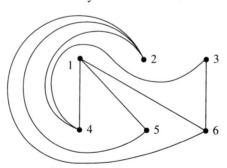

13. $n = 6$, $a = 7$, $r = 3$, and $6 - 7 + 3 = 2$

14. Without this condition on the arc, a figure such as the one below could result. Then the graph would be split into two disconnected subgraphs and the inductive hypothesis would not apply. Also the number of regions would not change.

15. In $K_{3,3}$, $a = 9$, $n = 6$, and $9 \le 3(6) - 6$.

16.
$$A = \begin{bmatrix} 1 & 1 & 0 & 1 \\ 1 & 0 & 1 & 0 \\ 0 & 1 & 0 & 2 \\ 1 & 0 & 2 & 0 \end{bmatrix}$$

17.

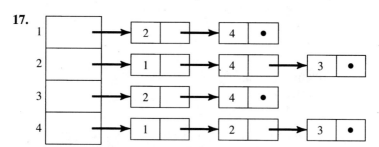

18. a. 2   **b.** 4   **c.** 2

19.

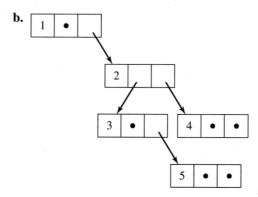

20. a.

	Left child	Right child
1	0	2
2	3	4
3	0	5
4	0	0
5	0	0

b.

21. $a, b, e, f, c, d, g, i, h$
$e, b, f, a, e, i, g, d, h$
$e, f, b, c, i, g, h, d, a$

22.

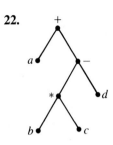

Prefix notation: $+\,a\,-\,*\,b\,c\,d$
Postfix notation: $a\,b\,c\,*\,d\,-\,+$

23. For the base case, $n = 1$, the tree consists of a single node and no arcs; therefore no arc ends. The number of arc ends is $0 = 2(1) - 2$. Assume that any tree with k nodes has a total number of arc ends of $2k - 2$. Consider a tree with $k + 1$ nodes, and show that the number of arc ends is $2(k + 1) - 2$. In this tree, remove a leaf node and the arc to that node's parent. This leaves a tree with k nodes and, by the inductive hypothesis, $2k - 2$ arc ends. The original graph had one more arc, and two more arc ends, so it had $2k - 2 + 2 = 2k = 2(k + 1) - 2$ arc ends.

24. a.

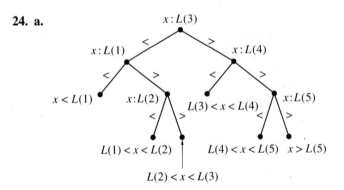

b. Depth of tree $= 3 = 1 + \lfloor \log 5 \rfloor$

25. a.

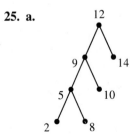

b. $d = 3$

26.

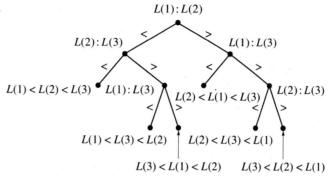

L(1):L(2)

< >

L(2):L(3)   L(1):L(3)

< >   < >

L(1) < L(2) < L(3)   L(1):L(3)   L(2) < L(1) < L(3)   L(2):L(3)

< >   < >

L(1) < L(3) < L(2)   L(2) < L(3) < L(1)

L(3) < L(1) < L(2)   L(3) < L(2) < L(1)

27. a. *ppca?*   **b.** *cagak*   **c.** *?kac?*

28.

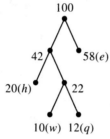

100

42   58(*e*)

20(*h*)   22

10(*w*) 12(*q*)

29. *w*: 010
q: 011
h: 00
e: 1

Chapter 6

1. {(2, 1), (2, 2), (3, 1), (3, 4)}

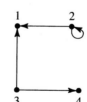

1   2

3   4

2. There are two distinct nodes, 3 and 4, with 3 ρ 4 and 4 ρ 3.

3.

$$\mathbf{A} = \begin{bmatrix} 0 & 0 & 0 & 1 \\ 0 & 0 & 1 & 1 \\ 0 & 0 & 0 & 0 \\ 1 & 0 & 0 & 0 \end{bmatrix} \quad \mathbf{A}^{(2)} = \begin{bmatrix} 1 & 0 & 0 & 0 \\ 1 & 0 & 0 & 0 \\ 0 & 0 & 0 & 0 \\ 0 & 0 & 0 & 1 \end{bmatrix}$$

$\mathbf{A}^{(2)}[2, 1] = 1$ because there is a path from 2 to 1 of length 2 (2–4–1).

4. There is a length-4 path (2–4–1–4–1) from 2 to 1, so $\mathbf{A}^{(4)}[2, 1]$ should be 1.

$$\mathbf{A}^{(3)} = \begin{bmatrix} 0 & 0 & 0 & 1 \\ 0 & 0 & 0 & 1 \\ 0 & 0 & 0 & 0 \\ 1 & 0 & 0 & 0 \end{bmatrix} \qquad \mathbf{A}^{(4)} = \begin{bmatrix} 1 & 0 & 0 & 0 \\ 1 & 0 & 0 & 0 \\ 0 & 0 & 0 & 0 \\ 0 & 0 & 0 & 1 \end{bmatrix}$$

5.

$$\mathbf{R} = \begin{bmatrix} 1 & 0 & 0 & 1 \\ 1 & 0 & 1 & 1 \\ 0 & 0 & 0 & 0 \\ 1 & 0 & 0 & 1 \end{bmatrix}$$

Column 2 is all 0s, so 2 is not reachable from any node.

6.

$$\mathbf{M}_0 = \begin{bmatrix} 0 & 0 & 0 & 1 \\ 0 & 0 & 1 & 1 \\ 0 & 0 & 0 & 0 \\ 1 & 0 & 0 & 0 \end{bmatrix} \qquad \mathbf{M}_1 = \begin{bmatrix} 0 & 0 & 0 & 1 \\ 0 & 0 & 1 & 1 \\ 0 & 0 & 0 & 0 \\ 1 & 0 & 0 & 1 \end{bmatrix}$$

$$\mathbf{M}_2 = \begin{bmatrix} 0 & 0 & 0 & 1 \\ 0 & 0 & 1 & 1 \\ 0 & 0 & 0 & 0 \\ 1 & 0 & 0 & 1 \end{bmatrix} \qquad \mathbf{M}_3 = \begin{bmatrix} 0 & 0 & 0 & 1 \\ 0 & 0 & 1 & 1 \\ 0 & 0 & 0 & 0 \\ 1 & 0 & 0 & 1 \end{bmatrix}$$

$$\mathbf{M}_4 = \mathbf{R} = \begin{bmatrix} 1 & 0 & 0 & 1 \\ 1 & 0 & 1 & 1 \\ 0 & 0 & 0 & 0 \\ 1 & 0 & 0 & 1 \end{bmatrix}$$

7. a. No   **b.** Yes

8. a. No, four odd nodes   **b.** Yes, no odd nodes

9. No

10.

	A	B	C	D
A	0	2	2	1
B	2	0	0	1
C	2	0	0	1
D	1	1	1	0

After row C, *total* = 3, the loop terminates, and there is no path.

11. a. No   **b.** Yes

12.

IN = {x}					
	x	1	2	3	*y*
d	0	1	∞	4	∞
s	–	*x*	*x*	*x*	*x*

$p = 1$
IN = {x, 1}					
	x	1	2	3	*y*
d	0	1	4	2	6
s	–	*x*	1	1	1

$p = 3$
IN = {x, 1, 3}					
	x	1	2	3	*y*
d	0	1	4	2	5
s	–	*x*	1	1	3

$p = 2$
IN = {x, 1, 3, 2}					
	x	1	2	3	*y*
d	0	1	4	2	5
s	–	*x*	1	1	3

$p = y$
IN = {x, 1, 3, 2, y}					
	x	1	2	3	*y*
d	0	1	4	2	5
s	–	*x*	1	1	3

Path: *x*, 1, 3, *y*   Distance: 5

13.
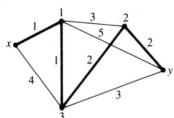

14. *a, e, d, b, c, i, f, g, h, l, k, m, j*

15. *a, e, f, d, i, b, c, g, h, j, k, m, l*

16. a.
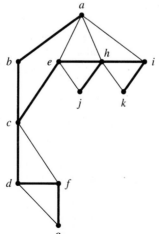

b.
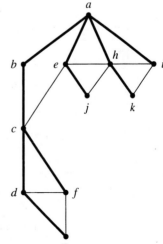

17. a (5) c (2)

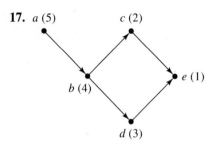

b (4)

e (1)

d (3)

Topological sort: a, b, d, c, e

18. The depth-first search progresses from node a to b and then c, with *TreeNumbers* and *BackNumbers* assigned in sequence. At node c, the back arc to node a causes *BackNumber* of c to be changed to *TreeNumber* of a. The search progresses to d, which has a sequential *TreeNumber* and *BackNumber* assigned. The search then backs up to node c. Because $BackNumber(d) > TreeNumber(c)$, c is recognized as an articulation point. The search backs up to node b, and b's *BackNumber* is reduced to that of c. The search backs up to a, but a is not an articulation point because it is the starting node with only one tree arc.

Chapter 7

1. $0 \cdot 1 = 0$
   $1 \cdot 1 = 1$

2. a. $A \vee A = A$
   b. $A \cup A = A$

3. a. $x + 1 = x + (x + x')$ (5a, complement property)
          $= (x + x) + x'$ (2a, associative property)
          $= x + x'$ (idempotent property)
          $= 1$ (5a, complement property)

   b. $x \cdot 0 = 0$

4. $0 + 1 = 1$ (Practice 3a)
       $0 \cdot 1 = 1 \cdot 0$ (1b)
       $= 0$ (Practice 3b)

   Therefore $1 = 0'$ by the theorem on the uniqueness of complements.

5. a. $f(x \cdot y) = f(x) * f(y)$
   b. $f(x') = [f(x)]''$

6. Property 2:

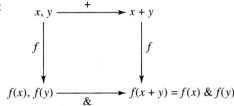

Property 3:

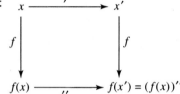

Property 4:

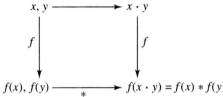

7. a. $f(0 + a) = f(a) = \{1\} = \emptyset \cup \{1\} = f(0) \cup f(a)$
  b. $f(a + a') = f(1) = \{1, 2\} = \{1\} \cup \{2\} = f(a) \cup f(a')$
  c. $f(a \cdot a') = f(0) = \emptyset = \{1\} \cap \{2\} = f(a) \cap f(a')$
  d. $f(1') = f(0) = \emptyset = \{1, 2\}' = (f(1))'$

8. a. A single truth function on $\{0, 1\}^n$ must map each of the elements in the domain to a 0 or a 1, and there are 2^n n-tuples in the domain $\{0, 1\}^n$. Hence the table for the function will have 2^n rows.
  b. Any truth function must fill 4 "slots" (corresponding to the $2^2 = 4$ domain elements) with one of 2 values, a 0 or a 1. There are $2^4 = 16$ different ways to do this.
  c. Any truth function must fill 2^n "slots" (corresponding to the 2^n domain elements) with one of 2 values, a 0 or a 1. There are 2^{2^n} different ways to do this.

9. a.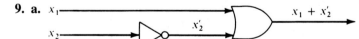

  b.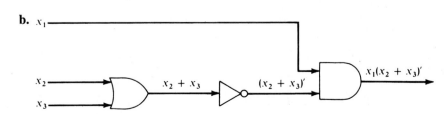

10. a. $(x_1' + x_2)x_3'$

b.

x_1	x_2	x_3	$(x_1' + x_2)x_3'$
1	1	1	0
1	1	0	1
1	0	1	0
1	0	0	0
0	1	1	0
0	1	0	1
0	0	1	0
0	0	0	1

11. a. $x_1x_2x_3 + x_1x_2'x_3 + x_1x_2'x_3' + x_1'x_2'x_3 + x_1'x_2'x_3'$

b.

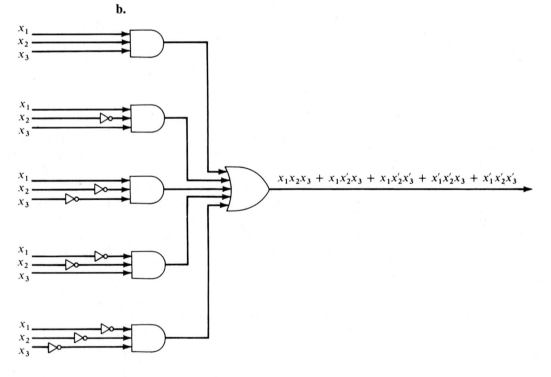

$x_1x_2x_3 + x_1x_2'x_3 + x_1x_2'x_3' + x_1'x_2'x_3 + x_1'x_2'x_3'$

12.   101
$\underline{\quad 111}$
(1)100

13. a.

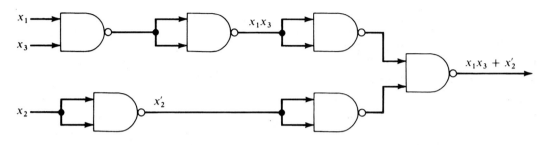

b. $x_1x_2 + x_2' = ((x_1x_3)'x_2)'$

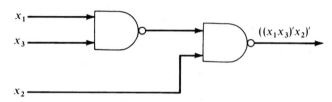

14. a.

x_1	x_2	$f(x_1, x_2)$
1	1	0
1	0	1
0	1	1
0	0	0

b. One possibility is the canonical sum-of-products form, $x_1x_2' + x_1'x_2$

c.

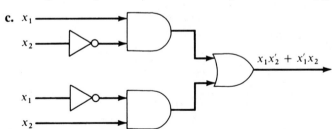

15. a. $x_1x_2 + x_1'x_2 = x_2x_1 + x_2x_1'$
$\qquad\qquad\quad = x_2(x_1 + x_1')$
$\qquad\qquad\quad = x_2 \cdot 1$
$\qquad\qquad\quad = x_2$

b. $x_1 + x_1'x_2 = x_1 \cdot 1 + x_1'x_2$
$\qquad\qquad\quad = x_1(1 + x_2) + x_1'x_2 \qquad$ (see Practice 3)
$\qquad\qquad\quad = x_1 + x_1x_2 + x_1'x_2$
$\qquad\qquad\quad = x_1 + x_2(x_1 + x_1')$
$\qquad\qquad\quad = x_1 + x_2 \cdot 1$
$\qquad\qquad\quad = x_1 + x_2$

16.

The reduced expression is x_1'.

17. x_1x_3 (4 squares) and $x_1'x_2x_3'$ (2 squares)

18. $x_1x_2'x_4 + x_1x_3'x_4 + x_2'x_3'$ (see following figure)

19. The reduction table follows.

Number of 1s	x_1	x_2	x_3	
Three	1	1	1	1
Two	1	1	0	1,2
One	1	0	0	2,3
	0	0	1	4
None	0	0	0	3,4

Number of 1s	x_1	x_2	x_3
Two	1	1	–
One	1	–	0
None	–	0	0
	0	0	–

The comparison table follows.

	111	110	100	001	000
11–	✓	✓			
1–0		✓	✓		
–00			✓		✓
00–				✓	✓

Essential terms are 11– and 00–. Either 1–0 or –00 can be used as the third reduced term. The minimal sum-of-products form is

$$x_1x_2 + x_1'x_2' + x_1x_3' \quad \text{or} \quad x_1x_2 + x_1'x_2' + x_2'x_3'$$

Chapter 8

1. $[\mathbb{R}, \cdot]$ is not a commutative group because $0 \in \mathbb{R}$ does not have an inverse with respect to multiplication; there is no real number y such that $0 \cdot y = y \cdot 0 = 1$.

2. See Practice 45 of Chapter 4.

3. Many elements of $M_2(\mathbb{Z})$ do not have inverses under matrix multiplication. For example, if

$$\begin{bmatrix} 2 & 0 \\ 0 & 2 \end{bmatrix}$$

is to have an inverse

$$\begin{bmatrix} a & b \\ c & d \end{bmatrix}$$

under multiplication, then

$$\begin{bmatrix} 2 & 0 \\ 0 & 2 \end{bmatrix} \cdot \begin{bmatrix} a & b \\ c & d \end{bmatrix} = \begin{bmatrix} a & b \\ c & d \end{bmatrix} \cdot \begin{bmatrix} 2 & 0 \\ 0 & 2 \end{bmatrix} = \begin{bmatrix} 1 & 0 \\ 0 & 1 \end{bmatrix}$$

By the definition of matrix multiplication, the only matrix that satisfies this equation would have $2a = 1$ or $a = \frac{1}{2}$ and thus not be a member of $M_2(\mathbb{Z})$.

4. Subtraction is not associative; for example, $5 - (3 - 1) = 3$ but $(5 - 3) - 1 = 1$.

5. No, S is not closed under multiplication; for example, $\dfrac{2}{3} \cdot \dfrac{3}{2} = 1$.

6. All except $[\mathbb{R}, +]$ (which has no identity) are monoids; the identities are, respectively, 0, 1, 1, 0, $\varnothing$, S.

7. $[\mathbb{R}, +]$

8. **a.** $f(x) + g(x) = g(x) + f(x)$
   $[f(x) + g(x)] + h(x) = f(x) + [g(x) + h(x)]$
   b. The zero polynomial, 0
   c. $-7x^4 + 2x^3 - 4$

9. **a.**

$+_5$	0	1	2	3	4
0	0	1	2	3	4
1	1	2	3	4	0
2	2	3	4	0	1
3	3	4	0	1	2
4	4	0	1	2	3

$\cdot_5$	0	1	2	3	4
0	0	0	0	0	0
1	0	1	2	3	4
2	0	2	4	1	3
3	0	3	1	4	2
4	0	4	3	2	1

   b. 0; 1   **c.** 3   **d.** All except 0

10. a.

$\cdot_6$	0	1	2	3	4	5
0	0	0	0	0	0	0
1	0	1	2	3	4	5
2	0	2	4	0	2	4
3	0	3	0	3	0	3
4	0	4	2	0	4	2
5	0	5	4	3	2	1

b. 1 and 5

11. Let $f, g, h \in S$. Then for any $x \in A$, $[(f \circ g) \circ h](x) = (f \circ g)(h(x)) = f(g[h(x)])$ and $[f \circ (g \circ h)](x) = f[(g \circ h)(x)] = f(g[h(x)])$. Hence, $(f \circ g) \circ h = f \circ (g \circ h)$.

12. a.

$\circ$	α_1	α_2	α_3	α_4	α_5	α_6
α_1	α_1	α_2	α_3	α_4	α_5	α_6
α_2	α_2	α_1	α_6	α_5	α_4	α_3
α_3	α_3	α_5	α_1	α_6	α_2	α_4
α_4	α_4	α_6	α_5	α_1	α_3	α_2
α_5	α_5	α_3	α_4	α_2	α_6	α_1
α_6	α_6	α_4	α_2	α_3	α_1	α_5

b. No

13. a. No   **b.** No

14. $i_1 = i_1 i_2$ because i_2 is an identity $i_1 i_2 = i_2$ because i_1 is an identity

15. Let y and z both be inverses of x. Let i be the identity. Then $y = y \cdot i = y \cdot (x \cdot z) = (y \cdot x) \cdot z = i \cdot z = z$.

16. $7^{-1} = 5, 3^{-1} = 9$; so $10^{-1} = (7 +_{12} 3)^{-1} = 3^{-1} +_{12} 7^{-1} = 9 +_{12} 5 = 2$

17. $z \cdot x = z \cdot y$ implies
$z^{-1} \cdot (z \cdot x) = z^{-1} \cdot (z \cdot y)$
$(z^{-1} \cdot z) \cdot x = (z^{-1} \cdot z) \cdot y$
$i \cdot x = i \cdot y$
$x = y$

18. $x = 1 +_8 (3)^{-1} = 1 +_8 5 = 6$

19.

*	1	a	b	c	d
1	1	a	b	c	d
a	a	b	c	d	1
b	b	c	d	1	a
c	c	d	1	a	b
d	d	1	a	b	c

20. a. $[\mathbb{Z}_{18}, +_{18}]$ **b.** $[S_3, \circ]$

21. Requirement 2, $i \in A$, ensures that $A \neq \emptyset$.

22. a. Closure holds; $0 \in \{0, 2, 4, 6\}$; $0^{-1} = 0$, $4^{-1} = 4$, and 2 and 6 are inverses of each other.
   b. Closure holds; $1 \in \{1, 2, 4\}$; $1^{-1} = 1$, and 2 and 4 are inverses of each other.

23. $[\{\alpha_1, \alpha_5, \alpha_6\}, \circ]$, $[\{\alpha_1, \alpha_2\}, \circ]$, $[\{\alpha_1, \alpha_3\}, \circ]$, $[\{\alpha_1, \alpha_4\}, \circ]$

24. To show f is one-to-one, let α and β belong to A_n and suppose $f(\alpha) = f(\beta)$. Then $\alpha \circ (1, 2) = \beta \circ (1, 2)$. By the cancellation law available in the group S_n, $\alpha = \beta$. To show f is onto, let $\gamma \in O_n$. Then $\gamma \circ (1, 2) \in A_n$ and $f(\gamma \circ (1, 2)) = \gamma \circ (1, 2) \circ (1, 2) = \gamma$.

25. For $nz_1, nz_2 \in n\mathbb{Z}$, $nz_1 + nz_2 = n(z_1 + z_2) \in n\mathbb{Z}$, so closure holds; $0 = n \cdot 0 \in n\mathbb{Z}$; for $nz \in n\mathbb{Z}$, $-nz = n(-z) \in n\mathbb{Z}$.

26.

$$x, y \xrightarrow{\quad \cdot \quad} x \cdot y$$

$$f \downarrow \qquad\qquad \downarrow f$$

$$f(x), f(y) \xrightarrow[+]{\quad} f(x \cdot y) = f(x) + f(y)$$

27. $f(s) + f(s^{-1}) = f(s \cdot s^{-1}) = f(i_S) = i_T$. Similarly, $f(s^{-1}) + f(s) = i_T$. Therefore $f(s^{-1})$ acts like the inverse of $f(s)$ in T, and since inverses are unique, $f(s^{-1}) = -f(s)$.

28. Let t_1 and t_2 be members of T. Because f is onto, $t_1 = f(s_1)$ and $t_2 = f(s_2)$ for some $s_1, s_2 \in S$. Then

$$t_1 + t_2 = f(s_1) + f(s_2) = f(s_1 \cdot s_2) = f(s_2 \cdot s_1) \quad \text{(because } [S, \cdot] \text{ is commutative)}$$
$$= f(s_2) + f(s_1) = t_2 + t_1$$

so $[T, +]$ is commutative.

29. Clearly f is onto. f is also one-to-one: Let $f(x) = f(y)$, then $5x = 5y$ and $x = y$. f is a homomorphism: For $x, y \in \mathbb{Z}$, $f(x + y) = 5(x + y) = 5x + 5y = f(x) + f(y)$.

30. a. Composition of bijections is a bijection, and for $x, y \in S$, $(g \circ f)(x \cdot y) =$
$g(f(x \cdot y)) = g(f(x) + f(y)) = g(f(x)) * g(f(y)) = (g \circ f)(x) * (g \circ f)(y)$.
b. $S \simeq S$ by the identity mapping. If f is an isomorphism from S to T, then f^{-1}
is an isomorphism from T to S. If $S \simeq T$ and $T \simeq V$, then by part (a), $S \simeq V$.

31. To show that α_g is an onto function, let $y \in G$. Then $g^{-1} \cdot y$ belongs to G and
$\alpha_g(g^{-1} \cdot y) = g(g^{-1} \cdot y) = (g \cdot g^{-1})y = y$. To show that α_g is one-to-one, let
$\alpha_g(x) = \alpha_g(y)$. Then $g \cdot x = g \cdot y$, and by cancellation, $x = y$.

32. a. For $\alpha_g \in P$, $\alpha_g \circ \alpha_1 = \alpha_{g \cdot 1} = \alpha_g$ and $\alpha_1 \circ \alpha_g = \alpha_{1 \cdot g} = \alpha_g$
b. $\alpha_g \circ \alpha_{g^{-1}} = \alpha_{g \cdot g^{-1}} = \alpha_1$ and $\alpha_{g^{-1}} \circ \alpha_g = \alpha_{g^{-1} \cdot g} = \alpha_1$

33. a. Let $f(g) = f(h)$. Then $\alpha_g = \alpha_h$ and, in particular, $\alpha_g(1) = \alpha_h(1)$, or $g \cdot 1 = h \cdot 1$ and $g = h$.
b. For $g, h \in G, f(g \cdot h) = \alpha_{g \cdot h} = \alpha_g \circ \alpha_h = f(g) \circ f(h)$

34. 000110

35.

Present State	Next State		Output
	Present Input		
	0	1	
s_0	s_0	s_3	0
s_1	s_0	s_2	1
s_2	s_3	s_3	1
s_3	s_1	s_3	2

36. a.

b. 01011

37. a. s_1 **b.** s_1

38. 11001011

39.

40. a. Set consisting of a single 0
   b. Set consisting of any number of 0s (including none) followed by 10
   c. Set consisting of a single 0 or a single 1
   d. Set consisting of any number (including none) of pairs of 1s

41. The string in part (b) does not belong.

42. a. 0   **b.** 0*10   **c.** 0 ∨ 1   **d.** (11)*

43. s_2, s_3

44. A state s produces the same output as itself for any input. If s_i produces the same output as s_j, then s_j produces the same output as s_i. Transitivity is equally clear.

45. Property 1 is satisfied because all states in the same class have the same output strings for any input string, including the empty input string. To see that property 2 is satisfied, assume s_i and s_j are equivalent states proceeding under the input symbol i to states s_i' and s_j' that are not equivalent. Then there is an input string α such that $f_0(s_i', \alpha) \neq f_0(s_j', \alpha)$. Thus, for the input string $i\alpha$, s_i and s_j produce different output strings, contradicting the equivalence of s_i and s_j.

46. Equivalent states of M in Table 8.9 are $A = \{0, 1, 3\}$, $B = \{2\}$, and $C = \{4\}$. The reduced machine is

Present State	Next State		Output
	Present Input		
	0	1	
A	B	A	1
B	C	A	0
C	A	A	0

Equivalent states of M in Table 8.10 are $\{0\}$, $\{1\}$, $\{2\}$, and $\{3\}$. M is already minimal.

47. a. ... $b\,0\,0\,b$ ...
   b. The machine cycles endlessly over the two nonblank tape squares.
   c. The machine changes the two nonblank squares to 0 1 and then moves endlessly to the right.

48. a. $b\,X\,X\,1\,X\,X\,b$ halts without accepting
         ↑
         7
   b. $b\,X\,X\,X\,X\,X\,b$ halts without accepting
         ↑
         2
   c. $b\,X\,X\,X\,0\,X\,X\,b$ halts without accepting
         ↑
         2

49. State 3 is the only final state.

$(0, 1, 1, 0, R)$
$(0, 0, 0, 1, R)$
$(1, 0, 0, 2, R)$
$(1, 1, 1, 0, R)$
$(2, 0, 0, 2, R)$
$(2, 1, 1, 0, R)$
$(2, b, b, 3, R)$

50. Change $(2, 1, X, 3, L)$ to $(2, 1, X, 9, L)$ and add $(9, 1, X, 3, L)$.

51. One machine that works, together with a description of its actions:

$(0, 1, 1, 1, R)$	reads first 1
$(1, b, 1, 6, R)$	$n = 0$, changes to 1 and halts
$(1, 1, 1, 2, R)$	reads second 1
$(2, b, b, 6, R)$	$n = 1$, halts
$(2, 1, 1, 3, R)$	$n \geq 2$

$\left.\begin{array}{l}(3, 1, 1, 3, R)\\(3, b, b, 4, L)\end{array}\right\}$ finds right end of $\bar{n}$

$\left.\begin{array}{l}(4, 1, b, 5, L)\\(5, 1, b, 6, L)\end{array}\right\}$ erases two 1s from $\bar{n}$ and halts

52. $0S \Rightarrow 00S \Rightarrow 000S \Rightarrow 0000S \Rightarrow 00001$ **53.** $L = \{0^n 1 \mid n \geq 0\}$

54. For example:

 a. $G(V, V_T, S, P)$ where $V = \{0, 1, S\}$, $V_T = \{0, 1\}$, and $P = \{S \rightarrow 1, S \rightarrow 0S0\}$
 b. $G = (V, V_T, S, P)$ where $V = \{0, 1, S, M\}$, $V_T = \{0, 1\}$, and $P = \{S \rightarrow 0M0,$
 $M \rightarrow 0M0, M \rightarrow 1\}$

55. In G_1: $S \Rightarrow ABA \Rightarrow 00A \Rightarrow 0000A \Rightarrow 00000$
 In G_2: $S \Rightarrow 00A \Rightarrow 0000A \Rightarrow 00000$
 In G_3: $S \Rightarrow 0A \Rightarrow 00B \Rightarrow 000C \Rightarrow 0000B \Rightarrow 00000$

56.

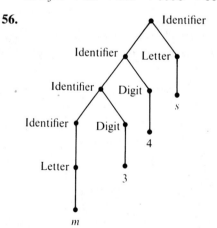

Answers to Selected Exercises

Chapter 1

Exercises 1.1

1. (a), (c), (e), (f)

4. a. Antecedent: sufficient water
Consequent: healthy plant growth
b. Antecedent: further technological advances
Consequent: increased availability of information
c. Antecedent: errors will be introduced
Consequent: there is a modification of the program
d. Antecedent: fuel savings
Consequent: good insulation or storm windows throughout

6. a. The food is good but the service is poor.
b. The food is poor and so is the service.

9. a. A: prices go up; B: housing will be plentiful; C: housing will be expensive
$[A \rightarrow B \wedge C] \wedge (C' \rightarrow B)$
b. A: going to bed; B: going swimming; C: changing clothes
$[(A \vee B) \rightarrow C] \wedge (C \rightarrow B)'$
c. A: it will rain; B: it will snow
$(A \vee B) \wedge (A \wedge B)'$
d. A: Janet wins; B: Janet loses; C: Janet will be tired
$(A \vee B) \rightarrow C$
e. A: Janet wins; B: Janet loses; C: Janet will be tired
$A \vee (B \rightarrow C)$

11. a.

A	B	$A \rightarrow B$	A'	$A' \vee B$	$(A \rightarrow B) \leftrightarrow A' \vee B$
T	T	T	F	T	T
T	F	F	F	F	T
F	T	T	T	T	T
F	F	T	T	T	T

Tautology

b.

A	B	C	$A \wedge B$	$(A \wedge B) \vee C$	$B \vee C$	$A \wedge (B \vee C)$	$(A \wedge B) \vee C \rightarrow A \wedge (B \vee C)$
T	T	T	T	T	T	T	T
T	T	F	T	T	T	T	T
T	F	T	F	T	T	T	T
T	F	F	F	F	F	F	T
F	T	T	F	T	T	F	F
F	T	F	F	F	T	F	T
F	F	T	F	T	T	F	F
F	F	F	F	F	F	F	T

12. $2^{2^4} = 2^{16}$

15. a.

A	A'	$A \vee A'$
T	F	T
F	T	T

c.

A	B	$A \wedge B$	$A \wedge B \rightarrow B$
T	T	T	T
T	F	F	T
F	T	F	T
F	F	F	T

19. a. Assign $B' \wedge (A \rightarrow B)$ true
               A' false

From the second assignment, A is true. From the first assignment, B' is true (so B is false), and $A \rightarrow B$ is true. If $A \rightarrow B$ is true and A is true, then B is true. B is thus both true and false, and $[B' \wedge (A \rightarrow B)] \rightarrow A'$ is a tautology.

25. $A \wedge B$ is equivalent to $(A|B)|(A|B)$.

A	B	$A \wedge B$	$A|B$	$(A|B)|(A|B)$	$A \wedge B \leftrightarrow (A|B)|(A|B)$
T	T	T	F	T	T
T	F	F	T	F	T
F	T	F	T	F	T
F	F	F	T	F	T

A' is equivalent to $A|A$.

A	A'	$A|A$	$A' \leftrightarrow A|A$
T	F	F	T
F	T	T	T

27. If Percival is a liar, then his statement is false. Therefore it is false that there is at least one liar, and both Percival and Llewellyn must be truthtellers. But this is impossible, since we assumed Percival is a liar. Therefore Percival is a truthteller, and his statement is true. Because he said, "At least one of us is a liar," Llewellyn must be a liar. Therefore Percival is a truthteller and Llewellyn is a liar.

Exercises 1.2

1.
1.	$A \rightarrow (B \vee C)$	hyp
2.	B'	hyp
3.	C'	hyp
4.	$B' \wedge C'$	2, 3, con
5.	$(B \vee C)'$	4, De Morgan
6.	A'	1, 5, mt

4. $(A \rightarrow B) \wedge [A \rightarrow (B \rightarrow C)] \rightarrow (A \rightarrow C)$

1.	$A \rightarrow B$	hyp
2.	$A \rightarrow (B \rightarrow C)$	hyp
3.	A	hyp
4.	B	1, 3, mp
5.	$B \rightarrow C$	2, 3, mp
6.	C	4, 5, mp

6. $A' \wedge (A \vee B) \rightarrow B$

1.	A'	hyp
2.	$A \vee B$	hyp
3.	$(A')' \vee B$	2, dn
4.	$A' \rightarrow B$	3, imp
5.	B	1, 4, mp

12. $(P \vee Q) \wedge P' \rightarrow Q$

1.	$P \vee Q$	hyp
2.	P'	hyp
3.	$(P')' \vee Q$	1, dn
4.	$P' \rightarrow Q$	3, imp
5.	Q	2, 4, mp

18. $P \wedge P' \rightarrow Q$

1.	P	hyp
2.	P'	hyp
3.	$P \vee Q$	1, add
4.	$Q \vee P$	3, comm
5.	$(Q')' \vee P$	4, dn
6.	$Q' \rightarrow P$	5, imp
7.	$(Q')'$	2, 6, mt
8.	Q	7, dn

22. $(P \rightarrow Q) \wedge (P' \rightarrow Q) \rightarrow Q$

1.	$P \rightarrow Q$	hyp
2.	$P' \rightarrow Q$	hyp
3.	$Q' \rightarrow P'$	1, cont
4.	$Q' \rightarrow Q$	2, 3, hs
5.	$(Q')' \vee Q$	imp
6.	$Q \vee Q$	5, dn
7.	Q	6, self

27. $(A \wedge B)' \wedge (C' \wedge A)' \wedge (C \wedge B')' \to A'$

1. $(A \wedge B)'$	hyp
2. $(C' \wedge A)'$	hyp
3. $(C \wedge B')'$	hyp
4. $A' \vee B'$	1, De Morgan
5. $B' \vee A'$	4, comm
6. $B \to A'$	5, imp
7. $(C')' \vee A'$	2, De Morgan
8. $C' \to A'$	7, imp
9. $C' \vee (B')'$	3, De Morgan
10. $(B')' \vee C'$	9, comm
11. $B' \to C'$	10, imp
12. $B' \to A'$	8, 11, hs
13. $(B \to A') \wedge (B' \to A')$	6, 12, con
14. A'	Exercise 22

33. The argument is $(R \wedge (F' \vee N)) \wedge N' \wedge (A' \to F) \to (A \wedge R)$
A proof sequence is

1. $R \wedge (F' \vee N)$	hyp
2. N'	hyp
3. $A' \to F$	hyp
4. R	1, sim
5. $F' \vee N$	1, sim
6. $N \vee F'$	5, comm
7. F'	2, 6, ds
8. $F' \to (A')'$	3, cont
9. $(A')'$	7, 8, mp
10. A	9, dn
11. $A \wedge R$	4, 10, con

36. a.

A	B	C	$B \to C$	$A \to (B \to C)$	$A \wedge B$	$(A \wedge B) \to C$	$A \to (B \to C) \leftrightarrow (A \wedge B) \to C$
T	T	T	T	T	T	T	T
T	T	F	F	F	T	F	T
T	F	T	T	T	F	T	T
T	F	F	T	T	F	T	T
F	T	T	T	T	F	T	T
F	T	F	F	T	F	T	T
F	F	T	T	T	F	T	T
F	F	F	T	T	F	T	T

b. $A{\rightarrow}(B{\rightarrow}C) \Leftrightarrow A{\rightarrow}(B' \lor C) \Leftrightarrow A' \lor (B' \lor C) \Leftrightarrow (A' \lor B') \lor C \Leftrightarrow (A \land B)' \lor C \Leftrightarrow (A \land B) \rightarrow C$

c. By part (a) (or (b)),

$$[P_1 \land P_2 \land \cdots \land P_n] \rightarrow (R{\rightarrow}S) \Leftrightarrow (P_1 \land P_2 \land \cdots \land P_n \land R) {\rightarrow}S$$

which says to take each of $P_1, P_2, \ldots, P_n, R$ as hypotheses and deduce S.

Exercises 1.3

2. a. True (pick $y = 0$)
 b. True (pick $y = 0$)
 c. True (pick $y = -x$)
 d. False (no one y works for all x's)

4. a. True: domain is the integers, $A(x)$ is "x is even," $B(x)$ is "x is odd"
  False: domain is the positive integers, $A(x)$ is "$x > 0$," $B(x)$ is "$x \geq 1$"

6. a. $(\forall x)(D(x) \rightarrow S(x))$
 b. $(\exists x)[D(x) \land (R(x))'] $ or $[(\forall x)(D(x) \rightarrow R(x))]'$
 c. $(\forall x)[D(x) \land S(x) \rightarrow (R(x))']$

8. b. $(\forall x)[W(x) \rightarrow (L(x) \land C(x))']$
 c. $(\exists x)[L(x) \land (\forall y)(A(x, y) \rightarrow J(y))]$ or $(\exists x)(\forall y)[L(x) \land (A(x, y) \rightarrow J(y))]$

9. a. $(\forall x)(C(x) \land F(x))'$
 b. $(\exists x)[P(x) \land (\forall y)(S(x, y) \rightarrow F(y))]$ or $(\exists x)(\forall y)[P(x) \land (S(x, y) \rightarrow F(y))]$

12. a. John is handsome and Kathy loves John.
 b. All men are handsome.

13. a. 2

15. c. If there is a single x that is in relation P to all y, then for every y an x exists (this same x) that is in relation P to y.

16. a. Domain is the integers, $A(x)$ is "x is even," $B(x)$ is "x is odd."

Exercises 1.4

2. a. The domain is the set of integers, $P(x, y)$ is "$x < y$," and $Q(x, y)$ is "$x > y$." For every integer x, there is some integer that is larger and there is some integer that is smaller. But it is false that for every integer x there is some one integer that is both larger and smaller than x.
 b. To get to step 2, ei was performed on two different existential quantifiers, neither of which was in front with the whole rest of the wff as its scope. Also, both existential quantifiers were removed at once, with the same constant a substituted for the variable in each case; this should be done in two steps, and the second would then have to introduce a new constant not previously used in the proof. And at step 3, the existential quantifier was not inserted at the front of the wff.

4. $(\forall x)P(x) \rightarrow (\forall x)[P(x) \vee Q(x)]$

1. $(\forall x)P(x)$ hyp
2. $P(x)$ 1, ui
3. $P(x) \vee Q(x)$ 2, add
4. $(\forall x)(P(x) \vee Q(x))$ 3, ug (note that $P(x) \vee Q(x)$ was deduced from $(\forall x)P(x)$ in which x is not free)

9. $(\exists x)[A(x) \wedge B(x)] \rightarrow (\exists x)A(x) \wedge (\exists x)B(x)$

1. $(\exists x)(A(x) \wedge B(x))$	hyp	5. $(\exists x)A(x)$	3, eg
2. $A(a) \wedge B(a)$	1, ei	6. $(\exists x)B(x)$	4, eg
3. $A(a)$	2, sim	7. $(\exists x)A(x) \wedge (\exists x)B(x)$	5, 6, con
4. $B(a)$	2, sim		

13. $(\exists x)(\forall y)Q(x, y) \rightarrow (\forall y)(\exists x)Q(x, y)$

1. $(\exists x)(\forall y)Q(x, y)$	hyp	4. $(\exists x)Q(x, y)$	3, eg
2. $(\forall y)Q(a, y)$	1, ei	5. $(\forall y)(\exists x)Q(x, y)$	4, ug
3. $Q(a, y)$	2, ui		

17. $[P(x) \rightarrow (\exists y)Q(x, y)] \rightarrow (\exists y)[P(x) \rightarrow Q(x, y)]$

1. $P(x) \rightarrow (\exists y)Q(x, y)$ hyp
2. $P(x)$ temporary hyp
3. $(\exists y)Q(x, y)$ 1, 2, mp
4. $Q(x, a)$ 3, ei
5. $P(x) \rightarrow Q(x, a)$ temporary hyp discharged
6. $(\exists y)(P(x) \rightarrow Q(x, y))$ 5, eg

20. The argument is

$$(\forall x)(\forall y)[C(x) \wedge A(y) \rightarrow B(x, y)] \wedge C(s) \wedge (\exists x)(S(x) \wedge [B(s, x)]') \rightarrow$$
$$(\exists x)[A(x)]'$$

A proof sequence is

1. $(\forall x)(\forall y)[C(x) \wedge A(y) \rightarrow B(x, y)]$ hyp
2. $C(s)$ hyp
3. $(\forall y)[C(s) \wedge A(y) \rightarrow B(s, y)]$ 1, ui
4. $(\exists x)(S(x) \wedge [B(s, x)]')$ hyp
5. $S(a) \wedge [B(s, a)]'$ 4, ei
6. $C(s) \wedge A(a) \rightarrow B(s, a)$ 3, ui
7. $[B(s, a)]'$ 6, sim
8. $[C(s) \wedge A(a)]'$ 6, 7, mt
9. $[C(s)]' \vee [A(a)]'$ 8, De Morgan
10. $[[C(s)]']'$ 2, dn
11. $[A(a)]'$ 9, 10, ds
12. $(\exists x)[A(x)]'$ 11, eg

22. The argument is

$$(\forall x)(M(x) \rightarrow I(x) \vee G(x)) \wedge (\forall x)(G(x) \wedge L(x) \rightarrow F(x)) \wedge (I(j))' \wedge L(j) \rightarrow (M(j) \rightarrow F(j))$$

A proof sequence is

1. $(\forall x)(M(x) \rightarrow I(x) \vee G(x))$	hyp
2. $(\forall x)(G(x) \wedge L(x) \rightarrow F(x))$	hyp
3. $M(j) \rightarrow I(j) \vee G(j)$	1, ui
4. $G(j) \wedge L(j) \rightarrow F(j)$	2, ui
5. $M(j)$	hyp
6. $I(j) \vee G(j)$	3, 5, mp
7. $(I(j))'$	hyp
8. $G(j)$	6, 7, ds
9. $L(j)$	hyp
10. $G(j) \wedge L(j)$	8, 9, con
11. $F(j)$	4, 10, mp

Exercises 1.5

3. fish

7. *herbivore*(x) **if** *eat*(x, y) **and** *plant*(y)

11. a. is(*author-of*(mark-twain, hound-of-the-baskervilles))
    b. which(x: *author-of*(faulkner, x))
    c. *nonfiction-author*(x) **if** *author-of*(x, y) **and not**(*fiction*(y))
    d. which(x: *nonfiction-author*(x))

Exercises 1.6

1. $x + 1 = y - 1$,   or $x = y - 2$

5. The desired postcondition is $y = x(x - 1)$. Working back from the postcondition, using the assignment rule, gives

$$\{x(x - 1) = x(x - 1)\}$$
$$y = x - 1$$
$$\{xy = x(x - 1)\}$$
$$y = x * y$$
$$\{y = x(x - 1)\}$$

Because the precondition is always true, so is each subsequent assertion, including the postcondition.

7. The two implications to prove are

$$\{y = 0 \text{ and } y < 5\} \ y = y + 1 \ \{y = 1\}$$

and

$$\{y = 0 \text{ and } y \geq 5\} \ y = 5 \ \{y = 1\}$$

The first implication holds because

$$\{y = 0\} \ y = y + 1 \ \{y = 1\}$$

is true by the assignment rule, and the second is true because the antecedent is false. The program segment is correct by the conditional rule.

Chapter 2

Exercises 2.1

1. a. Converse: Healthy plant growth implies sufficient water.
Contrapositive: If there is not healthy plant growth, then there is not sufficient water.
b. Converse: Increased availability of information implies further technological advances.
Contrapositive: If there is not increased availability of information, then there are no further technological advances.
c. Converse: If there is a modification of the program, then errors will be introduced.
Contrapositive: No modification of the program implies that errors will not be introduced.
d. Converse: Good insulation or storm windows throughout implies fuel savings.
Contrapositive: Poor insulation and some windows are not storm windows implies no fuel savings.

3. $25 = 5^2 = 9 + 16 = 3^2 + 4^2$
$100 = (10)^2 = 36 + 64 = 6^2 + 8^2$
$169 = (13)^2 = 25 + 144 = 5^2 + (12)^2$

9. Let $x = 2m + 1$, $y = 2n + 1$, where m and n are integers. Then $x + y = (2m + 1) + (2n + 1) = 2m + 2n + 2 = 2(m + n + 1)$, where $m + n + 1$ is an integer, so $x + y$ is even.

13. Let $x = 2m$ where m is an integer. Then $x^2 = (2m)^2 = 4m^2$, where m^2 is an integer, so x^2 is divisible by 4.

17. If $x < y$, then multiplying both sides of the inequality by the positive numbers x and y in turn gives $x^2 < xy$ and $xy < y^2$; therefore $x^2 < xy < y^2$ or $x^2 < y^2$. For the other direction, if $x^2 < y^2$, then

$$y^2 - x^2 > 0 \qquad \text{(definition of } < \text{)}$$
$$(y + x)(y - x) > 0 \qquad \text{(factoring)}$$
$$y + x < 0 \text{ and } y - x < 0 \qquad \text{(a positive number is the product}$$
$$\text{of two negatives or two positives)}$$
or
$$y + x > 0 \text{ and } y - x > 0$$

But it cannot be that $y + x < 0$ because y and x are both positive; therefore $y + x > 0$ and $y - x > 0$ and $y > x$.

20. Let x and y be divisible by n. Then $x = k_1 n$ and $y = k_2 n$, where k_1 and k_2 are integers, and $x + y = k_1 n + k_2 n = (k_1 + k_2)n$, where $k_1 + k_2$ is an integer. Therefore $x + y$ is divisible by n.

23. Let $x = 2n + 1$. Then $x^2 = (2n + 1)^2 = 4n^2 + 4n + 1 = 4n(n + 1) + 1$. But $n(n + 1)$ is even (Exercise 11), so $n(n + 1) = 2k$ for some integer k. Therefore $x^2 = 4(2k) + 1 = 8k + 1$.

26. $m^2 n^2 = (mn)^2$

34. Proof: If x is even, then $x = 2n$ and

$$x(x + 1)(x + 2) = (2n)(2n + 1)(2n + 2) = 2[(n)(2n + 1)(2n + 2)]$$

which is even. If x is odd, then $x = 2n + 1$ and

$$x(x + 1)(x + 2) = (2n + 1)(2n + 2)(2n + 3) = 2[(2n + 1)(n + 1)(2n + 3)]$$

which is even.

37. Proof: If x is even, then $x = 2n$ and

$$2n + (2n)^3 = 2n + 8n^3 = 2(n + 4n^3)$$

which is even. If x is odd, then $x = 2n + 1$ and

$$(2n + 1) + (2n + 1)^3 = (2n + 1) + (8n^3 + 12n^2 + 6n + 1) =$$
$$8n^3 + 12n^2 + 8n + 2 = 2(4n^3 + 6n^2 + 4n + 1)$$

which is even.

44. Proof: Let x and y be rational numbers, $x = p/q$, $y = r/s$ with p, q, r, s integers and q, $s \neq 0$. Then $x + y = p/q + r/s = (ps + rq)/qs$, where $ps + rq$ and qs are integers with $qs \neq 0$. Thus $x + y$ is rational.

48. Angle 6 plus angle 5 plus the right angle sum to $180°$ by the first fact. The right angle is $90°$ by the fourth fact. Therefore angle 6 plus angle 5 sum to $90°$. Angle 6 is the same size as angle 3 by the second fact. Therefore angle 3 plus angle 5 sum to $90°$.

Exercises 2.2

1. $P(1)$: $4(1) - 2 = 2(1)^2$ or $2 = 2$ true
Assume $P(k)$: $2 + 6 + 10 + \cdots + (4k - 2) = 2k^2$
Show $P(k + 1)$: $2 + 6 + 10 + \cdots + [4(k + 1) - 2] = 2(k + 1)^2$

$$2 + 6 + 10 + \cdots + [4(k + 1) - 2] \qquad \text{left side of } P(k + 1)$$
$$= 2 + 6 + 10 + \cdots + (4k - 2) + [4(k + 1) - 2]$$
$$= 2k^2 + 4(k + 1) - 2 \qquad \text{using } P(k)$$
$$= 2k^2 + 4k + 2$$
$$= 2(k^2 + 2k + 1)$$
$$= 2(k + 1)^2 \qquad \text{right side of } P(k + 1)$$

3. $P(1)$: $1 = 1(2(1) - 1)$ true

Assume $P(k)$: $1 + 5 + 9 + \cdots + (4k - 3) = k(2k - 1)$

Show $P(k + 1)$: $1 + 5 + 9 + \cdots + [4(k + 1) - 3] = (k + 1)[2(k + 1) - 1]$

$$1 + 5 + 9 + \cdots + [4(k + 1) - 3] \qquad \text{left side of } P(k + 1)$$
$$= 1 + 5 + 9 + \cdots + (4k - 3) + [4(k + 1) - 3]$$
$$= k(2k - 1) + 4(k + 1) - 3 \qquad \text{using } P(k)$$
$$= 2k^2 - k + 4k + 1$$
$$= 2k^2 + 3k + 1$$
$$= (k + 1)(2k + 1)$$
$$= (k + 1)[2(k + 1) - 1] \qquad \text{right side of } P(k + 1)$$

5. $P(1)$: $6 - 2 = 1[3(1) + 1]$ true

Assume $P(k)$: $4 + 10 + 16 + \cdots + (6k - 2) = k(3k + 1)$

Show $P(k + 1)$: $4 + 10 + 16 + \cdots + [6(k + 1) - 2] = (k + 1)[3(k + 1) + 1]$

$$4 + 10 + 16 + \cdots + [6(k + 1) - 2] \qquad \text{left side of } P(k + 1)$$
$$= 4 + 10 + 16 + \cdots + (6k - 2) + [6(k + 1) - 2]$$
$$= k(3k + 1) + 6(k + 1) - 2 \qquad \text{using } P(k)$$
$$= 3k^2 + k + 6k + 4$$
$$= 3k^2 + 7k + 4$$
$$= (k + 1)(3k + 4)$$
$$= (k + 1)[3(k + 1) + 1] \qquad \text{right side of } P(k + 1)$$

9. $P(1)$: $1^2 = 1(2 - 1)(2 + 1)/3$ true

Assume $P(k)$: $1^2 + 3^2 + \cdots + (2k - 1)^2 = k(2k - 1)(2k + 1)/3$

Show $P(k + 1)$: $1^2 + 3^2 + \cdots + [2(k + 1) - 1]^2$
$$= (k + 1)(2(k + 1) - 1)(2(k + 1) + 1)/3$$

$$1^2 + 3^2 + \cdots + [2(k + 1) - 1]^2 \qquad \text{left side of } P(k + 1)$$
$$= 1^2 + 3^2 + \cdots + (2k - 1)^2 + [2(k + 1) - 1]^2$$
$$= \frac{k(2k - 1)(2k + 1)}{3} + [2(k + 1) - 1]^2 \qquad \text{using } P(k)$$
$$= \frac{k(2k - 1)(2k + 1)}{3} + (2k + 1)^2$$
$$= (2k + 1)\left[\frac{k(2k - 1)}{3} + 2k + 1\right] \qquad \text{factoring}$$
$$= (2k + 1)\left[\frac{2k^2 - k + 6k + 3}{3}\right]$$
$$= \frac{(2k + 1)(2k^2 + 5k + 3)}{3}$$
$$= (2k + 1)(k + 1)(2k + 3)/3$$
$$= (k + 1)(2(k + 1) - 1)(2(k + 1) + 1)/3 \qquad \text{right side of } P(k + 1)$$

13. $P(1)$: $\dfrac{1}{1 \cdot 2} = \dfrac{1}{1 + 1}$ true

Assume $P(k)$: $\dfrac{1}{1 \cdot 2} + \dfrac{1}{2 \cdot 3} + \cdots + \dfrac{1}{k(k + 1)} = \dfrac{k}{k + 1}$

Show $P(k + 1)$: $\dfrac{1}{1 \cdot 2} + \dfrac{1}{2 \cdot 3} + \cdots + \dfrac{1}{(k + 1)(k + 2)} = \dfrac{k + 1}{k + 2}$

$\dfrac{1}{1 \cdot 2} + \dfrac{1}{2 \cdot 3} + \cdots + \dfrac{1}{(k + 1)(k + 2)}$ left side of $P(k + 1)$

$= \dfrac{1}{1 \cdot 2} + \dfrac{1}{2 \cdot 3} + \cdots + \dfrac{1}{k(k + 1)} + \dfrac{1}{(k + 1)(k + 2)}$

$= \dfrac{k}{k + 1} + \dfrac{1}{(k + 1)(k + 2)}$ using $P(k)$

$= \dfrac{k(k + 2) + 1}{(k + 1)(k + 2)}$

$= \dfrac{k^2 + 2k + 1}{(k + 1)(k + 2)}$

$= \dfrac{(k + 1)^2}{(k + 1)(k + 2)}$

$= \dfrac{k + 1}{k + 2}$ right side of $P(k + 1)$

15. $P(1)$: $1^2 = (-1)^2(1)(2)/2$ true

Assume $P(k)$: $1^2 - 2^2 + \cdots + (-1)^{k+1}k^2 = (-1)^{k+1}(k)(k + 1)/2$

Show $P(k + 1)$: $1^2 - 2^2 + \cdots + (-1)^{k+2}(k + 1)^2 = (-1)^{k+2}(k + 1)(k + 2)/2$

$1^2 - 2^2 + \cdots + (-1)^{k+2}(k + 1)^2$ left side of $P(k + 1)$

$= 1^2 - 2^2 + \cdots + (-1)^{k+1}k^2 + (-1)^{k+2}(k + 1)^2$

$= (-1)^{k+1}(k)(k + 1)/2 + (-1)^{k+2}(k + 1)^2$ using $P(k)$

$= [(-1)^{k+1}(k)(k + 1) + 2(-1)^{k+2}(k + 1)^2]/2$

$= (-1)^{k+2}(k + 1)[k(-1)^{-1} + 2(k + 1)]/2$

$= (-1)^{k+2}(k + 1)[-k + 2k + 2]/2$

$= (-1)^{k+2}(k + 1)(k + 2)/2$ right side of $P(k + 1)$

21. $P(1)$: $a = \dfrac{a - ar}{1 - r} = \dfrac{a(1 - r)}{1 - r}$ true

Assume $P(k)$: $a + ar + \cdots + ar^{k-1} = \dfrac{a - ar^k}{1 - r}$

Show $P(k + 1)$: $a + ar + \cdots + ar^k = \dfrac{a - ar^{k+1}}{1 - r}$

$a + ar + \cdots + ar^k$ right side of $P(k + 1)$

$= a + ar + \cdots + ar^{k-1} + ar^k$

$= \dfrac{a - ar^k}{1 - r} + ar^k$ using $P(k)$

$= \dfrac{a - ar^k + ar^k(1 - r)}{1 - r}$

$= \dfrac{a - ar^{k+1}}{1 - r}$ right side of $P(k + 1)$

25. $P(2)$: $2^2 > 2 + 1$ true
Assume $P(k)$: $k^2 > k + 1$
Show $P(k + 1)$: $(k + 1)^2 > k + 2$

$(k + 1)^2$
$= k^2 + 2k + 1$
$> (k + 1) + 2k + 1$ using $P(k)$
$= 3k + 2$
$> k + 2$

29. $P(4)$: $2^4 < 4!$ or $16 < 24$ true
Assume $P(k)$: $2^k < k!$
Show $P(k + 1)$: $2^{k+1} < (k + 1)!$

2^{k+1}
$= 2 \cdot 2^k$
$< 2 \cdot k!$ using $P(k)$
$< (k + 1)k!$ because $k \geq 4$
$= (k + 1)!$

34. $P(2)$: $1 + 2 < 2^2$ or $3 < 4$ true
Assume $P(k)$: $1 + 2 + \cdots + k < k^2$
Show $P(k + 1)$: $1 + 2 + \cdots + (k + 1) < (k + 1)^2$

$1 + 2 + \cdots + (k + 1)$
$= 1 + 2 + \cdots + k + (k + 1)$
$< k^2 + k + 1$ using $P(k)$
$< k^2 + 2k + 1 = (k + 1)^2$

36. $P(1)$: $2^3 - 1 = 8 - 1 = 7$ and $7|7$ true
Assume $P(k)$: $7|2^{3k} - 1$ so $2^{3k} - 1 = 7m$ or $2^{3k} = 7m + 1$ for some integer m
Show $P(k + 1)$: $7|2^{3(k+1)} - 1$

$2^{3(k+1)} - 1 = 2^{3k+3} - 1 = 2^{3k} \cdot 2^3 - 1$
$= (7m + 1)2^3 - 1$ using $P(k)$
$= 7(2^3 m) + 8 - 1$
$= 7(2^3 m + 1)$ where $2^3 m + 1$ is an integer, so $7|2^{3(k+1)} - 1$

40. $P(1)$: $2 + (-1)^2 = 2 + 1 = 3$ and $3|3$ true
Assume $P(k)$: $3|2k + (-1)^{k+1}$ so $2^k + (-1)^{k+1} = 3m$ or $2^k = 3m - (-1)^{k+1}$ for some integer m
Show $P(k + 1)$: $3|2^{k+1} + (-1)^{k+2}$

$2^{k+1} + (-1)^{k+2} = 2 \cdot 2^k + (-1)^{k+2}$
$= 2(3m - (-1)^{k+1}) + (-1)^{k+2}$ using $P(k)$
$= 3(2m) - 2(-1)^{k+1} + (-1)^{k+2}$
$= 3(2m) + (-1)^{k+1}(-2 + (-1))$
$= 3(2m) + (-1)^{k+1}(-3)$
$= 3(2m - (-1)^{k+1})$ where $2m - (-1)^{k+1}$ is an integer, so $3|2^{k+1} + (-1)^{k+2}$

44. $P(1)$: $10 + 3 \cdot 4^3 + 5 = 10 + 192 + 5 = 207 = 9 \cdot 23$ true
Assume $P(k)$: $9 | 10k + 3 \cdot 4^{k+2} + 5$ so $10^k + 3 \cdot 4^{k+2} + 5 = 9m$ or
$10^k = 9m - 3 \cdot 4^{k+2} - 5$ for some integer m
Show $P(k + 1)$: $9 | 10^{k+1} + 3 \cdot 4^{k+3} + 5$

$$10^{k+1} + 3 \cdot 4^{k+3} + 5 = 10 \cdot 10^k + 3 \cdot 4^{k+3} + 5$$
$$= 10(9m - 3 \cdot 4^{k+2} - 5) + 3 \cdot 4^{k+3} + 5 \qquad \text{using } P(k)$$
$$= 9(10m) - 30 \cdot 4^{k+2} - 50 + 3 \cdot 4^{k+2} \cdot 4 + 5$$
$$= 9(10m) - 45 - 3 \cdot 4^{k+2}(10 - 4) = 9(10m - 5) - 18 \cdot 4^{k+2}$$
$$= 9(10m - 5 - 2 \cdot 4^{k+2}) \text{ where } 10m - 5 - 2 \cdot 4^{k+2} \text{ is an integer,}$$
so $9 | 10^{k+1} + 3 \cdot 4^{k+3} + 5$

48. $P(1)$: $\cos \theta + i \sin \theta = \cos \theta + i \sin \theta$ true
Assume $P(k)$: $(\cos \theta + i \sin \theta)^k = \cos k\theta + i \sin k\theta$
Show $P(k + 1)$: $(\cos \theta + i \sin \theta)^{k+1} = \cos (k+1)\theta + i \sin (k+1)\theta$

$$(\cos \theta + i \sin \theta)^{k+1} = (\cos \theta + i \sin \theta)^k (\cos \theta + i \sin \theta)$$
$$= (\cos k\theta + i \sin k\theta)(\cos \theta + i \sin \theta) \qquad \text{using } P(k)$$
$$= \cos k\theta \cos \theta + i \sin k\theta \cos \theta + i \cos k\theta \sin \theta + i^2 \sin k\theta \sin \theta$$
$$= \cos k\theta \cos \theta - \sin k\theta \sin \theta + i(\sin k\theta \cos \theta + \cos k\theta \sin \theta)$$
$$= \cos(k\theta + \theta) + i \sin(k\theta + \theta)$$
$$= \cos(k + 1)\theta + i \sin(k + 1)\theta$$

50. The statement to be proved is that $n(n + 1)(n + 2)$ is divisible by 3 for $n \geq 1$.
$P(1)$: $1(1 + 1)(1 + 2) = 6$ is divisible by 3 true
Assume $P(k)$: $k(k + 1)(k + 2) = 3m$ for some integer m
Show $P(k + 1)$: $(k + 1)(k + 2)(k + 3)$ is divisible by 3

$$(k + 1)(k + 2)(k + 3) = (k + 1)(k + 2)k + (k + 1)(k + 2)3$$
$$= 3m + (k + 1)(k + 2)3 \qquad \text{using } P(k)$$
$$= 3[m + (k + 1)(k + 2)]$$

56. $P(1)$ is $1 = 1 + 1$ which is not true.

59. For the base case, the simplest such wff is a single statement letter, which has 1 symbol; 1 is odd. Assume that for any such wff with r symbols, $1 \leq r \leq k$, r is odd. Consider a wff with $k + 1$ symbols. It must have the form $(P) \wedge (Q)$, $(P) \vee (Q)$, or $(P) \rightarrow (Q)$, where P has r_1 symbols, $1 \leq r_1 < k$, and Q has r_2 symbols, $1 \leq r_2 < k$. By the inductive hypothesis, both r_1 and r_2 are odd. The number of symbols in the original wff is then $r_1 + r_2 + 5$ (four parentheses plus one connective), which is odd.

61. $P(2)$ and $P(3)$ are true by the equations $2 = 2$ and $3 = 3$. Now assume that $P(r)$ is true for any r, $2 \leq r \leq k$, and consider $P(k + 1)$. We may assume that $k + 1 \geq 4$, so that $(k + 1) - 2 \geq 2$ and by the inductive hypothesis can be written as a sum of 2s and 3s. Adding an additional 2 gives $k + 1$ as a sum of 2s and 3s.

64. $P(64)$, $P(65)$, $P(66)$, $P(67)$, and $P(68)$ are true by the equations

$$64 = 6(5) + 2(17)$$
$$65 = 13(5)$$
$$66 = 3(5) + 3(17)$$
$$67 = 10(5) + 17$$
$$68 = 4(17)$$

Now assume that $P(r)$ is true for any r, $64 \le r \le k$, and consider $P(k + 1)$. We may assume that $k + 1 \ge 69$, so that $(k + 1) - 5 \ge 64$ and by the inductive hypothesis can be written as a sum of 5s and 17s. Adding an additional 5 gives $k + 1$ as a sum of 5s and 17s.

Exercises 2.3

2. $Q(0)$: $j_0 = (i_0 - 1)!$ true since $j = 1$, $i = 2$ before loop is entered and $1 = 1!$
Assume $Q(k)$: $j_k = (i_k - 1)!$
Show $Q(k + 1)$: $j_{k+1} = (i_{k+1} - 1)!$

$j_{k+1} = j_k \cdot i_k = (i_k - 1)! i_k = (i_k)! = (i_{k+1} - 1)!$
At loop termination, $j = (i - 1)!$ and $i = x + 1$ so $j = x!$

6. $735 = 8 * 90 + 15$, $90 = 6 * 15 + 0$, so $\gcd(735, 90) = 15$

10. Q: $j = x * y^i$
$Q(0)$: $j_0 = x * y^{i_0}$ true since $j = x$, $i = 0$ before loop is entered
Assume $Q(k)$: $j_k = x * y^{i_k}$
Show $Q(k + 1)$: $j_{k+1} = x * y^{i_{k+1}}$

$j_{k+1} = j_k * y = x * y^{i_k} * y = x * y^{i_k+1} = x * y^{i_{k+1}}$
At loop termination, $j = x * y^i$ and $i = n$, so $j = x * y^n$

12. Q: $j = x * i!$
$Q(0)$: $j_0 = x * i_0!$ true since $j = x$, $i = 1$ before loop is entered
Assume $Q(k)$: $j_k = x * (i_k)!$
Show $Q(k + 1)$: $j_{k+1} = x * (i_{k+1})!$

$j_{k+1} = j_k * (i_k + 1) = x * (i_k)!(i_k + 1) = x * (i_k + 1)! = x * (i_{k+1})!$
At loop termination, $j = x * i!$ and $i = n$ so $j = x * n!$

Exercises 2.4

1. $10, 20, 30, 40, 50$

4. $1, 1 + \dfrac{1}{2}, 1 + \dfrac{1}{2} + \dfrac{1}{3}, 1 + \dfrac{1}{2} + \dfrac{1}{3} + \dfrac{1}{4}, 1 + \dfrac{1}{2} + \dfrac{1}{3} + \dfrac{1}{4} + \dfrac{1}{5}$

7. $2, 2, 6, 14, 34$

11. $F(n + 1) + F(n - 2) = F(n - 1) + F(n) + F(n - 2)$
$= [F(n - 2) + F(n - 1)] + F(n)$
$= F(n) + F(n) = 2F(n)$

16. $n = 1: F(1) = F(3) - 1$ or $1 = 2 - 1$ ⠀⠀⠀⠀⠀⠀true
Assume true for $n = k: F(1) + \cdots + F(k) = F(k + 2) - 1$
Show true for $n = k + 1: F(1) + \cdots + F(k + 1) = F(k + 3) - 1$

$F(1) + \cdots + F(k + 1)$
$= F(1) + \cdots + F(k) + F(k + 1)$
$= F(k + 2) - 1 + F(k + 1)$ ⠀⠀⠀⠀⠀⠀inductive hypothesis
$= F(k + 3) - 1$ ⠀⠀⠀⠀⠀⠀recurrence relation

20. $n = 1: F(4) = 2F(2) + F(1)$ or $3 = 2(1) + 1$ ⠀⠀true
$n = 2: F(5) = 2F(3) + F(2)$ or $5 = 2(2) + 1$ ⠀⠀true
Assume for all $r, 1 \le r \le k, F(r + 3) = 2F(r + 1) + F(r)$
Show that $F(k + 4) = 2F(k + 2) + F(k + 1)$ where $k + 1 \ge 3$

$F(k + 4) = F(k + 2) + F(k + 3)$ ⠀⠀⠀recurrence relation
$= 2F(k) + F(k - 1) + 2F(k + 1) + F(k)$ ⠀⠀inductive hypothesis
$= 2[F(k) + F(k + 1)] + [F(k - 1) + F(k)]$
$= 2F(k + 2) + F(k + 1)$ ⠀⠀⠀recurrence relation

26. Yes, this sequence continues to generate all of the Fibonacci numbers; this is **true** because of Exercise 16: $F(1) + F(2) + \cdots + F(n) = F(n + 2) - 1$, or $1 + F(1) + F(2) + \cdots + F(n) = F(n + 2)$.

30. $S(1) = a$
$S(n) = rS(n - 1)$ for $n \ge 2$

32. a. $A(1) = 50{,}000$
$A(n) = 3A(n - 1)$ for $n \ge 2$
b. 4

36. (a), (b), and (e)

38. 1. Any unary predicate in x is a wff.
2. If P and Q are unary predicate wffs in x, so are $(P \wedge Q)$, $(P \vee Q)$, $(P \to Q)$, (P'), $(P \leftrightarrow Q)$, $(\forall x)P$, and $(\exists x)P$.

42. 1. $\lambda^R = \lambda$
2. If x is a string with a single character, $x^R = x$.
3. If $x = yz$, then $x^R = z^R y^R$.

44. $\langle\text{positive digit}\rangle ::= 1|2|3|4|5|6|7|8|9$
$\langle\text{digit}\rangle ::= 0|\langle\text{positive digit}\rangle$
$\langle\text{positive integer}\rangle ::= \langle\text{positive digit}\rangle|\langle\text{positive integer}\rangle\langle\text{digit}\rangle$

46. $1! = 1$
$n! = n(n - 1)!$ for $n \ge 2$

50. $A \vee (B_1 \wedge B_2) \Leftrightarrow (A \vee B_1) \wedge (A \vee B_2)$ ⠀⠀⠀by equivalence 3a

Assume that
$A \vee (B_1 \wedge \cdots \wedge B_k) \Leftrightarrow (A \vee B_1) \wedge \cdots \wedge (A \vee B_k)$

Then

$$A \lor (B_1 \land \cdots \land B_{k+1})$$
$$= A \lor [(B_1 \land \cdots \land B_k) \land B_{k+1}] \qquad \text{by Exercise 49}$$
$$\Leftrightarrow (A \lor (B_1 \land \cdots \land B_k)) \land (A \lor B_{k+1}) \qquad \text{by equivalence 3a}$$
$$\Leftrightarrow [(A \lor B_1) \land \cdots \land (A \lor B_k)] \land (A \lor B_{k+1}) \qquad \text{by inductive hypothesis}$$
$$\Leftrightarrow (A \lor B_1) \land \cdots \land (A \lor B_{k+1}) \qquad \text{by Exercise 49}$$

The proof of the other statement is similar.

54. **if** $n = 1$ **then**
    return 1

else
    return $S(n - 1) + (n - 1)$

57. **if** $n = 1$ **then**
    return p

else
    if odd(n) **then**
        return $S(n - 1) + (n - 1) * q$
    else
        return $S(n - 1) - (n - 1) * q$

60. If the list has 1 element or 0 elements, then we are done, else exchange the first and last element in the list and invoke the algorithm on the list minus its first and last elements.

63.   4, 10, −6, 2, 5
       4, 5, −6, 2, 10
       4, 2, −6, 5, 10
       −6, 2, 4, 5, 10

68. The recurrence relation matches equation (6) with $c = 1$ and $g(n) = 5$. From equation (8), the solution is

$$S(n) = 5 + \sum_{i=2}^{n} 5 = 5 + (n - 1)5 = n(5)$$

71. The recurrence relation matches equation (6) with $c = 1$ and $g(n) = n$. From equation (8), the solution is

$$A(n) = 1 + \sum_{i=2}^{n} i = 1 + 2 + 3 + \cdots + n = n(n + 1)/2$$

74. $F(n) = nF(n - 1)$
    $= n[(n - 1)F(n - 2)] = n(n - 1)F(n - 2)$
    $= n(n - 1)[(n - 2)F(n - 3)] = n(n - 1)(n - 2)F(n - 3)$
In general, $F(n) = n(n - 1)(n - 2) \cdots (n - (k - 1))F(n - k)$
When $n - k = 1, k = n - 1,$
    $F(n) = n(n - 1)(n - 2) \cdots (2)F(1) = n(n - 1)(n - 2) \cdots (2)(1) = n!$

Now prove by induction that $F(n) = n!$

$F(1): F(1) = 1! = 1$ true

Assume $F(k): F(k) = k!$

Show $F(k + 1): F(k + 1) = (k + 1)!$

$F(k + 1) = (k + 1)F(k) = (k + 1)k! = (k + 1)!$

77. The recurrence relation is $T(n) = 0.95T(n - 1)$ with a base case $T(1) = X$. This is a linear, first-order recurrence relation with constant coefficients, so equation (8) applies and gives the solution $T(n) = (0.95)^{n-1}(X)$. At the end of 20 years (the beginning of the 21st year), the amount is $T(21) = (0.95)^{20}(X) = 0.358(X)$, which is slightly more than one-third the original amount X.

80. The recurrence relation is $P(n) = P(n - 1) + n$, with $P(1) = 1$. Equation (8) applies and gives the solution $P(n) = 1 + \sum_{i=2}^{n} i = 1 + 2 + 3 + \cdots + n = n(n+1)/2$.

Exercises 2.5

2. If the list is empty, write "not found," else compare x to the first item in the current list segment. If that item equals x, write "found," otherwise remove that first item and search the remainder of the list segment. Initially, the list segment is the entire list. More formally:

SequentialSearch(list L; integer i, n; itemtype x)
//searches list L from $L(i)$ to $L(n)$ for item x

if $i > n$ **then**
    write("not found")
else
    if $L(i) = x$ **then**
        write("found")
    else
        SequentialSearch(L, $i + 1$, n, x)
    end if
end if

5. This is in the form of equation (1) with $c = 2$ and $g(n) = 3$. By equation (6), the solution is

$$2^{\log n}(1) + \sum_{i=1}^{\log n} 2^{(\log n)-i}(3) = 2^{\log n} + 3[2^{(\log n)-1} + 2^{(\log n)-2} + \cdots + 2^0]$$
$$= 2^{\log n} + 3[2^{\log n} - 1] = n + 3(n - 1) = 4n - 3$$

8. $n - 1$ compares are always needed—every element after the first must be considered a potential new maximum.

11. a. The merged list is 1, 4, 5, 6, 8, 9
   3 comparisons—6 vs. 1, 6 vs. 4, and 6 vs. 5
b. The merged list is 1, 2, 3, 4, 5, 8
   4 comparisons—1 vs. 2, 5 vs. 2, 5 vs. 3, 5 vs. 4
c. The merged list is 0, 1, 2, 3, 4, 7, 8, 9, 10
   8 comparisons—0 vs. 1, 2 vs. 1, 2 vs. 8, 3 vs. 8, 4 vs. 8, 7 vs. 8, 10 vs. 8, 10 vs. 9

13. $C(1) = 0$ (no comparisons are required; a 1-element list is already sorted)

$C(n) = 2C(n/2) + (n - 1)$ ($C(n/2)$ comparisons are required
   for $n = 2^m, n \geq 2$ for each half, and $n - 1$
   comparisons are required to
   merge the two sorted halves.)

17. From Exercise 16, $a = n \geq F(m + 2)$. By Exercise 23 of Section 2.4,
$F(m + 2) > \left(\dfrac{3}{2}\right)^{m+1}$. Therefore $\left(\dfrac{3}{2}\right)^{m+1} < F(m + 2) \leq n$.

Chapter 3

Exercises 3.1

1. a. T   **b.** F   **c.** F   **d.** F

4. a. $\{0, 1, 2, 3, 4\}$
b. $\{4, 6, 8, 10\}$
c. {Washington, Adams, Jefferson}
d. $\varnothing$
e. {Maine, Vermont, New Hampshire, Massachusetts, Connecticut, Rhode Island}
f. $\{-3, -2, -1, 0, 1, 2, 3\}$

8. If $A = \{x \mid x = 2^n$ for n a positive integer$\}$, then $16 \in A$.
But if $A = \{x \mid x = 2 + n(n - 1)$ for n a positive integer$\}$, then $16 \notin A$.

10. a. F; $\{1\} \in S$ but $\{1\} \notin R$ **b.** T   **c.** F; $\{1\} \in S$, not $1 \in S$
d. F; 1 is not a set; the correct statement is $\{1\} \subseteq U$ **e.** T   **f.** F; $1 \notin S$

13. Let $(x, y) \in A$. Then (x, y) lies within 3 units of the point $(1, 4)$, so by the distance formula, $\sqrt{(x - 1)^2 + (y - 4)^2} \leq 3$, or $(x - 1)^2 + (y - 4)^2 \leq 9$, which means $(x - 1)^2 + (y - 4)^2 \leq 25$, so $(x, y) \in B$. The point $(6, 4)$ satisfies the inequality $(x - 1)^2 + (y - 4)^2 \leq 25$, so $(6, 4) \in B$, but $(6, 4)$ is not within 3 units of $(1, 4)$, so $(6, 4)$ does not belong to A.

16. a. T   **b.** F   **c.** F   **d.** T   **e.** T   **f.** F

20. $\wp(S) = \{\varnothing, \{1\}, \{2\}, \{3\}, \{4\}, \{1, 2\}, \{1, 3\}, \{1, 4\}, \{2, 3\}, \{2, 4\}, \{3, 4\}, \{1, 2, 3\}, \{1, 2, 4\}, \{1, 3, 4\}, \{2, 3, 4\}, \{1, 2, 3, 4\}\}$; $2^4 = 16$ elements

24. $A = \{x, y\}$

27. Let $x \in A$. Then $\{x\} \in \wp(A)$, so $\{x\} \in \wp(B)$ and $x \in B$. Thus $A \subseteq B$. A similar argument shows that $B \subseteq A$ so that $A = B$.

30. a. Binary operation   **b.** No; $0 \circ 0 \notin \mathbb{N}$ **c.** Binary operation

34. a. 13   **b.** 2   **c.** 28

36. a. $\{1, 2, 4, 5, 6, 8, 9\}$ **b.** $\{4, 5\}$ **c.** $\{2, 4\}$ **h.** $\{0, 1, 2, 3, 6, 7, 8, 9\}$

40. a. B' **b.** $B \cap C$ **c.** $A \cap B$ **d.** $B' \cap C$ **e.** $B' \cap C'$ or $(B \cup C)'$ or $B' - C$

42. a. C' **b.** $B \cap D$ **c.** $A \cap B$

44. a. T
   c. F (Let $A = \{1, 2, 3\}$, $B = \{1, 3, 5\}$, $S = \{1, 2, 3, 4, 5\}$. Then $(A \cap B)' = \{2, 4, 5\}$ but $A' \cap B' = \{4, 5\} \cap \{2, 4\} = \{4\}$.)
   e. F (Take A, B, and S as in (c), then $A - B = \{2\}$, $(B - A)' = \{1, 2, 3, 4\}$.)

45. a. $B \subseteq A$

48. Let $C \in \wp(A) \cap \wp(B)$. Then $C \in \wp(A)$ and $C \in \wp(B)$, from which $C \subseteq A$ and $C \subseteq B$, so $C \subseteq A \cap B$ or $C \in \wp(A \cap B)$. Therefore $\wp(A) \cap \wp(B) \subseteq \wp(A \cap B)$. The same argument works in reverse.

51. Suppose $A \cap B \neq \emptyset$. Let $x \in A \cap B$. Then $x \in A$ and $x \in B$ so $x \in A \cup B$, but $x \notin A - B$ and $x \notin B - A$ so $x \notin (A - B) \cup (B - A)$, which contradicts the equality of $A \cup B$ and $(A - B) \cup (B - A)$.

56. c. $x \in A \cup (B \cap A) \leftrightarrow x \in A$ or $x \in (B \cap A) \leftrightarrow x \in A$ or $(x \in B$ and $x \in A) \leftrightarrow x \in A$

57. a.
$$
\begin{aligned}
(A \cup B) \cap (A \cup B') &= A \cup (B \cap B') &&\text{(3a)} \\
&= A \cup \emptyset &&\text{(5b)} \\
&= A &&\text{(4a)}
\end{aligned}
$$
The dual is $(A \cap B) \cup (A \cap B') = A$.

58. e. $x \in (A')' \leftrightarrow x \notin A' \leftrightarrow x \notin \{y \mid y \notin A\} \leftrightarrow x \in A$

59. a.
$$
\begin{aligned}
A \cap (B \cup A') &= (A \cap B) \cup (A \cap A') &&\text{(3b)} \\
&= (A \cap B) \cup \emptyset &&\text{(5b)} \\
&= A \cap B &&\text{(4a)} \\
&= B \cap A
\end{aligned}
$$

61. The proof is by induction on n. For $n = 3$,
$$
\begin{aligned}
(A_1) \cup (A_2 \cup A_3) &= (A_1 \cup A_2) \cup A_3 \\
&= A_1 \cup A_2 \cup A_3
\end{aligned}
$$
set identity 2a
by the answer
to Exercise 60(b)

Assume that for $n = k$ and $1 \leq p \leq k - 1$,
$$(A_1 \cup \cdots \cup A_p) \cup (A_{p+1} \cup \cdots \cup A_k) = A_1 \cup \cdots \cup A_k$$
Then for $1 \leq p \leq k$,
$$
\begin{aligned}
&(A_1 \cup \cdots \cup A_p) \cup (A_{p+1} \cup \cdots \cup A_{k+1}) \\
&= (A_1 \cup \cdots \cup A_p) \cup [(A_{p+1} \cup \cdots \cup A_k) \cup A_{k+1}] &&\text{by the answer to 60(b)} \\
&= [(A_1 \cup \cdots \cup A_p) \cup (A_{p+1} \cup \cdots \cup A_k)] \cup A_{k+1} &&\text{by set identity 2a} \\
&= (A_1 \cup \cdots \cup A_k) \cup A_{k+1} &&\text{by inductive hypothesis} \\
&= A_1 \cup \cdots \cup A_{k+1} &&\text{by the answer to 60(b)}
\end{aligned}
$$

64. a. Proof is by induction on n.

For $n = 2$, $B \cup (A_1 \cap A_2) = (B \cup A_1) \cap (B \cup A_2)$ by identity 3a

Assume that $B \cup (A_1 \cap \cdots \cap A_k) = (B \cup A_1) \cap \cdots \cap (B \cup A_k)$

Then $B \cup (A_1 \cap \cdots \cap A_{k+1})$

$$= B \cup ((A_1 \cap \cdots \cap A_k) \cap A_{k+1}) \qquad \text{by Exercise 62b}$$
$$= (B \cup (A_1 \cap \cdots \cap A_k)) \cap (B \cup A_{k+1}) \qquad \text{by identity 3a}$$
$$= ((B \cup A_1) \cap \cdots \cap (B \cup A_k)) \cap (B \cup A_{k+1}) \qquad \text{by inductive hyp}$$
$$= (B \cup A_1) \cap \cdots \cap (B \cup A_{k+1}) \qquad \text{by Exercise 62b}$$

69. If T is a nonempty set, then by the principle of well-ordering, T has a smallest member t_0. Then $P(t_0)$ is not true, so by statement $1'$, $t_0 \neq 1$. Also $P(r)$ is true for all r, $1 \leq r \leq t_0 - 1$. This contradicts the implication in $2'$, so T is the empty set and therefore $P(n)$ is true for all positive integers n.

70. An enumeration of the set is 1, 3, 5, 7, 9, 11, ...

72. An enumeration of the set is a, aa, aaa, $aaaa$, ...

77. Let A and B be denunmerable sets with enumerations $A = a_1, a_2, a_3, \ldots$ and $B = b_1, b_2, b_3, \ldots$. Then use the list $a_1, b_1, a_2, b_2, a_3, b_3, \ldots$ and eliminate any duplicates. This will be an enumeration of $A \cup B$, which is therefore denumerable.

Exercises 3.2

1. $5 \cdot 3 \cdot 2 = 30$

2. $4 \cdot 2 \cdot 2 = 16$

7. $45 \cdot 13 = 585$

9. 10^9

11. $26 \cdot 26 \cdot 26 \cdot 1 \cdot 1 = 17{,}576$

14. $26 + 26 \cdot 10 = 286$

17. $5 \cdot 3 \cdot 4 \cdot 3 = 180$

21. $9 \cdot 10 \cdot 26 \cdot 10 \cdot 10 + 9 \cdot 10 \cdot 26 \cdot 10 \cdot 10 \cdot 10 + 9 \cdot 10 \cdot 26 \cdot 10 \cdot 10 \cdot 10 = 25{,}974{,}000$

23. $5 \cdot 3 = 15$

26. $900 - 180 = 720$

31. $2^8 = 256$

33. $1 \cdot 2^7$ (begin with 0) $+ 1 \cdot 2^6 \cdot 1$ (begin with 1, end with 0) $= 2^7 + 2^6 = 192$

39. 8 (This is the same problem as Exercise 36.)

42. $6 \cdot 1 = 6$

51. $52 \cdot 52 = 2704$

54. $4 \cdot 4 = 16$ ways to get 2 of one kind; there are 13 distinct "kinds," so by the Addition Principle, the answer is $16 + 16 + \cdots + 16 = 13 \cdot 16 = 208$.

59. $12 \cdot 52$ (flower face card, any bird card) $+ 40 \cdot 12$ (flower nonface card, bird face card) $= 1104$

or

$52 \cdot 52$ (total number of hands—Exercise 41) $- 40 \cdot 40$ (hands with no face cards—Exercise 48) $= 1104$

62.

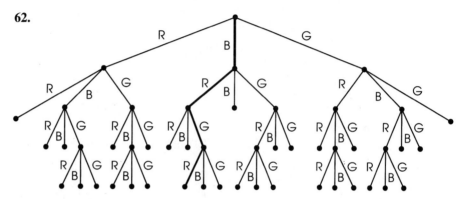

33 ways; the outcome that is highlighted is BRGR.

Exercises 3.3

2. Let A = guests who drink coffee
B = guests who drink tea
Then $|A| = 13$, $|B| = 10$, and $|A \cap B| = 4$
$|A \cup B| = |A| + |B| - |A \cap B| = 13 + 10 - 4 = 19$

5. Let A = breath set, B = gingivitis set, C = plaque set
a. $|A \cap B \cap C| = 2$
b. $|A - C| = |A| - |A \cap C| = 12 - 6 = 6$

8. Let A = auto set, B = bike set, C = motorcycle set
a. $|B - (A \cup C)| = |B| - |B \cap (A \cup C)|$ by Example 29
$\qquad = |B| - |(B \cap A) \cup (B \cap C)|$
$\qquad = |B| - (|B \cap A| + |B \cap C| - |B \cap A \cap C|)$
$\qquad = 97 - (53 + 7 - 2) = 39$
b. $|A \cup B \cup C| = 97 + 83 + 28 - 53 - 14 - 7 + 2 = 136$, so $150 - 136 = 14$ do not own any of the three.

14. No. There are 13 different denominations (bins), so 12 cards could all be different.

19. There are 3 pairs—1 and 6, 2 and 5, 3 and 4—that add up to 7. Each element in the set belongs to one of these pairs. Apply the Pigeonhole Principle, where the pairs are the bins and the numbers are the items.

Exercises 3.4

1. a. 42

5. 5! (total permutations)
3! (arrangement of the 3 R's for each distinguished permutation)
= 5·4 = 20

7. $P(15, 3) = \dfrac{15!}{12!} = 15 \cdot 14 \cdot 13 = 2730$

10. $(2!)(11!)(8!) = 2(39,916,800)(40,320)$

men first or
women first

14. a. 120

16. $C(300,25) = \dfrac{300!}{25!275!}$

18. $C(17, 5) \cdot C(23, 7) = (6,188)(245,157)$

20. $C(7, 1) \cdot C(14, 1) \cdot C(4, 1) \cdot C(5, 1) \cdot C(2, 1) \cdot C(3, 1) = 7 \cdot 14 \cdot 4 \cdot 5 \cdot 2 \cdot 3 = 11,750$

23. all committees $-$ (none or 1 from manufacturing) $= C(35, 6) - [C(21, 6) + C(14, 1) \cdot C(21, 5)]$

25. $C(13, 3) \cdot C(13, 2)$ **27.** $C(13, 5) + C(13, 5) + C(13, 5) + C(13, 5) = 4C(13, 5)$

38. $C(32, 2) \cdot C(16, 12)$ (Choose the two processors from B, then the remaining 12 modules are assigned to cluster A.)

39. $C(12, 4) = 495$

44. $C(60, 1) + C(60, 2)$

47. $C(12, 3) = 220$

49. no Democrats + no Republicans $-$ all independents (so as not to count twice) $= C(7, 3) + C(9, 3) - C(4, 3) = 115$

51. $C(14, 6) = 3003$

55. $C(25, 5) - C(23, 5)$ (all committees—those with neither)
(not $1 \cdot C(24, 4) + 1 \cdot C(24, 4)$—this number is too big, it counts some combinations more than once)

59. a. $\dfrac{8!}{3!2!}$ **b.** $\dfrac{7!}{3!2!}$

63. $C(7, 5)$

67. a. $C(8, 6)$
b. $C(7, 6)$ (choose 6 from 2 with repetitions)
c. $C(5, 3)$ (3 of the 6 shipments are fixed, choose the remaining 3 from among 3, with repetition)

69. a. $C(16, 10)$

  b. $C(9, 3)$ (7 of the 10 assignments are fixed, choose the remaining 3 from among 7, with repetitions)

71. $C(13, 10)$

Exercises 3.5

1. a. $(a + b)^5 = C(5, 0)a^5 + C(5, 1)a^4b + C(5, 2)a^3b^2 + C(5, 3)a^2b^3 + C(5, 4)ab^4 + C(5, 5)b^5 = a^5 + 5a^4b + 10a^3b^2 + 10a^2b^3 + 5ab^4 + b^5$

  c. $(a + 2)^5 = C(5, 0)a^5 + C(5, 1)a^4(2) + C(5, 2)a^3(2)^2 + C(5, 3)a^2(2)^3 + C(5, 4)a(2)^4 + C(5, 5)(2)^5 = a^5 + 10a^4 + 40a^3 + 80a^2 + 80a + 32$

  e. $(2x + 3y)^3 = C(3, 0)(2x)^3 + C(3, 1)(2x)^2(3y) + C(3, 2)(2x)(3y)^2 + C(3, 3)(3y)^3 = 8x^3 + 36x^2y + 54xy^2 + 27y^3$

4. $C(9, 5)(2x)^4(-3)^5 = +489{,}888x^4$

6. $C(8, 8)(-3y)^8 = 6561y^8$

8. $C(5, 2)(4x)^3(-2y)^2 = 2560x^3y^2$

12. $C(8, 1)(2x - y)^{7}5^1 = C(8, 1)(5)[C(7, 4)(2x)^3(-y)^4] = C(8, 1)C(7, 4)(2)^3(5)x^3y^4 = 11{,}200x^3y^4$

17. From the Binomial Theorem with $a = 1, b = 2$:

$$C(n, 0) + C(n, 1)2 + C(n, 2)2^2 + \cdots + C(n, n)2^n = (1 + 2)^n = 3^n$$

Chapter 4

Exercises 4.1

1. a. $(1, 3), (3, 3)$ **b.** $(4, 2), (5, 3)$
  c. $(5, 0), (2, 2)$ **d.** $(1, 1), (3, 9)$

3. a.

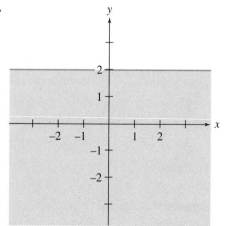

5. a. Many-to-many   **b.** Many-to-one   **c.** One-to-one   **d.** One-to-many

7. a. (2, 6), (3, 17), (0, 0)   **b.** (2, 12)   **c.** none   **d.** (1, 1), (4, 8)

10. a. Reflexive, transitive   **b.** Reflexive, symmetric, transitive   **c.** Symmetric

15. a. $x \rho^* y \leftrightarrow x$ is some number of years older than y

   b. $x \rho^* y \leftrightarrow x$ is a male ancestor of y

   c. $x \rho^* y \leftrightarrow$ you can drive from x to y in some number of days

16. a. $\rho = \{(1, 1)\}$

20. b.

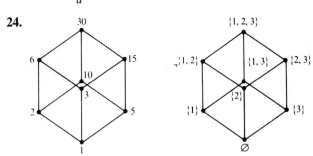

24.

The two graphs are identical in structure.

25. a. $\rho = \{(1, 1), (2, 2), (3, 3), (4, 4), (5, 5), (1, 3), (3, 5), (1, 5), (2, 4), (4, 5), (2, 5)\}$

28. Assume that ρ is reflexive and transitive on S. Then for all $x \in S$, $(x, x) \in \rho$, which means $(x, x) \in \rho^{-1}$, so $(x, x) \in \rho \cap \rho^{-1}$ and $\rho \cap \rho^{-1}$ is reflexive.

Let $(x, y) \in \rho \cap \rho^{-1}$. Then $(x, y) \in \rho$ and $(x, y) \in \rho^{-1}$, which means $(x, y) \in \rho$ and $(y, x) \in \rho$. This implies $(y, x) \in \rho^{-1}$ and $(y, x) \in \rho$, so $(y, x) \in \rho \cap \rho^{-1}$ and $\rho \cap \rho-1$ is symmetric.

Let $(x, y) \in \rho \cap \rho^{-1}$ and $(y, z) \in \rho \cap \rho^{-1}$. Then $(x, y) \in \rho$ and $(x, y) \in \rho^{-1}$ and $(y, z) \in \rho$ and $(y, z) \in \rho^{-1}$, so that $(x, y) \in \rho$ and $(y, x) \in \rho$ and $(y, z) \in \rho$ and $(z, y) \in \rho$. Because ρ is transitive, this means $(x, z) \in \rho$ and $(z, x) \in \rho$ or $(x, z) \in \rho$ and $(x, z) \in \rho^{-1}$, so $(x, z) \in \rho \cap \rho^{-1}$ and $\rho \cap \rho^{-1}$ is transitive.

31. a. "when;" no; all but the last

   b.

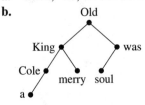

Maximal elements: "a," "merry," "soul"

33. a. $[a] = \{a, c\} = [c]$
   b. $[3] = \{1, 2, 3\}$
      $[4] = \{4, 5\}$
   c. $[1] = \{..., -5, -3, -1, 1, 3, 5, ...\}$
   d. $[-3] = \{..., -13, -8, -3, 2, 7, 12, ...\}$

36. Reflexive—$(x, y)\ \rho\ (x, y)$ because $y = y$
Symmetric—if $(x, y)\ \rho\ (z, w)$, then $y = w$ so $w = y$ and $(z, w)\ \rho\ (x, y)$
Transitive—if $(x, y)\ \rho\ (z, w)$ and $(z, w)\ \rho\ (s, t)$, then $y = w$ and $w = t$ so $y = t$
and $(x, y)\ \rho\ (s, t)$
The equivalence classes are sets of ordered pairs with the same second
components.

39. Clearly $P \leftrightarrow P$ is a tautology. If $P \leftrightarrow Q$ is a tautology, then P and Q have the same
truth values everywhere, so $Q \leftrightarrow P$ is a tautology. If $P \leftrightarrow Q$ and $Q \leftrightarrow R$ are tautologies, then P, Q, and R have the same truth values everywhere, and $P \leftrightarrow R$ is a
tautology. The equivalence classes are sets consisting of wffs with the same truth
values everywhere.

42. a. Partitions of 3 elements into 2 blocks can only be done with 2 elements in
one block and 1 in the other, so the answer is the number of ways to select
the 2 elements, or $C(3, 2) = 3$.
   b. Partitions of 4 elements into 2 blocks can be done with 3 elements in one
block and 1 in the other (pick the 3 elements out of 4), or two elements in
each block (pick 2 elements out of 4 but this determines the other two
elements). The answer is $C(4, 3) + C(4, 2)/2 = 7$.

45. a. $S(3, 2) = S(2, 1) + 2S(2, 2) = 1 + 2 \cdot 1 = 3$
   b. $S(4, 2) = S(3, 1) + 2S(3, 2) = 1 + 2 \cdot 3 = 7$

47. $S(4, 3) = 6$

Exercises 4.2

2.

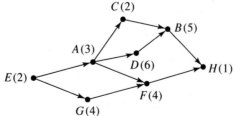

4. Minimum time to completion is 17 time units. Critical path: E, A, D, B, H

7. For example: E, A, C, D, G, F, B, H

Exercises 4.3

1.

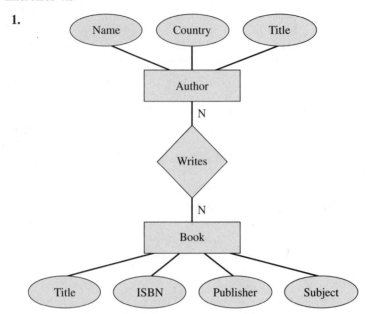

The "writes" relation is many-to-many; that is, one author can write many books, and one book can have more than one author.

4.

Results1		
Name	*Country*	*Title*
Bert Kovalsco	U.S.	Baskets for Today
Jane East	U.S.	Springtime Gardening

8.

Results5	
Name	*Title*
Dorothy King	Springtime Gardening
Jon Nkoma	Birds of Africa
Won Lau	Early Tang Paintings
Bert Kovalsco	Baskets for Today
Jimmy Chan	Early Tang Paintings
Dorothy King	Autumn Annuals
Jane East	Springtime Gardening

12.

Results9

Title	ISBN	Publisher	Subject	Name
Springtime Gardening	816-35421-8	Harding	Nature	Dorothy King
Early Tang Paintings	364-87547-8	Bellman	Art	Jimmy Chan
Early Tang Paintings	364-87547-8	Bellman	Art	Won Lau
Birds of Africa	115-67813-3	Loraine	Nature	Jon Nkoma
Springtime Gardening	816-89335-8	Swift-Key	Nature	Jane East
Baskets for Today	778-53705-7	Harding	Art	Bert Kovalsco
Autumn Annuals	414-88506-9	Harding	Nature	Dorothy King

15. a. project(restrict Author **where** Country = "U.S.")**over** Title **giving** Results11.
   b. SELECT Title **FROM** Author
      WHERE Author.Country = "U.S."
   c. Range of x is Author
      $\{x.\text{title} | x.\text{Country} = \text{"U.S."}\}$
   d. Baskets for Today
      Springtime Gardening

Exercises 4.4

1. a. Domain = $\{4, 5, 6, 7, 8\}$, codomain = $\{8, 9, 10, 11\}$, range = $\{8, 9, 10\}$
   b. 8, 10
   c. 6, 7
   d. No, no

4. a. $f(A) = \{3, 9, 15\}$
   b. $f(A) = \{x | x \in \mathbb{Z} \text{ and } (\exists y)(y \in \mathbb{Z} \text{ and } x = 6y)\}$

8. a. Function
   b. Not a function; undefined at $x = 0$
   c. Function; onto
   d. Bijection; $f^{-1} : \{p, q, r\} \rightarrow \{1, 2, 3\}$ where $f^{-1} = \{(q, 1), (r, 2), (p, 3)\}$
   e. Function; one-to-one
   f. Bijection; $h^{-1} : \mathbb{R}^2 \rightarrow \mathbb{R}^2$ where $h^{-1}(x, y) = (y - 1, x - 1)$

13. Let $k \leq x < k + 1$ where k is an integer. Then $\lfloor x \rfloor = k$. Also, $-k \geq -x > -k - 1$ so $\lceil -x \rceil = -k$ and $-\lceil -x \rceil = k$.

15. If $2^k < n < 2^{k+1}$, then $\log(2^k) < \log n < \log(2^{k+1})$ or $k < \log n < k + 1$ and $\lfloor \log n \rfloor = k, \lceil \log n \rceil = k + 1$.

20. a. $(g \circ f)(5) = g(f(5)) = g(6) = 18$
   b. $(f \circ g)(5) = f(g(5)) = f(15) = 16$
   c. $(g \circ f)(x) = g(f(x)) = g(x + 1) = 3(x + 1) = 3x + 3$
   d. $(f \circ g)(x) = f(g(x)) = f(3x) = 3x + 1$
   e. $(f \circ f)(x) = f(f(x)) = f(x + 1) = (x + 1) + 1 = x + 2$
   f. $(g \circ g)(x) = g(g(x)) = g(3x) = 3(3x) = 9x$

24. a. $f^{-1}(x) = x/2$
   b. $f^{-1}(x) = \sqrt[3]{x}$
   c. $f^{-1}(x) = 3x - 4$

27. a. $(1, 3, 5, 2)$
   b. $(1, 4, 3, 2, 5)$

30. a. $(1, 2, 5, 3, 4)$ **b.** $(1, 7, 8) \circ (2, 4, 6)$
   c. $(1, 5, 2, 4) \circ (3, 6)$ **d.** $(2, 3) \circ (4, 8) \circ (5, 7)$

34. a. 3^4 **b.** 36

37. a. n^n **b.** $n!$ **c.** $n!$ **d.** $n!$

$$\textbf{e.} \quad n!\left[1 - \frac{1}{1!} + \frac{1}{2!} - \frac{1}{3!} + \cdots + (-1)^n \frac{1}{n!}\right] = n!\left[\frac{1}{2!} - \frac{1}{3!} + \cdots + (-1)^n \frac{1}{n!}\right]$$

$$< n!\left[\frac{1}{2!}\right] = n!\cdot\frac{1}{2} < n!$$

   f. Number of derangements $< n! < n^n$. The total number of functions, with no restrictions, is the maximum. Only some of these functions are one-to-one and onto, but this is the definition of a permutation as well. Not all permutations are derangements, so the number of derangements is smaller still.

41. This is the number of derangements of 7 items, which is 1854.

47. a. $\{m, n, o, p\}$
   b. $\{n, o, p\}; \{o\}$

49. For example, $n_0 = c_2 = 1$, $c_1 = 1/34$

$$\textbf{54.} \quad \lim_{x \to \infty} \frac{x}{17x + 1} = \lim_{x \to \infty} \frac{1}{17 + \dfrac{1}{x}} = \frac{1}{17}$$

$$\textbf{56.} \quad \lim_{x \to \infty} \frac{x}{x^2} = \lim_{x \to \infty} \frac{1}{2x} = 0$$

Exercises 4.5

1. $2, -4$

3. $x = 1, y = 3, z = -2, w = 4$

5. a. $\begin{bmatrix} 6 & -5 \\ 0 & 3 \\ 5 & 3 \end{bmatrix}$ **e.** $\begin{bmatrix} 14 & -17 \\ 2 & 9 \\ 9 & 1 \end{bmatrix}$ **i.** $\begin{bmatrix} 21 & -23 \\ 33 & -44 \\ 11 & 1 \end{bmatrix}$ **m.** $\begin{bmatrix} 28 & 4 \\ 6 & 25 \end{bmatrix}$

6. c. $\mathbf{A}(\mathbf{B} + \mathbf{C}) = \begin{bmatrix} 3 & -1 \\ 2 & 5 \end{bmatrix}\begin{bmatrix} 10 & -4 \\ 4 & 3 \end{bmatrix} = \begin{bmatrix} 26 & -9 \\ 40 & -23 \end{bmatrix}$

$\mathbf{A} \cdot \mathbf{B} + \mathbf{A} \cdot \mathbf{C} = \begin{bmatrix} 10 & 4 \\ 18 & -3 \end{bmatrix} + \begin{bmatrix} 16 & -13 \\ 22 & -20 \end{bmatrix} = \begin{bmatrix} 26 & -9 \\ 40 & -23 \end{bmatrix}$

11. a. $\begin{bmatrix} 1 & 3 \\ 2 & 2 \end{bmatrix}\begin{bmatrix} -1/2 & 3/4 \\ 1/2 & -1/4 \end{bmatrix} = \begin{bmatrix} 1 & 0 \\ 0 & 1 \end{bmatrix} = \begin{bmatrix} -1/2 & 3/4 \\ 1/2 & -1/4 \end{bmatrix}\begin{bmatrix} 1 & 3 \\ 2 & 2 \end{bmatrix}$

b. For $\begin{bmatrix} 1 & 2 \\ 2 & 4 \end{bmatrix}\begin{bmatrix} b_{11} & b_{12} \\ b_{21} & b_{22} \end{bmatrix} = \begin{bmatrix} 1 & 0 \\ 0 & 1 \end{bmatrix}$

$$b_{11} + 2b_{21} = 1 \qquad b_{12} + 2b_{22} = 0$$
$$2b_{11} + 4b_{21} = 0 \qquad 2b_{12} + 4b_{22} = 1$$

which is an inconsistent system of equations with no solution.

15. First find $\mathbf{A}^{-1}$ by the method of Exercise 14:

$$\mathbf{A} = \begin{bmatrix} 1 & 1 \\ 24 & 14 \end{bmatrix} \qquad \mathbf{I} = \begin{bmatrix} 1 & 0 \\ 0 & 1 \end{bmatrix}$$

Multiply row 1 by -24 and add to row 2:

$$\begin{bmatrix} 1 & 1 \\ 0 & -10 \end{bmatrix}\begin{bmatrix} 1 & 0 \\ -24 & 1 \end{bmatrix}$$

Multiply row 2 by $-1/10$:

$$\begin{bmatrix} 1 & 1 \\ 0 & 1 \end{bmatrix}\begin{bmatrix} 1 & 0 \\ 24/10 & -1/10 \end{bmatrix}$$

Multiply row 2 by -1 and add to row 1:

$$\begin{bmatrix} 1 & 0 \\ 0 & 1 \end{bmatrix}\begin{bmatrix} -14/10 & 1/10 \\ 24/10 & -1/10 \end{bmatrix} = \mathbf{A}^{-1}$$

Now multiply $\mathbf{A}^{-1} \cdot \mathbf{B}$:

$$\begin{bmatrix} -14/10 & 1/10 \\ 24/10 & -1/10 \end{bmatrix}\begin{bmatrix} 70 \\ 1180 \end{bmatrix} = \begin{bmatrix} 20 \\ 50 \end{bmatrix} = \begin{bmatrix} x \\ y \end{bmatrix}$$

so the solution is $x = 20$, $y = 50$.

18. For example, $\begin{bmatrix} 1 & 1 \\ -1 & -1 \end{bmatrix}\begin{bmatrix} 1 & 1 \\ -1 & -1 \end{bmatrix} = \begin{bmatrix} 0 & 0 \\ 0 & 0 \end{bmatrix}$

22. $\mathbf{A} \wedge \mathbf{B} = \begin{bmatrix} 1 & 0 & 0 \\ 0 & 1 & 0 \\ 0 & 1 & 1 \end{bmatrix} \qquad \mathbf{A} \vee \mathbf{B} = \begin{bmatrix} 1 & 0 & 1 \\ 1 & 1 & 1 \\ 1 & 1 & 1 \end{bmatrix}$

$\mathbf{A} \times \mathbf{B} = \begin{bmatrix} 1 & 0 & 1 \\ 1 & 1 & 1 \\ 1 & 1 & 1 \end{bmatrix} \qquad \mathbf{B} \times \mathbf{A} = \begin{bmatrix} 1 & 1 & 1 \\ 1 & 1 & 1 \\ 1 & 1 & 1 \end{bmatrix}$

26. The i, j entry of $\mathbf{A}^2$ is $\displaystyle\sum_{k=1}^{n} a_{ik}a_{kj}$.

The j, i entry of $\mathbf{A}^2$ is $\displaystyle\sum_{k=1}^{n} a_{jk}a_{ki}$.

But these are the same because $a_{ik} = a_{ki}$ and $a_{kj} = a_{jk}$ ($\mathbf{A}$ is symmetric).

Chapter 5

Exercises 5.1

3. a. **b.** For example, **c.**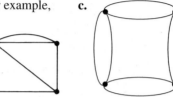

4. a. 4, 5, 6
  b. length 2
  c. for example (naming the nodes), 1–2–1–2–2–1–4–5–6

7. (b), because there is no node of degree 0.

9. $f_1: 1 \rightarrow a$ $f_2: a_1 \rightarrow e_2$
       $2 \rightarrow b$ $a_2 \rightarrow e_7$
       $3 \rightarrow c$ $a_3 \rightarrow e_6$
       $4 \rightarrow d$ $a_4 \rightarrow e_1$
                          $a_5 \rightarrow e_3$
                          $a_6 \rightarrow e_4$
                          $a_7 \rightarrow e_5$

13. Not isomorphic; graph in (b) has a node of degree 5, graph in (a) does not.

16.

21.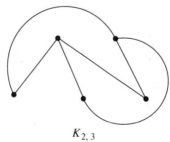

$K_{2,3}$

23. 5 (by Euler's formula)

27. Planar

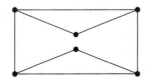

29. Nonplanar—subgraph below can be obtained from $K_{3,3}$ by elementary subdivisions.

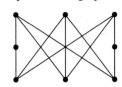

31. $\begin{bmatrix} 1 & 1 & 0 & 0 & 2 \\ 1 & 1 & 1 & 1 & 1 \\ 0 & 1 & 0 & 1 & 0 \\ 0 & 1 & 1 & 0 & 0 \\ 2 & 1 & 0 & 0 & 0 \end{bmatrix}$

35. $\begin{bmatrix} 0 & 1 & 0 & 0 \\ 0 & 0 & 1 & 1 \\ 0 & 0 & 0 & 1 \\ 0 & 0 & 1 & 0 \end{bmatrix}$

39.

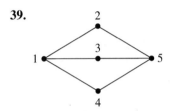

41.

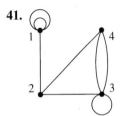

43. The graph consists of n disconnected nodes with a loop at each node.

48.

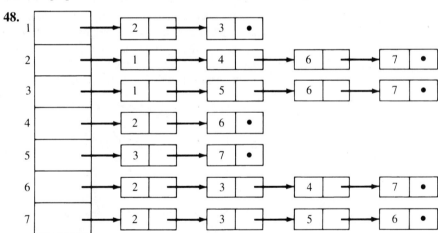

53.

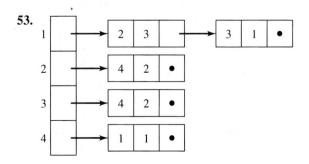

54.

Node		Pointer
1		5
2		7
3		11
4		0
5	2	6
6	3	0
7	1	8
8	2	9
9	3	10
10	4	0
11	4	0

58.

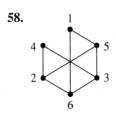

62. a. If G is not connected, then G consists of two or more connected subgraphs that have no paths between them. Let x and y be distinct nodes. If x and y are in different subgraphs, there is no x–y arc in G; hence there is an x–y arc in G', and a path exists from x to y in G'. If x and y are in the same subgraph, then pick a node z in a different subgraph. There is an arc x–z in G' and an arc z–y in G'; hence there is a path from x to y in G'.

b.

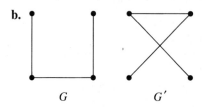

G G'

66. Let a simple connected graph have n nodes and m arcs. The original statement is $m \geq n - 1$, which says $n \leq m + 1$, or that the number of nodes is at most $m + 1$. For the base case, let $m = 0$. The only simple connected graph with 0 arcs consists of a single node, and the number of nodes, 1, is $\leq 0 + 1$. Now assume that any simple connected graph with r arcs, $0 \leq r \leq k$, has at most $r + 1$ nodes. Consider a simple connected graph with $k + 1$ arcs, and remove one arc. If the remaining graph is connected, it has k arcs and, by the inductive hypothesis, the number n of nodes satisfies $n \leq k + 1$. Therefore in the original graph (with the same nodes), $n \leq k + 1 < (k + 1) + 1$. If the remaining graph is not connected, it consists of two connected subgraphs with r_1 and r_2 arcs, $r_1 \leq k$ and $r_2 \leq k$, $r_1 + r_2 = k$. By the inductive hypothesis, one subgraph has at most $r_1 + 1$ nodes and the other has at most $r_2 + 1$ nodes. The original graph therefore had at most $r_1 + 1 + r_2 + 1 = (k + 1) + 1$ nodes.

71. a. 3

73. If this result is not true, then every node in such a graph has degree greater than 5, that is, degree 6 or higher. The total number of arc ends in the graph is then at least $6n$, where n is the number of nodes. But the number of arc ends is exactly twice the number a of arcs, so

$$6n \le 2a$$

By inequality (2) of the theorem on the number of nodes and arcs, $a \le 3n - 6$ or $2a \le 6n - 12$. Combining these inequalities, we obtain

$$6n \le 6n - 12$$

which is a contradiction.

75. a. A graph with parallel arcs would contain

Temporarily create small countries at the nodes of parallel arcs, giving

which is a simple graph. Any coloring satisfactory for this graph is satisfactory for the original graph (let the new regions shrink to points).

Exercises 5.2

1. a.  **b.** **c.**

4.

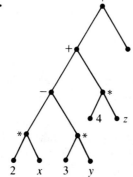

7.

	Left child	Right child
1	2	3
2	0	4
3	5	6
4	7	0
5	0	0
6	0	0
7	0	0

9.

12.

16. Preorder: $a\,b\,d\,e\,h\,f\,c\,g$
Inorder: $d\,b\,h\,e\,f\,a\,g\,c$
Postorder: $d\,h\,e\,f\,b\,g\,c\,a$

20. Preorder: $a\,b\,c\,e\,f\,d\,g\,h$
Inorder: $e\,c\,f\,b\,g\,d\,h\,a$
Postorder: $e\,f\,c\,g\,h\,d\,b\,a$

22. Prefix: $+\,/\,3\,4\,-\,2\,y$
Postfix: $3\,4\,/\,2\,y\,-\,+$

26. Prefix: $+\,*\,4\,-\,7\,x\,z$
Infix: $(4\,*\,(7\,-\,x))\,+\,z$

30. a

b

c

d

Both inorder and postorder traversal give $d\,c\,b\,a$.

32. Consider a simple graph that is a nonrooted tree. A tree is an acyclic and connected graph, so for any two nodes x and y, a path from x to y exists. If the path
is not unique, then the two paths diverge at some node n_1 and converge at some
node n_2, and there is a cycle from n_1 through n_2 and back to n_1, which is a contradiction.

Now consider a simple graph that has a unique path between any two nodes. The
graph is clearly connected. Also, there are no cycles because the presence of a
cycle produces a nonunique path between two nodes on the cycle. The graph is
thus acyclic and connected and is a nonrooted tree.

36. Proof is by induction on d. For $d = 0$, the only node is the root, and $2^0 = 1$.
Assume that there are at most 2^d nodes at depth d, and consider depth $d + 1$.
There are at most two children for each node at depth d, so the maximum num-
ber of nodes at depth $d + 1$ is $2 \cdot 2^d = 2^{d+1}$.

40. By Exercise 37, a full tree of height $h - 1$ has $2^h - 1$ nodes. When $n = 2^h$, this is the beginning of level h. The height h remains the same until $n = 2^{h+1}$, when it increases by 1. Therefore for $2^h \le n < 2^{h+1}$, the height of the tree remains the same, and is given by $h = \lfloor \log n \rfloor$. For example, for $2^2 \le n < 2^3$, that is, $n = 4$, 5, 6, or 7, the tree height is 2, and $2 = \lfloor \log n \rfloor$.

44.

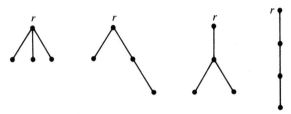

Exercises 5.3

1.

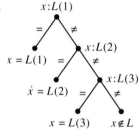

$x:L(1)$

$=$ $\neq$

$x = L(1)$ $=$ $x:L(2)$
$\neq$

$x = L(2)$ $=$ $x:L(3)$
$\neq$

$x = L(3)$ $x \notin L$

5.

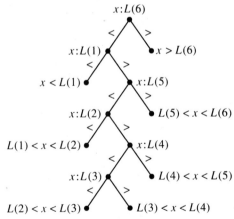

$x:L(6)$

$<$ $>$

$x:L(1)$ $x > L(6)$

$<$ $>$

$x < L(1)$ $x:L(5)$

$<$ $>$

$x:L(2)$ $L(5) < x < L(6)$

$<$ $>$

$L(1) < x < L(2)$ $x:L(4)$

$<$ $>$

$x:L(3)$ $L(4) < x < L(5)$

$<$ $>$

$L(2) < x < L(3)$ $L(3) < x < L(4)$

depth $= 6$; algorithm is not optimal

7. a.

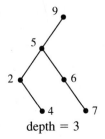

9

5

2   6

4   7

depth $= 3$

b. 2.83

11.

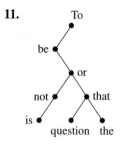

be is not or question that the To

13. a. $\lceil \log 4! \rceil = \lceil \log 24 \rceil = 5$
   b. $\lceil \log 8! \rceil = \lceil \log 40{,}320 \rceil = 16$
   c. $\lceil \log 16! \rceil = \lceil \log 2.09 \times 10^{13} \rceil = 45$

16. a. 10 (any of the five coins can be heavy or light)
   b. 3 (the minimum depth for a ternary tree with 10 leaves)
   c.

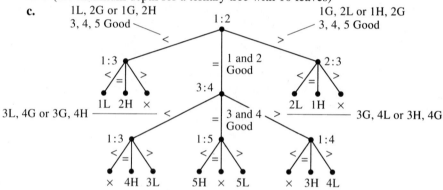

21. The three-way comparison would be done something like

> **if** (x = node element)
>   write "found"
> **else**
>   **if** ($x <$ node element)
>   ⋮
>   **else**
>   ⋮

so that in the worst case, where x is not equal to the node element, 2 comparisons are done at the node. The number of comparisons in the worst case for binary search of n elements is therefore 2 times the depth of the tree, or $2 * (1 + \lfloor \log n \rfloor)$.

Exercises 5.4

1. a. *ooue*
   b. *iaou*
   c. *eee*

7. a.

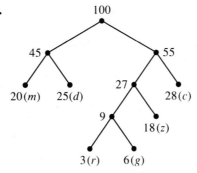

b.

m	00
g	1001
z	101
c	11
d	01
r	1000

10. Every single character *z* has been replaced by the two-character string *sh*. The frequencies of occurrence of all the characters must be recomputed. For example, if there were 100 characters in the original file, there are now 112. If there were 27 *a*'s per 100 characters in the original file, there are now 27 *a*'s per 112 characters, so the frequency of occurrence of an *a* is $(27/112) * 100 = 24$. The new frequencies are

a	*s*	*h*	*t*	*e*	*c*
24	11	11	13	28	13

The new Huffman tree is

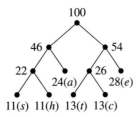

and a new encoding (one of several possibilities) is

s	000
h	001
a	01
t	100
c	101
e	11

13. a. Because we are assuming that $f(x) < f(p)$, we can write $f(x) + j = f(p)$ for some positive quantity j. Because x is above p in tree T (Figure 5.57a), we can write $d(x) + k = d(p)$ for some positive integer k. The contributions to $E(T)$ from nodes x and p are given by

$$f(x)d(x) + f(p)d(p) = f(x)d(x) + [(f(x) + j)(d(x) + k)]$$
$$= 2f(x)d(x) + jd(x) + kf(x) + jk$$

In tree T' (Figure 5.57b), the contributions to $E(T')$ from nodes x and p are given by the following (using the original $d(x)$ and $d(p)$ values):

$$f(x)d(p) + f(p)d(x) = f(x)(d(x) + k) + (f(x) + j)d(x)$$
$$= 2f(x)d(x) + kf(x) + jd(x)$$

which is jk smaller than the previous expression.

b. In Figure 5.58d, the contribution to $E(B)$ from the node with frequency $f(x) + f(y)$ is

$$[f(x) + f(y)] * r$$

where r is the depth of that node. The corresponding contribution to $E(B')$ (Figure 5.58c) is

$$f(x) * (r + 1) + f(y) * (r + 1) = f(x) * r + f(y) * r + f(x) + f(y)$$
$$= [f(x) + f(y)] * r + f(x) + f(y)$$

so $E(B')$ exceeds $E(B)$ by $f(x) + f(y)$.

Chapter 6

Exercises 6.1

1. $A = \begin{bmatrix} 1 & 0 & 0 \\ 1 & 0 & 1 \\ 0 & 1 & 0 \end{bmatrix}$ $\rho = \{(1, 1), (2, 1), (2, 3), (3, 2)\}$

3.

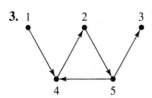

$\rho = \{(1, 4), (2, 5), (4, 2), (5, 3), (5, 4)\}$

8. The graph is a "star" with node 1 at the center, i.e., 1 is adjacent to every node and every node is adjacent to 1, but no other nodes are adjacent.

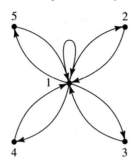

12.

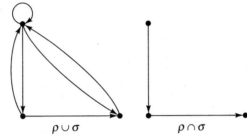

$\rho \cup \sigma$ $\rho \cap \sigma$

15. $\mathbf{R} = \begin{bmatrix} 1 & 1 & 1 & 1 \\ 1 & 1 & 1 & 1 \\ 1 & 1 & 1 & 1 \\ 0 & 0 & 0 & 0 \end{bmatrix}$

19. $\mathbf{R} = \begin{bmatrix} 0 & 1 & 1 & 1 & 1 \\ 0 & 1 & 1 & 1 & 1 \\ 0 & 0 & 0 & 0 & 0 \\ 0 & 1 & 1 & 1 & 1 \\ 0 & 1 & 1 & 1 & 1 \end{bmatrix}$

23. $\mathbf{R} = \begin{bmatrix} 1 & 0 & 0 \\ 1 & 1 & 1 \\ 1 & 1 & 1 \end{bmatrix}$

27. $\mathbf{R} = \begin{bmatrix} 0 & 1 & 1 & 1 & 1 & 1 \\ 0 & 0 & 0 & 0 & 0 & 0 \\ 0 & 0 & 0 & 0 & 0 & 0 \\ 0 & 0 & 0 & 0 & 0 & 0 \\ 0 & 0 & 0 & 0 & 0 & 0 \\ 0 & 1 & 1 & 0 & 1 & 0 \end{bmatrix}$

29. $A^2[i, j] = \sum_{k=1}^{n} a_{ik}a_{kj}$

If a term such as $a_{i2}a_{2j}$ in this sum is 0, then either $a_{i2} = 0$ or $a_{2j} = 0$ (or both) and there is either no path of length 1 from n_i to n_2 or no path of length 1 from n_2 to n_j (or both). Thus there are no paths of length 2 from n_i to n_j passing through n_2. If $a_{i2}a_{2j} \neq 0$, then $a_{i2} = p$ and $a_{2j} = q$, where p and q are positive integers. Then there are p paths of length 1 from n_i to n_2 and q paths of length 1 from n_2 to n_j. By the Multiplication Principle, there are pq possible paths of length 2 from n_i to n_j through n_2. By the Addition Principle, the sum of all such terms gives all possible paths of length 2 from n_i to n_j.

Exercises 6.2

2. Yes

7. No

10.
$$\begin{bmatrix} 0 & 1 & 1 & 1 & 1 & 0 \\ 1 & 0 & 0 & 0 & 0 & 1 \\ 1 & 0 & 0 & 0 & 0 & 1 \\ 1 & 0 & 0 & 0 & 0 & 1 \\ 1 & 0 & 0 & 0 & 0 & 1 \\ 0 & 1 & 1 & 1 & 1 & 0 \end{bmatrix}$$
total after row 2 is 0

12.
$$\begin{bmatrix} 0 & 1 & 0 & 1 & 1 & 0 & 0 \\ 1 & 0 & 1 & 0 & 1 & 1 & 0 \\ 0 & 1 & 0 & 0 & 0 & 1 & 1 \\ 1 & 0 & 0 & 0 & 1 & 0 & 0 \\ 1 & 1 & 0 & 1 & 0 & 1 & 0 \\ 0 & 1 & 1 & 0 & 1 & 0 & 1 \\ 0 & 0 & 1 & 0 & 0 & 1 & 0 \end{bmatrix}$$
$i = 8$

14. No

19. Yes

23. Any two nodes must be part of the Hamiltonian circuit; therefore there is a path between them, namely that part of the circuit that is between them.

26. a. $n = 2$ or $n =$ any odd number
b. $n > 2$

29. Begin at any node and take one of the arcs out from that node. Each time a new node is entered on an arc, there is exactly one unused arc on which to exit that node; because the arc is unused, it will lead to a new node or to the initial node. Upon return to the initial node, if all nodes have been used, we are done. If there is an unused node, because the graph is connected, there is an unused path from that node to a used node, which means the used node has degree ≥ 3, a contradiction.

Exercises 6.3

1. $IN = \{2\}$

	1	2	3	4	5	6	7	8
d	3	0	2	∞	∞	∞	1	∞
s	2	–	2	2	2	2	2	2

$p = 7, IN = \{2, 7\}$

	1	2	3	4	5	6	7	8
d	3	0	2	∞	∞	6	1	2
s	2	–	2	2	2	7	2	7

$p = 3, IN = \{2, 7, 3\}$

	1	2	3	4	5	6	7	8
d	3	0	2	3	∞	6	1	2
s	2	–	2	3	2	7	2	7

$p = 8, IN = \{2, 7, 3, 8\}$

	1	2	3	4	5	6	7	8
d	3	0	2	3	3	6	1	2
s	2	–	2	3	8	7	2	7

$p = 5, IN = \{2, 7, 3, 8, 5\}$

	1	2	3	4	5	6	7	8
d	3	0	2	3	3	6	1	2
s	2	–	2	3	8	7	2	7

path: 2, 7, 8, 5   distance $= 3$

5. $IN = \{1\}$

	1	2	3	4	5	6	7
d	0	2	∞	∞	3	2	∞
s	–	1	1	1	1	1	1

$p = 2, IN = \{1, 2\}$

	1	2	3	4	5	6	7
d	0	2	3	∞	3	2	∞
s	–	1	2	1	1	1	1

$p = 6$, $IN = \{1, 2, 6\}$

	1	2	3	4	5	6	7
d	0	2	3	∞	3	2	5
s	–	1	2	1	1	1	6

$p = 3$, $IN = \{1, 2, 6, 3\}$

	1	2	3	4	5	6	7
d	0	2	3	4	3	2	5
s	–	1	2	3	1	1	6

$p = 5$, $IN = \{1, 2, 6, 3, 5\}$

	1	2	3	4	5	6	7
d	0	2	3	4	3	2	5
s	–	1	2	3	1	1	6

$p = 4$, $IN = \{1, 2, 6, 3, 5, 4\}$

	1	2	3	4	5	6	7
d	0	2	3	4	3	2	5
s	–	1	2	3	1	1	6

$p = 7$, $IN = \{1, 2, 6, 3, 5, 4, 7\}$

	1	2	3	4	5	6	7
d	0	2	3	4	3	2	5
s	–	1	2	3	1	1	6

path: 1, 6, 7   distance = 5

9.

	1	2	3	4	5	6	7	8
d	3	0	2	∞	∞	∞	1	∞
s	2	–	2	2	2	2	2	2

	1	2	3	4	5	6	7	8
d	3	0	2	3	11	4	1	2
s	2	–	2	3	1	1	2	7

	1	2	3	4	5	6	7	8
d	3	0	2	3	3	4	1	2
s	2	–	2	3	8	1	2	7

No further changes in d or s.

13. Initial **A** and after $k = x$:

	x	1	2	3	y
x	0	1	∞	4	∞
1	1	0	3	1	5
2	∞	3	0	2	2
3	4	1	2	0	3
y	∞	5	2	3	0

after $k = 1$ and $k = 2$:

	x	1	2	3	y
x	0	1	4	2	6
1	1	0	3	1	5
2	4	3	0	2	2
3	2	1	2	0	3
y	6	5	2	3	0

after $k = 3$ and $k = y$:

	x	1	2	3	y
x	0	1	4	2	5
1	1	0	3	1	4
2	4	3	0	2	2
3	2	1	2	0	3
y	5	4	2	3	0

17. $IN = \{1, 8, 5, 6, 2, 7, 4, 3\}$

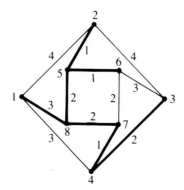

19. Kruskal's algorithm develops a minimal spanning tree as follows.

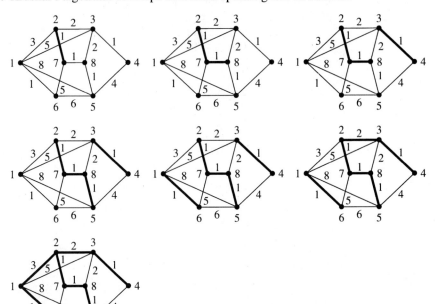

24. a. weight = 300

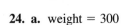
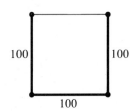

b. $d = 50\sqrt{2} = 70.750$ $4d = 282.8$

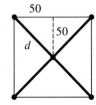

Exercises 6.4

1. $a, b, c, e, f, d, h, g, j, i$

5. $e, b, a, c, f, d, h, g, j, i$

7. $a, b, c, f, j, g, d, e, h, k, i$

9. $f, c, a, b, d, e, h, k, i, g, j$

11. $a, b, c, d, e, g, f, h, j, i$

17. $a, b, c, d, e, f, g, h, i, j, k$

21. a, b, c, e, g, d, f, h

24. a, b, c, d, e, g, h, f

27. Begin a dfs at node a: a, c, f, g, e, b, d

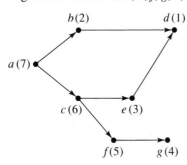

Exercises 6.5

1.

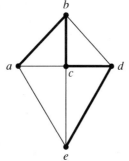

Back arcs: a–c, a–e, b–d, c–e

4.

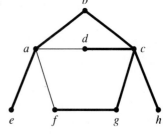

Back arcs: a–d, a–f

7.

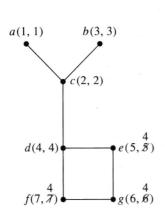

Articulation points: c, d

Biconnected components

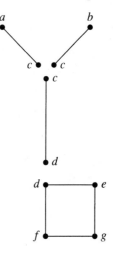

★10.

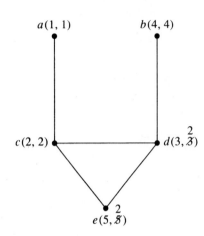

a(1, 1)   b(4, 4)

c(2, 2)   d(3, $\frac{2}{3}$)

e(5, $\frac{2}{3}$)

Articulation points: c, d

Biconnected components

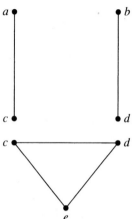

Chapter 7

Exercises 7.1

1.

+	0	1	a	a'
0	0	1	a	a'
1	1	1	1	1
a	a	1	a	1
a'	a'	1	1	a'

·	0	1	a	a'
0	0	0	0	0
1	0	1	a	a'
a	0	a	a	0
a'	0	a'	0	a'

4. a.
$$x' + x = x + x' \quad\quad\quad\quad (1a)$$
$$= 1 \quad\quad\quad\quad (5a)$$

and

$$x' \cdot x = x \cdot x' \quad\quad\quad\quad (1b)$$
$$= 0 \quad\quad\quad\quad (5b)$$

Therefore $x = (x')'$ by the theorem on the uniqueness of complements.

5. a.
$$x + (x \cdot y) = x \cdot 1 + x \cdot y \quad\quad\quad (4b)$$
$$= x(1 + y) \quad\quad\quad (3b)$$
$$= x(y + 1) \quad\quad\quad (1a)$$
$$= x \cdot 1 \quad\quad\quad (\text{Practice 3a})$$
$$= x \quad\quad\quad (4b)$$
$$x \cdot (x + y) = x \text{ follows by duality}$$

6. a. $x + (x' \cdot y + x \cdot y)' = x + (y \cdot x' + y \cdot x)'$ (1b)

$\qquad\qquad\qquad\qquad\quad = x + (y \cdot (x' + x))'$ (3b)

$\qquad\qquad\qquad\qquad\quad = x + (y \cdot (x + x'))'$ (1a)

$\qquad\qquad\qquad\qquad\quad = x + (y \cdot 1)'$ (5a)

$\qquad\qquad\qquad\qquad\quad = x + y'$ (4b)

8. a. Let $x = 0$. Then

$\qquad x \cdot y' + x' \cdot y = 0 \cdot y' + x' \cdot y$ $x = 0$

$\qquad\qquad\qquad\quad = y' \cdot 0 + x' \cdot y$ 1b

$\qquad\qquad\qquad\quad = 0 + x' \cdot y$ Practice 3b

$\qquad\qquad\qquad\quad = x' \cdot y + 0$ 1a

$\qquad\qquad\qquad\quad = x' \cdot y$ 4a

$\qquad\qquad\qquad\quad = 1 \cdot y$ Practice 4

$\qquad\qquad\qquad\quad = y \cdot 1$ 1b

$\qquad\qquad\qquad\quad = y$ 4b

b. Let $x \cdot y' + x' \cdot y = y$. Then

$\qquad x \cdot x' + x' \cdot x = x$ letting y in the hypothesis have the value x

$\qquad x \cdot x' + x \cdot x' = x$ 1b

$\qquad\qquad\quad 0 + 0 = x$ 5b

$\qquad\qquad\qquad\quad 0 = x$ 4a

9. a. $x \oplus y = x \cdot y' + y \cdot x'$ (definition of $\oplus$)

$\qquad\qquad = y \cdot x' + x \cdot y'$ (1a)

$\qquad\qquad = y \oplus x$ (definition of $\oplus$)

13. a.   **b.**

c.

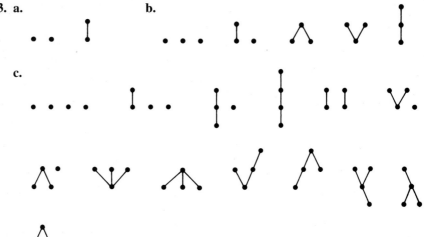

16

16. a. $f: \mathbb{R} \to \mathbb{R}^+$

   f is onto; given $y \in \mathbb{R}^+$, let $x = \log y$; then $x \in \mathbb{R}$ and $f(x) = 2^x = 2^{\log y} = y$.
   f is one-to-one; if $f(x) = f(w)$, then $2^x = 2^w$ and (taking the log of both sides)
   $x = w$.

   b. For $x, y \in \mathbb{R}$, $g(x + y) = g(x) \cdot g(y)$.

   c. f is a bijection from $\mathbb{R}$ to $\mathbb{R}^+$ and for $x, y \in \mathbb{R}$, $f(x + y) = 2^{x+y} = 2^x \cdot 2^y = f(x) \cdot f(y)$.

   d. f^{-1} is the function $f^{-1}(y) = \log y$.

   e. f^{-1} is a bijection from $\mathbb{R}^+$ to $\mathbb{R}$ and for any $x, y \in \mathbb{R}^+$, $f^{-1}(x \cdot y) = \log(x \cdot y) = \log x + \log y = f^{-1}(x) + f^{-1}(y)$.

19. a. For any $y \in b$, $y = f(x)$ for some $x \in B$. Then $y \& f(0) = f(x) \& f(0) = f(x + 0) = f(x) = y$, and $f(0) = \phi$ because the zero element in any Boolean algebra is unique (see Exercise 12).

   b. $f(1) = f(0') = [f(0)]'' = \phi'' = \dagger$

20. a. (i) If $x \leqslant y$, then $x \leqslant y$ and $x \leqslant x$ so x is a lower bound of x and y. If $w^* \leqslant x$ and $w^* \leqslant y$, then $w^* \leqslant x$ so x is a greatest lower bound, and $x = x \cdot y$. If $x = x \cdot y$, then x is a greatest lower bound of x and y, so $x \leqslant y$.

   (ii) Similar to i.

Exercises 7.2

1. a.

3. $(x_1 x_2)'(x_2 + x_3')$

x_1	x_2	x_3	$f(x_1, x_2, x_3)$
1	1	1	0
1	1	0	0
1	0	1	0
1	0	0	1
0	1	1	1
0	1	0	1
0	0	1	0
0	0	0	1

7. $x_1'x_2'$

13. a. $x_1x_2x_3' + x_1x_2'x_3'$

b.

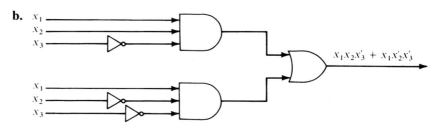

$$x_1x_2x_3' + x_1x_2'x_3'$$

c. $x_1x_2x_3' + x_1x_2'x_3' = x_1x_3'x_2 + x_1x_3'x_2' = x_1x_3'(x_2 + x_2') = x_1x_3'\cdot 1 = x_1x_3'$

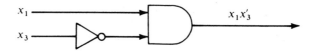

$$x_1x_3'$$

16. a. $(x_1' + x_2')(x_1' + x_2)(x_1 + x_2')$
   b. $(x_1' + x_2)(x_1 + x_2)$
   c. $(x_1' + x_2' + x_3')(x_1' + x_2 + x_3)(x_1 + x_2' + x_3)(x_1 + x_2 + x_3')$
   d. $(x_1' + x_2' + x_3')(x_1' + x_2' + x_3)(x_1 + x_2' + x_3')(x_1 + x_2 + x_3')(x_1 + x_2 + x_3)$

21. Network is represented by $(x_1'(x_2'x_3)')'$ and $(x_1'(x_2'x_3)')' = x_1 + x_2'x_3$

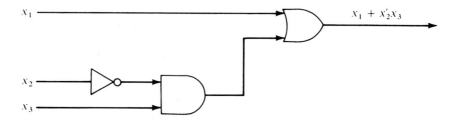

$$x_1 + x_2'x_3$$

23. The truth function for $|$ is that of the NAND gate, the truth function for $\downarrow$ is that of the NOR gate. In Section 1.1, we learned that every compound statement is equivalent to one using only $|$ or to one using only $\downarrow$; therefore any truth function can be realized by using only NAND gates or only NOR gates.

25. $x_1 = $ neutral $x_2 = $ park $x_3 = $ seat belt

x_1	x_2	x_3	$f(x_1, x_2, x_3)$
1	1	1	—
1	1	0	—
1	0	1	1
1	0	0	0
0	1	1	1
0	1	0	0
0	0	1	0
0	0	0	0

$$(x_1 + x_2)x_3$$

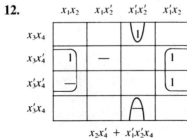

Exercises 7.3

1.

	x_1x_2	x_1x_2'	$x_1'x_2'$	$x_1'x_2$
x_3			1	1
x_3'	1	1		1

$$x_1'x_3 + x_1x_3' + x_1'x_2 \quad \text{or} \quad x_1'x_3 + x_1x_3' + x_2x_3'$$

3.

	x_1x_2	x_1x_2'	$x_1'x_2'$	$x_1'x_2$
x_3	1	1	1	1
x_3'	1			1

$$x_3 + x_2$$

6.

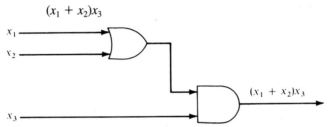

	x_1x_2	x_1x_2'	$x_1'x_2'$	$x_1'x_2$
x_3	1			1
x_3'	1			

$$x_1x_2 + x_2x_3$$

12.

	x_1x_2	x_1x_2'	$x_1'x_2'$	$x_1'x_2$
x_3x_4		1		
x_3x_4'	1	—		1
$x_3'x_4'$	—			1
$x_3'x_4$				

$$x_2x_4' + x_1'x_2'x_4$$

14.

x_1	x_2	x_3	
1	1	1	1,2,3
1	0	1	1,4
0	1	1	2,5,6
1	1	0	3,7
0	0	1	4,5
0	1	0	6,7

x_1	x_2	x_3	
1	–	1	1
–	1	1	1,2
1	1	–	2
–	0	1	1
0	–	1	1
0	1	–	2
–	1	0	2

x_1	x_2	x_3
–	–	1
–	1	–

	111	101	011	110	001	010
– –1	✓	✓	✓		✓	
–1–	✓		✓	✓		✓

– –1 and –1– are essential. The minimal form is $x_3 + x_2$.

18. a.

x_1	x_2	x_3	x_4	
0	1	1	1	1
1	0	1	0	
0	0	1	1	1,2
0	1	0	0	3
0	0	0	1	2,4
0	0	0	0	3,4

x_1	x_2	x_3	x_4
0	–	1	1
0	0	–	1
0	–	0	0
0	0	0	–

	0111	1010	0011	0100	0001	0000
1010		✓				
0–11	✓		✓			
00–1			✓		✓	
0–00				✓		✓
000–					✓	✓

1010, 0–11, 0–00 are essential. Either 00–1 or 000– can be used as the fourth term. The minimal sum-of-products form is

$$x_1x_2'x_3x_4' + x_1'x_3x_4 + x_1'x_3'x_4' + x_1'x_2'x_4$$

or

$$x_1x_2'x_3x_4' + x_1'x_3x_4 + x_1'x_3'x_4' + x_1'x_2'x_3'$$

Chapter 8

Exercises 8.1

1. a. Associative   **b.** Commutative

2. a. Not commutative: $a \cdot b \neq b \cdot a$
Not associative: $a \cdot (b \cdot d) \neq (a \cdot b) \cdot d$

b.

·	p	q	r	s
p	p	q	r	s
q	q	r	s	p
r	r	s	p	q
s	s	p	q	r

Commutative

5. a. Semigroup
b. Not a semigroup—not associative
c. Not a semigroup—S not closed under ·
d. Monoid; $i = 1 + 0\sqrt{2}$
e. Group; $i = 1 + 0\sqrt{2}$
f. Group; $i = 1$
g. Monoid; $i = 1$

6. a. $f_0 =$ identity function $f_1(1) = 2$ $f_1(2) = 1$
$f_2(1) = 1$ $f_2(2) = 1$ $f_3(1) = 2$ $f_3(2) = 2$

○	f_0	f_1	f_2	f_3
f_0	f_0	f_1	f_2	f_3
f_1	f_1	f_0	f_3	f_2
f_2	f_2	f_2	f_2	f_2
f_3	f_3	f_3	f_3	f_3

b. f_0 and f_1 are the elements.

○	f_0	f_1
f_0	f_0	f_1
f_1	f_1	f_0

7.

○	R_1	R_2	R_3	F_1	F_2	F_3
R_1	R_2	R_3	R_1	F_3	F_1	F_2
R_2	R_3	R_1	R_2	F_2	F_3	F_1
R_3	R_1	R_2	R_3	F_1	F_2	F_3
F_1	F_2	F_3	F_1	R_3	R_1	R_2
F_2	F_3	F_1	F_2	R_2	R_3	R_1
F_3	F_1	F_2	F_3	R_1	R_2	R_3

Identity element is R_3; inverse for F_1 is F_1; inverse for R_2 is R_1.

9. a. No, not the same operation
   b. No, zero polynomial (identity) does not belong to P.
   c. No, not every element of $\mathbb{Z}*$ has an inverse in $\mathbb{Z}*$.

10. $[\{0\}, +_{12}]$, $[\mathbb{Z}_{12}, +_{12}]$, $[\{0, 2, 4, 6, 8, 10\}, +_{12}]$, $[\{0, 4, 8\}, +_{12}]$,
   $[\{0, 3, 6, 9\}, +_{12}]$, $[\{0, 6\}, +_{12}]$

12. $4!/2 = 24/2 = 12$ elements
   $\begin{array}{llll}
   \alpha_1 = i & \alpha_2 = (1, 2) \circ (3, 4) & \alpha_3 = (1, 3) \circ (2, 4) & \alpha_4 = (1, 4) \circ (2, 3) \\
   \alpha_5 = (1, 3) \circ (1, 2) & \alpha_6 = (1, 2) \circ (1, 3) & \alpha_7 = (1, 3) \circ (1, 4) & \alpha_8 = (1, 4) \circ (1, 2) \\
   \alpha_9 = (1, 4) \circ (1, 3) & \alpha_{10} = (1, 2) \circ (1, 4) & \alpha_{11} = (2, 4) \circ (2, 3) & \alpha_{12} = (2, 3) \circ (2, 4)
   \end{array}$

13. a. No, $f(x + y) = 2, f(x) + f(y) = 2 + 2 = 4$
   b. No, $f(x + y) = |x + y|, f(x) + f(y) = |x| + |y|$
   c. Yes, $f(x \cdot y) = |x \cdot y|, f(x) \cdot f(y) = |x| \cdot |y|$

16. a. Closure: $\begin{bmatrix} 1 & z \\ 0 & 1 \end{bmatrix} \cdot \begin{bmatrix} 1 & w \\ 0 & 1 \end{bmatrix} = \begin{bmatrix} 1 & w + z \\ 0 & 1 \end{bmatrix} \in M_2^0(\mathbb{Z})$

   Matrix multiplication is associative.

   $\begin{bmatrix} 1 & 0 \\ 0 & 1 \end{bmatrix} \in M_2^0(\mathbb{Z})$

   The inverse of $\begin{bmatrix} 1 & z \\ 0 & 1 \end{bmatrix}$ is $\begin{bmatrix} 1 & -z \\ 0 & 1 \end{bmatrix}$, which belongs to $M_2^0(\mathbb{Z})$.

   b. f is a bijection and

   $f\left(\begin{bmatrix} 1 & z \\ 0 & 1 \end{bmatrix} \cdot \begin{bmatrix} 1 & w \\ 0 & 1 \end{bmatrix}\right) = f\left(\begin{bmatrix} 1 & w + z \\ 0 & 1 \end{bmatrix}\right) = w + z = z + w =$

   $f\left(\begin{bmatrix} 1 & z \\ 0 & 1 \end{bmatrix}\right) + f\left(\begin{bmatrix} 1 & w \\ 0 & 1 \end{bmatrix}\right)$

   c. $f\left(\begin{bmatrix} 1 & 7 \\ 0 & 1 \end{bmatrix}\right) = 7$ and $f\left(\begin{bmatrix} 1 & -3 \\ 0 & 1 \end{bmatrix}\right) = -3$

   $7 + (-3) = 4$

   $f^{-1}(4) = \begin{bmatrix} 1 & 4 \\ 0 & 1 \end{bmatrix}$

   d. $f^{-1}(2) = \begin{bmatrix} 1 & 2 \\ 0 & 1 \end{bmatrix}$ and $f^{-1}(3) = \begin{bmatrix} 1 & 3 \\ 0 & 1 \end{bmatrix}$

   $\begin{bmatrix} 1 & 2 \\ 0 & 1 \end{bmatrix} \cdot \begin{bmatrix} 1 & 3 \\ 0 & 1 \end{bmatrix} = \begin{bmatrix} 1 & 5 \\ 0 & 1 \end{bmatrix}$ and $f\left(\begin{bmatrix} 1 & 5 \\ 0 & 1 \end{bmatrix}\right) = 5$

19. a. $i_L = i_L \cdot i_R = i_R$ so $i_L = i_R$ and this element is an identity in $[S, \cdot]$.

   b. For example,

$\cdot$	a	b
a	a	b
b	a	b

   c. For example,

$\cdot$	a	b
a	a	a
b	b	b

   d. For example, $[\mathbb{R}^+, +]$.

22. a. $x \rho x$ because $i \cdot x \cdot i^{-1} = x \cdot i^{-1} = x \cdot i = x$.

If $x \rho y$, then for some $g \in G$, $g \cdot x \cdot g^{-1} = y$ or $g \cdot x = y \cdot g$
or $x = g^{-1} \cdot y \cdot g = (g^{-1}) \cdot y \cdot (g^{-1})^{-1}$ so $y \rho x$.
If $x \rho y$ and $y \rho z$, then for some $g_1, g_2 \in G$, $g_1 \cdot x \cdot g_1^{-1} = y$ and $g_2 \cdot y \cdot g_2^{-1}$
$= z$, so $g_2 \cdot g_1 \cdot x \cdot g_1^{-1} \cdot g_2^{-1} = z$ or $(g_2 \cdot g_1) \cdot x \cdot (g_2 \cdot g_1)^{-1} = z$ and $x \rho z$.

b. Suppose G is commutative and $y \in [x]$. Then for some $g \in G$,
$y = g \cdot x \cdot g^{-1} = x \cdot g \cdot g^{-1} = x \cdot i = x$. Thus $[x] = \{x\}$. Conversely
suppose $[x] = \{x\}$ for each $x \in G$, let $x, y \in G$, and denote the element
$y \cdot x \cdot y^{-1}$ by z. Then $x \rho z$, so $z = x$ and $y \cdot x \cdot y^{-1} = x$ or $y \cdot x = x \cdot y$.

28. a. $S \cap T \subseteq G$.

Closure: for $x, y \in S \cap T$, $x \cdot y \in S$ because of closure in S, $x \cdot y \in T$ because
of closure in T, so $x \cdot y \in S \cap T$. Identity: $i \in S$ and $i \in T$ so $i \in S \cap T$.
Inverses: for $x \in S \cap T$, $x^{-1} \in S$ and $x^{-1} \in T$ so $x^{-1} \in S \cap T$.

b. No. For example, $[\{0, 4, 8\}, +_{12}]$ and $[\{0, 6\}, +_{12}]$ are subgroups of
$[\mathbb{Z}_{12}, +_{12}]$ but $[\{0, 4, 6, 8\}, +_{12}]$ is not a subgroup of $[\mathbb{Z}_{12}, +_{12}]$ (not closed).

30. Closure: Let $x, y \in B_k$. Then $(x \cdot y)^k = x^k \cdot y^k = i \cdot i = i$, so $x \cdot y \in B_k$.
Identity: $i^k = i$ so $i \in B_k$.
Inverses: For $x \in B_k$, $(x^{-1})^k = (x^k)^{-1} = i^{-1} = i$, so $x^{-1} \in B_k$.

35. Let $x = a^{z_1}$, $y = a^{z_2} \in G$. Then $x \cdot y = a^{z_1} \cdot a^{z_2} = a^{z_1 + z_2} = a^{z_2 + z_1} = a^{z_2} \cdot a^{z_1} = y \cdot x$.

37. a. $[\text{Aut}(S), \circ]$ is closed because composition of isomorphisms is an isomorphism (Practice 30). Associativity always holds for function composition.
The identity function i_S is an automorphism on S. Finally, if f is an automorphism on S, so is f^{-1}.

b.

$i: 0 \to 0$ $f: 0 \to 0$
$ 1 \to 1$ $ 1 \to 3$
$ 2 \to 2$ $ 2 \to 2$
$ 3 \to 3$ $ 3 \to 1$

$\circ$	i	f
i	i	f
f	f	i

Exercises 8.2

1. a. 0001111110   **b.** *aaacaaaa*   **c.** 00100110

5.

Present State	Next State		Output
	Present Input		
	0	1	
s_0	s_1	s_2	a
s_1	s_2	s_3	b
s_2	s_2	s_1	c
s_3	s_2	s_3	b

Output is *abbcbb*.

9.

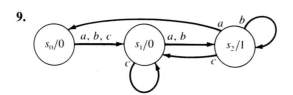

Output is 0001101.

13. a.

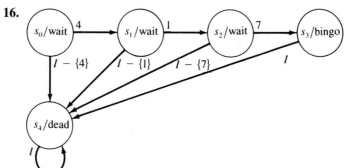

b. The length of time required to remember a given input grows without bound and eventually would exceed the number of states.

16.

17. a.

b.

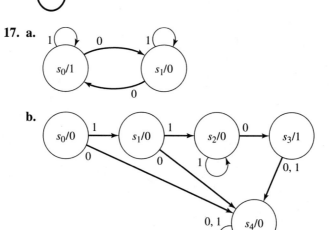

c.

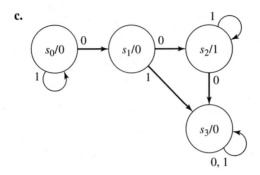

21. Once a state is revisited, behavior will be periodic since there is no choice of paths from a state. The maximum number of inputs which can occur before this happens is $n - 1$ (visiting all n states before repeating). The maximum length of a period is n (output from all n states, with the last state returning to s_0).

23. b. $01^* \lor (110)^*$

25. c. 100^*1

26. a. Yes   **b.** No   **c.** No   **d.** Yes

28. $dd^*(+ \lor -)dd^*$

33. a. s_1 **b.** s_2

34. $A = \{0\}, B = \{1, 2, 5\}, C = \{3, 4\}, D = \{6\}$

Present State	Next State		Output
	Present Input		
	0	1	
A	C	D	1
B	C	B	0
C	B	A	1
D	C	B	1

39. $A = \{0, 2\}, B = \{1, 3\}, C = \{4\}$

Present State	Next State			Output
	Present Input			
	a	b	c	
A	B	C	A	1
B	C	A	B	0
C	B	A	A	0

Exercises 8.3

1. a. Halts with final tape $\cdots$

b	0	0	0	0	0	b

$\cdots$

  b. Does not change the tape and moves forever to the left.

5. One answer: State 2 is a final state.

$(0, b, b, 2, R)$	blank tape or no more 1s
$(0, 1, 1, 1, R)$	has read odd number of 1s
$(1, 1, 1, 0, R)$	has read even number of 1s

8. One answer: State 9 is a final state.

$(0, b, b, 9, R)\}$	accepts blank tape
$\left.\begin{array}{l}(0, 0, 0, 0, R)\\(0, 1, X, 1, R)\end{array}\right\}$	finds first 1, marks with X
$\left.\begin{array}{l}(1, 1, 1, 1, R)\\(1, Y, Y, 1, R)\end{array}\right\}$	searches right for 2s
$\left.\begin{array}{l}(1, 2, Y, 3, R)\\(3, 2, Y, 4, L)\end{array}\right\}$	pair of 2s, marks with Y's
$\left.\begin{array}{l}(4, Y, Y, 4, L)\\(4, X, X, 4, L)\\(4, 1, 1, 4, L)\\(4, Z, Z, 4, L)\end{array}\right\}$	searches left for 0s
$\left.\begin{array}{l}(4, 0, Z, 5, L)\\(5, 0, Z, 6, R)\end{array}\right\}$	pair of 0s, marks with Z's
$\left.\begin{array}{l}(6, Z, Z, 6, R)\\(6, X, X, 6, R)\\(6, 1, X, 1, R)\end{array}\right\}$	passes right to next 1
$(6, Y, Y, 7, R)\}$	no more 1s
$\left.\begin{array}{l}(7, Y, Y, 7, R)\\(7, b, b, 8, L)\end{array}\right\}$	no more 2s
$\left.\begin{array}{l}(8, Y, Y, 8, L)\\(8, X, X, 8, L)\\(8, Z, Z, 8, L)\\(8, b, b, 9, L)\end{array}\right\}$	no more 0s, halts and accepts

12. One approach uses the following general plan: Put a marker at the right end of the original string X_1 and build a new string X_2 to the right of the marker. Working from the lower order end of X_1, for each new symbol in X_1, put a block of that symbol on the end of X_2 twice as long as the previous block of symbols added to X_2. When all symbols in X_1 have been processed, erase X_1 and the marker. Working from left to right in X_2, replace any 0s with 1s from the end of the string until there are no 0s left in X_2. (My implementation of this approach required 23 states and 85 quintuples; can you improve upon this solution?)

14. $f(n_1, n_2, n_3) = \begin{cases} n_2 + 1 & \text{if } n_2 \geq 0 \\ \text{undefined} & \text{if } n_2 = 0 \end{cases}$

16. $(0, 1, 1, 1, R)$

$\left. \begin{array}{l} (1, b, 1, 4, R)\} \\ (1, 1, 1, 2, R) \\ (2, b, 1, 4, R) \end{array} \right.$ $n = 0$, add 1 and halt

$\left. \begin{array}{l} (1, 1, 1, 2, R) \\ (2, b, 1, 4, R) \end{array} \right\}$ $n = 1$, add additional 1 and halt

$\left. \begin{array}{l} (2, 1, 1, 3, R) \\ (3, 1, b, 3, R) \\ (3, b, b, 4, R) \end{array} \right\}$ $n \geq 2$, erase extra 1s and halt

19. One answer:

$\left. \begin{array}{l} (0, 1, b, 1, R)\} \\ (1, *, b, 3, R)\} \end{array} \right.$ erases one extra 1
   $n_1 = 0$

$\left. \begin{array}{l} (1, 1, b, 2, R) \\ (2, 1, 1, 2, R) \\ (2, *, 1, 3, R) \end{array} \right\}$ $n_1 > 0$, replaces * with leftmost 1 of $\bar{n}_1$, halts

Exercises 8.4

1. a. $L(G) = \{a\}$
  b. $L(G) = \{010101, 010111, 011101, 011111, 110101, 110111, 111101, 111111\}$
  c. $L(G) = 0(10)*$
  d. $L(G) = 0*1111*$

3. $L(G) = aa*bb*$. G is context-sensitive. An example of a regular grammar that generates $L(G)$ is $G' = (V, V_T, S, P)$ where $V = \{a, b, A, B, S\}$, $V_T = \{a, b\}$, and P consists of the productions

$\begin{array}{ll} S \rightarrow aA & A \rightarrow aA \quad B \rightarrow bB \\ S \rightarrow aB & A \rightarrow aB \quad B \rightarrow b \end{array}$

6. a. $1*1(00)*$
  b. $S \rightarrow 1$
    $S \rightarrow 1S$
    $S \rightarrow 0A$
    $A \rightarrow 0$
    $A \rightarrow 0B$
    $B \rightarrow 0A$

9. For example, $G = (V, V_T, S, P)$ where $V = \{0, 1, A, S\}$, $V_T = \{0, 1\}$ and P consists of the productions

$\begin{array}{lll} S \rightarrow 01 & A \rightarrow A0 & A \rightarrow 0 \\ S \rightarrow A01 & A \rightarrow A1 & A \rightarrow 1 \end{array}$

14. For example, $G = (V, V_T, S, P)$ where $V = \{0, 1, S, S_1\}$, $V_T = \{0, 1\}$, and P consists of the productions

$$S \to \lambda \qquad\qquad S_1 \to 1S_11$$
$$S \to S_1 \qquad\qquad S_1 \to 00$$
$$S_1 \to 0S_10 \qquad\quad S_1 \to 11$$

17. a.

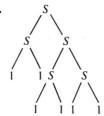

18

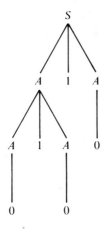 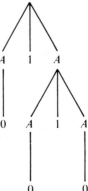

Answers to Self-Tests

Chapter 1

1. F   If a statement is not a tautology, it does not have values that are all true, but that does not make them all false.

2. T   Because of the truth table for disjunction, ( ) ∨ *T* is *T*.

3. F   The statement must have an implication as its main connective.

4. T

5. F   The negation of a disjunction is the *conjunction* of the negations.

6. T

7. F

8. T

9. F   It is one in which hypothesis → conclusion is always true.

10. T

11. F   In fact, $(\forall x)(P(x) \wedge [P(x)]')$ would be false in all interpretations.

12. T

13. T

14. F   There is no one predicate wff defined on an interpretation, nor is the domain at all determined by truth values.

15. T

16. T

17. F   Existential instantiation should be used early.

18. F   Universal instantiation would only strip off the leading universal quantifier.

19. F   Wffs in propositional logic are not even wffs in predicate logic.

20. T Predicate logic is correct—only valid wffs are provable.

21. T

22. F A single negated predicate is only one kind of Horn clause.

23. T

24. F A PROLOG recursive rule is not a rule of inference.

25. T

26. F It only guarantees that the output satisfies certain conditions, given that the input satisfies certain conditions.

27. F Nothing much can be said about the precondition without knowing the assignment, but at any rate the strict inequality will not go away.

28. F Program testing involves test data sets.

29. T

30. T

Chapter 2

1. F A conjecture that only asserts something about a finite number of cases can be proved by proving all cases.

2. T

3. F A universal quantifier is understood, because the formal statement of the theorem is $(\forall x)(x \text{ odd} \rightarrow 2 * x \text{ is even})$.

4. F The second statement is the converse of the first, not the contrapositive.

5. T

6. T

7. F The basis step need not be $n = 1$.

8. T

9. T

10. F This omits the first $k - 1$ terms of the series.

11. F A loop invariant remains true after loop termination.

12. F It means that correctness has been proved only given that the loop terminates.

13. F The first principle of induction is used because the values at pass $k + 1$ through the loop depend only on the values at pass k.

14. F But Q should give the desired result when the condition B' is true.

15. T (12 is the remainder when 42 is divided by 30)

16. T

17. F They are valuable because they represent natural ways of thinking about certain problems, but they typically use more storage and perform more operations than a corresponding iterative program.

18. T

19. F Induction may be used to verify a proposed closed-form solution, but not to determine such a solution from the recurrence relation.

20. F It is not linear because of the presence of the $S(n - 2)$ term.

21. T

22. T

23. T

24. F The recursive version looks at the first item in the list, and if that is not the target, it searches the rest of the list. The "input size" only goes down by 1 each time the algorithm is invoked.

25. T

Chapter 3

1. F It is not a proper subset of itself.

2. T

3. T

4. F This is the closure property.

5. F It is a way to prove that certain sets are uncountable.

6. T

7. F

8. T

9. T

10. F

11. F

12. F The number of elements in the union plus the number of elements in the intersection is the sum of the number of elements in each set.

13. F All must be finite.

14. F

15. T

16. T

17. T

18. F   Use $C(n, r)$.

19. F   It is $n!/n_1!n_2!n_3!$.

20. T

21. F   Combinations, not arrangements

22. T

23. F   All terms are found in row n.

24. T

25. T

Chapter 4

1. T

2. F (x, x) can belong.

3. T

4. F   The relation of equality is both a partial ordering and an equivalence relation.

5. F   An equivalence relation does this.

6. F

7. T

8. F   The converse is true.

9. F   The maximum value is used.

10. F   See Example 18 and Practice 19.

11. T

12. T

13. F

14. T

15. F   If the data satisfies data integrity to begin with, then data integrity will still be true after a delete. It is referential integrity that may be lost.

16. F It may not have an image for each member of the domain.

17. F Every element of the range has a preimage; begin with an element of the codomain.

18. T

19. T

20. F Other constants may work where these do not.

21. T

22. F

23. T

24. T

25. F See Practice 50.

Chapter 5

1. F A *complete* graph has an arc between any two nodes.

2. T

3. F A planar graph could still be drawn with arcs that cross.

4. F It means that nodes 2, 3, and 4 are all adjacent to some one node.

5. F It could be symmetric, it just doesn't have to be.

6. T

7. F

8. T

9. T

10. T

11. T

12. F This is the worst case; other cases could require fewer comparisons.

13. T

14. F The binary search tree depends on the order in which data elements are inserted.

15. F It must have at least $n!$ leaves.

16. T

17. F In a prefix code, NO code word is the prefix of another code word

18. F   Characters that occur most frequently have the shortest strings, and frequency does not affect the number of 0s versus the number of 1s.

19. T

20. F   The code table file (giving the code for each character) must be stored along with the encoded file.

Chapter 6

1. T

2. T

3. F

4. F

5. F

6. T

7. F   The graph can have at most two odd nodes.

8. T

9. F   Some arcs may go unused.

10. F   No efficient algorithm is known, but trial and error solves the problem.

11. F   This is how Prim's algorithm works.

12. F

13. T

14. T

15. F   It will generally not form a tree at all.

16. F

17. T

18. T

19. F   It is the equivalent of *depth-first* search, assuming that sibling nodes are labeled in order from left to right.

20. F   Use a succession of *depth-first* searches.

21. F

22. T

23. T

24. T

25. F   The root is a special case.

Chapter 7

1. F $x + x' = 1$

2. T

3. T
$$\begin{aligned}
x + (y + x \cdot z) &= x + (x \cdot z + y)\\
&= x \cdot 1 + (x \cdot z + y)\\
&= (x \cdot 1 + x \cdot z) + y\\
&= x \cdot (1 + z) + y\\
&= x \cdot (z + 1) + y\\
&= x \cdot 1 + y\\
&= x + y
\end{aligned}$$

4. T

5. T

6. T $(x + y)' = x' \cdot y'$

7. F   It has as many terms as the function has 1-values (or one term if the function has all 0-values).

8. F   Only one half-adder is needed (see Figure 7.14a).

9. T   Because $x + x = x$

10. F   This usually results in an unnecessarily large number of devices; start again from the truth function.

11. T

12. T

13. F   Look for two-square blocks first and work up from there.

14. T

15. F   The term is essential if the check for that row is the only one in some column.

Chapter 8

1. F   This describes a commutative operation.

2. T   Even though this is not the definition of the identity.

3. T

4. T   By Lagrange's theorem

5. F f must also be a bijection.

6. T

7. T $(0 \vee 1)*00$

8. F   See Figure 8.6a.

9. T

10. F

11. F   Entering a final state causes a Turing machine to halt, but it can also halt in a nonfinal state if there is no instruction for the current state–input pair.

12. F   There will be $n + 2$ 1s on the tape.

13. F

14. F   This is the version with the quantifiers reversed that is trivially true.

15. T

16. T

17. F   Depending on the productions selected, nonterminal symbols may never be removed.

18. F   There may be an equivalent grammar for the language that is context-free.

19. T

20. T

Index